Tech. Info. Ctr.
Hewlett Packard
Corvallis

1000098815

D1774294

DISCARDED

2 1996

H. P. Corvallis
Research Library

Linear Control Systems

A Computer-aided Approach

International Series on

SYSTEMS AND CONTROL, Volume 7

Editorial Board

Professor M. G. SINGH, UMIST, Manchester, England (Coordinating Editor)
Professor H. AKASHI, University of Kyoto, Japan
Professor Y. C. HO, Harvard University, USA
Academician B. PETROV, Moscow Aviation Institute, USSR

Other Titles of Interest

AKASHI
Control Science and Technology for the Progress of Society
(8th Triennial IFAC World Congress)

ANAND
Introduction to Control Systems, 2nd Edition

BROADBENT & MASUBUCHI
Multilingual Glossary of Automatic Control Technolgoy

ELLOY & PIASCO
Classical and Modern Control with Worked Examples

EYKHOFF
Trends and Progress in System Identification

GERTLER & KEVICZKY
A Bridge between Control Science and Technology
(9th Triennial IFAC World Congress)

GVISHIANI
Systems Research: A Methodological Approach

HASEGAWA
Real-Time Programming 1981

MAHALANABIS
Theory and Application of Digital Control

MAHMOUD & SINGH
Large Scale Systems Modelling

MILLER
Distributed Computer Control Systems

MORRIS
Communication for Command and Control Systems

PATEL & MUNRO
Multivariable System Theory and Design

SINGH *et al.*
Applied Industrial Control: An Introduction

SINGH & TITLI
Systems: Decomposition, Optimization and Control

TZAFESTAS
Distributed Parameter Control Systems: Theory and Application

Pergamon Related Journals (free specimen copies gladly sent on request)

AUTOMATICA
COMPUTERS & INDUSTRIAL ENGINEERING
COMPUTERS & OPERATIONS RESEARCH
JOURNAL OF THE OPERATIONAL RESEARCH SOCIETY

Linear Control Systems

A Computer-aided Approach

MOHAMMAD JAMSHIDI
University of New Mexico, USA

and

MANU MALEK-ZAVAREI
AT & T Bell Laboratories*

*Major part of this work done while with the
University of New Mexico

PERGAMON PRESS

OXFORD · NEW YORK · BEIJING · FRANKFURT
SÃO PAULO · SYDNEY · TOKYO · TORONTO

U.K.	Pergamon Press, Headington Hill Hall, Oxford OX3 0BW, England
U.S.A.	Pergamon Press, Maxwell House, Fairview Park, Elmsford, New York 10523, U.S.A.
PEOPLE'S REPUBLIC OF CHINA	Pergamon Press, Qianmen Hotel, Beijing, People's Republic of China
FEDERAL REPUBLIC OF GERMANY	Pergamon Press, Hammerweg 6, D-6242 Kronberg, Federal Republic of Germany
BRAZIL	Pergamon Editora, Rua Eça de Queiros, 346, CEP 04011, São Paulo, Brazil
AUSTRALIA	Pergamon Press Australia, P.O. Box 544, Potts Point, N.S.W. 2011, Australia
JAPAN	Pergamon Press, 8th Floor, Matsuoka Central Building, 1-7-1 Nishishinjuku, Shinjuku-ku, Tokyo 160, Japan
CANADA	Pergamon Press Canada, Suite 104, 150 Consumers Road, Willowdale, Ontario M2J 1P9, Canada

Copyright © 1986 Pergamon Books Ltd.

All Rights Reserved. No part of this publication may be reproduced, stored in a retrieval system or transmitted in any form or by any means: electronic, electrostatic, magnetic tape, mechanical, photocopying, recording or otherwise, without permission in writing from the publishers.

First edition 1986

Library of Congress Cataloging in Publication Data

Jamshidi, Mohammad.
Linear control systems.
(International series on systems and control; v. 7)
Includes bibliographical references.
I. Automatic control — Data processing.   2. Control theory — Data processing.   I. Malek-Zavarei,
Manu. II. Title. III. Series.
TJ213.J32 1984   629.8'312   84–6164

British Library Cataloguing in Publication Data

Jamshidi, Mohammad
Linear control systems. — (International series on systems and control ; v. 7).
1. Automatic control — Data processing.   I. Title
II. Malek-Zavarei, Manu. III. Series
629.8'32'02854   TJ213
ISBN 0–08–028701–8 (Hardcover)
ISBN 0–08–028702–6 (Flexicover)

Printed in Great Britain by A. Wheaton & Co. Ltd., Exeter

To our wives and children:
Jila, Ava and Nima Habib - M.J.
Lynn, Mitra, Majid and Cameron - M.M.-Z.

PREFACE

This text is intended to give an introduction to classical and modern control systems modeling, analysis and design. The bulk of the manuscript is based on two courses on classical/modern control and linear systems at the *University of New Mexico's Department of Electrical and Computer Engineering*. The book's three main themes are: control, systems and computer-aided design and analysis.

The text has evolved over several years of the author's combined effort at various locations including Shiraz University (Shiraz, Iran), IBM Thomas J. Watson Research Center (Yorktown Heights, NY) and the University of New Mexico (Albuquerque, NM). The book is intended to bridge the gap between a first course in classical control and a theoretical-oriented graduate course such as optimal control. Continuous-time and discrete-time as well as time-domain and frequency-domain presentations of linear systems are considered on a balanced basis. The intended audience are thus the advanced undergraduate and the first-year graduate electrical engineering students. There has been no attempt to teach computer programming in the text although there are over 100 BASIC subroutines and driver programs listed in the manuscript. All the computer programs have been originally written for an HP-9845A or B desk-top computers. These programs are now being put together in the form of two integrated and interconnected packages called FREDOM© and TIMDOM© to treat frequency-domain and time-domain techniques of analysis and design. The packages are also being tailored for a wide range of personal and other computers such as IBM/PC, HP-2816, Apple IIe and SUN workstation. For more information on the availability and acquisition of these packages, the interested reader may write directly to the Publisher.

The detailed plan of the book is shown on page 4. A typical chapter of the book is structured as follows. The chapter begins with an Introduction Section which is followed by a theoretical development or a mathematical description, which are further illustrated by numerical examples. Occasionally certain important techniques are summarized by an algorithm. The various system and control concepts are coded in extended BASIC and the respective main or driver programs are listed within the text. In the cases where a given subroutine is called once the subroutines is also listed along with the main program. All the subroutines called more than once within the programs and all the utility routines are listed in Appendix A. Indices to all programs and subroutines are given at the end of the book. Each chapter ends with a Problems Section and a list of cited references.

Chapter 1 serves as an introduction to the text. In this chapter a system is defined and important classes of systems are presented. Chapter 2 and 3 present a review of linear algebra and the transforms (Laplace and z) theories. A well familiar reader can skip these two chapters without any loss of continuity. System modeling utilizing both transfer functions (frequency domain) and statespace representation (time-domain) have been presented in Chapter 4. State transformations and linearization of nonlinear systems are among other topics discussed here.

In Chapter 5 model reduction of large-scale linear time-invariant systems is discussed. Both perturbation (singular and regular) and aggregation (time-domain and frequency-domain) are presented here. The solution of the state equations for both continuous-time and discrete-time systems are discussed in Chapter 6. Both time-invariant and time-varying cases are considered.

Chapter 7 is devoted to system stability. This chapter includes notions of stability in the classical control such as Routh-Hurwitz and Jury-Blanchard criteria as well as the root locus method, Bode and Nyquist diagrams and the Nyquist criterion. It also includes modern system theoretical stability concepts such as zero-input and zero-state stability, the Lyapunov method and the circle criterion are. The important notions of controllability and observability are discussed in Chapter 8. The canonical decomposition of an uncontrollable (unobservable) system into controllable (observable) and uncontrollable (unobservable) subsystem are discussed here. Minimal realization and duality are among other topics discussed in Chapter 8.

The next three chapters of the book are generally concerned with the design of linear single-input single-output (SISO) or multi-input multi-output (MIMO) systems. In Chapter 9, design of a SISO linear control system via series and feedback compensations is presented. The design using root locus and frequency responses (Bode and Nyquist diagrams) as design tools with the aid of available computer programs are illustrated by a number of detail design problems. The computer-aided design of a SISO system via feedback compensation is illustrated as well.

Design of linear MIMO systems via modern system concepts such as pole placement along with state estimation are taken up in Chapter 10. The design of a state estimator (observer) using full or reduced order formulations are presented here.

Chapter 11 is concerned with design of linear SISO systems through optimization of system criteria. Here functional minimization techniques are utilized to design a controller for a SISO control system subject to plant parameter variations. Parseval's theorems for both continuous-time and discrete-time systems have been utilized for this purpose. In short, Chapter 11 is concerned with design of a linear SISO system via parameter optimization.

There are two main schemes to utilize the text: classical control and modern system engineering. The suggested sequence of chapters and sections to be used for the above two schemes are given below.

Classical Control (a senior-level course): Chapters 1, 3, Sections 4.2, 4.3, 4.5, 7.4, 7.5, 7.6 and 7.7, Chapters 9 and 11.

Linear System (a first-year graduate course): Chapters 1, 2, Sections 3.6 through 3.9, 4.2 through 4.4, 4.6 and 4.7, Chapters 6, Sections 7.2, 7.3, and 7.9, Chapter 8, Chapter 10, Sections 5.3 and 5.4.

The manuscript has been used in two such courses at the Department of Electrical and Computer Engineering of the University of New Mexico. The text can be even more useful to the reader when it is used in conjunction with the packages FREDOM© and TIMDOM©.

The authors are indebted to many people for their various contributions. We would like to thank Dean Jerry May of the University of New Mexico's College of Engineering and Peter Dorato, Chairperson of the Electrical and Computer Engineering Department for their leadership and continuous support. The authors would like to thank Professor Madan Singh of the University of Manchester Institute of Technology for reviewing the manuscript and many helpful suggestions. We would like to thank Professor Singh in another capacity — the series editor of Pergamon's International Series on Systems and Control for suggesting to write the book. We would like to thank many of our former students who have contributed to the text in various forms. In particular we like to thank D. Behroozi, R.-W. Chang, S.-K. Chiang, G. Eisler, V. Gieri, M. Masukawa, R. Morel, S. Otis, R. Owen, P.-E. Tang and J. Wilcoxen for writing the original versions of some of the programs.

The manuscript was prepared using a DEC 11/780 VAX digital computer. We would like to thank Mrs. Gladys Ericksen for many hours of tireless and persistent text processing at the computer's terminals.

M. Jamshidi
Albuquerque, New Mexico
U.S.A.

M. Malek-Zavarei
Holmdel, New Jersey
U.S.A.

CONTENTS

1 PRELIMINARIES

1.1	INTRODUCTION	1
1.2	WHAT IS A SYSTEM?	1
1.3	NOTATIONS	1
	1.3.1 Numbering and Cross-Referencing	2
	1.3.2 Conventions	2
	1.3.3 Abbreviations and Symbols	2
1.4	CLASSIFICATION OF SYSTEMS	3
	1.4.1 Lumped-Parameter and Distributed-Parameter Systems	3
	1.4.2 Deterministic and Stochastic Systems	4
	1.4.3 Continuous-Time and Discrete-Time Systems	4
	1.4.4 Linear and Nonlinear Systems	4
	1.4.5 Time-Invariant and Time-Varying System	4
1.5	SCOPE OF THE BOOK	4
	REFERENCES	5

PART I
MATHEMATICAL BACKGROUND

2 REVIEW OF LINEAR ALGEBRA

2.1	INTRODUCTION	9
2.2	FIELDS AND VECTOR SPACES	9
2.3	LINEAR INDEPENDENCE AND RANK	11
2.4	BASES AND DIMENSION	14
2.5	INNER PRODUCT AND NORM	16
2.6	EIGENVALUES AND EIGENVECTORS	23
2.7	DIAGONALIZATION OF MATRICES	29
2.8	GENERALIZED EIGENVECTORS AND JORDAN CANONICAL FORM	31
2.9	FUNCTIONS OF A SQUARE MATRIX	37
	APPLICATIONS OF CAYLEY-HAMILTON THEOREM	40
2.10	QUADRATIC FORMS	45
2.11	PROBLEMS	51
	REFERENCES	55

3 REVIEW OF THE LAPLACE AND z-TRANSFORMS

3.1	INTRODUCTION	57

3.2	DEFINITION OF THE LAPLACE TRANSFORM	57
3.3	PROPERTIES OF THE LAPLACE TRANSFORM	59
3.4	INVERSION OF THE LAPLACE TRANSFORM	65
	3.4.1 Inversion of the Laplace Transform by Partial Fraction Expansion	65
	3.4.2 Numerical Inversion of the Laplace Transform	81
3.5	SOLUTION OF DIFFERENTIAL EQUATIONS USING THE LAPLACE TRANSFORM	88
3.6	DEFINITION OF THE z-TRANSFORM	89
3.7	PROPERTIES OF THE z-TRANSFORM	92
3.8	INVERSION OF THE z-TRANSFORM	93
	3.8.1 The Power Series Method	94
	3.8.2 The Residue Method	94
	3.8.3 The Partial Fraction Expansion Method	96
3.9	SOLUTION OF DIFFERENCE EQUATIONS USING THE z-TRANSFORM	106
	PROBLEMS	107
	REFERENCES	109

PART II
ANALYSIS

4 SYSTEM MODELING

4.1	INTRODUCTION	113
4.2	DEVELOPING SYSTEM MODELS	113
4.3	THE CONCEPT OF STATE	115
4.4	STATE-SPACE REPRESENTATION OF LINEAR SYSTEMS	117
	4.4.1 State and Output Equations	117
	4.4.2 Simulation Diagrams	120
	4.4.3 State Equations in Standard Forms	123
	4.4.3.1 The controllable companion form	123
	4.4.3.2 The observable companion form	124
4.5	STATE TRANSFORMATIONS	130
4.6	TRANSFER FUNCTION	138
	4.6.1 Continuous-Time Systems	139
	4.6.2 Discrete-Time Systems	140
	4.6.3 Transfer Function of a SISO System	145
	4.6.4 Signal Flow Graphs	150
4.7	LINEARIZATION OF NONLINEAR SYSTEMS	152
	4.7.1 Differentiation with Respect to a Vector	153
	4.7.2 Linearization Around an Operating Point	154

CONTENTS

	PROBLEMS	155
	REFERENCES	158

5 MODEL REDUCTION OF LARGE-SCALE LINEAR SYSTEMS

5.1	INTRODUCTION	161
5.2	PERTURBATION METHODS	164
	5.2.1 Regular Perturbation	164
	5.2.2 Singular Perturbation	167
	5.2.2.1 Boundary layer correction	167
	5.2.2.2 Time-scale separation	168
5.3	AGGREGATION METHODS – TIME DOMAIN	175
	5.3.1 Exact Aggregation	176
	5.3.2 Modal Aggregation	181
5.4	AGGREGATION METHODS – FREQUENCY DOMAIN	185
	5.4.1 Moment Matching	185
	5.4.2 Padé Approximation	187
	5.4.3 Mixed Methods and Multi-Variable Systems	191
	5.4.3.1 Padé-modal	191
	5.4.3.2 Matrix continued fraction	193
	PROBLEMS	196
	REFERENCES	199

6 SOLUTION OF THE STATE EQUATION

6.1	INTRODUCTION	201
6.2	THE HOMOGENEOUS CASE – TRANSITION MATRIX	202
	6.2.1 Continuous-Time Systems	202
	6.2.2 Discrete-Time Systems	203
6.3	PROPERTIES OF THE TRANSITION MATRIX	204
6.4	CALCULATION OF THE TRANSITION MATRIX	206
	6.4.1 Time-Invariant Continuous Case	206
	6.4.1.1 The eigenvalue method	206
	6.4.1.2 The Cayley-Hamilton method	207
	6.4.1.3 The Laplace transform method	207
	6.4.2 Time-Invariant Discrete Case	215
	6.4.2.1 The eigenvalue method	216
	6.4.2.2 The Cayley-Hamilton method	216
	6.4.2.3 The z-transform method	216
	6.4.3 Transition Matrix in the Time-Varying Case	223
6.5	RESOLVENT MATRIX	224
6.6	THE FORCED CASE – COMPLETE SOLUTION	234

CONTENTS

	6.6.1 Continuous-Time Systems	234
	6.6.2 Impulse Response Matrix	237
	6.6.3 Discrete-Time Systems	249
	6.6.4 Pulse Response Matrix	252
6.7	ADJOINT AND DUAL SYSTEMS	263
	PROBLEMS	266
	REFERENCES	272

7 SYSTEM STABILITY

7.1	INTRODUCTION	273
7.2	STABILITY CONCEPTS AND DEFINITIONS	274
	7.2.1 Zero-Input Stability	274
	7.2.2 Zero-State Stability	276
7.3	STABILITY CRITERIA	277
	7.3.1 Zero-Input Stability	277
	7.3.2 Zero-State Stability	288
7.4	ROUTH-HURWITZ STABILITY CRITERION	290
7.5	JURY-BLANCHARD STABILITY CRITERION	299
7.6	THE ROOT LOCUS METHOD	310
	7.6.1 Conventional Approach in Root Locus Plotting	310
	7.6.2 Computer-Aided Approach in Root Locus Plotting	316
7.7	FREQUENCY RESPONSE METHODS	325
	7.7.1 Nyquist Diagram	326
	7.7.1.1 Nyquist stability criterion	328
	7.7.1.2 Stability margins and conditional stability	335
	7.7.2 Bode Diagram	338
	7.7.3 Phase and Gain Margins	346
7.8	THE CIRCLE CRITERION	358
7.9	THE LYAPUNOV'S METHOD	363
	PROBLEMS	372
	REFERENCES	379

8 CONTROLLABILITY AND OBSERVABILITY

8.1	INTRODUCTION	381
8.2	LINEAR INDEPENDENCE OF FUNCTIONS	381
8.3	CONTROLLABILITY	384
	8.3.1 Definitions and Examples	384
	8.3.2 Characterization of Controllability	386
8.4	OBSERVABILITY	389
	8.4.1 Definitions and Examples	389
	8.4.2 Characterization of Observability	391

CONTENTS

8.5	DUALITY	396
8.6	CANONICAL DECOMPOSITION OF THE STATE SPACE	397
	8.6.1 Separation of the Controllable Part	398
	8.6.2 Separation of the Observable Part	403
	8.6.3 Canonical Structure of Systems	408
8.7	MINIMAL REALIZATIONS	409
	PROBLEMS	411
	REFERENCES	414

PART III
DESIGN

9 DESIGN OF LINEAR CONTROL SYSTEMS BY COMPENSATION

9.1	INTRODUCTION	419
9.2	DOMINANT POLE ASSUMPTION	420
9.3	STEADY-STATE ERRORS AND STATIC ERROR COEFFICIENTS	421
	9.3.1 Steady-State Error	421
	9.3.2 Static Error Coefficients	422
9.4	COMPENSATING NETWORKS	426
	9.4.1 Lead Network	426
	9.4.2 Lag Network	428
	9.4.3 Lag-Lead Network	429
9.5	PERFORMANCE SPECIFICATIONS	431
	9.5.1 Time-Domain Specifications	431
	9.5.2 Frequency-Domain Specifications	435
9.6	DESIGN VIA ROOT LOCUS	451
9.7	DESIGN VIA BODE AND NYQUIST DIAGRAMS	460
9.8	RULES FOR SELECTION OF A COMPENSATOR	468
	9.8.1 Lead Compensation and Root Locus Scheme	469
	9.8.2 Lag Compensation and Root Locus Scheme	470
	9.8.3 Lead-Lag Compensation and Root Locus Scheme	470
	9.8.4 Lead Compensation and Frequency-Response Schemes	471
	9.8.5 Lag Compensation and Frequency Response Schemes	472
	9.8.6 Lead-Lag Compensation and Frequency-Response Schemes	472
9.9	FEEDBACK COMPENSATION	472
	PROBLEMS	476
	REFERENCES	479

10 FEEDBACK CONTROL AND STATE ESTIMATION

10.1	INTRODUCTION	481
10.2	STATE AND OUTPUT FEEDBACK	481
10.3	EFFECT OF FEEDBACK ON SYSTEM PROPERTIES	483

10.4	POLE PLACEMENT	486
	10.4.1 Pole Placement Via State Feedback	487
	10.4.2 Pole Placement Via Output Feedback	493
10.5	STATE ESTIMATION – OBSERVERS	494
10.6	FULL-DIMENSION OBSERVERS	496
10.7	REDUCED-ORDER OBSERVER	499
	PROBLEMS	503
	REFERENCES	505

11 DESIGN VIA PARAMETER OPTIMIZATION

11.1	INTRODUCTION	507
11.2	PERFORMANCE INDICES	507
11.3	OPTIMIZATION BY PARSEVAL'S THEOREM	510
	11.3.1 Continuous-Time Systems	510
	11.3.2 Discrete-Time Systems	513
11.4	FUNCTIONAL MINIMIZATION TECHNIQUES	515
	11.4.1 Direct Search Method	515
	11.4.2 Variable Metric Method	525
	11.4.3 An Automatic Constrained Method	529
11.5	DESIGN OF OPTIMAL STATE FEEDBACK	537
	PROBLEMS	543
	REFERENCES	545
	APPENDIX	547
	INDEX	617

CHAPTER 1
PRELIMINARIES

1.1 INTRODUCTION

This chapter serves as a general introduction to the book. It is intended to provide a framework within which the analysis and design of linear systems can be explained. As such it may refer to some concepts which are not defined yet but will be defined in later chapters.

In the following sections, first the concept of a system will be introduced. Then the notation to be used in the book will be explained. A section on systems classification and a final section on the scope of the book will conclude this chapter.

1.2 WHAT IS A SYSTEM?

Webster's dictionary defines a *system* as "a regularly interacting or independent group of items forming a unified whole." From our point of view a system is an entity that can be characterized by a finite number of attributes. An example of a system is a lumped electric circuit. Some attributes of this system are the values of the elements used in the circuit. Another example is a weight resting on a spring. Some attributes of this system are the mass of the weight and the spring constant. Sufficient attributes should be given for each system to wholly characterize it.

A *control system* is a system capable of monitoring or regulating the operation of a process or a plant. An early example of a control system is the centrifugal steam engine governor invented by James Watt in the eighteenth century [1.1,1.2]. This control system controlled the flow of steam by the application of centrifugal force to a lever, to maintain the speed of a steam engine at a relatively constant level. A modern example of a control system is the autopilot system of an airplane.

1.3 NOTATIONS

In this section first the method of numbering and cross-referencing in the book will be explained. Then the notations and abbreviations used throughout the book will be compiled.

LINEAR CONTROL SYSTEMS

1.3.1 Numbering and Cross-Referencing

Chapters of the book are consecutively numbered. The sections in each chapter are numbered consecutively by two digits, the first of which is the chapter number. Subsections in each section are similarly numbered consecutively by three digits. The equations in each section are numbered on the right-hand side consecutively. Also, the figures in each section are numbered consecutively. For reference to an equation in the *same* section, only the number of the equation will be referred to. But to refer to an equation or a figure in *another* section, the two-digit section number will also precede the number of the equation or the figure.

Also in each section consecutive numbers are used to identify definitions, theorems and examples. In the *same* section, these are referred to by their numbers. In *other* sections, these are referred to by a three-digit number consisting of their number preceded by the two-digit number of the section in which they appear.

1.3.2 Conventions

Capital letters denote sets or vector spaces, e.g., S, V.

Lower case and Greek letters indicate scalars and scalar-valued functions, e.g., $m, \alpha, f(t)$.

Bold lower case letters indicate vectors, e.g., **x**, **y**.

Bold capital letters indicate matrices, e.g., **A**, **B**.

The Laplace transform of a function is denoted by the corresponding capital letter, e.g., $G(s) = L[g(t)]$.

The z-transform of a function is indicated by the corresponding capital letter, e.g., $H(z) = Z[h(t)]$.

Superscript "'" denotes the transpose of a vector or a matrix, e.g., $\mathbf{x}'$, $\mathbf{A}'$.

Superscript "*" denotes the conjugate transpose of a vector or a matrix, e.g., $\mathbf{x}^*$, $\mathbf{A}^*$.

Superscript "-1" denotes the inverse of a matrix or a transformation, e.g., $\mathbf{A}^{-1}$, L^{-1}.

Superscript "$\perp$" denotes the orthogonal complement of a subspace, e.g., $V^\perp$.

A dot over a time function denotes its derivative with respect to time, e.g., $\dot{x}$, $\dot{\mathbf{x}}$.

Two or three dots over a time function denote its second or third derivative with respect to time, respectively, e.g., $\ddot{x}$, $\dddot{x}$. For higher-order derivatives a superscript with the corresponding derivative order in parenthesis may also be used, e.g., $x^{(3)}$, $y^{(4)}$. A bar over a scalar or a vector denotes its complex conjugate, e.g., $\bar{\alpha}$, $\bar{\mathbf{x}}$.

Braces indicate sets, e.g., $\{x\}$.

1.3.3 Abbreviations and Symbols

$(.,.)$, $(.,.]$ and $[.,.]$ denote, respectively, open, semiclosed, and closed interval, respectively, e.g., t in the interval $(a,b]$ means $a < t \leq b$.

||.||: norm of a vector or a matrix, e.g., $\|x\|$, $\|A\|$.
$\triangleq$: equal by definition
$\Rightarrow$: implies
$\Leftarrow$: is implied by
$\Leftrightarrow$: implies and is implied by
$\exists$: there exists
$\forall$: for all
$\ni$: such that
ϵ: belongs to, e.g., $x \epsilon S$
$\notin$: does not belong to
$\supset$: contains, e.g., $S_1 \supset S_2$
$\subset$: is contained in
$\cup$: union
$\cap$: intersection
$\oplus$: direct sum
Δ: end of an example or discussion
$*$: convolution
$(.,.)$: inner product
j: the imaginary number $\sqrt{-1}$
adj: adjoint of a matrix, e.g., $adj\ A$
$det(.)$: determinant of a matrix, e.g., $det(A)$
$\rho(.)$: rank of a matrix, e.g., $\rho(A)$
$\gamma(.)$: nullity of a matrix, e.g. $\gamma(A)$
$tr(.)$: trace of a matrix, e.g., $tr(A)$
$D^{-n}(.)$: nth-order integration of a function, e.g., $D^{-2}(f)$
$diag$: diagonal matrix
$q.e.d.$: quod erat demonstrandum (which was to be proved)
$l.t.i.$: linear time-invariant
$w.r.t.$: with respect to
KCL: Kirchoff's current law
KVL: Kirchoff's voltage law
$r.h.p.$: right-half plane
$l.h.p.$: left-half plane
$exp(.)$: exponential of a scalar or a matrix, e.g., $exp(At) \triangleq e^{At}$
SISO: single-input single-output
MIMO: multi-input multi-output
AMRE: algebraic matrix Riccati equation
DMRE: discrete matrix Riccati equation
TPBVP: two-point boundary value problem

1.4 CLASSIFICATION OF SYSTEMS

From this point on the word "system" means a system model, i.e., a mathematical representation of a physical system. (See Chapter 4.) Systems can be classified according to the type of equations describing them. To this end there are five different ways of classifying systems.

1.4.1 Lumped-Parameter and Distributed-Parameter Systems

Lumped-parameter systems are those which can be described by ordinary differential (or difference) equations. Such systems are also referred to as finite-dimensional systems for reasons which will become clear later. In contrast, *distributed-parameter* systems, referred to as infinite-dimensional systems, are those which require partial differential equations for their characterization.

A lumped electric circuit, i.e., one in which the element sizes are negligible compared to the wavelength of the highest frequency of operation, is an example of a lumped-parameter system. A transmission line is an example of a distributed-parameter system. We will be concerned only with lumped-parameter systems in this book.

1.4.2 Deterministic and Stochastic Systems

In *deterministic* systems all the parameters can be described exactly. In *stochastic* systems, however, some (or all) parameters can be described only probabilistically, i.e., as random variables. We will deal only with deterministic systems in this book.

1.4.3 Continuous-Time and Discrete-Time Systems

In *continuous-time* systems, described by differential equations, all the variables are defined for all values of time in typically a semi-infinite interval [t_o,∞). In contrast, the variables in *discrete-time* systems are defined only at discrete instants of time. Discrete-time systems are described by difference equations.

1.4.4 Linear and Nonlinear Systems

A *linear* system is one to which the superposition principle applies. That is, a linear system exhibits proportionality of input and output. This, however, is an over-simplification: for a rigorous definition of system linearity will be given in Chapter 4. A linear system can be described by a linear differential or difference equation. A system which is not linear is called a *nonlinear* system.

1.4.5 Time-Invariant and Time-Varying System

A system is called *time-invariant* or *stationary* if all its parameters are constant. Such systems can be described by constant-coefficient differential or difference equations. If one or more of the system parameters vary with time, the system will be called *time-varying* or *nonstationary*. Such systems are described by differential or difference equations with time-varying coefficients.

It should be noted that a given system falls into one of the two categories in each of the above classifications. For example a system may be lumped, stochastic, continuous-time, linear and time-varying.

1.5 SCOPE OF THE BOOK

The main theme of the book is the application of computers in the analysis and design of linear systems. The book is divided into three parts:

Part I.   Mathematical Background
Part II.   Analysis
Part III.   Design

Part I provides the mathematical background on which Parts II and III are based. This part includes a review of linear algebra, Laplace and z transforms. Each topic is supported by several computer programs and accompanied by many numerical examples.

Part II deals with the analysis of linear systems. Topics such as system modeling, model reduction in large-scale linear systems as well as the solution of system equations and stability are included in this part. Also, controlability and observability of linear systems are addressed in Part II. Relevant ready-to-use computer programs and numerical examples support each topic.

Design of linear control systems from the classical and modern points of view is considered in Part III. This part treats the design of linear systems by compensation, state estimation in linear systems, parameter optimization and optimal control. As in Parts I and II, every topic in Part III is accompanied by computer programs as well as numerical examples.

SCOPE OF THE BOOK

The book is mainly concerned with lumped, deterministic, linear time-invariant systems. Both continuous-time and discrete-time systems are treated in each topic. At some points, digressions may be made to nonlinear and/or time-varying systems.

The computer programs are written in BASIC language. An appendix is included at the end of the book which contains the source code for various subroutines.

REFERENCES

[1.1] O. Mayr, *The Origins of Feedback Control*, M.I.T. Press, 1970 (Also in *Scientific American*, Vol. 223, pp. 110-118, 1970)

[1.2] J. C. Maxwell, "On Governors", *Proc. Royal Soc. of London*, Vol. 16, pp. 270-283, 1868. Also in *Selected Papers on Mathematical Trends in Control Theory*, R. Bellman and R. Kalaba (Eds.), pp. 3-17, Dover, 1964

[1.3] L. A. Zadeh and C. A. Desoer, *Linear System Theory*, McGraw-Hill, 1963

PART I

MATHEMATICAL BACKGROUND

CHAPTER 2

REVIEW OF LINEAR ALGEBRA

2.1 INTRODUCTION

In this chapter we review some results in linear algebra which are essential to the study of linear systems. In order to maintain a reasonable size for the chapter, we assume that the reader has some basic knowledge of linear algebra. Thus we dispense with elementary definitions such as sets, vectors, matrices and determinants, and basic operations such as matrix addition, multiplication and inversion. Also the proofs of some theorems may be omitted or may be left as exercises to the reader.

2.2 FIELDS AND VECTOR SPACES

1. **Definition**. A field F consists of a set F whose elements are called scalars, two operations called addition "+" and multiplication "." and two distinguished elements 0 and 1. The set is closed under addition and multiplication. Also, the two operations are cumulative and associative and multiplication is distributive w.r.t. addition. That is, we have the following six axioms

a) For all $\alpha \epsilon F$, $\alpha+0 = \alpha$ and $\alpha \cdot 1 = \alpha$.

b) For all $\alpha \epsilon F$ there exists $\beta \epsilon F$ such that $\alpha + \beta = 0$. Then β is indicated $-\alpha$.

c) For all $\alpha \epsilon F$, $\alpha \neq 0$ there exists $\gamma \epsilon F$ such that $\alpha \cdot \gamma = 1$. Then γ is indicated α^{-1}.

d) For $\alpha, \beta \epsilon F$, $\alpha+\beta = \beta+\alpha \epsilon F$ and $\alpha \cdot \beta = \beta \cdot \alpha \epsilon F$.

e) For $\alpha,\beta,\gamma \epsilon F$, $\alpha+(\beta+\gamma) = (\alpha+\beta) + \gamma$ and $\alpha \cdot (\beta \cdot \gamma) = (\alpha \cdot \beta) \cdot \gamma$.

f) For $\alpha,\beta,\gamma \epsilon F$, $\alpha \cdot (\beta+\gamma) = \alpha \cdot \beta + \alpha \cdot \gamma$.

2. **Example.** The set of all real numbers forms a field called the real field denoted by R.

3. **Example.** The set of all complex numbers forms a field called the complex field denoted by C.

4. **Example.** The set of all positive numbers does not form a field because axiom (b) is not satisfied.

5. **Definition.** A *vector space* (also referred to as a *linear space* or a *linear vector space*) V over a field F, denoted by (V,F), or simply by V when no ambiguity might arise, is a set of elements called *vectors* and two operations called *vector addition* and *scalar multiplication* satisfying the following nine axioms:

a) For $x,y \epsilon V$, $x+y = y+x \epsilon V$.

b) For $x,y,z \epsilon V$ $x+(y+z) = (x+y) + z$.

c) For all $x \epsilon V$ there exists a unique vector $0 \epsilon V$ such that $x+0 = x$.

d) For all $x \epsilon V$ there exists a unique vector $y \epsilon V$ such that $x+y = 0$.

e) For any $x \epsilon V$, $1 \cdot x = x$ where $1 \epsilon F$.

f) $\alpha \epsilon F$ and $x \epsilon V$, implies $\alpha \cdot x \epsilon V$.

g) $\alpha,\beta \epsilon F$ and $x \epsilon V$ implies $\alpha \cdot (\beta \cdot x) = (\alpha \cdot \beta) \cdot x$.

h) $\alpha \epsilon F$ and $x,y \epsilon V$ implies $\alpha \cdot (x+y) = \alpha \cdot x + \alpha \cdot y$.

i) $\alpha,\beta \epsilon F$ and $x \epsilon V$ implies $(\alpha+\beta) x = \alpha \cdot x + \beta \cdot x$.

6. **Example.** Consider a field F. Let F^n denote all n-tuples of the form

$$x = \begin{bmatrix} x_1 \\ x_2 \\ \cdot \\ \cdot \\ \cdot \\ x_n \\ \cdot \\ \cdot \end{bmatrix}, \quad x_i \epsilon F, \ i = 1,2, \cdots ,n.$$

LINEAR INDEPENDENCE AND RANK

If vector addition and scalar multiplication are defined as

$$\mathbf{x}+\mathbf{y} = \begin{bmatrix} x_1+y_1 \\ x_2+y_2 \\ \cdot \\ \cdot \\ \cdot \\ x_n+y_n \end{bmatrix}, \; x_i, y_i \epsilon F, \; i = 1,2,\cdots,n$$

and

$$\alpha \cdot \mathbf{x} = \begin{bmatrix} \alpha x_1 \\ \alpha x_2 \\ \cdot \\ \cdot \\ \cdot \\ \alpha x_n \end{bmatrix}, \; \alpha \epsilon F, \; x_i \epsilon F, \; i = 1,2,\cdots,n,$$

then F^n forms a vector space over F, (F^n, F). Examples of this vector space are (R^n, R), the space of n-dimensional real vectors (the n-dimensional Euclidean space) and (C^n, C), the space of n-dimensional complex vectors.

2.3 LINEAR INDEPENDENCE AND RANK

1. **Definition**. A set of vectors $v_1, v_2, \cdots, v_n$ belonging to a vector space (V,F) are said to be *linearly independent* if $\sum_{i=1}^{n} \alpha_i v_i = 0$, where $\alpha_1, \alpha_2, \ldots, \alpha_n \epsilon F$ implies that $\alpha_1 = \alpha_2 = \cdots = \alpha_n = 0$. If at least one $\alpha_i \epsilon F$ is nonzero, then the vectors are said to be *linearly dependent*. The sum $\sum_{i=1}^{n} \alpha_i v_i$ is referred to as a *linear combination* of vectors $v_1, v_2, \cdots, v_n$.

Note that if $v_i = 0$ for any $i=1,2,\cdots,n$, then the vectors $v_1, v_2, \cdots, v_n$ will be linearly dependent.

2. **Example**. The vectors $v_1 = [1,0,0]'$, $v_2 = [0,1,0]'$ and $v_3 = [0,0,1]'$ are linearly independent. But v_1, v_2, v_3 with $v_4 = [2,0,-3]'$ form a linearly dependent set because $-2v_1 + 0 \cdot v_2 + 3v_3 + v_4 = 0$.

Program "LININD" determines whether a set of m n-dimensional real vectors are linearly independent. The source listing of "LININD" followed by an example is given below.

3. "LININD".

```
10    !        PROGRAM NAME : "LININD" PROG
20    PRINT "'LININD' DETERMINES WHETHER A SET OF m REAL
      VECTORS OF DIMENSION"
30    PRINT "n ARE LINEARLY INDEPENDENT.";LIN(2)
40    INPUT "Give the DIMENSION of the vectors:",N
50    PRINT "DIMENSION of the vectors=";N
60    INPUT "Give the NUMBER of the vectors:",M
70    PRINT "NUMBER of the vectors=";M;LIN(2)
80    CALL R(M,N)
90    END
100   SUB R(M,N)
110   OPTION BASE 1
120   DIM A(M,N),V(1,N)
130   FOR I=1 TO M
140   PRINT "VECTOR number";I;":"
150   CALL Mat(V(*),1,N,"V")
160   FOR J=1 TO N
170   A(I,J)=V(1,J)
180   NEXT J
190   NEXT I
200   CALL Rank(A(*),M,N,R)
210   IF (R=M) OR (R=N) THEN Lind
220   PRINT "Only";R;"of the";M;" vectors in the set are
      linearly independent."
230   GOTO 250
240 Lind: PRINT "The vectors are LINEARLY INDEPENDENT."
250   SUBEND
```

4. Example.

'LININD' DETERMINES WHETHER A SET OF m REAL VECTORS OF DIMENSION n ARE LINEARLY INDEPENDENT.

DIMENSION of the vectors= 4
NUMBER of the vectors= 4

VECTOR number 1 :

Row Vector V(4):

  1.0000E+00   2.0000E+00   3.0000E+00   4.0000E+00

VECTOR number 2 :

Row Vector V(4):

  2.0000E+00   4.0000E+00   6.0000E+00   8.0000F+00

VECTOR number 3 :

Row Vector V(4):

  2.0000F+00   4.0000E+00   6.0000F+00   8.0000F+00

VECTOR number 4 :

Row Vector V(4):

  0.0000E+00   -1.0000E+00   2.0000E+00   3.0000F+00

Only 2 of the 4 vectors in the set are linearly independent.

5. Definition. An $m \times n$ matrix A has *rank r*, denoted $\rho(A) = r$, if and only if it has at least one nonsingular $r \times r$ submatrix but has no nonsingular submatrix of order higher than r.

6. Definition. An $m \times n$ matrix A is said to have *full rank* if $\rho(A) = \min(m, n)$.

△

Note that a square matrix has full rank if and only if it has a nonzero determinant. The following theorem relates the concepts of linear independence and rank.

7. Theorem. Let vectors $v_1, v_2, \cdots, v_k$ in $[\rho^n, \rho]$, where $k < n$, form columns of matrix A. Then these vectors are linearly independent if and only if $\rho(A) = k$.

Proof. The proof follows from the following two facts:

1) $\rho(A) = k \Leftrightarrow$ at least one $k \times k$ submatrix of A is nonsingular and all submatrices of higher order are singular.

2) The determinant of a matrix is zero if and only if the columns of the matrix are linearly dependent.

△

Program "RANK" determines the rank of any real $m \times n$ matrix. The source listing of this program with an example follows.

8. "RANK".

```
10    !    PROGRAM NAME : "RANK" PROG
20         PRINT " 'RANK' DETERMINES THE RANK OF ANY REAL MxN
           MATRIX.";LIN(1)
30         INPUT "Give number of ROWS , COLUMNS ",M,N
40         CALL R(M,N)
50         END
60         SUB R(M,N)
70         OPTION BASE 1
```

```
80      DIM A(M,N)
90      CALL Mat(A(*),M,N,"A")
100     CALL Rank(A(*),M,N,R)
110     PRINT LIN(1);"The RANK of matrix A =";R
120     SUBEND
```

9. Example.

'RANK' DETERMINES THE RANK OF ANY REAL MxN MATRIX.

Matrix A(3 x 4):

0.0000E+00	0.0000E+00	0.0000E+00	0.0000E+00
1.0000E+00	2.0000E+00	3.0000E+00	4.0000E+00
-2.0000E+00	-3.0000E+00	1.0000E+00	5.0000E+00

The RANK of matrix A = 2

2.4 BASES AND DIMENSION

1. Definition. A set of vectors is said to *span* a vector space V if any vector in V can be expressed as a linear combination of vectors in that set.

2. Definition. A set of vectors in a vector space V is said to form a *basis* of V if

   a) they span V, and

   b) they are linearly independent.

△

Consider a vector space (V,F) with a basis $\{e_1, e_2, \ldots, e_n\}$. Then for any vector $v \in (V,F)$, unique elements $\alpha_1, \alpha_2, \ldots, \alpha_n \in F$ exist such that v has a representation

$$v = \sum_{i=1}^{n} \alpha_i e_i. \qquad (1)$$

(See Problem 2-4). The scalars $\alpha_1, \alpha_2, \ldots, \alpha_n$ are then called the *components* of v w.r.t. basis $\{e_1, e_2, \ldots, e_n\}$.

3. Example. The vectors $e_1 = (2,0,0,)'$, $e_2 = (0,1,1,)'$, $e_3 = (-1,2,0)'$ and $x = (-3,0,4)'$ span R^3, however, they do not form a basis for R^3. The vectors e_1, e_2 and e_3 form a basis for R^3 since they are linearly independent. The vector x will then have components $\alpha_1 = -2.5$, $\alpha_2 = 4$ and $\alpha_3 = -2$ w.r.t. this basis because $x = \sum_{i=1}^{3} \alpha_i e_i$.

△

BASES AND DIMENSION

It can be proved [2.1] that all bases of a vector space have the same number of elements. This number is called the *dimension* of the vector space.

4. Definition. A vector space is said to be *finite dimensional* if its bases have a finite number of elements. Otherwise, it is an *infinite - dimensional* vector space.

5. Definition. Let (V,F) be a vector space and let $W \subset V$. Then (W,F) is called a *subspace* of (V,F) if (W,F) is a vector space.

6. Example.
1. The real vector space (R^n,R) is a subspace of the real vector space (R^{n+m},R) for any integers $m, n \geq 1$.
2. The real vector space (R^n,R) is a subspace of the complex vector space (C^n,C).

7. Definition. Consider a matrix A. Then vectors y such that $Ax = y$ for some $x \in (V,F)$ form a vector space $R(A)$ called the *range space* of matrix A, i.e.,

$$R(A) = \{Ax \ni x \in V\}. \qquad (2)$$

Note that $R(A)$ is a subspace of (V,F). (See Problem 2-6.)

8. Theorem. The rank of matrix A is equal to the dimension of its range space.

Proof. The proof follows from the definitions of rank and range space. It is left to the reader.

9. Definition. Consider a matrix A. Then vectors $x \in (V,F)$ such that $Ax = 0$ form a vector space $N(A)$, a subspace of (V,F), called the *null space* of matrix A, i.e.,

$$N(A) = \{x \in V \ni Ax = 0\} \qquad (3)$$

(See Problem 2-6.)

10. Definition. The dimension of the null space of matrix A is called the *nullity* of A denoted by $\gamma(A)$.

11. Definition. A subspace W is said to be *invariant* A if $x \in W$ implies $Ax \in W$.

Δ

In Problem (2-6) the reader is asked to show that the range space and the null space of a matrix A are invariant subspaces under A.

12. Definition. A vector space V is said to be the *direct sum* of two subspaces W_1 and W_2, denoted $V = W_1 \oplus W_2$, if W_1 and W_2 are subspaces of V and if every $x \in V$ can be uniquely represented as $x = x_1 + x_2$ where $x_1 \in W_1$, and $x_2 \in W_2$.

13. Example. Vector space R^3 is the direct sum of the two subspaces R^1 and R^2, i.e.,

$$R^3 = R^1 \oplus R^2 \qquad (4)$$

14. Theorem. Let $C^n = M \oplus N$ where M has dimension k and is invariant under $\mathbf{A}$. Then a basis for C^n exists w.r.t which $\mathbf{A}$ has a representation of the form

$$k \left\{ \begin{bmatrix} \overset{k}{A_{11}} & A_{12} \\ \cdots & \cdots \\ 0 & A_{22} \end{bmatrix} \right\} n-k \qquad (3)$$

$$n-k$$

Proof. Consider a basis $\{u_1, u_2, \cdots, u_n\}$ for C^n where $\{u_1, u_2, \cdots u_n\}$ is a basis for M and $\{u_{n+1}, u_{n+2}, \cdots u_n\}$ is a basis for N. We will show that w.r.t. this basis $\mathbf{A}$ has the representation given in (3). Since M is invariant under $\mathbf{A}$, $x \epsilon M$ implies $\mathbf{A}x \epsilon M$. But $x \epsilon M$ means that numbers α_i exist such that $x = \sum_{i=1}^{k} \alpha_i u_i$. Since $\mathbf{A}x \epsilon M$, numbers β_i must exist such that $\mathbf{A}x = \sum_{i=1}^{k} \beta_i u_i$. This implies that the $(n-k) \times k$ block shown in (3) must indeed be a zero matrix.

2.5 INNER PRODUCT AND NORM

1. Definition. The *inner product* of two vectors $x, y \epsilon (V, F)$ is a complex number denoted by (x, y) satisfying the following properties.

a) $(x, x) > 0$ for *all* $x \neq 0 \epsilon V$.

b) $(x, x) = 0 \Leftrightarrow x = 0 \epsilon V$.

c) $\overline{(x, y)} = (y, x)$ for *all* $x, y \epsilon V$.

d) $(\alpha x + \beta y, z) = \overline{\alpha}(x, z) + \overline{\beta}(y, z)$ for *all* $x, y, z \epsilon V$ and $\alpha, \beta \epsilon F$.

An inner product is sometimes also referred to as a *scalar product* or a *dot product*.

2. Example. An inner product for the vector space (C^n, C) is

$$(x, y) = \sum_{i=1}^{n} \overline{x}_i y_i, \qquad (1)$$

where x_i and y_i are the components of x and y, respectively, w.r.t. some basis. It can be easily verified that the above inner product satisfies the conditions of Definition 1. △

Program "INNERP" calculates the inner product of any two complex vectors in (C^n, C). The source listing of this program followed by an example is given below.

3. "INNERP".

```
10   !       PROGRAM NAME : "INNERP" PROG
20   PRINT "´INNERP´ CALCULATES THE INNER PRODUCT OF ANY
     TWO COMPLEX VECTORS"
30   OPTION BASE 1
40   DIM Ar(15,1),Ai(15,1),Br(15,1),Bi(15,1)
50   INPUT " Vector Dimension (<16)",N
60   PRINT " Vector Dimension n = ";N;LIN(1)
70   REDIM Ar(N,1),Ai(N,1),Br(N,1),Bi(N,1)
80   PRINT "Give real components of the first vector"
90   CALL Mat(Ar(*),N,1,"Ar")
100  PRINT "Give imaginary components of the first vector"
110  CALL Mat(Ai(*),N,1,"Ai")
120  PRINT "Give real components of the second vector"
130  CALL Mat(Br(*),N,1,"Br")
140  PRINT "Give imaginary components of the second vector"
150  CALL Mat(Bi(*),N,1,"Bi")
160  Cr=Ci=0
170  FOR I=1 TO N
180  Cr=Cr+Ar(I,1)*Br(I,1)+Ai(I,1)*Bi(I,1)
190  Ci=Ci+Ar(I,1)*Bi(I,1)-Ai(I,1)*Br(I,1)
200  NEXT I
210  PRINT "Real Part of Inner Product = ";Cr
220  PRINT "Imaginary Part of Inner Product = ";Ci
230  END
```

4. Example.

´INNERP´ CALCULATES THE INNER PRODUCT OF ANY TWO COMPLEX VECTORS
Vector Dimension n =   3.0000E+00

Give real components of the first vector

Column Vector Ar(3):

 1.0000E+00

 3.0000E+00

 0.0000E+00
Give imaginary components of the first vector

Column Vector Ai(3):

 6.0000E+00

-1.0000E+00

 4.0000E+00

Give real components of the second vector

Column Vector Br(3):

-1.0000E+00

2.0000E+00

3.0000E+00

Give imaginary components of the second vector

Column Vector Bi(3):

0.0000E+00

5.0000E+00

2.0000E+00

Real Part of Inner Product = 8.0000E+00
Imaginary Part of Inner Product = 1.1000E+01

5. Definition. : Two vectors x and y are said to be *orthogonal* if

$$(x,y) = 0 \qquad (2)$$

6. Definition. A set of vectors $x_1, x_2, \cdots, x_n$ are said to be *orthonormal* if

$$(x_i, x_j) = \delta_{ij}, \quad i,j=1,2,\ldots,n \qquad (3)$$

where δ_{ij} is the *Kronecker delta,* i.e.

$$\delta_{ij} = \begin{cases} 0 & \text{if } i \neq j \\ 1 & \text{if } i=j \end{cases} \qquad (4)$$

7. Example. Vectors $e_1, e_2, \cdots, e_n$ in (R^n, R) where

$$e_i = \begin{bmatrix} 0 \\ \cdot \\ \cdot \\ \cdot \\ 0 \\ 1 \\ 0 \\ \cdot \\ \cdot \\ 0 \end{bmatrix} \leftarrow \text{ith position} \qquad (5)$$

are orthonormal. Further, they are linearly independent and form a basis for (R^n,R), called an *orthonormal basis*.

8. Definition. Consider a matrix **A**. The *conjugate transpose* of **A**, denoted by **A***, is another matrix defined by

$$(Ax,y) = (x,A^*y) \quad \forall x,y \tag{6}$$

Note that if **A** is a real matrix, then its transpose and its conjugate transpose will be the same, i.e., $A' = A^*$. Some of the properties of conjugate transpose are given in Problem 2.9.

9. Definition. Matrix **A** is said to be *symmetric* if $A = A'$, *skew symmetric* if $A = -A'$, *hermitian* if $A = A^*$, *normal* if $A^*A = AA^*$, and *unitary* or *orthogonal* if $A^*A = I$ (identity matrix).

10. Definition. Consider a vector space (V,F) where $F = R$ or $F = C$. It is said to be a *normed vector space* if there exists a function, denoted by $\|.\|$ and called a *norm* on V, which maps V into R_+ (the set of nonnegative real numbers) and satisfies the following postulates:

a) $\|x\| = 0 \Leftrightarrow x = 0$

b) $\|\alpha x\| = |\alpha| \|x\| \quad \forall \alpha \in F, \ x \in V$

c) $\|x+y\| \leq \|x\| + \|y\| \quad \forall x,y \in V$ (*the triangle inequality*)

Then $\|x\|$ is called the *norm* of **x**.

△

The norm of a vector is a nonnegative real number which is a measure of its "length". Different norms may be defined or a given vector space. For example, the norms $\|\cdot\|_1$, $\|\cdot\|_2$, $\|\cdot\|_p$ and $\|\cdot\|_\infty$ are defined as follows. Let **x** have components $x_1, x_2, \cdots, x_n$ w.r.t. some basis. Then

$$\|x\|_1 = \sum_{i=1}^{n} |x_i| \tag{7}$$

$$\|x\|_2 = \left(\sum_{i=1}^{n} |x_i|^2\right)^{1/2} = (x,x)^{1/2} \tag{8}$$

$$\|x\|_p = \left(\sum_{i=1}^{n} |x_i|^p\right)^{1/p} \tag{9}$$

$$\|x\|_\infty = \max_i |x_i| \tag{10}$$

The norm $\|\cdot\|_2$ is also called the *Euclidean norm*. Note that if orthonormal basis is used, the Euclidean norm of a vector in the two-dimensional vector space is indeed the same as its length.

One of the major uses of norm is that it reduces the concept of convergence of vector sequences to that of real number sequences. More precisely, given a norm $\|\cdot\|$ on a vector space (V,F) we say a sequence of vectors $x_1, x_2, \cdots, x_n \in V$ converges to a vector $x \in V$ in that norm if and only if the sequence of nonnegative real numbers $\|x - x_i\|$ converges to zero.

11. Definition. Two norms are said to be *equivalent* if any sequence of vectors which converges in one norm, also converges in the other. (See Problem 2-10.)

△

Consider an $n \times n$ matrix A mapping (C^n, C) into itself. Then for any vector $x \in C^n$, Ax will also be a vector in C^n. The *norm of matrix* A can be defined in terms of the vector norms $\|x\|$ and $\|Ax\|$ as follows.

12. Definition. The norm of a matrix A, denoted by $\|A\|$, is the minimum value of k such that

$$\|Ax\| \leq k \|x\| \quad \forall x. \tag{11}$$

Using the above definition, the reader can verify that a matrix norm has the following properties:

a) $\|A\| = 0 \Leftrightarrow A = 0$. (12)

b) $\|Ax\| \leq \|A\| \|x\| \quad \forall x$. (13)

c) $\|A\| = \max \|Ax\|$ (14)

$\|x\| = 1$

d) $\|A + B\| \leq \|A\| + \|B\|$. (15)

e) $\|AB\| \leq \|A\| \|B\|$. (16)

INNER PRODUCT AND NORM

The *Euclidean norm* of an $n \times n$ matrix **A** is defined as

$$\|A\|_E = \left(\sum_{i=1}^{n} \sum_{j=1}^{n} |a_{ij}|^2 \right)^{1/2} \qquad (17)$$

This norm is easy to calculate and it provides a bound for $\|A\|$. (See Problem 2-13.)

△

Program "NORM" determines the Euclidean norm of any vector or square matrix. The source listing of "NORM" with two examples follow.

13. **"NORM"**.

```
10   !         PROGRAM NAME : "NORM" PROG
20   PRINT " 'NORM' FINDS THE EUCLIDEAN NORM OF ANY VECTOR
     'V' OR"
30   PRINT "    ANY SQUARE MATRIX 'A'.";LIN(2)
40   DIM A(20,20)
50   INPUT "Have SUB's <<Mat>> & <<Norm>> already been
     LINKED (Y/N)?",C$
60   IF (C$="Y") OR (C$="y") THEN 110
70   IF (C$="N") OR (C$="n") THEN 90
80   GOTO 50
90   LINK "Mat",400,100
100  LINK "Norm",2000,110
110  INPUT "Do you want the norm of a VECTOR or a MATRIX
     (V/M)?",A$
120  IF A$="V" THEN 210
130  INPUT "Order of the matrix (<=20) ?",N
140  PRINT "Order of the matrix = ";N
150  REDIM A(N,N)
160  FLOAT 4
170  CALL Mat(A(*),N,N,"A")
180  CALL Norm(A(*),N,M,Anorm)
190  PRINT "Norm of matrix A = ";Anorm
200  GOTO 270
210  INPUT "Dimension of the vector (<=20) ?",N
220  PRINT "Dimension of the vector =";N
230  CALL Mat(A(*),N,1,"V")
240  CALL Norm(A(*),N,1,Anorm)
250  FLOAT 4
260  PRINT "Norm of vector V = ";Anorm
270  END
```

14. Example.

`NORM` FINDS THE EUCLIDEAN NORM OF ANY VECTOR `V` OR ANY SQUARE MATRIX `A`.

Order of the matrix = 5

Matrix A(5 x 5):

```
 1.0000E+00   0.0000E+00   2.0000E+00  -3.0000E+00   4.0000E+00
-6.0000E+00   2.0000E+00   0.0000E+00   1.0000E+00   5.0000E+00
-8.0000E+00   6.0000E+00   0.0000E+00   2.0000E+00   0.0000E+00
 4.0000E+00   3.0000E+00   7.0000E+00  -9.0000E+00   2.0000E+00
-5.0000E+00   0.0000E+00   3.0000E+00   0.0000E+00   1.2000E+01
```

15. Example.

Norm of matrix A = 2.3173E+01

Dimension of the vector = 6

Column Vector V(6):

2.0000E+00

0.0000E+00

-4.0000E+00

6.0000E+00

5.0000E+00

-3.0000E+00

Norm of vector V = 9.4868E+00

2.6 EIGENVALUES AND EIGENVECTORS

1. Definition. The scalar λ is called an *eigenvalue* of the square matrix **A** if there exists a nonzero column vector **u** such that

$$\mathbf{Au} = \lambda \mathbf{u}. \tag{1}$$

The vector **u** is then called an *eigenvector* or a *right eigenvector* of **A** corresponding to eigenvalue λ. A row vector **v** satisfying

$$\mathbf{vA} = \lambda \mathbf{v} \tag{2}$$

is called a *left eigenvector* of **A** corresponding to eigenvalue λ. We will be concerned mainly with right eigenvector of matrices. Thus in order to find the eigenvalues and eigenvectors of a square matrix **A**, one has to solve

$$(\mathbf{A} - \lambda \mathbf{I})\mathbf{u} = 0. \tag{3}$$

Nontrivial solutions **u** to this equation exist only if

$$\det(\mathbf{A} - \lambda \mathbf{I}) = 0. \tag{4}$$

2. Definition. The *characteristic polynomial* of an $n \times n$ matrix **A**, denoted by $\Delta(\lambda)$, is $\det(\mathbf{A} - \lambda \mathbf{I})$, i.e.

$$\Delta(\lambda) = \det(\mathbf{A} - \lambda \mathbf{I}). \tag{5}$$

Δ

Note that the characteristic polynomial of an $n \times n$ matrix is an nth-order polynomial in λ. Thus there are n eigenvalues $\lambda_1, \lambda_2, ..., \lambda_n$ which are the roots of this polynomial. Also note that the eigenvalues may be distinct or some of them may be repeated.

3. Example.: Consider the square matrix

$$A = \begin{bmatrix} 1 & 2 \\ 2 & 1 \end{bmatrix} \tag{6}$$

The characteristic polynomial of this matrix is

$$\Delta(\lambda) = \det(A-\lambda-I) = \begin{bmatrix} 1-\lambda & 2 \\ 2 & 1-\lambda \end{bmatrix} = \lambda^2-2\lambda-3 \tag{7}$$

The eigenvalues of A are the roots of $\Delta(\lambda)$, i.e., $\lambda_1 = -1$, $\lambda_2 = 3$. To find the corresponding eigenvectors determine u_1 and u_2 where $(A-\lambda_i I)u_i = 0$, $i=1,2$. That is,

$$(A-\lambda_1 I)u_1 = 0 \implies \begin{bmatrix} 2 & 2 \\ 2 & 2 \end{bmatrix} u_1 = 0 \implies u_1 = \begin{bmatrix} 1 \\ -1 \end{bmatrix} \tag{8}$$

$$(A-\lambda_2 I)u_2 = 0 \implies \begin{bmatrix} -2 & 2 \\ 2 & -2 \end{bmatrix} u_2 = 0 \implies u_2 = \begin{bmatrix} 1 \\ 1 \end{bmatrix} \quad \Delta \tag{9}$$

Note that if **u** is an eigenvector of **A**, then so is αu for any scalar α.

Program "EIGEN" calculates the eigenvalues and eigenvectors of any real square matrix. The source listing of "EIGEN" with an example is given below.

4. "EIGEN".

```
10   !       PROGRAM NAME : "EIGEN" PROG
20   PRINT " 'EIGEN' FINDS THE EIGENVALUES AND EIGENVECTORS OF"
30   PRINT " ANY REAL SQUARE MATRIX.";LIN(2)
40   INPUT "Give ORDER of the matrix:",N
50   PRINT "Order of the matrix =";N;LIN(1)
60   IF (N<1) OR (N<>INT(N)) THEN 40
70   CALL Mjmmz(N)
80   END
90   SUB Mjmmz(N)
100  OPTION BASE 1
110  DIM A(N,N),Vr(N,N),Vi(N,N),Er(N),Ei(N),Indx(N)
120  CALL Mat(A(*),N,N,"A")
130  CALL Eigen(N,A(*),Er(*),Ei(*),Vr(*),Vi(*),Indx(*))
```

```
140    PRINT LIN(2),"Real Components of Eigenvalues:",LIN(1)
150    FLOAT 4
160    MAT PRINT Er;
170    PRINT LIN(2),"Imaginary Components of Eigenvalues:",LIN(1)
180    MAT PRINT Ei;
190    PRINT LIN(2),"Real Components of Eigenvectors πContained
       in Columns→",LIN(1)
200    MAT PRINT Vr;
210    PRINT LIN(2),"Imaginary Components of Eigenvectors
       πContained in Columns→:" IN(1)
220    MAT PRINT Vi;
230    SUBEND
```

5. Example..

```
'EIGEN' FINDS THE EIGENVALUES AND EIGENVECTORS OF
ANY REAL SQUARE MATRIX.


Order of the matrix = 3

Matrix A( 3 x 3 ):

-2.0000E+00  -2.0000E+00   0.0000E+00

 0.0000E+00   0.0000E+00   1.0000E+00

 0.0000E+00  -3.0000E+00  -4.0000E+00


Real Components of Eigenvalues:

-2.0000E+00  -3.0000E+00  -1.0000E+00



Imaginary Components of Eigenvalues:

 0.0000E+00   0.0000E+00   0.0000E+00
```

Real Components of Eigenvectors πContained in Columns→

1.0000E+00   5.3452E-01   -8.1650E-01

0.0000E+00   2.6726E-01   4.0825E-01

0.0000E+00   -8.0178E-01   -4.0825E-01

Imaginary Components of Eigenvectors πContained in Columns→:

0.0000E+00   0.0000E+00   0.0000E+00

0.0000E+00   0.0000E+00   0.0000E+00

0.0000E+00   0.0000E+00   0.0000E+00

Program "CHRPOL" determines the characteristic polynomial of any real square matrix. The source listing of "CHRPOL" with an example follows.

6. "CHRPOL".

```
10    !    PROGRAM NAME : "CHRPOL" PROG
20    PRINT " ´CHRPOL´ DETERMINES THE CHARACTERISTIC POLYNOMIAL
      P(s) OF ANY"
30    PRINT "REAL nxn MATRIX : ";LIN(1)
40    PRINT "P(s)=(-s)^n+SUM OF P(i)*s^i  ,    i=0 TO n-1";LIN(2)
50    INPUT "Give ORDER of the square matrix ",N
60    PRINT "Order of the square matrix =";N;LIN(1)
70    CALL R(N)
80    END
90    SUB R(N)
100   OPTION BASE 1
110   DIM A(N,N),P(N+1),Q(N,N,N)
120   CALL Mat(A(*),N,N,"A")
130   CALL Resmat(N,A(*),P(*),Q(*))
140   P(N+1)=1
150   L=(-1)^N
160   MAT P=(L)*P
170   PRINT "The coefficients of the CHARACTERISTIC POLYNOMIAL
      in ASCENDING order ;LIN(1)
180     FOR I=1 TO N+1
190     PRINT "P(";I-1;")=";P(I)
200     NEXT I
210   SUBEND
```

7. Example.

```
CHRPOL   DETERMINES THE CHARACTERISTIC POLYNOMIAL P(S) OF ANY
REAL nxn MATRIX :

P(s)=(-s)^n+SUM OF P(i)*s^i ,    i=0 TO n-1


Order of the square matrix = 4


Matrix A( 4 x 4 ):

1.0000E+00    3.0000E+00    0.0000E+00    0.0000E+00

5.0000E+00    2.0000E+00    1.0000E+00   -3.0000E+00

0.0000E+00    1.0000E+00    0.0000E+00   -5.0000E+00

-2.0000E+00    0.0000E+00   -3.0000E+00    1.0000E+00
```

The coefficients of the CHARACTERISTIC POLYNOMIAL in ASCENDING order:

```
P( 0 ) = 173
P( 1 ) =  33
P( 2 ) = -26
P( 3 ) = -4
P( 4 ) =  1
```

8. Definition. An eigenvector **u** is said to be *normalized* if $\|u\|_2 = 1$.   △

The eigenvectors u_1 and u_2 in Example 3 are normalized as follows:

$$u_1 = \frac{1}{\sqrt{2}} \begin{bmatrix} 1 \\ -1 \end{bmatrix}, \tag{10}$$

$$u_2 = \frac{1}{\sqrt{2}} \begin{bmatrix} 1 \\ 1 \end{bmatrix} \tag{11}$$

9. Theorem. If a matrix has distinct eigenvalues, then its eigenvectors will be linearly independent.
Proof. Consider an $n \times n$ matrix **A** with distinct eigenvalues $\lambda_1, \lambda_2, \cdots, \lambda_n$. Call the corresponding eigenvectors $u_1, u_2, \cdots, u_n$. That is

$$Au_i = \lambda_i u_i, \; i=1,2,...,n \tag{12}$$

Assume that the eigenvectors are not linearly independent. Then there exists scalars $\alpha_1, \alpha_2, \ldots, \alpha_n$, not all zero, such that

$$\alpha_1 u_1 + \alpha_2 u_2 + \cdots + \alpha_n u_n = 0 \tag{13}$$

Let

$$\alpha_k \neq 0. \tag{14}$$

Then from (13) we have

$$(A - \lambda_1 I)(A - \lambda_2 I) \cdots (A - \lambda_{k-1} I)(A - \lambda_{k+1} I) \cdots (A - \lambda_n I) \times$$

$$\sum_{i=1}^{n} \alpha_i u_i = 0 \tag{15}$$

Using (12), (15) yields

$$\alpha_k (\lambda_k - \lambda_1)(\lambda_k - \lambda_2) \cdots (\lambda_k - \lambda_{k-1})(\lambda_k - \lambda_{k+1}) \cdots (\lambda_k - \lambda_n) u_k = 0 \tag{16}$$

Since, $u_k \neq 0$ and $\lambda_k \neq \lambda_i$ for $i \neq k$, (16) yields $\alpha_k = 0$. This contradicts the original assumption. Therefore, the eigenvectors $u_1, u_2, \ldots, u_n$ are linearly independent.

△

Note that the class of matrices whose eigenvectors are linearly independent contains the class of matrices with distinct eigenvalues. For example the matrix

$$A = \begin{bmatrix} 1 & 0 \\ 0 & 1 \end{bmatrix} \tag{17}$$

has repeated eigenvalues $\lambda_1 = \lambda_2 = 1$ but its eigenvectors are linearly independent:

$$u_1 = \begin{bmatrix} 1 \\ 0 \end{bmatrix}, \tag{18}$$

$$u_2 = \begin{bmatrix} 0 \\ 1 \end{bmatrix} \tag{19}$$

In fact, if λ is a repeated eigenvalue of a square matrix $\mathbf{A}$, then the number of linearly independent eigenvectors corresponding to λ will be equal to the nullity of $\mathbf{A}-\lambda\mathbf{I}$. Thus for the matrix $\mathbf{A}$ in (17) we have

$$\mathbf{A}-\lambda\mathbf{I} = \begin{bmatrix} 0 & 0 \\ 0 & 0 \end{bmatrix} \tag{20}$$

and $\gamma(\mathbf{A}-\lambda\mathbf{I}) = 2$ which indicates the existence of two linearly independent eigenvectors $\mathbf{u}_1$ and $\mathbf{u}_2$ in (18), corresponding to $\lambda=1$. But for the matrix

$$\mathbf{B} = \begin{bmatrix} 1 & 1 \\ 0 & 1 \end{bmatrix} \tag{21}$$

with eigenvalues $\lambda_1=\lambda_2=1=\lambda$, we have

$$\mathbf{B}-\lambda\mathbf{I} = \begin{bmatrix} 0 & 1 \\ 0 & 0 \end{bmatrix} \tag{22}$$

whose nullity $\gamma(\mathbf{B}-\lambda\mathbf{I})=1$ indicates the existence of only one (linearly independent) eigenvector

$$\mathbf{u} = \begin{bmatrix} 1 \\ 0 \end{bmatrix} \tag{23}$$

2.7 DIAGONALIZATION OF MATRICES

Consider an $n \times n$ matrix $\mathbf{A}$. If $\mathbf{M}$ is any nonsingular matrix, then matrices $\mathbf{A}$ and $\mathbf{M}^{-1}\mathbf{A}\mathbf{M}$ will have the same eigenvalues. (See Problem 2-16.) These matrices are said to be *similar*, the transformation $\mathbf{M}^{-1}\mathbf{A}\mathbf{M}$ is called a *similarity transformation* and $\mathbf{M}$ is called a *similarity transformation matrix*.

In this section we will show that a square matrix can be diagonalized, i.e., can be transformed to a similar diagonal matrix, if its eigenvectors are linearly independent. The concept of diagonalization of a square matrix is useful in uncoupling an nth-order system into n first-order subsystems as we will see later.

1. **Theorem.** Any square matrix can be transformed to a similar diagonal matrix if and only if its eigenvectors are linearly independent.

Proof. Consider an $n \times n$ matrix $\mathbf{A}$ with eigenvalues $\lambda_1, \lambda_2, \ldots, \lambda_n$ and corresponding linearly independent eigenvectors $\mathbf{u}_1, \mathbf{u}_2, \ldots, \mathbf{u}_n$. Note that we have $\mathbf{A}\mathbf{u}_i = \lambda_i \mathbf{u}_i$, $i=1,2,\cdots,n$ or equivalently,

$$\mathbf{A}\{\mathbf{u}_1, \mathbf{u}_2, \ldots, \mathbf{u}_n\} = \{\lambda_1 \mathbf{u}_1, \lambda_2 \mathbf{u}_2, \ldots, \lambda_n \mathbf{u}_n\}$$

$$= \{\mathbf{u}_1, \mathbf{u}_2, \ldots, \mathbf{u}_n\} \Lambda \tag{1}$$

where Λ is the diagonal matrix formed by the eigenvalues, i.e.,

$$\Lambda = \text{diag}\left[\lambda_1, \lambda_2, \ldots, \lambda_n\right]. \tag{2}$$

Let

$$\mathbf{M} = \{\mathbf{u}_1, \mathbf{u}_2, \ldots, \mathbf{u}_n\} \tag{3}$$

$\mathbf{M}$ is called the *modal matrix* of $\mathbf{A}$. Note that $\mathbf{M}$ is nonsingular since the eigenvectors are linearly independent. Thus (1) implies that $\mathbf{AM} = \mathbf{M}\Lambda$ or

$$\mathbf{M}^{-1}\mathbf{A}\mathbf{M} = \Lambda, \tag{4}$$

which completes the necessity part of the proof. The sufficiency part of the proof is straightforward and is left to the reader.

2. **Example.** Consider the matrix of Example 2-6-3

$$A = \begin{bmatrix} 1 & 2 \\ 2 & 1 \end{bmatrix} \tag{5}$$

whose eigenvalues were found to be $\lambda_1 = -1$, $\lambda_2 = 3$ with the corresponding eigenvectors $u_1 = [1, -1]'$ and $u_2 = [1, 1]'$. Since the eigenvectors are linearly independent, the matrix can be diagonalized using the modal matrix

$$M = \{u_1, u_2\} = \begin{bmatrix} 1 & 1 \\ -1 & 1 \end{bmatrix}. \tag{6}$$

That is,

$$M^{-1}AM = \begin{bmatrix} 1 & 1 \\ -1 & 1 \end{bmatrix}^{-1} \begin{bmatrix} 1 & 2 \\ 2 & 1 \end{bmatrix} \begin{bmatrix} 1 & 1 \\ -1 & 1 \end{bmatrix} = \begin{bmatrix} -1 & 0 \\ 0 & 3 \end{bmatrix} = \begin{bmatrix} \lambda_1 & 0 \\ 0 & \lambda_2 \end{bmatrix} = \Lambda. \tag{7}$$

2.8 GENERALIZED EIGENVECTORS AND JORDAN CANONICAL FORM

If the eigenvectors of a square matrix are not linearly independent, it cannot be diagonalized. However, it can be transformed into a similar matrix which is in block diagonal form. Such a similar matrix is called a *Jordan canonical form*. In order to define Jordan canonical form, we first have to discuss the concept of generalized eigenvectors.

1. **Definition.** A nonzero vector u for which $(A-\lambda I)^k u = 0$ but $(A-\lambda I)^{k-1} u \neq 0$, $k=1,2,...$, is called a *generalized eigenvector* of order k of A associated with eigenvalue λ.

△

From the above definition, the generalized eigenvectors corresponding to eigenvalue λ can be found as follows:

$$k=1 \implies (A-\lambda I)u_1 = 0$$

$$k=2 \implies (A-\lambda I)u_2 = 0 = (A-\lambda I)u_1$$

$$\implies (A-\lambda I)u_2 = u_1.$$

Similarly,

$$(A-\lambda I)^k u_k = 0 \implies (A-\lambda I) u_k = u_{k-1} \qquad (1)$$

Note that the generalized eigenvector of order 1, i.e., u_1, is the same as an *eigenvector* of A associated with eigenvalue λ.

It can be shown that the generalized eigenvectors of A associated with each eigenvalue are linearly independent. (See Problem 2-23.) Note that the total number of generalized eigenvectors (including eigenvectors) associated with each eigenvalue is equal to the multiplicity of that eigenvalue. (See Example 3 below.) It can be proved that the generalized eigenvectors associated with different eigenvalues form a linearly independent set [2.3].

2. **Example.** Consider the matrix

$$A = \begin{bmatrix} 2 & 1 & 0 \\ 1 & 2 & 1 \\ 0 & -1 & 2 \end{bmatrix} \qquad (2)$$

with characteristic polynomial $\Delta(\lambda) = \det(A-\lambda I) = (2-\lambda)^3$. Thus $\lambda_1=\lambda_2=\lambda_3=2$. The only nonzero vector resulting from $(A-2I)u = 0$ is $u = [1,0,-1]'$ (an eigenvector). Thus A has two generalized eigenvectors u_2 and u_3 which are found as follows:

$$(A-2I)u_2 = u_1 \implies u_2 = [1,1,-1]'$$

$$(A-2I)u_3 = u_2 \implies u_3 = [1,1,0]'$$

Note that u_1, u_2 and u_3 form a linearly independent set.

3. **Example.** Consider the matrix

$$B = \begin{bmatrix} 2 & 1 & 0.5 \\ 0 & 3 & 0.5 \\ 0 & -2 & 1 \end{bmatrix} \qquad (3)$$

GENERALIZED EIGENVECTORS AND JORDAN CANONICAL FORM

with characteristic polynomial $\Delta(\lambda)=\det(\mathbf{B}-\lambda\mathbf{I}) = -(\lambda-2)^3$. Thus again $\lambda_1 = \lambda_2 = \lambda_3 = 2$. However, in this example two linearly independent vectors result from $(\mathbf{B}-2\mathbf{I})\mathbf{u}=0$: $\mathbf{u}_1 = [1,0,0]'$ and $\mathbf{u}_2 = [1,1,-2]'$ (eigenvectors). Thus $\mathbf{B}$ has one generalized eigenvector $\mathbf{u}_3$ which can be found from $(\mathbf{B}-2\mathbf{I})\mathbf{u}_3 = \mathbf{u}_2$ as $\mathbf{u}_3 = [0,0,2]'$. Note that it appears that the choice of $\mathbf{u}_3$ is not unique. However, a construction procedure exists for $\mathbf{u}_3$ which will result in $\mathbf{u}_3$ as shown. This procedure is complex and will not be explained here.

4. **Example.** Consider the matrix

$$\mathbf{C} = \begin{bmatrix} 1 & 0 & 0 \\ 0 & 2 & 1 \\ 0 & 0 & 2 \end{bmatrix} \tag{4}$$

with eigenvalues $\lambda_1=1$ *and* $\lambda_2=\lambda_3=2$. The eigenvector $\mathbf{u}_1$ corresponding to λ_1 is found from $(\mathbf{C}-\lambda_1\mathbf{I})\mathbf{u}_1 = 0$ which yields $\mathbf{u}_1 = [1,0,0]'$. For eigenvalues λ_2 and λ_3, the only nonzero vector resulting from $(\mathbf{C}-2\mathbf{I})\mathbf{u} = 0$ is $\mathbf{u}_2 = [0,1,0]'$. Thus a generalized eigenvector $\mathbf{u}_3$ corresponding to $\lambda_2=\lambda_3=2$ exists which is found from $(\mathbf{C}-2\mathbf{I})\mathbf{u}_3=\mathbf{u}_2$ as $\mathbf{u}_3=[0,0,1]'$. Note that $\mathbf{u}_1$, $\mathbf{u}_2$ and $\mathbf{u}_3$ are linearly independent.

5. **Theorem.** Consider an $n \times n$ matrix $\mathbf{A}$ with eigenvalues λ_i, $i=1,2,\cdots,p$. Let each λ_i have multiplicity m_i. Then a similarity transformation matrix $\mathbf{M}$ exists which transforms $\mathbf{A}$ into its *Jordan Canonical form* $\mathbf{J}$; that is,

$$\mathbf{M}^{-1}\mathbf{A}\mathbf{M} = \mathbf{J} = \begin{bmatrix} \mathbf{J}_1 & & & 0 \\ & \mathbf{J}_2 & & \\ & & \ddots & \\ 0 & & & \mathbf{J}_p \end{bmatrix} \tag{5}$$

where each $\mathbf{J}_i$ is an $m_i \times m_i$ matrix corresponding to λ_i and has the general form

$$\mathbf{J}_i = \begin{bmatrix} \begin{matrix} \lambda_i & 1 & & \\ & \ddots & 1 & 0 \\ & & \ddots & 1 \\ 0 & & & \lambda_i \end{matrix} & 0 \\ 0 & \begin{matrix} \lambda_i & & \\ & \ddots & \\ & & \lambda_i \end{matrix} \end{bmatrix} \tag{6}$$

Each block of J_i is called a *Jordan block*. The proof of this theorem may be found in Reference [2.4]. It can also be shown [2.3] that the columns of **M** are the generalized eigenvectors of **A**.

Δ

Note that $\sum_{i=1}^{p} m_i = n$. Also note that the Jordan form of an $n \times n$ matrix **A** is always an nxn matrix which is in an upper triangular form. If **A** has m linearly independent eigenvectors, then its Jordan form **J** will have $n-m$ ones above the diagonal. In the special case where all the eigenvectors of **A** are linearly independent, then $p=n$; $m_i=1$, $i=1,2,\cdots,p$ and **J** reduces to $\Lambda = diag(\lambda_1, \lambda_2, \cdots, \lambda_n)$.

6. **Example.** Consider the matrices **A, B** and **C** in (2), (3) and (4). For **A** the modal matrix is

$$\mathbf{M}_a = [\mathbf{u}_1, \mathbf{u}_2, \mathbf{u}_3] = \begin{bmatrix} 1 & 1 & 1 \\ 0 & 1 & 1 \\ -1 & -1 & 0 \end{bmatrix} \quad (7)$$

and the Jordan form of **A** is

$$\mathbf{J}_a = \mathbf{M}_a^{-1}\mathbf{A}\mathbf{M}_a = \begin{bmatrix} 2 & 1 & 0 \\ 0 & 2 & 1 \\ 0 & 0 & 2 \end{bmatrix} \quad (8)$$

Similarly, for **B** the modal matrix is

$$\mathbf{M}_b = \begin{bmatrix} 1 & 1 & 0 \\ 0 & 1 & 0 \\ 0 & -2 & 2 \end{bmatrix} \quad (9)$$

and

$$\mathbf{J}_b = \mathbf{M}_b^{-1}\mathbf{B}\mathbf{M}_b = \begin{bmatrix} 2 & 0 & 0 \\ 0 & 2 & 1 \\ 0 & 0 & 2 \end{bmatrix} \quad (10)$$

Also,

$$\mathbf{M}_c = \begin{bmatrix} 1 & 0 & 0 \\ 0 & 1 & 0 \\ 0 & 0 & 1 \end{bmatrix} = I \quad (11)$$

and $\mathbf{J}_c = \mathbf{M}_c^{-1}\mathbf{C}\mathbf{M}_c = \mathbf{C}$ as expected, because **C** is already in Jordan canonical form.

7. **Example.** Consider a 4 × 4 matrix **A** with eigenvalues $\lambda_1=\lambda_2=\lambda_3=3$ and $\lambda_4=4$. Then its Jordan form will be

$$\mathbf{J} = \left[\begin{array}{c|c} \mathbf{J}_1 & 0 \\ \hline 0 & 4 \end{array}\right] \tag{12}$$

where the 3 × 3 matrix $\mathbf{J}_1$ will have one of the following forms:

a) $\mathbf{J}_1 = \begin{bmatrix} 3 & 0 & 0 \\ 0 & 3 & 0 \\ 0 & 0 & 3 \end{bmatrix}$ if **A** has 4 *linearly independent eigenvectors.*

b) $\mathbf{J}_1 = \begin{bmatrix} 3 & 1 & 0 \\ 0 & 3 & 0 \\ 0 & 0 & 3 \end{bmatrix}$ if **A** has 3 *linearly independent eigenvectors.*

c) $\mathbf{J}_1 = \begin{bmatrix} 3 & 1 & 0 \\ 0 & 3 & 1 \\ 0 & 0 & 3 \end{bmatrix}$ if **A** has 2 *linearly independent eigenvectors.*

Δ

For any $n \times n$ real matrix **A**, the characteristic polynomial (2-6-4) will have real coefficients. Thus complex eigenvalues of **A** will occur in conjugate pairs, i.e., if

$$\lambda_i \sigma_i + j\omega_i, \quad \omega_i \neq 0 \tag{13}$$

is an eigenvalue of **A**, then

$$\lambda_{i+1} = \bar{\lambda}_i = \sigma_i - j\omega_i \tag{14}$$

will also be an eigenvalue of **A**. Further, if

$$\mathbf{u}_i = \mathbf{v}_i + j\mathbf{w}_i \qquad (15)$$

is an eigenvector of **A** associated with λ_{i+1}, then

$$\mathbf{u}_{i+1} = \bar{\mathbf{u}}_i = \mathbf{v}_i - j\mathbf{w}_i \qquad (16)$$

will be an eigenvector of **A** associated with λ_{i+1}. Thus in such a case, the modal matrix **M** and the diagonal or the Jordan canonical form of **A** will have some complex entries. However, it is also possible to find a block diagonal *real* matrix which is similar to **A**. The following theorem, whose proof may be found in Reference [2.3], establishes this fact.

8. **Theorem.** Consider an $n \times n$ real matrix **A** with eigenvalues

$$\left.\begin{array}{l}\lambda_i = \sigma_i + j\omega_i \\ \lambda_{i+1} = \bar{\lambda}_i = \sigma_i - j\omega_i\end{array}\right\} i = 1, 3, \cdots, m-1 \qquad (17)$$

$$\lambda_i = \bar{\lambda}_i, \quad i = m+1, m+2, \ldots, p. \qquad (18)$$

Let each λ_i have multiplicity m_i. Then a similarity transformation matrix **M** exists which transforms **A** into its *block-Jordan form*, i.e.

$$\mathbf{M}^{-1}\mathbf{A}\mathbf{M} = \mathbf{J} = \begin{bmatrix} \mathbf{J}_1 & & & & & & & \\ & \mathbf{J}_3 & & & & & & \\ & & \cdot & & & & & \\ & & & \cdot & & & 0 & \\ & & & & \mathbf{J}_{m-1} & & & \\ & & & & & \mathbf{J}_{m+1} & & \\ & & & & & & \mathbf{J}_{m+2} & \\ & & 0 & & & & & \cdot \\ & & & & & & & & \cdot \\ & & & & & & & & & \mathbf{J}_p \end{bmatrix} \qquad (19)$$

For $i = 1, 3, \ldots, m - 1$, each $\mathbf{J}_i$ is a $2m_i \times 2m_i$ real matrix corresponding to λ_i which has the general form

$$\mathbf{J}_i = \begin{bmatrix} \Lambda_i & \mathbf{I} & & & & & & \\ & \Lambda_i & \mathbf{I} & 0 & & & & \\ & & \cdot & \cdot & & & 0 & \\ & & & \cdot & \mathbf{I} & & & \\ & & & & \Lambda_i & & & \\ \hline & & & & & \Lambda_i & & \\ & & & & & & \cdot & \\ & 0 & & & & & \cdot & \\ & & & & & & & \Lambda_i \end{bmatrix} \quad (20)$$

where

$$\Lambda_i = \begin{bmatrix} \sigma_i & \omega_i \\ -\omega_i & \sigma_i \end{bmatrix} \quad (21)$$

For $i = m + 1, m + 2, \ldots, p$ each $\mathbf{J}_i$ is an $m_i \times m_i$ real matrix as in (6). The modal matrix $\mathbf{M}$ is an $n \times n$ real matrix whose columns are the real and imaginary parts of the eigenvectors and the generalized eigenvectors of $\mathbf{A}$:

$$\mathbf{M} = \begin{bmatrix} \mathbf{v}_1, \mathbf{w}_1, \mathbf{v}_3, \mathbf{w}_3, \ldots, \mathbf{v}_{m-1}, \mathbf{w}_{m-1}, \mathbf{u}_{m+1}, \mathbf{u}_{m+2}, \ldots, \mathbf{u}_n \end{bmatrix} \quad (22)$$

2.9 FUNCTIONS OF A SQUARE MATRIX

Consider a finite polynomial $f(\cdot)$ of a scalar x:

$$f(x) = a_m x^m + a_{m-1} x^{m-1} + \cdots + a_1 x + a_0 \quad (1)$$

Then a corresponding polynomial of a square matrix $\mathbf{A}$ may be defined as

$$f(\mathbf{A}) = a_m \mathbf{A}^m + a_{m-1} \mathbf{A}^{m-1} + \cdots + a_1 \mathbf{A} + a_0 \mathbf{I} \quad (2)$$

If function $f(x)$ is analytic, then it can be uniquely expressed in a convergent Maclaurin series

$$f(x) = \sum_{k=0}^{\infty} a_k x^k / k! \qquad (3)$$

Where

$$a_k = \frac{d^k f(x)}{dx^k} \bigg|_{x=0} \qquad (4)$$

The concept of the function of a matrix can be extended to any analytic function as formalized by the following definition.

1. **Definition.** Let **A** be an $n \times n$ matrix with eigenvalues $\lambda_1, \lambda_2, \cdots, \lambda_p$ ($p \leq n$) and let $f(x)$ be a function which is analytic in an open set containing $\lambda_1, \lambda_2, \cdots, \lambda_p$ with a Maclaurin series (3). Then $f(\mathbf{A})$ is defined as

$$f(\mathbf{A}) = \sum_{k=0}^{\infty} a_k \mathbf{A}^k / k! \qquad (5)$$

2. **Example.** It is known that

$$e^{ax} = 1 + ax + \frac{a^2 x^2}{2!} + \cdots + \frac{a^K x^k}{k!} + \cdots \qquad (6)$$

Thus

$$e^{\mathbf{A}t} = \mathbf{I} + \mathbf{A}t + \frac{\mathbf{A}^2 t^2}{k!} + \cdots + \frac{\mathbf{A}^k t_k}{2!} + \cdots \qquad (7)$$

3. **Theorem.** Given an $n \times n$ matrix **A**, if **T** is any nonsingular $n \times n$ matrix then

$$\hat{\mathbf{A}} = \mathbf{T}^{-1} \mathbf{A} \mathbf{T} \implies f(\hat{\mathbf{A}}) = \mathbf{T}^{-1} f(\mathbf{A}) \mathbf{T} \qquad (8)$$

FUNCTIONS OF A SQUARE MATRIX

Proof. The proof follows from the following observation:

$$\hat{A}^k = \underbrace{\hat{A}.\hat{A} \cdots \hat{A}}_{k \text{ times}} = \underbrace{\left[T^{-1}AT\right]\left[T^{-1}AT\right] \cdots \left[T^{-1}AT\right]}_{k \text{ times}} = T^{-1}A^kT \qquad (9)$$

4. Theorem. *(Cayley-Hamilton Theorem).* Let A be a square matrix with characteristic polynomial $\Delta(\lambda) = \det(A - \lambda I)$. Then $\Delta(A) = 0$.

Proof. We will prove the theorem for the case where A has distinct eigenvalues. For the general case see Problem 2.25.

Let the $n \times n$ matrix A have distinct eigenvalues $\lambda_1, \lambda_2, ..., \lambda_n$ and the characteristic polynomial

$$\Delta(\lambda) = \alpha_n \lambda^n + \alpha_{n-1} \lambda^{n-1} + \cdots + \alpha_1 \lambda + \alpha_0 \qquad (10)$$

Thus A is diagonalizable and

$$M^{-1}AM = \Lambda \text{ or } A = M\Lambda M^{-1} \qquad (11)$$

(See Section 2.7.) where Λ is given by (2.7.2). From Theorem 3 we have

$$A^k = M\Lambda^k M^{-1}, \quad k = 0,1,2,...,n \qquad (12)$$

Therefore,

$$\Delta(A) = \alpha_n A^n + \alpha_{n-1} A^{n-1} + ... + \alpha_1 A + \alpha_0 I$$

$$= M\left[\alpha_n \Lambda^n + \alpha_{n-1} \Lambda^{n-1} + \cdots + \alpha_1 \Lambda + \alpha_0 I\right] M^{-1} \qquad (13)$$

Since Λ is a diagonal matrix it is easily seen that

$$\Delta(\mathbf{A}) = \mathbf{M} \begin{bmatrix} \Delta(\lambda_1) & & & 0 \\ & \Delta(\lambda_2) & & \\ & & \cdot & \\ & & & \cdot \\ 0 & & & \Delta(\lambda_n) \end{bmatrix} \mathbf{M}^{-1} = \mathbf{M}0\mathbf{M}^{-1} = 0 \qquad (14)$$

5. **Example.** Let

$$\mathbf{A} = \begin{bmatrix} 3 & 1 \\ 1 & 2 \end{bmatrix} \qquad (15)$$

Then

$$\Delta(\lambda) = \det(\mathbf{A} - \lambda \mathbf{I}) = \lambda^2 - 5\lambda + 5 \qquad (16)$$

and

$$\Delta(\mathbf{A}) = \mathbf{A}^2 - 5\mathbf{A} + 5\mathbf{I} = \begin{bmatrix} 10 & 5 \\ 5 & 5 \end{bmatrix} - 5\begin{bmatrix} 3 & 1 \\ 1 & 2 \end{bmatrix} + 5\begin{bmatrix} 1 & 0 \\ 0 & 1 \end{bmatrix} = \begin{bmatrix} 0 & 0 \\ 0 & 0 \end{bmatrix}$$

2.10 APPLICATIONS OF CALEY-HAMILTON THEOREM

Some important applications of Cayley-Hamilton Theorem will be discussed in this section. Using the theorem it can be shown that any mth-order polynomial

$$p(\mathbf{A}) = b_m \mathbf{A}^m + b_{m-1}\mathbf{A}^{m-1} + \cdots + b_1 \mathbf{A} + b_0 \mathbf{I} \qquad (1)$$

of an $n \times n$ matrix $\mathbf{A}$ ($m \geq n$) can be equivalently expressed as

$$p(\mathbf{A}) = a_{n-1}\mathbf{A}^{n-1} + a_{n-2}\mathbf{A}^{n-2} + \cdots + a_1 \mathbf{A} + a_0 \mathbf{I} \qquad (2)$$

This is true because for $m \geq n$ we can write

APPLICATIONS OF CAYLEY-HAMILTON THEOREM

$$p(\lambda) = \Delta(\lambda)D(\lambda) + R(\lambda) \qquad (3)$$

where $R(\lambda)$ is the remainder polynomial of order $n-1$ resulting from the division of $p(\lambda)$ by the characteristic polynomial $\Delta(\lambda)$ of **A**. Thus

$$p(\mathbf{A}) = \Delta(\mathbf{A})D(\mathbf{A}) + R(\mathbf{A}) = \Delta(\mathbf{A})0 + R(\mathbf{A}) = R(\mathbf{A}) \qquad (4)$$

Δ

Program "MATPOL" uses the above procedure to evaluate any mth order polynomial $p(\mathbf{A})$ of an $n \times n$ real matrix **A**. The source listing of "MATPOL" is given below followed by an example of a $4th$ order polynomial of a 2×2 matrix.

1. "MATPOL"

```
10    !    PROGRAM NAME  :  "MATPOL" PROG
20    PRINT " ´MATPOL´ EVALUATES ANY mTH-ORDER POLYNOMIAL p(A)
      OF ANY nxn REAL"
30    PRINT "MATRIX ´A´ USING CAYLEY-HAMILTON THEOREM:";LIN(2)
40    PRINT "p(A)= SUMπB(i)*A^i→   (i=0 TO m) = f(A)
      = SUMπR(j)*A^j→   (j=0 TO n-1)"
50    INPUT "Give ORDER of the matrix:",N
60    PRINT "Order of the matrix=";N
70    INPUT "Give ORDER of the matrix POLYNOMIAL:",M
80    PRINT "Order of the matrix polynomial=";M;LIN(1)
90    IF M<N THEN 130
100   N1=M-N
110   N2=M-1
120   GOTO 140
130   N2=M
140   CALL R(N,M,N1,N2)
150   END
160   SUB R(N,M,N1,N2)
170   DIM A(N,N),B(M),A1(N-1,N-1),R(N2)
180   CALL Mat(A(*),N,N,"A")
190   PRINT "Give COEFFICIENTS of the matrix polynomial in
      ASCENDING Order:"
200   CALL Vec(B(*),M,"B")
210   CALL Matpol(A(*),B(*),N,M,N1,A1(*),R(*))
220   PRINT LIN(1),"COEFFICIENTS of the resulting matrix
      polynomial in DESCENDING Order are:";LIN(1)
230     FOR I=N-1 TO 0 STEP -1
240       PRINT "R(";I;")=";R(I)
250     NEXT I
260   PRINT LIN(1),"The RESULTING matrix f(A) is:",LIN(1)
270   FLOAT 4
280   MAT PRINT A1;
290   SUBEND
```

2. Example.

```
'MATPOL' EVALUATES ANY mTH-ORDER POLYNOMIAL p(A)
OF ANY nxn REAL MATRIX 'A' USING CAYLEY-HAMILTON THEOREM:


p(A)= SUMπB(i)*A^i→   (i=0 TO m)  = f(A)
    = SUMπR(j)*A^j→   (j=0 TO n-1)

Order of the matrix= 2
Order of the matrix polynomial= 4
Matrix A( 2 x 2 ):

3.0000E+00    1.0000E+00

1.0000E+00    2.0000E+00

Give COEFFICIENTS of the matrix polynomial in ASCENDING Order:
B( 0 )= 1.0000E+00
B( 1 )= 1.0000E+00
B( 2 )= 2.0000E+00
B( 3 )= 3.0000E+00
B( 4 )= 1.0000E+00


COEFFICIENTS of the resulting matrix polynomial in
DESCENDING order are:

R( 1 )= 146
R( 0 )=-184

The RESULTING matrix f(A) is:

  2.5400E+02           1.4600E+02

  1.4600E+02           1.0800E+02
```

If the $n \times n$ matrix **A** has an inverse, it can similarly be shown that

$$\mathbf{A}^{-1} = c_{n-1}\mathbf{A}^{n-1} + c_{n-2}\mathbf{A}^{n-2} + \cdots + c_1\mathbf{A} + c_0\mathbf{I} \tag{5}$$

This follows from the fact that

$$\Delta(\mathbf{A}) = \alpha_n \mathbf{A}^n + \alpha_{n-1}\mathbf{A}^{n-1} + \cdots + \alpha_1\mathbf{A} + \alpha_0\mathbf{I} = 0 \tag{6}$$

APPLICATIONS OF CAYLEY-HAMILTON THEOREM

Thus

$$A^{-1}\Delta(A) = \alpha_n A^{n-1} + \alpha_{n-1} A^{n-2} + \cdots + \alpha I + \alpha_o A^{-1} = 0 \qquad (7)$$

which implies that

$$A^{-1} = -\frac{1}{\alpha_o}\left[\alpha_n A^{n-1} + \alpha_{n-1} A^{n-2} + \cdots + \alpha_1 I\right] \qquad (8)$$

Note that $\alpha_o = \det(A) \neq 0$ if A^{-1} exists.

Program "INVMAT" implements the above procedure. The source listing of this program followed by an example is given below.

3. "INVMAT"

```
10    !         PROGRAM NAME : "INVMAT" PROG
20    PRINT " 'INVMAT' COMPUTES THE INVERSE OF A MATRIX THROUGH"
30    PRINT " THE USE OF CAYLEY HAMILTON THEOREM.";LIN(2)
40       OPTION BASE 1
50       DIM A(9,9),B(9,9),A1(9,9),I1(9,9),P(9),Q(9,9,9),C(9,9)
60       INPUT "Give the order of the matrix (<10):",N
70       PRINT "The order of the matrix = ";N
80       REDIM A(N,N),B(N,N),A1(N,N),I1(N,N),P(N),Q(N,N,N),C(N,N)
90       CALL Mat(A(*),N,N,"A")
100      MAT I1=IDN
110         MAT B=A
120         IF N=1 THEN GOTO 260
130         CALL Resmat(N,A(*),P(*),Q(*))
140         REDIM P(N+1)
150         P(N+1)=1
160         IF P(1)=0 THEN GOTO 300
170         MAT A1=(P(2))*I1
180         FOR K=1 TO N-1
190         MAT C=(P(K+2))*A
200         MAT A1=A1+C
210         MAT C=A*B
220         MAT A=C
230         NEXT K
240         MAT A1=(-1/P(1))*A1
250         GOTO 270
260         A1(N,N)=1/A(N,N)
270         PRINT "Inverse of Matrix A: ",LIN(1)
280      CALL Prtmat(A1(*),N,N)
290         GOTO 310
300         PRINT "Matrix A is SINGULAR"
310         END
```

4. Example.

```
´INVMAT´ COMPUTES THE INVERSE OF A MATRIX THROUGH
THE USE OF CAYLEY HAMILTON THEOREM.

The order of the matrix =   3

Matrix A( 3 x 3 ):

1.0000E+00   0.0000E+00  -6.0000E+00

0.0000E+00   5.0000E+00  -2.0000E+00

3.0000E+00   2.0000E+00   1.0000E+00

Inverse of Matrix A:

 9.0909E-02  -1.2121E-01   3.0303E-01

-6.0606E-02   1.9192E-01   2.0202E-02

-1.5152E-01  -2.0202E-02   5.0505E-02
```

The above results can be extended to the case where the polynomial $p(\cdot)$ in (1) is of infinite order as long as it is analytic [2.5]. (See Section 2.9.) This can be exploited for the computation of $exp(\mathbf{A}t)$ which, as we will see in Chapter 6, is important in the analysis of l.t.i. continuous-time systems. That is, we have

$$exp(\mathbf{A}t) = \sum_{k=0}^{\infty} \frac{t^k}{k!} \mathbf{A}^k = \sum_{i=0}^{n-1} c_i \mathbf{A}^i \qquad (9)$$

Where c_i, $i=0,1,...,n-1$ are linearly independent functions of time [2.6]. Further, for any eigenvalue λ_k of $\mathbf{A}$ we have

$$exp(\lambda_k t) = \sum_{i=0}^{n-1} c_i \lambda_k^i, \quad k=1,2,...,n \qquad (10)$$

which can be used to calculate c_i, $i=0,1,...,n-1$. Note that if $\mathbf{A}$ is an $n \times n$ matrix, so is $exp(\mathbf{A}t)$.

APPLICATIONS OF CAYLEY-HAMILTON THEOREM

5. **Example.** Let

$$A = \begin{bmatrix} -3 & 1 \\ 0 & -2 \end{bmatrix} \tag{11}$$

Then

$$e^{At} = c_0 I + c_1 A \tag{12}$$

Note that scalars c_0 and c_1 are functions of time. Also

$$e^{\lambda_i t} = c_0 + c_1 \lambda_i, \quad i = 1,2 \tag{13}$$

where $\lambda_1 = -3$ and $\lambda_2 = -2$ are eigenvalues of **A**. Equations (13) imply that

$$c_0 = -2e^{-3t} + 3e^{-2t}, \quad c_1 = -e^{-3t} + e^{-2t} \tag{14}$$

Thus

$$e^{At} = \left(-2e^{-3t} + 3e^{-2t}\right)I + \left(-e^{-3t} + e^{-2t}\right)A$$

$$= \begin{bmatrix} e^{-3t} & -e^{-3t} + e^{-2t} \\ 0 & e^{-2t} \end{bmatrix} \tag{15}$$

Note that $e^{At}\big|_{t=0} = I$ and $\dfrac{d}{dt} e^{At}\big|_{t=0} = A$.

2.11 QUADRATIC FORMS

1. **Definition.** Consider the n-dimensional column vector $\mathbf{x} = [x_1, x_2, \ldots, x_n]'$ and the $n \times n$ matrix **Q**. Then the scalar function

$$Q\left[x_1, x_2, \ldots, x_n\right] = \left[x, Qx\right] \tag{1}$$

is called a *quadratic form*.

The matrix Q in quadratic form can be assumed hermitian with no loss of generality. (See Problem 2.31.) Note that a quadratic form is always real. (See Problem 2-11.)

2. **Definition.** A hermitian matrix Q (or its associated quadratic form) is said to be *positive definite*, denoted $Q > 0$, if $(x, Qx) > 0$ for all $x \neq 0$, *negative definite*, denoted $Q < 0$, if $(x, Qx) < 0$ for all $x \neq 0$, and *non-negative definite* or *positive semi-definite*, denoted $Q \geq 0$, if $(x, Qx) \geq 0$. Note that if $Q > 0$, then $-Q < 0$.

3. **Theorem.** A hermitian matrix Q, with eigenvalues $\lambda_1, \lambda_2, \ldots, \lambda_n$, can always be diagonalized by a unitary (or orthogonal) matrix P, i.e., unitary matrix $P(P^{-1} = P^*)$ always exists such that

$$P^{-1}QP = P^*QP = \Lambda\Lambda = \text{diag}\left[\lambda_1, \lambda_2, \ldots, \lambda_n\right] \tag{2}$$

Proof. The proof will be given for the case of distinct eigenvalues. For the general proof the reader is referred to Reference [2.1] or Reference [2.4]. If $\lambda_1, \lambda_2, \ldots, \lambda_n$ are distinct, then the eigenvectors $u_1, u_2, \ldots, u_n$ of Q are orthonormal (Problem 2-18) and linearly independent. Thus the matrix $P = [u_1, u_2, \ldots, u_n]$ diagonalizes Q. To show that P is a unitary matrix, note that

$$P^*P = \begin{bmatrix} u_1^* \\ u_2^* \\ \vdots \\ u_n^* \end{bmatrix} [u_1, u_2, \ldots, u_n] = \begin{bmatrix} u_1^*u_1 & u_1^*u_2 & \cdots & u_1^*u_n \\ u_2^*u_1 & u_2^*u_2 & \cdots & u_2^*u_n \\ \vdots & & & \vdots \\ u_n^*u_1 & u_n^*u_2 & \cdots & u_n^*u_n \end{bmatrix} \tag{3}$$

But $u_i^*u_j = (u_i, u_j) = \delta_{ij}$ because the eigenvectors are orthonormal. Thus $P^*P = I$ and since P^{-1} exists due to linear independence of the eigenvectors, the proof is complete.

4. **Theorem.** A hermitian matrix is positive definite if and only if all its eigenvalues are positive.

Proof: Using Theorem 3, we can write

$$(x, Qx) = x^*Qx = x^*P\Lambda\Lambda P^*x = y^*\Lambda\Lambda y = \sum_{i=1}^{n} \lambda_i |y_i|^2 \tag{4}$$

where $y = P^*x$. Note that $\lambda_1, \lambda_2, \ldots, \lambda_n$ are real (Problem 2-18). Thus

QUADRATIC FORMS

$$x^*Qx > 0 \text{ for all } x \neq 0 \iff \lambda_i > 0, \ i=1,2,...,n \tag{5}$$

△

It can similarly be proved that

$$Q < 0 \iff \lambda_i < 0, \ i=1,2,...,n \tag{6}$$

and

$$Q \geq 0 \iff \lambda_i \geq 0, \ i=1,2,...,n \tag{7}$$

5. **Definition.** The *minors* of an $n \times n$ matrix Q are the determinants of submatrices of Q formed by deleting an equal number of its rows and columns. The *principal minors* of Q are the minors which result from deleting pairs of identically numbered rows and columns of Q. The *mth leading principal minor* of Q, denoted $\det(Q_m)$, is the principal minor formed by deleting the last $\{n-m\}$ pairs of rows and columns of Q.

6. **Theorem.** (Sylvester's Theorem). Let Q be a hermitian matrix.

   (a) Q is positive definite if and only if all its leading principal minors $\det(Q_m)$, $m=1,2,...,n$, are positive.

   (b) Q is nonnegative definite if and only if all its principal minors are nonnegative.

Proof. For the proofs of parts (a) and (b), see References [2.1] and [2.7], respectively. Note that $\det(Q_m) \geq 0$, $m=1,2,...,n$ does *not* imply $Q \geq 0$.

7. **Example.** Consider the hermitian matrix

$$Q = \begin{bmatrix} q_1 & q_2 \\ \bar{q}_2 & q_3 \end{bmatrix} \tag{8}$$

a) $Q > 0 \iff q_1 > 0$ and $\det(Q) = q_1 q_3 - |q_2|^2 > 0$.

b) $Q \geq 0 \iff q_1 \geq 0, q_3 \geq 0$ and $\det(Q) \geq 0$.

Thus for $q_1 = q_2 = 0$, $q_3 < 0$, matrix Q will not be nonnegative definite.

△

Program "POSDEF" determines whether a symmetric square matrix is positive definite, positive semidefinite, negative definite or negative semidefinite. The source listing of this program followed by two examples are given below.

8. "POSDEF"

```
10    !       PROGRAM NAME : "POSDEF" PROG
20    PRINT "'POSDEF' DETERMINES WHETHER A SYMMETRIC SQUARE
      MATRIX 'A'"
30    PRINT "    OF ORDER N IS POSITIVE-DEFINITE, POSITIVE
      SEMI-DEFINITE,"
40    PRINT "    NEGATIVE-DEFINITE OR NEGATIVE SEMI-DEFINITE.";
      LIN(2)
50      OPTION BASE 1
60      DIM A(20,20),Evr(20),Evi(20),Vecr(20,20),Veci(20,20),
        Indic(20)
70      INPUT "What is the order n of the symmetric matrix?",N
80      PRINT "The order of the matrix = ";N;LIN(1)
90      REDIM A(N,N),Evr(N),Evi(N),Vecr(N,N),Veci(N,N),Indic(N)
100     CALL Mat(A(*),N,N,"A")
110     FOR I=1 TO N
120     FOR J=1 TO N
130     IF A(I,J)<>A(J,I) THEN 410
140     NEXT J
150     NEXT I
160     CALL Eigen(N,A(*),Evr(*),Evi(*),Vecr(*),Veci(*),Indic(*))
170       PRINT "The eigenvalues of the matrix are:",LIN(1)
180     MAT PRINT Evr;
190       FOR I=1 TO N
200       IF Evr(I)<=0 THEN 240
210       NEXT I
220       PRINT LIN(1),"Matrix A is POSITIVE DEFINITE."
230       GOTO End
240       FOR I=1 TO N
250       IF Evr(I)>=0 THEN 290
260       NEXT I
270       PRINT "Matrix A is NEGATIVE DEFINITE."
280       GOTO End
290       FOR I=1 TO N
300       IF Evr(I)<0 THEN 340
310       NEXT I
320       PRINT "Matrix A is POSITIVE SEMI-DEFINITE."
330       GOTO End
340       FOR I=1 TO N
350       IF Evr(I)>0 THEN 390
360       NEXT I
370       PRINT "Matrix A is NEGATIVE SEMI-DEFINITE."
380       GOTO End
390       PRINT "Matrix A does not have a definite sign."
400       GOTO End
410       PRINT "Matrix A is not symmetric!"
420 End:   END
```

9. Example.

'POSDEF' DETERMINES WHETHER A SYMMETRIC SQUARE MATRIX 'A' OF ORDER N IS POSITIVE-DEFINITE, POSITIVE SEMI-DEFINITE, NEGATIVE-DEFINITE OR NEGATIVE SEMI-DEFINITE.

The order of the matrix = 2

Matrix A(2 x 2):

-2.0000E+00   3.0000E+00

 3.0000E+00 -6.0000E+00

The eigenvalues of the matrix are:

-7.60555127546   -.394448724552

Matrix A is NEGATIVE DEFINITE.

10. Example.

'POSDEF' DETERMINES WHETHER A SYMMETRIC SQUARE MATRIX 'A' OF ORDER N IS POSITIVE-DEFINITE, POSITIVE SEMI-DEFINITE, NEGATIVE-DEFINITE OR NEGATIVE SEMI-DEFINITE.

The order of the matrix = 3

Matrix A(3 x 3):

 1.0000E+00   2.0000E+00 -1.0000E+00

 2.0000E+00   5.0000E+00   0.0000E+00

-1.0000E+00   0.0000E+00   6.0000E+00

The eigenvalues of the matrix are:

.028039231612   5.57653473371   6.39542603485

Matrix A is POSITIVE DEFINITE.

PROBLEMS

2.1 a) Determine whether the following vectors are linearly independent (directly and by using "LININD").

$$\begin{bmatrix}1\\1\\1\\0\end{bmatrix}, \begin{bmatrix}1\\3\\4\\1\end{bmatrix}, \begin{bmatrix}0\\0\\1\\0\end{bmatrix}, \begin{bmatrix}0\\2\\3\\0\end{bmatrix}, \begin{bmatrix}2\\6\\8\\2\end{bmatrix}$$

b) Find the rank of the matrix formed by these vectors.

c) Determine the largest number of linearly independent vectors in the above set.

2.2 For an $m \times n$ matrix A and an $n \times k$ matrix B show that

a) $\rho(A) = \rho(AA') = \rho(A'A) = \rho(A')$.

b) $\rho(AB) \leq \rho(A)$, $\rho(AB) \leq \rho(B)$.

Also verify these for arbitrary matrices A and B by using "RANK".

2.3 If B and C are nonsingular matrices, show that $\rho(A) = \rho(BA) = \rho(AB) = \rho(BAC)$. Also use "RANK" for arbitrary matrices A, B and C to verify this.

2.4 Let the vectors $e_1, e_2, \ldots, e_n$ in (V,F) form a basis for that vector space. Show that any vector $v \in (V,F)$ has unique components $\alpha_1, \alpha_2, \ldots, \alpha_n$ in F w.r.t. this basis.

2.5 For an $m \times n$ matrix A show that $\rho(A) + \gamma(A) = n$. Verify this for A and A' where

$$A = \begin{bmatrix}2 & 1 & -1 & -4\\0 & 3 & -2 & -3\\2 & 4 & -3 & -4\end{bmatrix}$$

2.6 Show that

a) The range space of any matrix is a vector space.

b) The null space of any matrix is a vector space.

c) $R(A)$ and $N(A)$ are invariant subspaces under A.

d) $N(A - \lambda_i I)$ is an invariant subspace of A for any eigenvalue λ_i of A.

e) Dimension of $N(A - \lambda_i I)$ is equal to the multiplicity of λ_i, an eigenvalue of A.

2.7 Consider an inner product $(.,.)$ for a vector space (V,F). Show that
   a) (x,x) is real *for all* $x \in V$
   b) $(x,0) = 0$ *for all* $x \in V$
   c) $(x,\alpha y) = \bar{\alpha}(x,y)$ *for all* $x,y \in V$, *for all* $\alpha \in F$

2.8 Show that for any inner product $(.,.)$, the Schwarz inequality $|(x,y)| \leq (x,x) \cdot (y,y)$ holds. Furthermore, show that the equality holds if and only if $x=0$ or $y=0$ or $x=\alpha y$ for some scalar α.

2.9 Show that for arbitrary matrices **A** and **B**
   a) $(A^*)^* = A$.
   b) $(AB)^* = B^*A^*$.
   c) $(A+B)^* = A^*+B^*$.
   d) $(\alpha A)^* = \bar{\alpha}A^*$ for *any complex scalar* α.

2.10 Verify that the norms $\|\cdot\|_1$, $\|\cdot\|_2$, $\|\cdot\|_p$ and $\|\cdot\|_\infty$ indeed satisfy the postulates of Definition 2.5.9. Further, show that in finite-dimensional vector spaces, these norms are equivalent.

2.11 Show that if **A** is hermitian, then (x,Ax) is real.

   Use "INNERP" to verify this for some arbitrary vectors **x** and hermitian matrices **A**.

2.12 If $\|A\|<1$, show that $(I-A)^{-1} = \sum_{n=0}^{\infty} A^n$.

2.13 For a matrix $A = (a_{ij})$ show that

   a) $\|A\|_1 = \max_j \left(\sum_{i=1}^{n} |a_{ij}| \right)$

   b) $\|A\|_\infty = \max_i \left(\sum_{j=1}^{n} |a_{ij}| \right)$

   c) $\|A\|_2 \leq \left(\sum_{i=1}^{n} \sum_{j=1}^{n} |a_{ij}|^2 \right)^{1/2}$

2.14 Prove that
   a) $\|A\|_2 = \lambda_{max}$
   b) $\lambda_{min}^2 \|x\|_2 \leq \|Ax\|_2 \leq \lambda_{max}^2 \|x\|_2$ where λ_{min}^2 and λ_{max}^2 are, respectively, the minimum and the maximum eigenvalues of A^*A. For arbitrary square matrices **A** and vectors **x** use "EIGEN" and "NORM" to verify (a) and (b).

2.15 Show that $|\lambda| \leq \|A\|$ for any eigenvalue λ of **A**. Also use EIGEN" and "NORM" to verify this for arbitrary square matrices **A**.

2.16 Let λ be an eigenvalue of **A**. Prove that
   a) λ is also an eigenvalue of A' and $P^{-1}AP$ for any nonsingular matrix **P**.
   b) $\bar{\lambda}$ is an eigenvalue of $\bar{A}$ and A^*.

c) λ^k is an eigenvalue of $\mathbf{A}^k$ for any positive or negative integer k.

d) $f(\lambda)$ is an eigenvalue of $f(\mathbf{A})$ where $f(\cdot)$ is any polynomial.

2.17 Use "EIGEN" to find the eigenvalues and eigenvectors of the following matrices and diagonalize them:

$$\mathbf{A} = \begin{bmatrix} 0 & 0 & 4 \\ 4 & 1 & 0 \\ 1 & 4 & 0 \end{bmatrix}, \quad \mathbf{B} = \begin{bmatrix} 1 & 0 & -1 \\ 0 & 1 & 0 \\ 0 & 0 & 2 \end{bmatrix}$$

2.18 Show that the eigenvalues of a hermitian matrix are real and the eigenvectors associated with distinct eigenvalues are orthogonal.

2.19 Show that the eigenvalues of a triangular matrix are the elements of its main diagonal.

2.20 If $\Delta(\lambda) = \lambda^n + \alpha_{n-1}\lambda^{n-1} + ... + \alpha_1\lambda + \alpha_0$ is the characteristic polynomial of $\mathbf{A}$, show that

a) $\alpha_0 = \det(\mathbf{A}) = \lambda_1\lambda_2 \cdots \lambda_n$

b) $-\alpha_{n-1} = tr(\mathbf{A}) = \lambda_1 + \lambda_2 + ... + \lambda_n$

where $tr(\mathbf{A})$, the *trace* of the square matrix $\mathbf{A}$, is the sum of the elements on the main diagonal of $\mathbf{A}$.

2.21 Show that the characteristic polynomial of the matrix

$$\mathbf{A} = \begin{bmatrix} 0 & 1 & 0 & \cdots & 0 \\ 0 & 0 & 1 & \cdots & \\ \vdots & & & \cdots & \\ & & & & 0 \\ & & & & 1 \\ -\alpha_0 & -\alpha_1 & -\alpha_2 & & -\alpha_{n-1} \end{bmatrix}$$

is $\Delta(\lambda) = \lambda^n + \alpha_{n-1}\lambda^{n-1} + ... + \alpha_1\lambda_1 + \alpha_0$. For an arbitrary 5×5 matrix use "CHRPOL" to verify this.

2.22 Show that a normal matrix $\mathbf{A}$ can always be diagonalized by a similarity transformation $\mathbf{Q}^{-1}\mathbf{A}\mathbf{Q}$ where $\mathbf{Q}$ is unitary.

2.23 Prove that the generalized eigenvectors $\mathbf{u}_i$, $i=1,2,...,k$ of matrix $\mathbf{A}$ associated with each eigenvalue λ are linearly independent. (Hint: Assume $\sum_{i=1}^{k}\alpha_i\mathbf{u}_i = 0$. Premultiply by $(\mathbf{A}-\lambda\mathbf{I})^{j-1}$ to show $\alpha_j = 0$, $j=k-1,k-2,...,1$.)

LINEAR CONTROL SYSTEMS

2.24 Find the eigenvalues, eigenvectors, generalized eigenvectors and the Jordan canonical form for the matrix

$$A = \begin{bmatrix} 1 & 1 & 0 \\ 0 & 1 & 1 \\ 0 & 0 & 2 \end{bmatrix}$$

Also calculate A^{-1} directly, by the application of Cayley-Hamilton theorem and by using "INVMAT".

2.25 Prove Cayley-Hamilton theorem for the case of matrix A with repeated eigenvalues. (Hint: Use the Jordan canonical form of A to show that $\Delta(A) = M\Delta(J)M^{-1}$ and $\Delta(J) = 0$.)

2.26 If

$$J = \begin{bmatrix} \lambda & 1 & 0 \\ 0 & \lambda & 1 \\ 0 & 0 & \lambda \end{bmatrix}$$

show that

$$e^{Jt} = \begin{bmatrix} e^{\lambda t} - te^{\lambda t} - \dfrac{t^2}{2}e^{\lambda t} \\ 0 - e^{\lambda t} - te^{\lambda t} - \\ 0 - 0 - e^{\lambda t} - \end{bmatrix}$$

2.27 Prove that $(e^{At})^* = e^{A^*t}$ and $(e^{At})^{-1} = e^{-At}$.

2.28 If A is a skew symmetric matrix $(A = -A')$, show that $exp(At)$ will be unitary, i.e., $exp(A't)exp(At) = I$.

2.29 Use "MATPOL" to find $A^3 - A^2 + 2A + I$ where

$$A = \begin{bmatrix} 1 & -2 & 0 \\ 1 & 0 & 2 \\ 3 & -2 & 1 \end{bmatrix}$$

2.30 Show that $exp(\mathbf{A}t)exp(\mathbf{B}t) = exp(\mathbf{A}+\mathbf{B})t$ if and only if **A** and **B** commute, i.e., **AB=BA**.

2.31 Show that in a quadratic form **x*Qx** there is no loss of generality in assuming that **Q** is hermitian, i.e., show that hermitian matrix **P** always exists such that **x*Qx = x*Px**.

2.32 Determine α such that the following matrix will be a) positive definite, b) positive semidefinite.

$$\begin{bmatrix} 1 & -1 & 0 \\ -1 & 4 & 1 \\ 0 & 1 & \alpha \end{bmatrix}$$

substitute a scalar in the determined range in each case for α and verify the result by using "POSDEF".

REFERENCES

[2.1] R. Bellman, *Introduction to Matrix Analysis*, McGraw-Hill, 1960

[2.2] D. M. Wiberg, *State Space and Linear Systems*, McGraw-Hill, 1971

[2.3] C. T. Chen, *Introduction to Linear System Theory*, Holt, Rinehart, and Winston, 1970

[2.4] F. R. Gantmacher, *The Theory of Matrices*, Volume I, Chelsea, 1959

[2.5] P. DeRusso, R. Roy and C. Close, *State Variables for Engineers*, Wiley, 1965

[2.6] M. Vidyasagar, "A Characterization of $e^{\mathbf{A}t}$ and a constructive proof of the controllability criterion," *IEEE Trans. Auto. Control*, Vol. AC-16, pp. 370-371, 1971

[2.7] H. W. Turnbull and A. C. Aitken, *An Introduction to the Theory of Canonical Matrices*, Dover, 1961

CHAPTER 3

REVIEW OF THE LAPLACE AND z-TRANSFORMS

3.1 INTRODUCTION

The Laplace transform and the z-transform are fundamental tools for studying linear time-invariant systems. These transforms convert a differential or a difference equation describing a l.t.i. system into an algebraic equation. They thus considerably reduce the computational effort required for the analysis or the design of such systems.

In order to keep the size of this chapter reasonable, we assume that the reader has had a previous introduction to these transforms. We will review only those aspects of these vast topics which have immediate relevance to the study of l.t.i. systems. For a more detailed and rigorous treatment of these subjects the reader is referred to references [3.1-3.9].

3.2 DEFINITION OF THE LAPLACE TRANSFORM

The Laplace transform is useful in transforming functions in the time domain to those in the complex frequency domain. We will deal only with the *unilateral* or *one-sided* Laplace transform. We assume that all time functions of interest vanish for $t < t_0$. Further, we assume $t_0 = 0$ with no loss of generality, since in the following treatment t can always be replaced by $t - t_0$.

1. Definition: Let $f(t)$ be an integrable function of time defined on the interval $[0,\infty]$. Then the (unilateral) Laplace transform of $f(t)$, denoted by $L[f(t)] \triangleq F(s)$, is defined by the integral

$$F(s) = \int_{0-}^{\infty} f(t) e^{-st} dt \qquad (1)$$

for all values of the complex frequency s for which the integral exists.

Δ

Note that the lower limit of integration is 0−. Thus, if $f(t)$ includes an impulse at $t = 0$, its contribution at $t = 0$ will be included in the integral (1).

It is clear from (1) that if $F(s)$ exists for some $s = \sigma_c + j\omega_c$, then it will also exist for all s such that $Re(s) > \sigma_c$. The greatest lower bound on σ_c for which integral (1) exists is called the *abscissa of convergence* for the function $f(t)$. The region in the complex plane to the right of the abscissa of convergence is called the *region* or the *domain* of convergence of $F(s)$. (See Fig. 1.)

The Laplace integral (1) can be extended to other values of $s = \sigma + j\omega$. That is, $F(s)$ can be considered a well-defined function for all values of s except for those which cause integral (1) to diverge.

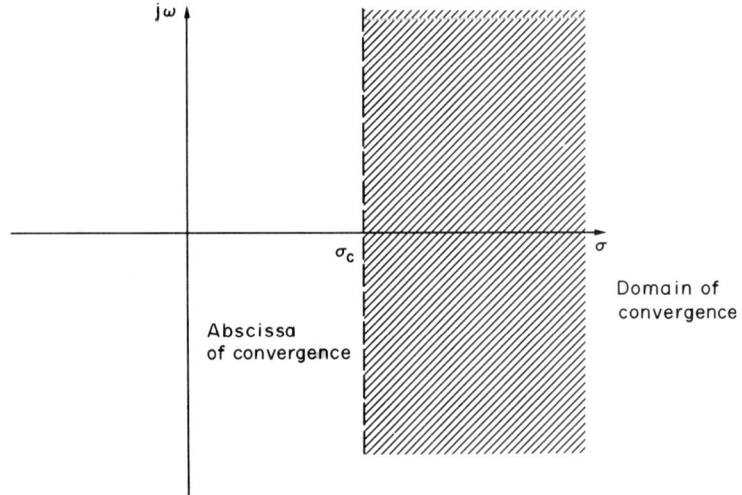

Fig. 3.2.1. The complex s-plane

2. Example. Consider the function $f(t) = e^{\alpha t}$ where α is a real or a complex number. Then

$$L[f(t)] \triangleq F(s) = \int_{0-}^{\infty} e^{\alpha t} e^{-st} dt = \frac{e^{-(s-\alpha)t}}{-(s-\alpha)} \Big|_{0-}^{\infty} \qquad (2)$$

Thus $F(s)$ exists, i.e. the above integral is finite when $Re(s-\alpha) > 0$. That is,

$$F(s) = L[e^{\alpha t}] = \frac{1}{s-\alpha} \quad \text{for } Re(s) > Re(\alpha) \qquad (3)$$

and $Re(\alpha)$ is the abscissa of convergence for $e^{\alpha t}$.

3.3 PROPERTIES OF THE LAPLACE TRANSFORM

In this section we will discuss the important properties of the Laplace transform. Further properties will be introduced in the problems at the end of this chapter.

1. Uniqueness. If two time functions $f_1(t)$ and $f_2(t)$ have the same function of the complex frequency s, say $F(s)$, as their Laplace transform, then $f_1(t)$ and $f_2(t)$ can differ only trivially.

△

Uniqueness is a fundamental property of the Laplace transform. Given a time function $f(t)$, it allows us to use a unique function $F(s)$ as its Laplace transform. Conversely, this property implies that given a Laplace transform $F(s)$, there is a *unique* time function $f(t)$ over the interval $[0,\infty]$, (except for trivialities) such that $F(s) = L[f(t)]$. This is written as

$$f(t) = L^{-1}[F(s)] \tag{1}$$

meaning that $f(t)$ is the *inverse Laplace transform* of $F(s)$.

2. Example. Consider the following time functions:

$$f_1(t) = \begin{cases} 1 & \text{for } t>0 \\ 0.5 & \text{for } t=0 \\ 0 & \text{for } t<0 \end{cases}, \quad f_2(t) = \begin{cases} 1 & \text{for } t>0 \\ 1 & \text{for } t=0 \\ 0 & \text{for } t<0 \end{cases} \tag{2}$$

For our purposes, the difference between these functions is trivial. Both functions are referred to as the *unit step function* and have the same Laplace transform:

$$F_1(s) = F_2(s) \triangleq \int_{0-}^{\infty} e^{-st} dt = \frac{1}{s} \tag{3}$$

3. Linearity. If $F_1(s)$ and $F_2(s)$ are the Laplace transforms of the time functions $f_1(t)$ and $f_2(t)$, respectively, then

$$L[\alpha_1 f_1(t) + \alpha_2 f_2(t)] = \alpha_1 F_1(s) + \alpha_2 F_2(s) \tag{4}$$

for arbitrary constants α_1 and α_2.

△

LINEAR CONTROL SYSTEMS

The linearity property of the Laplace transform follows immediately from the linearity of the Laplace integral (3.2.1). The reader is asked to verify this in Problem 3.3.

4. Example. Consider

$$f(t) = \sin \alpha t = \frac{1}{2j}(e^{j\alpha t} - e^{-j\alpha t}) \tag{5}$$

Thus, using (3.2.3) and the linearity property of the Laplace transform yields

$$L[\sin \alpha t] = \frac{1}{2j}\left[\frac{1}{s-j\alpha} - \frac{1}{s+j\alpha}\right] = \frac{\alpha}{s^2+\alpha^2} \tag{6}$$

5. Differentiation Rule. Let $F(s)$ be the Laplace transform of $f(t)$. Then

$$L[\frac{d}{dt}f(t)] = sF(s) - f(0-) \tag{7}$$

and

$$L[\frac{d^n}{dt^n}f(t)] = s^n F(s) - s^{n-1}f(0-) - s^{n-2}f^{(1)}(0-)$$

$$- \cdots - sf^{(n-2)}(0-) - f^{(n-1)}(0-) \tag{8}$$

where $f^{(i)}$ indicates the ith derivative of f w.r.t. t.

△

The differentiation rule can be verified by performing the Laplace integral (3.2.1) for $\dot{f}(t)$ by parts:

$$L[\dot{f}(t)] = f(t)e^{-st}\Big|_{0-}^{\infty} - \int_{0-}^{\infty} f(t)(-se^{-st})dt$$

$$= -f(0-) + s\int_{0-}^{\infty} f(t)e^{-st}dt = sF(s) - f(0-) \tag{9}$$

Application of (9) n-1 times implies (8).

PROPERTIES OF THE LAPLACE TRANSFORM

6. Example. Consider $\frac{d}{dt} \sin \alpha t = \alpha \cos \alpha t$. Thus, (7) and (6) yield

$$L[\frac{d}{dt} \sin \alpha t] = L[\alpha \cos \alpha t] = s[\frac{\alpha}{s^2 + \alpha^2}] \tag{10}$$

and using the linearity of the Laplace transforms results

$$L[\cos \alpha t] = \frac{s}{s^2 + \alpha^2} \tag{11}$$

7. Integration Rule. If $L[f(t)] = F(s)$, then

$$L[\int_{0-}^{t} f(\tau) d\tau] = \frac{1}{s} F(s) \tag{12}$$

and

$$L[\int_{0-}^{t} \int_{0-}^{\tau_1} \cdots \int_{0-}^{\tau_{n-1}} f(\tau_n) d\tau_n d\tau_{n-1} \cdots d\tau_1] = \frac{1}{s^n} F(s) \tag{13}$$

Δ

Equation (12) can be verified again by using integration by parts:

$$L[\int_{0-}^{t} f(\tau) d\tau] \triangleq \int_{0-}^{\infty} [\int_{0-}^{t} f(\tau) d\tau] e^{-st} dt$$

$$= [\int_{0-}^{t} f(\tau) d\tau] \frac{e^{-st}}{-s} \Big|_{0-}^{\infty} - \int_{0-}^{\infty} f(t) \left[\frac{e^{-st}}{-s}\right] dt$$

$$= 0 + \frac{1}{s} \int_{0-}^{\infty} f(t) e^{-st} dt = \frac{1}{s} F(s) \tag{14}$$

Repeated application of (14) implies (13).

8. Time-Function Translation. If $L[f(t)] = F(s)$, then for any positive scalar α,

$$L[f(t-\alpha)] = e^{-\alpha s} F(s) \tag{15}$$

Further, the scalar α may be negative if $f(t-\alpha)$ vanishes for $t<0$.

△

This property follows easily from the Laplace transform defining integral:

$$L[f(t-\alpha)] \triangleq \int_{0-}^{\infty} f(t-\alpha)e^{-st}dt = \int_{\alpha-}^{\infty} f(t-\alpha)e^{-st}dt$$

$$= \int_{0-}^{\infty} f(\tau)e^{-s(\tau+\alpha)}dt = e^{-s\alpha}\int_{0-}^{\infty} f(\tau)e^{-s\tau}d\tau = e^{-s\alpha}F(s) \qquad (16)$$

where $\tau = t - \alpha$. Note that $f(t-\alpha)$ is the original function delayed by α as illustrated in Fig. 2.

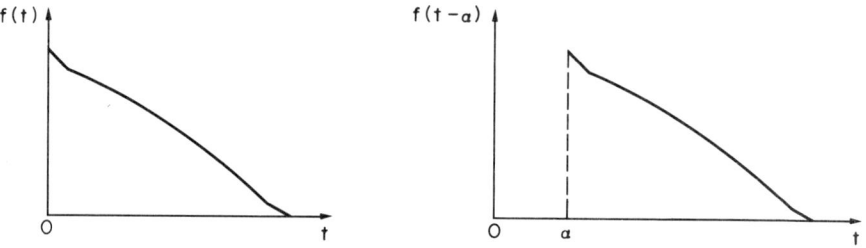

Fig. 3.2.2. A delayed function

9. Example. Consider a pulse function defined by

$$p(t) = \begin{cases} \dfrac{1}{a} & 0 < t \leqslant a \\ 0 & \text{elsewhere} \end{cases} \qquad (17)$$

as shown in Fig. 3.

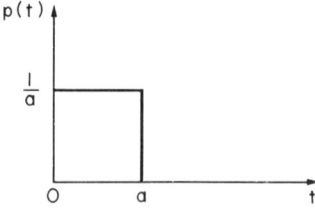

Fig. 3.2.3 A pulse function

Note that

$$p(t) = \frac{1}{a}[u(t) - u(t-a)] \qquad (18)$$

where $u(t)$ is the unit step function. Using (3), the linearity and the time-function translation properties of the Laplace transform we obtain

$$L[p(t)] = \frac{1}{a}\left\{L[u(t)] - L[u(t-a)]\right\}$$

$$= \frac{1}{a}\left[\frac{1}{s} - e^{-sa}\frac{1}{s}\right] = \frac{1-e^{-sa}}{sa} \qquad (19)$$

A *unit impulse* or a *Dirac delta function* at $t = 0$, denoted by $\delta(t)$, is defined as

$$\delta(t) = \lim_{a \to 0} p(t) \qquad (20)$$

Thus

$$L[\delta(t)] = \lim_{a \to 0} \frac{1-e^{-sa}}{sa} = 1 \qquad (21)$$

Also, by the definition of the derivative, (20) implies that

$$\delta(t) = \frac{d}{dt} u(t) \qquad (22)$$

and by the differentiation rule of the Laplace transform,

$$L[\delta(t)] = sL[u(t)] - u(0-) = s\left(\frac{1}{s}\right) - 0 = 1 \qquad (23)$$

which verifies (21).

10. Laplace Transform Translation. If $L[f(t)] = F(s)$, then

$$L[e^{\alpha t} f(t)] = F(s-\alpha) \tag{24}$$

where α is any complex scalar.

△

The reader is asked to prove (24) in Problem (3.5).

11. Initial Value Property. If $f(t)$ approaches a limit as t approaches zero from the right, then

$$f(0+) = \lim_{s \to \infty} s\, L[f(t)] \tag{25}$$

See Problem 3.6. Note that the limit on the right side of (25) may exist without the existence of $f(0+)$.

12. Final Value Property. If $f(t)$ approaches a limit as t approaches ∞, then

$$\lim_{t \to \infty} f(t) = \lim_{s \to 0} s\, L[f(t)] \tag{26}$$

See Problem 3.7. Again note that the limit on the right side of (26) may exist without the existence of the limit on the left side.

13. Example. Consider $f(t) = e^{-\alpha t} \cos \omega t$ where α is a real positive number. Using (11) and (24) we have

$$L[f(t)] \triangleq F(s) = \frac{s+\alpha}{(s+\alpha)^2 + \omega^2} \tag{27}$$

Now $\lim_{s \to \infty} sF(s) = 1$ which is equal to $f(0+)$. Also $\lim_{s \to 0} sF(s) = 0$ which is equal to $\lim_{t \to \infty} f(t)$.

14. Definition. Let time functions $f_1(t)$ and $f_2(t)$ be defined on the interval $(-\infty, \infty)$. Then the *convolution* of f_1 and f_2, denoted $f_1 * f_2$, is a time function defined by

$$f_1 * f_2(t) = \int_{-\infty}^{\infty} f_1(\tau) f_2(t-\tau) d\tau \tag{28}$$

provided that the integral exists for all t.

Note that if f_1 and f_2 vanish for $t<0$, then their convolution will be

$$f_1*f_2(t) = \int_0^t f_1(\tau)f_2(t-\tau)d\tau \qquad (29)$$

See Problem 3.8.

15. Convolution Property. If $F_1(s)$ and $F_2(s)$ are the Laplace transforms of $f_1(t)$ and $f_2(t)$, respectively, then

$$L[f_1*f_2(t)] = F_1(s)F_2(s) \qquad (30)$$

See Problem 3.9. This is an important property of the Laplace transform often used in the study of l.t.i. continuous-time systems.

For some further properties of the Laplace transform see Problem 3.10.

3.4 INVERSION OF THE LAPLACE TRANSFORM

The importance of the Laplace transform lies in the fact that there is a unique time function $f(t)$ for each Laplace transform $F(s)$. The following theorem provides a procedure for determining the inverse Laplace transform of $F(s)$.

1. Theorem. Let $F(s) = L[f(t)]$ and let σ_c be the abscissa of convergence for $f(t)$. Then

$$f(t) \triangleq L^{-1}[F(s)] = \frac{1}{2\pi j} \lim_{\omega \to \infty} \int_{\sigma-j\omega}^{\sigma+j\omega} F(s)e^{st}ds \qquad (1)$$

for all $\sigma > \sigma_c$.

The proof of this theorem may be found in reference [3.7]. Note that by assumption $f(t) = 0$ for $t<0$; thus the integral (1) yields $f(t)$ for $t>0$. Also, $f(\cdot)$ could be discontinuous at t; in fact the left-hand side of (1) may be replaced by $\frac{1}{2}[f(t+0) + f(t-0)]$.

3.4.1 Inversion of the Laplace Transform by Partial Fraction Expansion

Integral (1) is, in general, difficult to compute and is rarely used in practice directly for the inversion of the Laplace transform. Rather, the inverser Laplace transform is usually found by expanding it into easily invertible components and applying the linearity property of the Laplace transform. This procedure will be formalized below.

LINEAR CONTROL SYSTEMS

2. Definition. A function $F(s)$ is said to be *rational* if it is a ratio of two polynomials. Further, if the degree of the numerator polynomial is m and that of the denominator polynomial is n, it is called *proper* if $m \leq n$, *strictly proper* if $m < n$ and *improper* if $m > n$.

3. Theorem. Let $F(s)=L[f(t)]$ be a strictly proper rational function with real coefficients, i.e.

$$F(s) = \frac{Q(s)}{P(s)} = \frac{q_m s^m + q_{m-1} s^{m-1} + \cdots + q_1 s + q_0}{p_n s^n + p_{n-1} s^{n-1} + \cdots + p_1 s + p_0} \qquad (2)$$

$$= \frac{q_m}{p_n} \frac{\prod_{j=1}^{m}(s-z_j)}{\prod_{i=1}^{n}(s-p_i)}$$

where the coefficients q_j, $j = 1,2,\ldots,m$ and p_i, $i=1,2,\ldots,n$ are real. Then complex numbers K_{ij} exist such that $F(s)$ can be written as

$$F(s) = \sum_{i=1}^{p} \sum_{j=1}^{m_i} \frac{K_{ij}}{(s-p_i)^j} \qquad (3)$$

where m_i is the multiplicity of root p_i of $P(s)$ and p is the number of distinct roots of $P(s)$. The complex numbers K_{ij} are determined as follows:

$$K_{i,m_i-j+1} = \frac{1}{(j-1)!} \frac{d^{j-1}}{ds^{j-1}} (s-p_i)^{m_i} F(s) \Big|_{s=p_i},$$

$$j = 1,2,\ldots,m_i \quad ; \quad i = 1,2,\ldots,p \qquad (4)$$

The inverse Laplace transform of $F(s)$ is

$$f(t) = \sum_{i=1}^{p} \sum_{j=1}^{m_i} \frac{K_{ij} t^{j-1}}{(j-1)!} e^{\lambda_i t} \qquad (5)$$

The proof of this theorem is straightforward but tedious. It is left to the reader. (See reference [3.7].)

The roots z_j, $j = 1,2,\ldots,m$ of $Q(s)$ are called the *zeros* of $F(s)$ and the roots p_i, $i=1,2,\ldots,n$ of $P(s)$ are called the *poles* of $F(s)$. Clearly, the abscissa of convergence for $f(t)$ is $\max_i \{\text{Re}(p_i)\}$. The complex numbers K_{ij}, $j = 1,2,\ldots,m_i$ are called the *residues* corresponding to the pole p_i. The poles, the zeros and the residues either are real or they occur in complex conjugate pairs because the coefficients p_i and q_j are assumed to be real. (See Problem 3.12.) Note that $\sum_{i=1}^{p} m_i = n$. Also note that in the case where all the poles of $F(s)$ are distinct, i.e. where $m_i = 1$, $i = 1, 2, \ldots, p$, we have $p = n$ and (3) becomes

$$F(s) = \sum_{i=1}^{n} \frac{K_i}{s - p_i} \tag{6}$$

where

$$K_i = (s - p_i) F(s) \big|_{s = p_i}, \quad i = 1, 2, \ldots, n \tag{7}$$

and (5) becomes

$$f(t) = \sum_{i=1}^{n} K_i e^{\lambda_i t}. \tag{8}$$

4. Example. Consider

$$F(s) = \frac{s^2 + 3s + 5}{s^3 + 6s^2 + 11s + 6} = \frac{s^2 + 3s + 5}{(s+1)(s+2)(s+3)}. \tag{9}$$

It can be written as

$$F(s) = \frac{K_1}{s+1} + \frac{K_2}{s+2} + \frac{K_3}{s+3} \tag{10}$$

where

$$K_1 = (s+1) F(s) \big|_{s = -1} = 1.5 \tag{11}$$

$$K_2 = (s+2) F(s) \big|_{s = -2} = -3 \tag{12}$$

$$K_3 = (s+3) F(s) \big|_{s = -3} = 2.5 \tag{13}$$

Thus

$$L^{-1}[F(s)] = 1.5 e^{-t} - 3 e^{-2t} + 2.5 e^{-3t}, \quad t \geq 0. \tag{14}$$

5. Example. Consider

$$F(s) = \frac{s}{(s+1)^3(s+2)} \tag{15}$$

It can be written as

$$F(s) = \frac{K_{11}}{s+1} + \frac{K_{12}}{(s+1)^2} + \frac{K_{13}}{(s+1)^3} + \frac{K_{21}}{s+2} \tag{16}$$

where

$$K_{13} = (s+1)^3 F(s)\big|_{s=-1} = -1 \tag{17}$$

$$K_{12} = \frac{d}{ds}(s+1)^3 F(s)\big|_{s=-1} = \frac{2}{(s+2)^2}\big|_{s=-1} = 2 \tag{18}$$

$$K_{11} = \frac{1}{2}\frac{d^2}{ds^2}(s+1)^3 F(s)\big|_{s=-1} = \frac{-2}{(s+1)^3}\big|_{s=-1} = -2 \tag{19}$$

$$K_{21} = (s+2)F(s)\big|_{s=-2} = 2. \tag{20}$$

Thus

$$F(s) = \frac{-2}{s+1} + \frac{2}{(s+1)^2} - \frac{1}{(s+1)^3} + \frac{2}{s+2} \tag{21}$$

and

$$L^{-1}[F(s)] = -2e^{-t} + 2te^{-t} - \frac{t^2}{2}e^{-t} + 2e^{-2t}, \quad t \geq 0 \tag{22}$$

△

If $p_i = \sigma_1 + j\omega_i$ and $p_{i+1} = \bar{p}_1 = \sigma_i - j\omega_j$ are a pair of complex conjugate poles of $F(s)$, then the corresponding pair of residues K_i and K_{i+1} will also be complex conjugate (Problem 3.12). This arrangement can be combined into a more compact form as follows:

INVERSION OF THE LAPLACE TRANSFORM

$$\frac{K_i}{s-p_i} + \frac{K_i+1}{s-p_{i+1}} = \frac{K_i}{s-\sigma_i-j\omega_i} + \frac{\overline{K_i}}{s-\sigma_i+j\omega_i}$$

$$= \frac{A_i(s-\sigma_i) + B_i\omega_i}{(s-\sigma_i)^2 + \omega_i^2} \tag{23}$$

where $A_i = 2Re(K_i)$ and $B_i = -2Im(K_i)$. Then A_i and B_i may be found from

$$B_i + jA_i = 2jK_i = \frac{1}{\omega_i}[(s-\sigma_i)^2 + \omega_i^2]F(s)\Big|_{s=\sigma_i+j\omega_i} \tag{24}$$

The inverse Laplace transform of (23) is

$$e^{\sigma_i t}(A_i \cos \omega_i t + B_i \sin \omega_i t), \quad t \geq 0 \tag{25}$$

or

$$2|K_i|e^{\sigma_i t}\cos(\omega_i t + <K_i), \quad t \geq 0. \tag{26}$$

(See Problem 3.14.)

6. Example. Consider $F(s) = \dfrac{2s^2+25s+50}{s^3+10s^2+50s}$. It can be written as

$$\frac{2s^2+25s+50}{s[(s+5)^2+25]} = \frac{K_1}{s} + \frac{A_2(s+5)+B_2(5)}{(s+5)^2+5^2} \tag{27}$$

where

$$K_1 = sF(s)\Big|_{s=0} = 1 \tag{28}$$

and

$$B_2 + jA_2 = \frac{1}{5}[(s+5)^2 + 5^2]F(s)\Big|_{s=-5+j5} = \frac{2s^2+25s+50}{5s}\Big|_{s=-5+j5} \tag{29}$$

which yields $A_2 = 1$ and $B_2 = 2$. Thus

$$L^{-1}[F(s)] = 1 + e^{-5t}(\cos 5t + 2 \sin 5t), \quad t \geq 0 \quad \triangle \qquad (30)$$

If $F(s)$ is not a strictly proper rational function, it can always be written as a polynomial in s plus a strictly proper rational function. Then the linearity property of the Laplace transform can be applied to find the inverse Laplace transform. Note that

$$L^{-1}[a_k s^k + a_{k-1} s^{k-1} + \cdots + a_1 s + a_0] =$$

$$a_k \delta^{(k)}(t) + a_{k-1} \delta^{(k-1)}(t) + \cdots + a_1 \delta^{(1)}(t) + a_0 \delta(t) \qquad (31)$$

where $\delta(t)$ is the unit impulse and

$$\delta^{(i)}(t) \triangleq \frac{d^i}{dt^i} \delta(t).$$

Program "PFEXP" determines the partial fraction expansion of any strictly proper rational function $F(s)$. The source listing of this program followed by an example is given below.

7. "PFEXP".

```
10    !      PROGRAM NAME : "PFEXP" PROG
20    PRINT "This progam performs PARTIAL FRACTION expansion on "
30    PRINT "a rational function of Laplace transform operator s"
40    !   "Nn = Order of Denominator = Order of the System"
50    !   "Np = Number of distinct poles"
60    !   "P(*),Pj(*) contain the Real and Imaginary parts of
      the distinct poles"
70    !   "M(*) contains the multiplicities of the distinct poles"
80    !   "Rcoef(*),Icoef(*) contain the Real and Imaginary
      parts of the
90    !   "      denominator polynomial coefficients"
100   !   "Nm = Order of Numerator"
110   !   "Nz = Number of distinct zeroes"
120   !   "Z(*),Zj(*) contain the Real and Imaginary parts of
      the distinct zeroes"
130   !   "Mz(*) contains the multiplicities of the distinct
      zeroes"
140   !   "C(*),Cj(*) contain the Real and Imaginary parts of the
150   !   "      numerator polynomial coefficients"
160   DIM P(100),Pj(100),C(100),Cj(100),Cw(100),Cwj(100),
      Rcoefw(100),Icoefw(100)
```

```
170   DIM A(100),Aj(100),M(100),Mz(100),Rcoef(100),Icoef(100),
      Z(100),Zj(100)
180   DIM Rroot(100),Iroot(100)
190   N=Nz=Nn=Nd=Np=Nm=0
200   Gain=1
210 Options: INPUT "DENOM in POLE OR COEFFICIENT FORM (P/C)?
      (NO DEFAULT)",P$
220   IF P$="C" THEN Poles.coeff
230   IF P$="P" THEN Poles
240   GOTO Options
250 Poles: IF N<>0 THEN DISP "Enter the no. of DISTINCT poles
      (DEFAULT =";N;")";
260   IF N<>0 THEN INPUT N
270   IF N=0 THEN INPUT "Enter the no. of DISTINCT POLES
      (NO DEFAULT)",N
280   N=ABS(INT(N))
290   IF N=0 THEN Poles
300   FOR I=1 TO N
310   DISP "ENTER POLE(";I;"):REAL,IMAG,MULTIPL. (DEFAULT="
      ;P(I);Pj(I);M(I);")";
320   INPUT P(I),Pj(I),M(I)
330   IF Pj(I)>=0 THEN PRINT "POLE(";I;") = ";P(I);"+";Pj(I);
      "j","  MULTIPL.";M(I)
340   IF Pj(I)<0 THEN PRINT "POLE(";I;") = ";P(I);"+";Pj(I);
      "j","  MULTIPL.";M(I)
350   NEXT I
360   A$="N"
370   INPUT "ANY CHANGES (Y/N, DEFAULT = N)",A$
380   PRINT LIN(1)
390   IF A$="Y" THEN Poles
400   Nn=0
410   FOR I=1 TO N
420   Nn=M(I)+Nn
430   NEXT I
440 Numerator: REDIM C(Nn),Cj(Nn),Cw(Nn),Cwj(Nn),A(Nn),
      Aj(Nn),M(N),P(N),Pj(N)
450   INPUT "Numerator in ZERO or COEFFICIENT form (Z/C)?
      (NO DEFAULT)",N$
460   IF N$="C" THEN Numeratorcoeff
470   IF N$="Z" THEN Zeroes
480 Numeratorcoeff: IF Nm<>0 THEN DISP "Enter Numerator Order:
      (DEFAULT=";Nm;")";
490   IF Nm<>0 THEN INPUT Nm
500   IF Nm=0 THEN INPUT "ENTER ORDER OF THE NUMERATOR:
      (DEFAULT NOT ALLOWED",Nm
510   FOR I=1 TO Nm+1
520   DISP "COEFFICIENT n(";I-1;"): (DEFAULT=";C(I);Cj(I);")";
530   INPUT C(I),Cj(I)
540   IF Cj(I)>=0 THEN PRINT "n(";I-1;")=";C(I);"+";Cj(I);"j"
```

```
550    IF Cj(I)<0 THEN PRINT C(I);"-";ABS(Cj(I));"j"
560    NEXT I
570    B$="N"
580    INPUT "ANY CHANGES (Y/N), DEFAULT = N)",B$
590    PRINT LIN(1)
600    IF B$="Y" THEN Numeratorcoeff
610 List input:  IF P$="C" THEN Listdenomcoeff
620    PRINT LIN(2),"POLES OF THE DENOMINATOR IN ORDER OF
       ENTRY",LIN(1)
630    FOR I=1 TO N
640    IF Pj(I)>=0 THEN PRINT "POLE(";I;") = ";P(I);"+";Pj(I);
       "j"," MULTIPL.";M(I)
650    IF Pj(I)>0 THEN 670
660    PRINT "POLE(";I;") = ";P(I);"-";ABS(Pj(I));"j",
       " MULTIPLTIPLICITY";M(I)
670    NEXT I
680    PRINT LIN(2)
690    GOTO List numerator
700 Listdenomcoeff: PRINT LIN(4),"Coefficeints of the Denom.
       Polynomial",LIN(1)
710    FOR I=0 TO Nn
720    IF Rcoef(I)>=0 THEN PRINT "n(";I;")=";Rcoef(I);"+";
       Icoefj(I);"j"
730    IF Rcoef(I)<0 THEN PRINT "n(";I;")=";Rcoef(I);"-";
       ABS(Icoefj(I));"j"
740    NEXT I
750 List numerator: IF N$="Z" THEN Listnumzeroes
760    PRINT LIN(4),"COEFFICIENTS OF THE NUMERATOR IN
       ASCENDING ORDER",LIN(1)
770    FOR I=0 TO Nm
780    IF Cj(I)>=0 THEN PRINT "n(";I;")=";C(I+1);"+";Cj(I+1);"j"
790    IF Cj(I)<0 THEN PRINT "n(";I;")=";C(I+1);"-";
       ABS(Cj(I+1));"j"
800    NEXT I
810    GOTO Pfexp
820    PRINT LIN(2)
830 Listnumzeroes: PRINT "Zeroes in order of entry w.r.t
       MULTIPLICITIES",LIN(1)
840    FOR I=1 TO Nz
850    IF Zj(I)>=0 THEN PRINT "ZERO(";I;")=";Z(I);"+";
       Zj(I);"j"," MULTIPL.";Mz(I)
860    IF Zj(I)>0 THEN 880
870    PRINT "ZERO(";I;") = ";Z(I);"-";ABS(Zj(I));"j",
       " MULTIPLICITY ";Mz(I)
880    NEXT I
890 Pfexp: PRINT LIN(4),"PARTIAL FRACTION EXPANSION FOR
       GIVEN DATA",LIN(2)
900    FOR I=1 TO N
910    MAT Cw=C
```

```
920   MAT Cwj=Cj
930   PRINT LIN(2),"PCLE(";I;") = ";P(I);CHR$(43+(Pj(I)<0)*2);
      ABS(Pj(I));"j"
940   FOR J=1 TO M(I)
950   Prod=Cw(Nn)
960   Prodj=Cwj(Nn)
970   FOR K=Nn-1 TO 1 STEP -1
980   R=P(I)*Prod-Pj(I)*Prodj+Cw(K)
990   Prodj=P(I)*Prodj+Pj(I)*Prod+Cwj(K)
1000  Prod=R
1010  NEXT K
1020  FOR K=1 TO N
1030  IF K=I THEN Nextk1
1040  FOR L=1 TO M(K)
1050  D=(P(I)-P(K))^2+(Pj(I)-Pj(K))^2
1060  R=(Prod*(P(I)-P(K))+Prodj*(Pj(I)-Pj(K)))/D
1070  Prodj=(Prodj*(P(I)-P(K))-Prod*(Pj(I)-Pj(K)))/D
1080  Prod=R
1090  NEXT L
1100  Nextk1:   NEXT K
1110  IF Prodj<0 THEN 1140
1120  PRINT "K(";I;",";M(I)-J+1;") = ";Gain*Prod;" + ";
      Gain*Prodj;"j";
1130  GOTO 1150
1140  PRINT "K(";I;",";M(I)-J+1;") = ";Gain*Prod;" - ";
      ABS(Gain*Prodj);"j";
1150  MAT A=ZER
1160  MAT Aj=ZER
1170  A(1)=-Prod
1180  Aj(1)=-Prodj
1190  FOR K=1 TO N
1200  IF K=I THEN Nextk2
1210  FOR L=1 TO M(K)
1220  CALL Coeff(A(*),Aj(*),P(K),Pj(K),Nn)
1230  NEXT L
1240  Nextk2:NEXT K
1250  MAT Cw=Cw+A
1260  MAT A=Cw
1270  MAT Cwj=Cwj+Aj
1280  MAT Aj=Cwj
1290  Cw(Nn)=Cwj(Nn)=0
1300  FOR K=Nn-1 TO 1 STEP -1
1310  R=Cw(K+1)*P(I)-Cwj(K+1)*Pj(I)+A(K+1)
1320  Cwj(K)=Cw(K+1)*Pj(I)+Cwj(K+1)*P(I)+Aj(K+1)
1330  Cw(K)=R
1340  NEXT K
1350  NEXT J
1360  NEXT I
1370  PRINT LIN(2)
1380  GOTO End
```

```
1390 ! "This calculates numerator coefficients when operator
     inputs zeroes"
1400 Zeroes: IF Nz<>0 THEN DISP "Enter no. of DISTINCT
     ZEROES = (NO DEFAULT)";
1410 IF Nz<>0 THEN INPUT Nz
1420 IF Nz=0 THEN INPUT "Enter no. of DISTINCT ZEROES =
     (NO DEFAULT)",Nz
1430 Nz=ABS(INT(Nz))
1440 IF Nz=0 THEN GOTO Zeroes
1450 DISP "ENTER GAIN OF THE SYSTEM  (DEFAULT =";Gain;")";
1460 INPUT Gain
1470 PRINT LIN(1),"SYSTEM GAIN =";Gain;LIN(2)
1480 FOR I=1 TO Nz
1490 DISP "ENTER ZERO(";I;"):REAL,IMAG,MULTIPLI.(DEFAULT=";
     Z(I);Zj(I);Mz(I);")";
1500 INPUT Z(I),Zj(I),Mz(I)
1510 PRINT "ZERO(";I;") =";Z(I);Zj(I);Mz(I)
1520 Mz(I)=ABS(INT(Mz(I)))
1530 NEXT I
1540 A$="N"
1550 INPUT "ANY CHANGES (Y/N, DEFAULT=N)",A$
1560 IF A$="Y" THEN Zeroes
1570 C(1)=1
1580 Cj(1)=0
1590 FOR I=1 TO Nz
1600 FOR J=1 TO Mz(I)
1610 CALL Coeff(C(*),Cj(*),Z(I),Zj(I),Nz+1)
1620 NEXT J
1630 NEXT I
1640 GOTO List_input
1650 ! "This calculates distinct poles when operator inputs
     denominator coeff."
1660 Polescoeff:INPUT "ORDER OF DENOMINATOR",Nn
1670 PRINT LIN(2),"DEGREE OF POLYNOMIAL=";Nn
1680 REDIM Rroot(Nn),Iroot(Nn),Rcoef(Nn),Icoef(Nn),Rcoefw(Nn),
     Icoefw(Nn)
1690 MAT Rcoefw=Rcoef
1700 MAT Icoefw=Icoef
1710 CALL Sroot(Nn,Rroot(*),Iroot(*),Rcoefw(*),Icoefw(*),
     Rcoef(*),Icoef(*))
1720 FOR I=1 TO Nn
1730 Rroot(I)=DROUND(Rroot(I),4)
1740 IF ABS(Rroot(I))<.0001 THEN Rroot(I)=0
1750 Iroot(I)=DROUND(Iroot(I),4)
1760 IF ABS(Iroot(I))<.0001 THEN Iroot(I)=0
1770 NEXT I
1780 MAT M=ZER
1790 Np=Nn
1800 FOR I=1 TO Nn
1810 IF I>Np THEN Subend
```

INVERSION OF THE LAPLACE TRANSFORM

```
1820 FOR J=I TO Np
1830 IF J>Np THEN Nexti
1840 IF SGN(Rroot(I))<>SGN(Rroot(J)) THEN Nextj
1850 IF SGN(Iroot(I))<>SGN(Iroot(J)) THEN Nextj
1860 IF (Rroot(I)=0) AND (ABS(Rroot(J))>.0001) THEN Nextj
1870 IF (Iroot(I)=0) AND (ABS(Iroot(J))>.0001) THEN Nextj
1880 B=Rroot(I)
1890 A=Rroot(J)
1900 IF (B=0) AND (ABS(A)>.0001) THEN Nextj
1910 IF (B=0) AND (ABS(A)<.0001) THEN Shift
1920 IF (ABS(A/B)>1.1) OR (ABS(A/B)<.9) THEN Nextj
1930 B=Iroot(I)
1940 A=Iroot(J)
1950 IF (B=0) AND (ABS(A)>.0001) THEN Nextj
1960 IF (B=0) AND (ABS(A)<.0001) THEN Shift
1970 IF (ABS(A/B)>1.1) OR (ABS(A/B)<.9) THEN Nextj
1980 Shift:M(I)=M(I)+1
1990 IF J=I THEN Nextj
2000 FOR K=J TO Np-1
2010 Rroot(K)=Rroot(K+1)
2020 Iroot(K)=Iroot(K+1)
2030 M(K)=M(K+1)
2040 NEXT K
2050 Rroot(K)=Iroot(K)=M(K)=0
2060 Np=Np-1
2070 J=J-1
2080 Nextj:NEXT J
2090 Nexti: NEXT I
2100 Subend: PRINT LIN(2)
2110 REDIM Rroot(Np),Iroot(Np),P(Np),Pj(Np),M(Np)
2120 N=Np
2130 MAT P=Rroot
2140 MAT Pj=Iroot
2150 GOTO Numerator
2160 End: END
2170 SUB Sroot(N,Rroot(*),Iroot(*),Rcoefw(*),Iccefw(*),
     Rcoef(*),Iccef(*))
2180 Itmax=1000000000
2190 PRINT "MAX # OF ITERATIONS=";Itmax
2200 Tola=.000001
2210 PRINT "TOLERANCE FOR ROOTS=";Tola
2220 Tolf=.00000001
2230 PRINT "TOLERANCE FOR FUNCTIONAL EVALUATIONS=";Tolf
2240 PRINT "COEFF:(Rcoef(0)+Iccef(0)*j)*s^0 +
     (Rcoef(1)+Iccef(1)*j)*s^1 + ...."
2250 PRINT LIN(1),SPA(8),"REAL","    IMAGINARY",LIN(1)
2260 FOR I=0 TO N
2270 DISP "ENTER Rcoef(";I;"), Iccef(";I;")  (DEFAULT =";
     Rcoef(I);Iccef(I);")";
```

```
2280     INPUT Rcoef(I),Iccef(I)
2290      PRINT USING 2350;Rcoef(I),Iccef(I)
2300 NEXT I
2310 C$="N"
2320 INPUT "ANY CHANGES (Y/N), DEFAULT = N",C$
2330 PRINT LIN(1)
2340 IF C$="Y" THEN 2260
2350 IMAGE 3X,MZ.6DE,5X,MZ.6DE
2360 PRINT LIN(2)
2370 MAT Rcoefw=Rcoef
2380 MAT Iccefw=Iccef
2390 CALL Roctfd(N,Rcoefw(*),Icoefw(*),Tcla,Tolf,Itmax,
     Rroct(*),Iroct(*))
2400 SUBEXIT
2410 SUBEND
```

This progam performs PARTIAL FRACTION expansion on
a rational function of Laplace transform operator s

```
DEGREE OF POLYNOMIAL= 4
MAX # OF ITERATIONS= 1000000000
TOLERANCE FOR ROOTS= .000001
TOLERANCE FOR FUNCTIONAL EVALUATIONS= .00000001
COEFF:(Rcoef(0)+Icoef(0)*j)*s^0 + (Rcoef(1)+Icoef(1)*j)*s^1 + ..
```

REAL	IMAGINARY
0.000000E+00	0.000000E+00
5.000000E+00	0.000000E+00
3.700000E+01	0.000000E+00
1.000000E+01	0.000000E+00
1.000000E+00	0.000000E+00

```
n( 0 )=-3 + 0 j
n( 1 )= 47 + 0 j
n( 2 )= 12 + 0 j
n( 3 )= 1 + 0 j
```

Coefficeints of the Denom. Polynomial

```
n( 0 )= 0 + 0 j
n( 1 )= 5 + 0 j
n( 2 )= 37 + 0 j
n( 3 )= 10 + 0 j
n( 4 )= 1 + 0 j
```

COEFFICIENTS OF THE NUMERATOR IN ASCENDING ORDER

```
n( 0 )=-3 + 0 j
n( 1 )= 47 + 0 j
n( 2 )= 12 + 0 j
n( 3 )= 1 + 0 j
```

PARTIAL FRACTION EXPANSION FOR GIVEN DATA

```
POLE( 1 )  =  -4.93 - 3.363 j
K( 1 , 1 ) = -.173775067775  +  8.32008362522E-02 j

POLE( 2 )  =  -4.93 + 3.363 j
K( 2 , 1 ) = -.173775067775  -  8.32008362522E-02 j

POLE( 3 )  =   0 + 0 j
K( 3 , 1 ) = -.59996405884  +  0 j

POLE( 4 )  =  -.1404 + 0 j
K( 4 , 1 ) =   1.9475141944  +  2.07956186711E-12 j
```

This progam performs PARTIAL FRACTION expansion on
a rational function of Laplace transform operator s
```
POLE( 1 ) =  0 + 0 j MULTIPL. 2
POLE( 2 ) = -2 + 0 j MULTIPL. 2
POLE( 3 ) = -3 + 3 j MULTIPL. 1
POLE( 4 ) = -3 +-3 j MULTIPL. 1
```

```
n( 0 )= 36 + 0 j
n( 1 )= 12 + 0 j
```

POLES OF THE DENOMINATOR IN ORDER OF ENTRY

```
POLE( 1 ) =  0 + 0 j MULTIPL. 2
POLE( 1 ) =  0 - 0 j MULTIPLTIPLICITY 2
POLE( 2 ) = -2 + 0 j MULTIPL. 2
POLE( 2 ) = -2 - 0 j MULTIPLTIPLICITY 2
POLE( 3 ) = -3 + 3 j MULTIPL. 1
POLE( 4 ) = -3 - 3 j MULTIPLTIPLICITY 1
```

LINEAR CONTROL SYSTEMS

COEFFICIENTS OF THE NUMERATOR IN ASCENDING ORDER

n(0) = 36 + 0 j
n(1) = 12 + 0 j

PARTIAL FRACTION EXPANSION FOR GIVEN DATA

POLE(1) =   0 + 0 j
K(1 , 2) = .5 + 0 jK(1 , 1) = -.5 + 0 j

POLE(2) = -2 + 0 j
K(2 , 2) = .3 + 0 jK(2 , 1) = .54 + 0 j

POLE(3) = -3 + 3 j
K(3 , 1) = -.02 - 2.66666666667E-02 j

POLE(4) = -3 - 3 j
K(4 , 1) = -.02 + 2.66666666667E-02 j

This progam performs PARTIAL FRACTION expansion on
a rational function of Laplace transform operator s
POLE(1) = 0 + 0 j MULTIPL. 3
POLE(2) = -1 + 0 j MULTIPL. 1
POLE(3) = 2 + 0 j MULTIPL. 1

n(0) = 8 + 0 j

POLES OF THE DENOMINATOR IN ORDER OF ENTRY

POLE(1) = 0 + 0 j MULTIPL. 3
POLE(1) = 0 - 0 j MULTIPLTIPLICITY 3
POLE(2) = -1 + 0 j MULTIPL. 1
POLE(2) = -1 - 0 j MULTIPLTIPLICITY 1
POLE(3) = 2 + 0 j MULTIPL. 1
POLE(3) = 2 - 0 j MULTIPLTIPLICITY 1

INVERSION OF THE LAPLACE TRANSFORM

COEFFICIENTS OF THE NUMERATOR IN ASCENDING ORDER

n(0) = 8 + 0 j

PARTIAL FRACTION EXPANSION FOR GIVEN DATA

POLE(1) = 0 + 0 j
K(1,3) = -4 + 0 jK(1,2) = 2 + 0 jK(1, 1) = -3 + 0 j

POLE(2) = -1 + 0 j
K(2 , 1) = 2.66666666667 + 0 j

POLE(3) = 2 + 0 j
K(3 , 1) = .333333333333 + 0 j

The Laplace transforms of some common functions of time as well as the inverse Laplace transforms of some common functions of s are collected in Table 1 for convenience. Note that the time function $f(t)$ is assumed to vanish for $t < 0$.

TABLE 1. Laplace Transform Pairs

No.	f(t)	F(s)
1	$\delta(t)$	1
2	$u(t)$ (unit step function)	$\dfrac{1}{s}$
3	$u(t-\alpha)$	$\dfrac{e^{-\alpha s}}{s}$
4	$r(t) = tu(t)$ (unit ramp)	$\dfrac{1}{s^2}$
5	t^n	$\dfrac{n!}{s^{n+1}}$
6	$t^n e^{\alpha t}$	$\dfrac{n!}{(s-\alpha)^{n+1}}$
7	$-1-\alpha t + e^{\alpha t}$	$\dfrac{\alpha^2}{s^2(s-\alpha)}$
8	$1-e^{\alpha t} + \alpha t e^{\alpha t}$	$\dfrac{\alpha^2}{s(s-\alpha)^2}$

No.	f(t)	F(s)
9	$-\dfrac{\beta}{\alpha^2} + \dfrac{1+\alpha t-\beta t}{\alpha} e^{\alpha t}$	$\dfrac{s-\beta}{s(s-\alpha)^2}$
10	$\sin \omega t$	$\dfrac{\omega}{s^2+\omega^2}$
11	$\cos \omega t$	$\dfrac{s}{s^2+\omega^2}$
12	$e^{\alpha t} \sin \omega t$	$\dfrac{\omega}{(s-\alpha)^2+\omega^2}$
13	$e^{\alpha t} \cos \omega t$	$\dfrac{s-\alpha}{(s-\alpha)^2+\omega^2}$
14	$\dfrac{1}{\omega^2}(1-\cos \omega t)$	$\dfrac{1}{s(s^2+\omega^2)}$
15	$e^{-\zeta\omega_n t} \sin\left(\omega_n \sqrt{1-\zeta^2}\, t\right)$	$\dfrac{\omega_n \sqrt{1-\zeta^2}}{s^2+2\zeta\omega_n s+\omega_n^2}$
16	$\dfrac{e^{\alpha t}}{\alpha^2+\omega^2} + \dfrac{\sin(\omega t-\phi)}{\omega\sqrt{\alpha^2+\omega^2}}$ where $\phi = -\tan^{-1}\dfrac{\omega}{\alpha}$	$\dfrac{1}{(s-\alpha)(s^2+\omega^2)}$
17	$e^{\alpha t}(A \cos \omega t + B \sin \omega t)$	$\dfrac{A(s-\alpha)+B\omega}{(s-\alpha)^2+\omega^2}$
18	$e^{\alpha t}\left(A \cos \omega t + \dfrac{B+\alpha A}{\omega}\sin \omega t\right)$	$\dfrac{As+B}{(s-\alpha)^2+\omega^2}$
19	$\dfrac{\sqrt{(\alpha+\beta)^2+\omega^2}}{\omega} e^{\alpha t} \sin(\omega t+\delta)$ where $\delta = \tan^{-1}\dfrac{\omega}{\alpha+\beta}$	$\dfrac{s+\beta}{(s-\alpha)^2+\omega^2}$
20	$2\lvert K \rvert e^{\alpha t} \cos(\omega t + <K)$	$\dfrac{K}{s-\alpha-j\omega} + \dfrac{\overline{K}}{s-\alpha+j\omega}$
21	$\dfrac{1}{\alpha^2+\omega^2} + \dfrac{e^{\alpha t}}{\omega\sqrt{\alpha^2+\omega^2}}\sin(\omega t-\delta)$ where $\delta = \tan^{-1}\dfrac{\omega}{\alpha}$	$\dfrac{1}{s[(s-\alpha)^2+\omega^2]}$
22	$\dfrac{1}{\omega_n^2} - \dfrac{e^{-\zeta\omega_n t}}{\omega_n\sqrt{1-\zeta^2}}\sin(\omega_n\sqrt{1-\zeta^2}\,t+\delta)$ where $\delta = \cos^{-1}\zeta$	$\dfrac{1}{s(s^2+2\zeta\omega_n s+\omega_n^2)}$
23	$\dfrac{\beta}{\alpha^2+\omega^2} + \dfrac{e^{\alpha t}\sqrt{(\alpha+\beta)^2+\omega^2}}{\omega\sqrt{\alpha^2+\omega^2}}\sin(\omega t+\delta)$ where $\delta = \tan^{-1}\dfrac{\omega}{\alpha+\beta} - \tan^{-1}\dfrac{\omega}{\alpha}$	$\dfrac{s+\beta}{s[(s-\alpha)^2+\omega^2]}$

3.4.2 Numerical Inversion of the Laplace Transform

The inversion of the Laplace transform via numerical techniques will be considered here. The subject has been well treated in literature. An extensive bibliography can be found in references [3.10, 3.11] where some 300 works have been cited as early as 1934 and as late as 1976. An efficient method due to Liou [3.12] will be presented here. In order to illustrate this method, consider a third-order linear differential equation,

$$\dddot{x}(t) + a\ddot{x}(t) + b\dot{x}(t) + cx(t) = 0 \tag{32}$$

with initial conditions $x(0-)$, $\dot{x}(0-)$ and $\ddot{x}(0-)$. If the Laplace transform of (32) is taken and terms are rearranged, one obtains,

$$X(s) = x(0-)s^2 + [\dot{x}(0-) + ax(0-)]s +$$

$$\frac{[\ddot{x}(0-) + a\dot{x}(0-) + bx(0-)]}{(s^3 + as^2 + bs + c)} \tag{33}$$

This indicates that the third-order differential equation (32) can be represented by a strictly proper rational function whose coefficients depend directly on the initial conditions. The general discussion in this section is thus focused on the following rational function,

$$X(s) = \frac{b_{n-1}s^{n-1} + b_{n-2}s^{n-2} + \cdots + b_1 s + b_0}{s^n + a_{n-1}s^{n-1} + \cdots + a_1 s + a_0} \tag{34}$$

where a_i and b_i are real constant coefficients and n is a positive integer. The differential equation corresponding to (34) is

$$x^{(n)}(t) + a_{n-1}x^{(n-1)}(t) + \cdots + a_1\dot{x}(t) + a_0 x(t) = 0 \tag{35}$$

with initial conditions:

$$x(0) = b_{n-1} \tag{36a}$$

$$\dot{x}(0) = b_{n-2} - a_{n-1}x(0) \tag{36b}$$

$$\ddot{x}(0) = b_{n-3} - a_{n-1}\dot{x}(0) - a_{n-2}x(0) \tag{36c}$$

$$\vdots$$

$$x^{(n-1)}(0) = b_o - a_{n-1}x^{(n-2)}(0) - a_{n-2}x^{(n-3)}(0) - \cdots - a_1 x(0) \tag{36d}$$

The above equations can be put in vector form by letting $x_1 = x$, $x_2 = \dot{x}$, $x_3 = \ddot{x}$, ..., $x_n = x^{(n-1)}$ and $\mathbf{x}' = (x_1,...,x_n)$, then (35) can be rewritten as

$$\dot{\mathbf{x}}(t) = \mathbf{A}\mathbf{x}(t) \tag{37}$$

where

$$\mathbf{A} = \begin{bmatrix} 0 & 1 & 0 & \cdots & 0 & 0 \\ 0 & 0 & 1 & \cdots & 0 & 0 \\ \cdot & & & & & \\ \cdot & & & & & \\ \cdot & & & & & \\ 0 & 0 & 0 & \cdots & 0 & 1 \\ -a_0 & -a_1 & -a_2 & \cdots & -a_{n-2} & -a_{n-1} \end{bmatrix} \tag{38}$$

and the initial condition vector $\underline{x}(0)$ is given by (36). Therefore, the rational function (34) can be represented by a vector differential equation (37). The latter equation can be solved numerically by any standard technique such as Euler, Runge-Kutta [3.16], etc. In this presentation, (37) is solved by the Euler's method, i.e.

$$\mathbf{x}(t+\Delta t) = \mathbf{A}\,\mathbf{x}(t)\Delta t + \mathbf{x}(t) \tag{39}$$

where Δt is the step size and the initial vector $\underline{x}(0)$ is obtained from (36). The following example illustrates the use of this method. It is solved by a BASIC program called "INVLT".

8. Example. For the third order transfer function,

$$X(s) = \frac{s^2+s}{s^3+5s^2+5.25s+5} \tag{40}$$

it is desired to find $x(t)$ versus time for $0 \leqslant t \leqslant 5$. The program "INVLT" and a sample terminal session for the numerical inversion of (40) are given below. Figure 1 shows a plot of $x(t)$ versus time.

INVERSION OF THE LAPLACE TRANSFORM

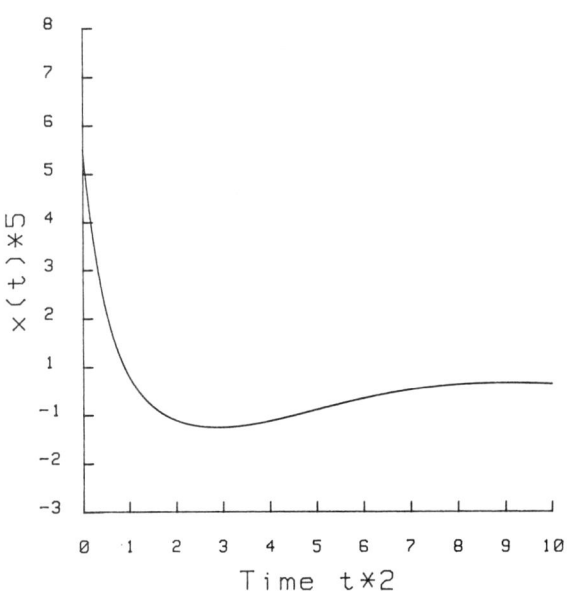

Fig. 3.4.1 Pictorial representation of $x(t)$ for Example 3.4.8

9. "INVLT".

```
10   !       PROGRAM NAME : "INVLT" PROG
20   PRINT " This program numerically inverts a LAPLACE.
     TRANSFORM:"
30   PRINT " X(s)=(ao + al.s + ... +an-1.s^n-1)/(bo + bl.s +
     ... +bn.s^n) and"
40   PRINT "  Provides x(t) vs t  for a set of known parameters"
50   DIM X(1000),Y(1000),Xlabel$(50),Ylabel$(50),T(1000)
60   DIM Z(100,0),W(100,0),Dy(100)
70   DIM A(20),B(20),M(20,20),Minv(20,20)
80   Iplot=0
90   INPUT "System Order?",N
100  INPUT "Number of Data Points e.g. 100,500,1000",M
110  REDIM Z(N-1,0),W(N-1,0),X(M),Y(M)
120  REDIM A(N),B(N-1),M(N-1,N-1),Minv(N-1,N-1),Dy(N-1),T(M)
130  INPUT "Inital Time to?",To
140  INPUT "Final Time tf?",Tf
150  S=(Tf-To)/M
160  PRINT "Give the coefficients of DENOMINATOR in ASCENDING
     order of s"
170  FIXED 0
180  FOR I=0 TO N
```

```
190     DISP "a";I;
200     INPUT A(I)
210     NEXT I
220     FIXED 2
230     MAT PRINT A
240     PRINT "Give the coefficients of NUMERATOR in ASCENDING
        order of s"
250     FIXED 0
260     FOR I=0 TO N-1
270     DISP "b";I;
280     INPUT B(I)
290     NEXT I
300     FIXED 2
310     MAT PRINT B
320     FOR I=0 TO N-1
330     FOR J=0 TO N-1
340     IF (I+J+1<N) OR (I+J+1=N) THEN M(I,J)=A(I+J+1)
350     IF I+J+1>N THEN M(I,J)=0
360     NEXT J
370     NEXT I
380     MAT Minv=INV(M)
390     MAT Dy=Minv*B
400     FOR J=0 TO M
410     T(J)=(Tf-To)/M*J+To
420     X(J)=T(J)
430     NEXT J
440     FOR I=0 TO N-2
450     FOR J=0 TO N-1
460     IF J=I+1 THEN M(I,J)=1
470     IF (J>I+1) OR (J<I+1) THEN M(I,J)=0
480     NEXT J
490     NEXT I
500     FOR J=0 TO N-1
510     M(N-1,J)=-A(J)/A(N)
520     NEXT J
530     Y(0)=Dy(0)
540     FOR J=0 TO N-1
550     Z(J,0)=Dy(J)
560     NEXT J
570     FOR J=1 TO M
580     MAT W=M*Z
590     Y(J)=Y(J-1)+W(0,0)*S
600     FOR I=0 TO N-1
610     Z(I,0)=Z(I,0)+W(I,0)*S
620     NEXT I
630     NEXT J
640     INPUT "Do you like to PRINT the values of x(t) vs t?",P$
650     IF (P$="y") OR (P$="Y") THEN 680
660     IF (P$="N") OR (P$="n") THEN 730
670     GOTO 640
```

```
680  PRINT " Time t   x(t) "
690  Step=INT(M/20)
700  FOR I=0 TO M STEP Step
710  PRINT X(I);Y(I)
720  NEXT I
730  INPUT "Do you like to PLOT the values of x(t) vs t?",L$
740  IF (L$="Y") OR (L$="y") THEN 770
750  IF (L$="N") OR (L$="n") THEN End
760  GOTO 730
770  Xmin=X(0)
780  FOR I=0 TO M
790  Xmin=MIN(Xmin,X(I))
800  NEXT I
810  Xmax=X(0)
820  FOR I=0 TO M
830  Xmax=MAX(Xmax,X(I))
840  NEXT I
850  Ymin=Y(0)
860  FOR I=0 TO M
870  Ymin=MIN(Ymin,Y(I))
880  NEXT I
890  Ymax=Y(0)
900  FOR I=0 TO M
910  Ymax=MAX(Ymax,Y(I))
920  NEXT I
930  Xrange=Xmax-Xmin
940  Yrange=Ymax-Ymin
950  PRINT "Trange=";Xrange
960  PRINT "Xrange=";Yrange
970  INPUT "Tscale?",Xscale
980  INPUT "Xscale?",Yscale
990  PRINT "Tscale=";Xscale
1000 PRINT "Xscale=";Yscale
1010 FOR I=0 TO M
1020 X(I)=Xscale*X(I)
1030 Y(I)=Yscale*Y(I)
1040 NEXT I
1050 Xmax=Xscale*Xmax
1060 Ymax=Yscale*Ymax
1070 Xmin=Xscale*Xmin
1080 Ymin=Yscale*Ymin
1090 PRINT "Tmin=";Xmin;"Tmax=";Xmax
1100 PRINT "XMIN=";Ymin;"Xmax=";Ymax
1110 INPUT "Initial t?",Xinitial
1120 INPUT "Final t?",Xfinal
1130 INPUT "Initial x?",Yinitial
1140 INPUT "Final x?",Yfinal
1150 Xinc=INT((Xfinal-Xinitial)/10)
1160 Yinc=INT((Yfinal-Yinitial)/10)
1170 LINPUT "X-axis Label?",Xlabel$
```

```
1180 LINPUT "Y-axis Label?",Ylabel$
1190 PLOTTER IS 13,"GRAPHICS"
1200 Graph:  GRAPHICS
1210 LOCATE 20,90,20,90
1220 SCALE Xinitial,Xfinal,Yinitial,Yfinal
1230 AXES Xinc,Yinc,Xinitial,Yinitial,100,100,6
1240 SETGU
1250 Xtic=Xinitial
1260 FOR Xposition=20 TO 90 STEP 7
1270 MOVE Xposition,15
1280 LORG 5
1290 CSIZE 3
1300 FIXED 0
1310 LABEL USING "K";VAL$(Xtic)
1320 FIXED 2
1330 Xtic=Xtic+Xinc
1340 NEXT Xposition
1350 Ytic=Yinitial
1360 FOR Yposition=20 TO 90 STEP 7
1370 MOVE 15,Yposition
1380 LORG 4
1390 CSIZE 3
1400 FIXED 0
1410 LABEL USING "K";VAL$(Ytic)
1420 FIXED 2
1430 Ytic=Ytic+Yinc
1440 NEXT Yposition
1450 MOVE 55,10
1460 LORG 5
1470 CSIZE 5
1480 LABEL USING "K";Xlabel$
1490 MOVE 10,55
1500 LORG 5
1510 DEG
1520 LDIR 90
1530 CSIZE 5
1540 LABEL USING "K";Ylabel$
1550 SETUU
1560 MOVE X(0),Y(0)
1570 FOR I=1 TO M
1580 DRAW X(I),Y(I)
1590 NEXT I
1600 IF Iplot=0 THEN 1630
1610 PENUP
1620 End:END
1630 INPUT "WANT TO DUMP GRAPHICS?",A$
1640 IF A$="N" THEN 1750
1650 DUMP GRAPHICS
1660 PRINTER IS 0
1670 PRINT LIN(2)
```

```
1680 FOR I=0 TO N
1690 PRINT "A(";I;")=";A(I)
1700 NEXT I
1710 FOR I=0 TO N-1
1720 PRINT "B(";I;")=";B(I)
1730 NEXT I
1740 PRINTER IS 16
1750 INPUT "WANT GRAPH ON 9872A?",A$
1760 IF A$="N" THEN End
1770 Hardcopy:   PLOTTER IS "9872A"
1780 Iplot=1
1790 INPUT "WHICH PEN?",Ipen
1800 PEN Ipen
1810 OUTPUT 705;"VS4;"
1820 GOTO Graph
```

This program numerically inverts a LAPLACE TRANSFORM:
$X(s) = (a_0 + a_1.s + \ldots + a_{n-1}.s^{n-1})/(b_0 + b_1.s + \ldots + b_n.s^n)$
and
Provides x(t) vs t for a set of known parameters
Give the coefficients of DENOMINATOR in ASCENDING order of s
5.00   5.25   5.00   1.00

Give the coefficients of NUMERATOR in ASCENDING order of s

0.00   1.00   1.00

```
Time t   x(t)
0.00    1.00
 .25     .33
 .50     .06
 .75    -.07
1.00    -.12
1.25    -.15
1.50    -.15
1.75    -.14
2.00    -.13
2.25    -.10
2.50    -.08
2.75    -.05
3.00    -.03
3.25    -.01
3.50     .01
3.75     .02
4.00     .03
4.25     .03
4.50     .03
4.75     .03
5.00     .03
```

```
Trange= 5.00
Xrange= 1.15
Tscale= 2.00
Xscale= 5.00
Tmin= 0.00 Tmax= 10.00
XMIN=-.76 Xmax= 5.00
```

3.5 SOLUTION OF DIFFERENTIAL EQUATIONS USING THE LAPLACE TRANSFORM

An important application of the Laplace transform is to solve linear constant-coefficient differential equations. By taking the Laplace transform of both sides of such a differential equation, the Laplace transform of the unknown function can be expressed as a rational function of s. This was demonstrated in Section 3.4.2. The inverse Laplace transform of this function then provides the solution. Thus the Laplace transform converts the solution of a linear constant-coefficient differential equation to that of an algebraic equation. The following example further illustrates this.

1. **Example.** Consider the differential equation

$$\dddot{f}(t) + 10\ddot{f}(t) + 37\dot{f}(t) + 52f(t) = -3u(t) + \dot{\delta}(t), \quad t \geq 0 \tag{1}$$

with initial conditions

$$f(0-) = \dot{f}(0-) = 1, \quad \ddot{f}(0-) = 0 \tag{2}$$

Taking the Laplace transform of both sides of (1) yields

$$[s^3F(s)-s^2f(0-)-s\dot{f}(0-)-\ddot{f}(0-)] + 10[s^2F(s)-sf(0-)-\dot{f}(0-)]$$

$$+ 37[sF(s)-f(0-)] + 52\,F(s) = \frac{-3}{s} + s$$

or

$$F(s) = \frac{s^3+12s^2+47s-3}{s(s^3+10s^2+37s+52)} \tag{3}$$

which can be written in partial fraction form as

$$F(s) = \frac{-3/52}{s} + \frac{63/20}{s+4} - \frac{\frac{136}{65}(s+3)+\frac{541}{130}}{(s+3)^2+4} \tag{4}$$

Thus, using entries 2, 6 and 17 of Table 3.4.1 the solution is

$$f(t) = L^{-1}[F(s)] = -\frac{3}{52} + \frac{63}{20}e^{-4t} - \frac{136}{65}e^{-3t}\cos 2t$$

$$+ \frac{541}{130}e^{-3t}\sin 2t, \quad t \geq 0 \tag{5}$$

3.6 DEFINITION OF THE z-TRANSFORM

In the previous section we illustrated how the Laplace transform can be used to convert the solution of a linear constant-coefficient differential equation into that of an algebraic equation. In a similar manner, the z-transform can be used to convert the solution of a linear difference equation with constant coefficients into that of an algebraic equation.

In order to study the z-transform, the idea of sampling a time function will be discussed first.

1. Ideal Sampling. Consider a function of continuous time $f(t)$ where $f(t) = 0$ for $t < 0$. An *ideal sampler* is one that takes samples of infinitesimal width of $f(t)$ at regular intervals of time (Fig. 1). Thus, an ideal sampling $f^*(t)$ of $f(t)$ can be viewed as a sequence of equally spaced impulses with magnitudes equal to the values of the function $f(\cdot)$ at the corresponding discrete times. That is,

$$f^*(t) = \sum_{k=0}^{\infty} f(kT)\delta(t-kT) \tag{1}$$

where T is called the *sampling period*.

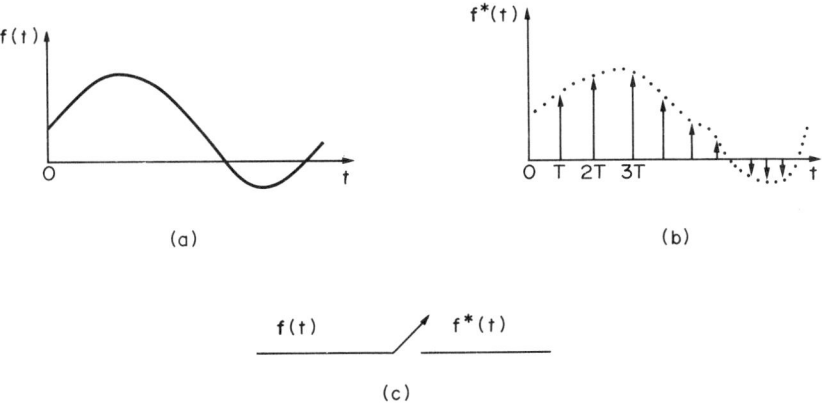

Fig. 3.6.1 Ideal sampling

Taking the Laplace transform of both sides of (1) yields

$$L[f^*(t)] \triangleq F^*(s) = \sum_{k=0}^{\infty} f(kT) L[\delta(t-kT)]$$

$$= \sum_{k=0}^{\infty} f(kT) e^{-kTs} \qquad (2)$$

Now define the complex variable z as

$$z = e^{Ts} \qquad (3)$$

Thus, (2) becomes

$$F^*(s) = \sum_{k=0}^{\infty} f(kT) z^{-k} \qquad (4)$$

which is defined as the *(one-sided) z-transform* of $f(t)$. More formally, we have the following definition.

2. Definition. Let T be a positive constant. The z-transform of a function $f(t)$ where $f(t) = 0$ for $t < 0$ and $f(t)$ is continuous at $t = kT$, $k = 0, 1, 2, ...$ is

$$Z[f(t)] \triangleq F(z) = \sum_{k=0}^{\infty} f(kT) z^{-k} \qquad (5)$$

Also, the z-transform of a sequence of scalars $f(k)$, $k = 0, 1, 2, ...$ is defined as

$$Z[f(k)] \triangleq F(z) = \sum_{k=0}^{\infty} f(k) z^{-k} \qquad (6)$$

$\triangle$

Note that the z-transform exists if the infinite sums in (5) and (6) converge. The *radius of convergence* r_c of the series in (6) is given by

$$r_c = \lim_{k \to \infty} |f(k)|^{1/k} \qquad (7)$$

DEFINITION OF THE z-TRANSFORM

The series in (6) is analytic for $|z|>r_c^*$. Thus it converges absolutely for all z in the domain $|z|>r_c$. For notational convenience, we will take $T=1$ in (5) from now on.

3. Example. Consider the unit step function $u(t)$. Then

$$Z[u(t)] \triangleq U(z) = \sum_{k=0}^{\infty} z^{-k} \tag{8}$$

which converges to

$$\frac{1}{1-\frac{1}{z}} = \frac{z}{z-1} \tag{9}$$

if $|z|>1$. Thus the radius of convergence of the series in (8) is 1. Note that (9) is also the z-transform of the sequence $u^*(t) = 1,1,1,\cdots$. Although (9) is the z-transform of this unique sequence, it does not correspond to a unique function of continuous time since there are many functions whose samples $f(k)$ at $k = 0, 1, 2, \ldots$ are 1 (Fig. 2).

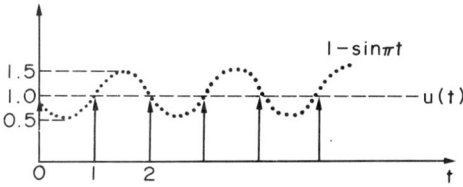

Fig. 3.6.2 Different time functions with the same samples

4. Example. Consider the sequence

$$f(k) = \alpha^k, \quad k = 0,1,2,\cdots \tag{10}$$

Then

$$Z[f(k)] \triangleq \sum_{k=0}^{\infty} f(k)z^{-k} = \sum_{k=0}^{\infty} \alpha^k z^{-k}$$

$$= \frac{1}{1-\frac{\alpha}{z}} = \frac{z}{z-\alpha} \tag{11}$$

---

* Some authors define radius of convergence of the series as $\dfrac{1}{r_c}$.

provided that $|\frac{\alpha}{z}|<1$ or $|z|>|\alpha|$. Thus the radius of convergence of the infinite series in (11) is $|\alpha|$.

3.7 PROPERTIES OF THE z-TRANSFORM

The z-transform is defined through the use of the Laplace transform. (See (3.6.2) to (3.6.4).) Thus the properties of the z-transform can be deduced from those of the Laplace transform. In this section we will list some useful properties of the z-transform without proof. The proofs follow the same lines as in Section 3.3. We assume that the functions considered in this section vanish for $t<0$ and that they have z-transforms with finite radii of convergence. We use $F(z)$ for the z-transform of $f(t)$.

1. Uniqueness. As discussed in Example 3.6.3, a given function $F(z)$ corresponds to a unique sequence $f(k)$. It does not, however, correspond to a unique function of continuous time $f(t)$.

2. Linearity. We have

$$Z[\alpha_1 f_1(t)+\alpha_2 f_2(t)] = \alpha_1 Z[f_1(t)] + \alpha_2 Z[f_2(t)] \qquad (1)$$

for arbitrary constants α_1 and α_2.

3. Advance. This property of the z-transform is analogous to the differentiation rule of the Laplace transform:

$$Z[f(k+1)] = zF(z) - zf(0) \qquad (2)$$

and, in general,

$$Z[f(k+m)] = z^m F(z) - \sum_{k=0}^{m-1} f(k) z^{m-k} \qquad (3)$$

for any positive integer m.

4. Delay. This property of the z-transform is analogous to the integration rule of the Laplace transform:

$$Z[f(k-m)] = z^{-m} F(z) \qquad (4)$$

for any positive integer m.

5. Initial Value Property. We have

$$f(0+) = \lim_{z \to \infty} F(z) \tag{5}$$

provided that the limit exists.

6. Final Value Property. If the sequence $f(k)$ tends to a limit as k tends to ∞, then

$$\lim_{k \to \infty} f(k) = \lim_{z \to 1} (z-1)F(z) \tag{6}$$

if the limit exists.

7. Convolution Property. If $F_1(z)$ and $F_2(z)$ are the z-transforms of $f_1(t)$ and $f_2(t)$, respectively, then

$$Z[\sum_{j=0}^{k} f_1(j)f_2(k-j)] = F_1(z)F_2(z) \tag{7}$$

For some further properties of the z-transform, see Problem 3.19.

3.8 INVERSION OF THE z-TRANSFORM

Given a function $F(z)$, of the complex variable z, one can find a function $f(\cdot)$ in the time domain such that

$$Z[f(t)] = F(z) \tag{1}$$

or equivalently

$$f(t) = Z^{-1}[F(z)] \tag{2}$$

Note, however, that unlike in the case of the Laplace transform, the inverse z-transform $f(t)$ is not unique. In fact an infinite number of functions $f(t)$ exist for which $F(z)$ is the z-transform. These functions have the same values only at the sampling instants, i.e. they can all be represented by the same ideal sampling $f^*(t)$. Three different methods of determining the inverse z-transform will be discussed below.

3.8.1 The Power Series Method

Given $F(z)$, the defining infinite series (3.6.6) may be used to determine the value of $Z^{-1}[F(z)]$ at the sampling instants. This is done by expanding the rational function $F(z)$ into an infinite series in z^{-k}. Then the coefficient of z^{-k} will be equal to $f(k)$. However, this procedure is, in general, tedious and does not yield a closed-form solution.

1. Example. Consider

$$F(z) = \frac{z}{z-2} \qquad (3)$$

Long division results in

$$F(z) = 1 + 2z^{-1} + 4z^{-2} + 8z^{-3} + \cdots \qquad (4)$$

Thus, by comparison with (3.6.6), we have

$$f(0) = 1, \quad f(1) = 2, \quad f(2) = 4, \quad f(3) = 8, \text{ etc.} \qquad (5)$$

3.8.2 The Residue Method

The following theorem is analogous to the Laplace transform inversion theorem.

2. Theorem. Let $F(z) = Z[f(k)]$. Then

$$f(k) = \frac{1}{2\pi j} \oint_\Gamma F(z) z^{k-1} dz, \quad k = 0, 1, 2, \cdots \qquad (6)$$

where integration is performed in counterclockwise direction and Γ is any closed curve enclosing the origin and lying outside the circle $|z| > r_c$ (the radius of convergence of $F(z)$); or, equivalently, Γ is any closed curve enclosing the poles of $F(z)$, $k = 0, 1, 2, \ldots$

△

The contour integral (6) can be evaluated by using Cauchy's residue theorem which will be given after the following definition.

3. Definition. The *residue* of a complex function $H(z)$ at a pole $z = \beta$ with multiplexity m of $H(z)$

is

$$\frac{1}{(m-1)!} \frac{d^{m-1}}{d_z{}^{m-1}} [(z-\beta)^m H(z)]\big|_{z=\beta} \qquad (7)$$

(Compare with (3.4.4) for $j = m_i$.)

4. Cauchy's Residue Theorem. [3.15] Let $H(z)$ be analytic in a region R except at a finite number of singularities. The if Γ is a closed curve in R not passing through these singularities,

$$\frac{1}{2\pi j} \oint_\Gamma H(z)\,dz = \text{sum of the residues of } H(z) \qquad (8)$$

where integration is performed in counterclockwise direction.

△

Thus from (6) we have

$$f(k) = Z^{-1}[F(z)] = \text{sum of the residues of } F(z)z^{k-1} \qquad (9)$$

5. Example. Again consider $F(z) = \dfrac{z}{z-\alpha}$ which has a simple poles at $z - \alpha$. Thus,

$$f(k) = Z^{-1}[F(z)] = \mathrm{Res}\left[\frac{z}{z-\alpha}z^{k-1}\right] = z^k\big|_{z=\alpha} = \alpha^k \qquad (10)$$

6. Example. Consider

$$G(z) = \frac{z}{(z-\beta)^2} \qquad (11)$$

$G(z)$ has a pole of multiplicity 2 at $z=\beta$. Thus,

$$g(k) = Z^{-1}[G(z)] = \mathrm{Res}\left[\frac{z}{(z-\beta)^2}z^{k-1}\right] = \frac{d}{dz}(z^k)\big|_{z=\beta} = \beta^{k-1}k \qquad (12)$$

3.8.3 The Partial Fraction Expansion Method

This method is similar to that used in determining the inverse Laplace transform of a rational function. Instead of $F(s)$ in the case of the Laplace transform, here we expand $\frac{F(z)}{z}$ is partial fractions. The reason for this is that it is usually more convenient to find the inverse z-transform of a fraction with a free z in its numerator. The following example illustrates this method.

7. Example. Consider

$$F(z) = \frac{z(-2z+1)}{z^2-3z+2} \tag{13}$$

The partial fraction expansion of $F(z)/z$ yields

$$\frac{F(z)}{z} = \frac{1}{z-1} - \frac{3}{z-2} \tag{14}$$

Thus

$$F(z) = \frac{z}{z-1} - 3\frac{z}{z-2} \tag{15}$$

and using (10), the inverse z-transform of $F(z)$ is

$$f(k) = 1 - 3(2^k) \tag{16}$$

Program "INVZT" uses the partial fraction expansion method to determine the inverse z-transform of any proper rational function $F(z)$. The source listing of this program followed by an example is given below.

8. "INVZT"

```
10      REM PROGRAM NAME : "INVZT" PROC
20      PRINT " 'INVZTR' FINDS THER INVERSE z-TRANSFORM OF A
        TRANSFER "
30      PRINT " FUNCTION BY THE RESIDUE METHOD"
40      DIM P(100),Pj(100),C(100),Cj(100),Cw(100),Cwj(100),
        Rccefw(100),Icoefw(100)
50      DIM A(100),Aj(100),M(100),Mz(100),Rcoef(100),Iccef(100),
        Z(100),Zj(100)
```

INVERSION OF THE z-TRANSFORM

```
60   DIM Rroot(100),Iroot(100)
70   DIM Kk(100)
80   N=Nz=Nn=Nd=Np=Nm=0
90   Gain=1
100  Options: INPUT "DENOM.IN POLE OR COEFF. FORM (P/C)?
     (NO DEFAULTS)",P$
110  IF P$="C" THEN Polescoeff
120  IF P$="P" THEN Poles
130  GOTO Options
140  Poles: IF N<>0 THEN DISP "ENTER NUM. OF DISTINCT POLES
     (DEFAULT =";N;")";
150  IF N<>0 THEN INPUT N
160  IF N=0 THEN INPUT "ENTER THE NUMBER OF DISTINCT POLES
     (NO DEFAULTS",N
170  N=ABS(INT(N))
180  IF N=0 THEN Poles
190  FOR I=1 TO N
200  DISP "ENTER POLE(";I;"):REAL,IMAG.,MULTIPL. (DEFAULT=";
     P(I);Pj(I);M(I);")";
210  INPUT P(I),Pj(I),M(I)
220  IF Pj(I)>=0 THEN PRINT "POLE(";I;") = ";P(I);"+";Pj(I);
     "j"," MULTIPL.";M(I)
230  IF Pj(I)<0 THEN PRINT "POLE(";I;") = ";P(I);"+";Pj(I);
     "j"," MULTIPL.";M(I)
240  NEXT I
250  A$="N"
260  INPUT "ANY CHANGES (Y/N, DEFAULT = N)",A$
270  PRINT LIN(1)
280  IF A$="Y" THEN Poles
290  Nn=0
300  FOR I=1 TO N
310  Nn=M(I)+Nn
320  NEXT I
330  Numerator: REDIM C(Nn),Cj(Nn),Cw(Nn),Cwj(Nn),A(Nn),Aj(Nn),
     M(N),P(N),Pj(N)
340  PRINT "PLEASE ENTER THE NUMERATOR IN COEFFICIENT FORM"
350  GOTO 390
360  INPUT "NUMERATOR IN ZERO OR COEFFICIENT FORM (Z/C)?
     (NO DEFAULTS)",N$
370  IF N$="C" THEN Numeratorcoeff
380  IF N$="Z" THEN Zeroes
390  Numeratorcoeff: IF Nm<>0 THEN DISP "ENTER ORDER OF NUMER.:
     (DEFAULT=";Nm;")";
400  IF Nm<>0 THEN INPUT Nm
410  IF Nm=0 THEN INPUT "ENTER ORDER OF THE NUMERATOR:
     (DEFAULT NOT ALLOWED",Nm
420  FOR I=1 TO Nm+1
430  DISP "COEFFICIENT n(";I-1;"): (DEFAULT=";C(I);Cj(I);")";
440  INPUT C(I),Cj(I)
450  IF Cj(I)>=0 THEN PRINT "n(";I-1;")=";C(I);"+";Cj(I);"j"
```

```
460   IF Cj(I)<0 THEN PRINT C(I);"-";ABS(Cj(I));"j"
470   NEXT I
480   B$="N"
490   INPUT "ANY CHANGES (Y/N), DEFAULT = N)",B$
500   PRINT LIN(1)
510   IF B$="Y" THEN Numeratorcoeff
520   List_input:  IF P$="C" THEN Listdenomcoeff
530   PRINT LIN(1),"POLES OF THE DENOMINATOR IN ORDER OF
      ENTRY",LIN(1)
540   FOR I=1 TO N
550   IF Pj(I)>=0 THEN PRINT "POLE(";I;") = ";P(I);"+";Pj(I);"j"
      " MULTIPL.";M(I)
560   PRINT "POLE(";I;") = ";P(I);"-";ABS(Pj(I));"j",
      " MULTIPL.";M(I)
570   NEXT I
580   PRINT LIN(2)
590   GOTO List_numerator
600   Listdenomcoeff: PRINT LIN(4),"COEFFICIENTS OF THE DENOM.
      POLYNOMIAL",LIN(1)
610   FOR I=0 TO Nn
620   IF Rcoef(I)>=0 THEN PRINT "n(";I;")=";Rcoef(I);"+";
      Icoefj(I);"j"
630   IF Rcoef(I)<0 THEN PRINT "n(";I;")=";Rcoef(I);"-";
640   NEXT I
650   List_numerator: IF N$="Z" THEN Listnumzeroes
      ABS(Icoefj(I));"j"
660   PRINT LIN(4),"COEFFICIENTS OF THE NUMERATOR IN ASCENDING
      ORDER",LIN(1)
670   FOR I=0 TO Nm
680   IF Cj(I)>=0 THEN PRINT "n(";I;")=";C(I+1);"+";Cj(I+1);"j"
690   IF Cj(I)<0 THEN PRINT "n(";I;")=";C(I+1);"-";
      ABS(Cj(I+1));"j"
700   NEXT I
710   GOTO Pfexp
720   Listnumzeroes: PRINT LIN(2),"ZEROES IN ORDER OF ENTRY
      W.R.T. MULTIPLICITIES"
730   FOR I=1 TO Nz
740   IF Zj(I)>=0 THEN PRINT "ZERO(";I;") = ";Z(I);"+";Zj(I);
      "j","MULTIPL.";Mz(I)
750   PRINT "ZERO(";I;") = ";Z(I);"-";ABS(Zj(I));"j",
      " MULTIPL.";Mz(I)
760   NEXT I
770   Pfexp: PRINT LIN(1),"INVERSE OF z-TRANSFORM FOR GIVEN
      DATA",LIN(1)
780   FOR I=1 TO N
790   MAT Cw=C
800   Zz=1
810   MAT Cwj=Cj
820   FOR Bb=M(I) TO 2 STEP -1
830   Zz=(Bb-1)*Zz
```

```
840  NEXT Bb
850  IF M(I)=1 THEN 890
860  PRINT LIN(1),"POLE(";I;") = ";P(I);CHR$(43+(Pj(I)<0)*2);
     ABS(Pj(I));"j";
870  PRINT "         Differential constant is";1/Zz
880  GOTO Diff
890  PRINT LIN(1),"POLE(";I;") = ";P(I);CHR$(43+(Pj(I)<0)*2);
     ABS(Pj(I));"j",
900  FOR J=1 TO M(I)
910  Prod=Cw(Nn)
920  Prodj=Cwj(Nn)
930  FOR K=Nn-1 TO 1 STEP -1
940  R=P(I)*Prod.-Pj(I)*Prodj+Cw(K)
950  Prodj=P(I)*Prodj+Pj(I)*Prod+Cwj(K)
960  Prod=R
970  NEXT K
980  FOR K=1 TO N
990  IF K=I THEN Nextkl
1000 FOR L=1 TO M(K)
1010 D=(P(I)-P(K))^2+(Pj(I)-Pj(K))^2
1020 R=(Prod*(P(I)-P(K))+Prodj*(Pj(I)-Pj(K)))/D
1030 Prodj=(Prodj*(P(I)-P(K))-Prod*(Pj(I)-Pj(K)))/D
1040 Prod=R
1050 NEXT L
1060 Nextkl:   NEXT K
1070 IF Prodj<0 THEN 1110
1080 PRINT LIN(1),"K(";I;",";M(I)-J+1;") = ";Gain*Prod;"+";
     Gain*Prodj;"j";
1090 PRINT " z power of k-1"
1100 GOTO 1130
1110 PRINT "K(";I;",";M(I)-J+1;") = ";Gain*Prod;"-";
     ABS(Gain*Prodj);"j";
1120 PRINT " z power of k-1"
1130 MAT A=ZER
1140 MAT Aj=ZER
1150 A(1)=-Prod
1160 Aj(1)=-Prodj
1170 FOR K=1 TO N
1180 IF K=I THEN Nextk2
1190 FOR L=1 TO M(K)
1200 CALL Coeff(A(*),Aj(*),P(K),Pj(K),Nn)
1210 NEXT L
1220 Nextk2:NEXT K
1230 MAT Cw=Cw+A
1240 MAT A=Cw
1250 MAT Cwj=Cwj+Aj
1260 MAT Aj=Cwj
1270 Cw(Nn)=Cwj(Nn)=0
1280 FOR K=Nn-1 TO 1 STEP -1
1290 R=Cw(K+1)*P(I)-Cwj(K+1)*Pj(I)+A(K+1)
1300 Cwj(K)=Cw(K+1)*Pj(I)+Cwj(K+1)*P(I)+Aj(K+1)
1310 Cw(K)=R
```

```
1320 NEXT K
1330 NEXT J
1340 Index: NEXT I
1350 PRINT LIN(2)
1360 GOTO End
1370 Polescoeff:INPUT "ORDER OF DENOMINATOR",Nn
1380 PRINT LIN(1),"DEGREE OF POLYNOMIAL=";Nn
1390 REDIM Rroot(Nn),Iroot(Nn),Rcoef(Nn),Iccef(Nn),Rcoefw(Nn),
     Icoefw(Nn)
1400 MAT Rcoefw=Rcoef
1410 MAT Icoefw=Iccef
1420 CALL Rootfdl(Nn,Rroot(*),Iroot(*),Rcoefw(*),Icoefw(*),
     Rcoef(*),Iccef(*))
1430 FOR I=1 TO Nn
1440 Rroot(I)=DROUND(Rroot(I),4)
1450 IF ABS(Rroot(I))<.0001 THEN Rroot(I)=0
1460 Iroot(I)=DROUND(Iroot(I),4)
1470 IF ABS(Iroot(I))<.0001 THEN Iroot(I)=0
1480 NEXT I
1490 MAT M=ZER
1500 Np=Nn
1510 FOR I=1 TO Nn
1520 IF I>Np THEN Subend
1530 FOR J=I TO Np
1540 IF J>Np THEN Nexti
1550 IF SGN(Rroot(I))<>SGN(Rroot(J)) THEN Nextj
1560 IF SGN(Iroot(I))<>SGN(Iroot(J)) THEN Nextj
1570 IF (Rroot(I)=0) AND (ABS(Rroot(J))>.0001) THEN Nextj
1580 IF (Iroot(I)=0) AND (ABS(Iroot(J))>.0001) THEN Nextj
1590 B=Rroot(I)
1600 A=Rroot(J)
1610 IF (B=0) AND (ABS(A)>.0001) THEN Nextj
1620 IF (B=0) AND (ABS(A)<.0001) THEN Shift
1630 IF (ABS(A/B)>1.1) OR (ABS(A/B)<.9) THEN Nextj
1640 B=Iroot(I)
1650 A=Iroot(J)
1660 IF (B=0) AND (ABS(A)>.0001) THEN Nextj
1670 IF (B=0) AND (ABS(A)<.0001) THEN Shift
1680 IF (ABS(A/B)>1.1) OR (ABS(A/B)<.9) THEN Nextj
1690 Shift:M(I)=M(I)+1
1700 IF J=I THEN Nextj
1710 FOR K=J TO Np-1
1720 Rroot(K)=Rroot(K+1)
1730 Iroot(K)=Iroot(K+1)
1740 M(K)=M(K+1)
1750 NEXT K
1760 Rroot(K)=Iroot(K)=M(K)=0
1770 Np=Np-1
1780 J=J-1
1790 Nextj:NEXT J
1800 Nexti:  NEXT I
```

INVERSION OF THE z-TRANSFORM

```
1810 Subend: PRINT LIN(2)
1820 REDIM Rroot(Np),Iroot(Np),P(Np),Pj(Np),M(Np)
1830 N=Np
1840 MAT P=Rroot
1850 MAT Pj=Iroot
1860 GOTO Numerator
1870 Diff: Y=M(I)-1
1880 Nmm=Nm-1
1890 FOR G=Nm+1 TO 1 STEP -1
1900 REDIM Kk(Y)
1910 IF Cj(G)>=0 THEN PRINT LIN(2),"COEFF. CONSTANT IS";C(G);
     "+";Cj(G);"j"
1920 IF Cj(G)<0 THEN PRINT LIN(2),"COEFF. CONSTANT IS";C(G);
     "-";ABS(Cj(G));"j"
1930 Kk(1)=1
1940 Kk(0)=G-2
1950 FOR Ii=2 TO Y
1960 Kk(Ii)=1
1970 FOR Jj=Ii-1 TO 1 STEP -1
1980 Kk(Jj)=Kk(Jj)*(G-2-Ii+1)+Kk(Jj-1)
1990 NEXT Jj
2000 Kk(0)=Kk(0)*(G-2-Ii+1)
2010 NEXT Ii
2020 FOR Aa=Y TO 0 STEP -1
2030 PRINT "K(";Aa;") = ";Kk(Aa)
2040 NEXT Aa
2050 IF G-1-M(I)>=0 THEN PRINT LIN(1),"z Power of","k +";
     G-1-M(I)
2060 IF G-1-M(I)<0 THEN PRINT LIN(1),"z Power of","k -";
     ABS(G-1-M(I))
2070 NEXT G
2080 GOTO Index
2090 End: END
2100 SUB Rootfdl(N,Rroot(*),Iroot(*),Rcoefw(*),Icoefw(*),
     Rcoef(*),Icoef(*))
2110 Itmax=1000000000
2120 PRINT "MAX # OF ITERATIONS=";Itmax
2130 Tola=.000001
2140 PRINT "TOLERANCE FOR ROOTS=";Tola
2150 Tclf=.00000001
2160 PRINT "TOLERANCE FOR FUNCTIONAL EVALUATIONS=";Tclf
2170 PRINT LIN(2),"COEFF: (Rcoef(0)+Icoef(0)*j)+(Rcoef(1)
     +Icoef(1)*j)*S+(Rcoef(2)+Icoef(2)*j)*S^2+..."
2180 PRINT LIN(1),SPA(8),"REAL","   IMAGINARY",LIN(1)
2190 FOR I=0 TO N
2200 DISP "ENTER Rcoef(";I;"), Icoef(";I;")   (DEFAULT =";
     Rcoef(I);Icoef(I);")";
2210    INPUT Rcoef(I),Icoef(I)
2220    PRINT USING 2280;Rcoef(I),Icoef(I)
2230 NEXT I
2240 C$="N"
```

```
2250 INPUT "ANY CHANGES (Y/N), DEFAULT = N)",C$
2260 PRINT LIN(1)
2270 IF C$="Y" THEN 2190
2280 IMAGE 3X,MZ.6DE,5X,MZ.6DE
2290 PRINT LIN(2)
2300 MAT Rcoefw=Rcoef
2310 MAT Icoefw=Icoef
2320 CALL Rootfd(N,Rcoefw(*),Icoefw(*),Tcla,Tclf,Itmax,
     Rroot(*),Iroot(*))
2330 SUBEXIT
2340 SUBEND
```

```
'INVZTR' FINDS THE INVERSE z-TRANSFORM OF A TRANSFER
FUNCTION BY THE RESIDUE METHOD
POLE( 1 ) = -2 + 0 j MULTIPL. 1
POLE( 2 ) = -1 + 0 j MULTIPL. 1
POLE( 3 ) =  1 + 0 j MULTIPL. 1


PLEASE ENTER THE NUMERATOR IN COEFFICIENT FORM
n( 0 )= 0 + 0 j
n( 1 )= 6 + 0 j
n( 2 )= 1 + 0 j


POLES OF THE DENOMINATOR IN ORDER OF ENTRY

POLE( 1 ) = -2 + 0 j MULTIPL. 1
POLE( 1 ) = -2 - 0 j MULTIPL. 1
POLE( 2 ) = -1 + 0 j MULTIPL. 1
POLE( 2 ) = -1 - 0 j MULTIPL. 1
POLE( 3 ) =  1 + 0 j MULTIPL. 1
POLE( 3 ) =  1 - 0 j MULTIPL. 1




COEFFICIENTS OF THE NUMERATOR IN ASCENDING ORDER

n( 0 )= 0 + 0 j
n( 1 )= 6 + 0 j
n( 2 )= 1 + 0 j
```

INVERSION OF THE z-TRANSFORM

INVERSE OF z-TRANSFORM FOR GIVEN DATA

POLE(1) = -2 + 0 j
K(1 , 1) = -2.66666666667 + 0 j z power of k-1

POLE(2) = -1 + 0 j
K(2 , 1) = 2.5 + 0 j z power of k-1

POLE(3) = 1 + 0 j
K(3 , 1) = 1.16666666667 + 0 j z power of k-1

'INVZTR' FINDS THE INVERSE z-TRANSFORM OF A TRANSFER
FUNCTION BY THE RESIDUE METHOD
POLE(1) = -5 + 0 j MULTIPL. 5

PLEASE ENTER THE NUMERATOR IN COEFFICIENT FORM
n(0)=-1 + 0 j
n(1)= 1 + 0 j

POLES OF THE DENOMINATOR IN ORDER OF ENTRY

POLE(1) = -5 + 0 j MULTIPL. 5
POLE(1) = -5 - 0 j MULTIPL. 5

COEFFICIENTS OF THE NUMERATOR IN ASCENDING ORDER

n(0)=-1 + 0 j
n(1)= 1 + 0 j

INVERSE OF z-TRANSFORM FOR GIVEN DATA

POLE(1) = -5 + 0 j   Differential constant is
                              4.16666666667E-02

COEFF. CONSTANT IS 1 + 0 j
K(4) =   1
K(3) = -6
K(2) = 11
K(1) = -6
K(0) =   0

z Power of   k - 4

COEFF. CONSTANT IS-1 + 0 j
K(4) =   1
K(3) = -10
K(2) = 35
K(1) = -50
K(0) = 24

z Power of   k - 5

'INVZTR' FINDS THEF INVERSE z-TRANSFCRM OF A TRANSFER
FUNCTION BY THE RESIDUE METHOD
POLE(1) =   2 + 0 j MULTIPL. 1
POLE(2) =   4 + 0 j MULTIPL. 1

PLEASE ENTER THE NUMERATOR IN COEFFICIENT FORM
n(0)= 0 + 0 j
n(1)= 1 + 0 j

POLES OF THE DENOMINATOR IN ORDER OF ENTRY

POLE(1) =   2 + 0 j MULTIPL. 1
POLE(1) =   2 - 0 j MULTIPL. 1
POLE(2) =   4 + 0 j MULTIPL. 1
POLE(2) =   4 - 0 j MULTIPL. 1

COEFFICIENTS OF THE NUMERATOR IN ASCENDING ORDER

n(0)= 0 + 0 j
n(1)= 1 + 0 j

INVERSE OF z-TRANSFORM FOR GIVEN DATA

POLE(1) =   2 + 0 j
K(1 , 1) = -1 + 0 j z power of k-1

POLE(2) =   4 + 0 j
K(2 , 1) = 2 + 0 j z power of k-1

The z-transforms of some common functions and sequences are collected in Table 1. Note that $f(k) = 0$ for $k < 0$.

TABLE 1. z-Transform Pairs

No.	$f(k)$	$F(z)$
1	$\delta(k)$*	1
2	1	$\dfrac{z}{z-1}$
3	α^k	$\dfrac{z}{z-\alpha}$
4	k	$\dfrac{z}{(z-1)^2}$
5	k^2	$\dfrac{z(z+1)}{(z-1)^3}$
6	k^3	$\dfrac{z(z^2+4z+1)}{(z-1)^4}$
7	k^n	$(-1)^n z^n \dfrac{d^n}{dz^n}\left(\dfrac{z}{z-1}\right)$
8	$\dfrac{1}{k},\ k>0$	$\ln \dfrac{z}{z-1}$
9	$e^{-\alpha k}$	$\dfrac{z}{z-e^{-\alpha}}$
10	$ke^{-\alpha k}$	$\dfrac{ze^{-\alpha}}{(z-e^{-\alpha})^2}$
11	$k\alpha^k$	$\dfrac{\alpha z}{(z-\alpha)^2}$
12	$k^2 \alpha^k$	$\dfrac{\alpha z(z+\alpha)}{(z-\alpha)^3}$
13	$\dfrac{\alpha^k}{k!}$	$e^{\alpha/z}$
14	$\sin \alpha k$	$\dfrac{z \sin \alpha}{z^2 - 2z\cos\alpha + 1}$
15	$\cos \alpha k$	$\dfrac{z(z-\cos \alpha)}{z^2 - 2z\cos \alpha + 1}$
16	$e^{-\alpha k} \sin \beta k$	$\dfrac{ze^{-\alpha}\sin \beta}{z^2 - 2ze^{-\alpha}\cos \beta + e^{-2\alpha}}$
17	$e^{-\alpha k} \cos \beta k$	$\dfrac{z(z-e^{-\alpha}\cos \beta)}{z^2 - 2ze^{-\alpha}\cos \beta + e^{-2\alpha}}$
18	$\dfrac{\alpha^k + (-\alpha^k)}{2\alpha^2}$	$\dfrac{1}{z^2 - \alpha^2}$
19	$\dfrac{\alpha^k - \beta^k}{\alpha - \beta}$	$\dfrac{z}{(z-\alpha)(z-\beta)}$

*$\delta(k) = \begin{cases} 1 & k=0 \\ 0 & k \neq 0 \end{cases}$ is the Kronecker delta.

3.9 SOLUTION OF DIFFERENCE EQUATIONS USING THE z-TRANSFORM

As mentioned before, the z-transform can be used to solve linear difference equations with constant coefficients. By taking the z-transform of both sides of such a difference equation, the z-transform of the unknown sequence can be expressed as a rational function of z. The inverse z-transform of this function then provides the solution. Thus, the z-transform converts the solution of a linear constant-coefficient difference equation to that of an algebraic equation. The following example illustrates the method.

1. Example. Consider the difference equation

$$f(k+2) + 3f(k+1) + 2f(k) = 0, \quad k = 0,1,2,\cdots \tag{1}$$

subject to

$$3f(0) + f(1) = 0 \tag{2}$$

Taking the z-transform of both sides of (1) yields

$$[z^2 F(z) - z^2 f(0) - zf(1)] + 3[zF(z) - zf(0)] + 2F(z) = 0$$

or, using (2),

$$F(z) = \frac{z^2 f(0)}{z^2 + 3z + 2} \tag{3}$$

which can be written as

$$\frac{F(z)}{z} = f(0)\left[\frac{-1}{z+1} + \frac{2}{z+2}\right]$$

Thus the solution is

$$f(k) = Z^{-1}[F(z)] = f(0)[-(-1)^k + 2(-2)^k] \tag{4}$$

PROBLEMS

3.1 Show that a sufficient condition for the Laplace transform of $f(t)$ to exist is that $f(t)$ be of *exponential order*, i.e.

$$\lim_{t \to \infty} f(t) e^{kt} = 0$$

for some real number k.

3.2 Show that $\exp(t^{ik})$ where $k > 1$ is not Laplace transformable.

3.3 If $L[f_1(t)] = F_1(s)$ and $L[f_2(t)] = F_s(s)$, show that

$$L[\alpha_1 f_1(t) + \alpha_2 f_2(t)] = \alpha_1 F_1(s) + \alpha_2 F_2(s)$$

for arbitrary constants α_1 and α_2.

3.4 Derive the integration rule of the Laplace transform from the differentiation rule.

3.5 If $L[f(t)] = F(s)$, show that $L[e^{\alpha t} f(t)] = F(s-\alpha)$ where α is any complex number.

3.6 Prove the initial value theorem of the Laplace transform, i.e. show that $f(0+) = \lim_{s \to \infty} sL[f(t)]$ provided that $f(0+)$ exists. Hint. Use the differentiation rule of the Laplace transform and take the limit as $s \to \infty$.

3.7 Prove the final value theorem of the Laplace transform, i.e. show that $\lim_{t \to \infty} f(t) = \lim_{s \to 0} sF(s)$ provided that the limit on the left exists.

3.8 Show that the convolution of two time functions is a commutative operation, i.e. prove that

$$f_1 * f_2(t) = f_2 * f_1(t) \quad \text{for all } t.$$

3.9 Prove the convolution theorem of the Laplace transform, i.e. show that $L[f_1 * f_2(t)] = L[f_1(t)]L[f_2(t)]$.
Hint. Since $f_1(t) = f_2(t) = 0$ for $t < 0$, then $f_1 * f_2(t) = \int_{-\infty}^{\infty} f_1(\tau) f_2(t-\tau) d\tau$. Use the definition of the Laplace transform and the Laplace transform translation property.

3.10 Show the following:

$$L[tf(t)] = -\frac{d}{ds} F(s)$$

$$L\left[\frac{f(t)}{t}\right] = \int_s^\infty F(\alpha) d\alpha$$

$$L[f(\tfrac{t}{\alpha})] = \alpha F(\alpha s) \quad \text{for any scalar } \alpha$$

3.11 Show that $L[t^n e^{\alpha t}] = \dfrac{n!}{(s-\alpha)^{n+1}}$ where n is any positive integer and α is any complex scalar.

3.12 If $F(s)$ is a strictly proper rational function with real coefficients show that

    a. the residues of real poles of $F(s)$ are real.

    b. the residues of a pair of complex conjugate poles of $F(s)$ are complex conjugate.

3.13 Show that

    a. $L^{-1}\left[\dfrac{s^2 - 3}{(s^2+1)(s^3+10s^2+37s+52)}\right] = \dfrac{13}{85} e^{-4t} + 0.047\cos t - 0.055\sin t$

$$+ 0.106 e^{-3t} \cos 2t - 0.119 e^{-3t} \sin 2t, \quad t \geqslant 0$$

    b. $L^{-1}\left[\dfrac{as^2 + (10a+b)s + 50}{s^3 + 10s^2 + 50s}\right] = 1 + (a-1) e^{-5t} \cos 5t$

$$+ \left(a + \dfrac{b}{5} - 1\right) e^{-5t} \sin 5t, \quad t \geqslant 0$$

3.14 Show that

$$L^{-1}\left[\dfrac{K}{s-\alpha-j\omega} + \dfrac{\bar{K}}{s-\alpha+j\omega}\right] = 2|K| e^{\alpha t} \cos(\omega t + <K) \quad t \geqslant 0$$

3.15 Determine the z-transform of

    a. the sequence $f(k) = \alpha^{k+1}$, $k = 0, 1, 2, \ldots$

    b. the sequence $g(k) = \alpha^{k-1}$, $k = 0, 1, 2, \ldots$

    c. the function $h(t) = e^{-\alpha t}$.

3.16 Show that

$$Z[t e^{-\alpha t}] = \dfrac{z e^{-\alpha}}{(z - e^{-\alpha})^2}$$

provided that $|z| > e^{-\alpha}$.

3.17 Prove the final value theorem of the z-transform:

$$\lim_{k \to \infty} f(k) = \lim_{z \to 1}(z-1)F(z)$$

provided that $F(z)$ is analytic for $|z|>1$.

3.18 Prove the convolution property of the z-transform:

$$Z[\sum_{j=0}^{k} f_1(j)f_2(k-j)] = Z[f_1(k)]Z[f_2(k)].$$

3.19 Show the following properties of the z-transform:

a) $Z[\dfrac{f(t)}{t}] = -\displaystyle\int_{z}^{\infty} \dfrac{F(z')}{z'} dz' + \lim_{t \to 0} \dfrac{f(t)}{t}$

b) $Z[tf(t)] = -z \dfrac{dF(z)}{dz}$

c) $Z[e^{-\alpha t}f(t)] = F(ze^{\alpha})$ provided that

$$|ze^{\alpha}| > \text{radius of convergence of } F(z).$$

REFERENCES

[3.1] G. Doetsch, *Guide to the Applications of the Laplace and z-Transforms*, Van Nostrand Reinhold, 1971

[3.2] G. Doetsch, *Introduction to the Theory and Application of the Laplace Transformation*, Springer, 1974

[3.3] H. Freeman, and O. Lowenschuss, "Bibliography of sampled-data control systems and z-transform applications," *I.R.E. Trans. Automatic Control (PGAC)*, pp. 28-30, March 1958

[3.4] H. A. Helm, "The z-transformation," *Bell System Technical Journal*, Vol. 38, No. 1 pp. 177-196, 1956

[3.5] E. I. Jury, *Theory and Application of the z-Transform Method*, John Wiley & Sons, New York, 1964

[3.6] G. V. Lago, "Additions to z-transformation theory for sampled-data systems," *Trans. A.I.E.E.*, Vol. 74, part II, pp. 403-408, 1955

[3.7] W. R. LaPage, *Complex Variables and the Laplace Transform for Engineers*, McGraw-Hill, New York, 1961

[3.8] M. R. Spiegel, *Theory and Problems of Laplace Transforms*, McGraw-Hill, 1965

[3.9] E. J. Watson, *Laplace Transforms and Applications*, Van Nostrand Reinhold, 1981

[3.10] R. Piessens, "A bibliography on numerical inversion of the Laplace transform and applications," *J. comput. and Appl. math.*, Vol. I, No. 2, pp. 115-128, 1975

[3.11] R. Piessens and N. D. P. Dang, "A bibliography on numerical inversion of the Laplace transform and applications: A supplement," ibid, Vol. 2, No. 3, pp. 225-227, 1976

[3.12] M. L. Liou, "A novel method of evaluating transient response," *Proc. IEEE*, Vol. 54, pp. 20-23, 1966

[3.13] A. Talbot, "The accurate numerical inversion of Laplace transforms," *J. Inst. Math. Appl.*, Vol. 23, pp. 97-120, 1979

[3.14] E. I. Jury, and C. A. Galtieri, "A Note on the Inverse z-Transform," *I.R.E. Trans. Circuit Theory* Vol. CT-9 pp. 371-374, 1961

[3.15] R. V. Churchill, J. W. Brown and R. F. Verhey, *Complex Variables and Applications*, McGraw-Hill, 1974

[3.16] A. Ralston, *A First Course in Numerical Analysis*, Mc-Graw-Hill, 1965

PART II

ANALYSIS

CHAPTER 4
SYSTEM MODELING

4.1 INTRODUCTION

The first step in studying a physical system is to develop a mathematical representation, or a *model,* for it. A model is developed so that, while it is computationally convenient, it adequately represents the system. In this chapter we will discuss procedures for developing models for physical systems. We will be concerned mainly with *l.t.i.* continuous- and discrete-time models. If only the terminal behavior of the system is of interest, then an input-output description or a transfer function will be adequate to represent the system. If, however, the internal behavior of the system is also of interest, a state-space representation of the system will be useful. These concepts will be discussed in the following sections. Other topics to be covered in this chapter are state transformations and linearization of nonlinear systems.

4.2 DEVELOPING SYSTEM MODELS

A model is an idealization of a physical system. It is used to reduce the computational effort in the analysis and design of the system.

In developing a model for a physical system, certain parameters and/or variables of the system or relationships between its components may be neglected. However, one must be careful not to neglect parameters or relationships that are crucial to the accuracy of the model. This immediately implies that a physical system may have different models depending on the questions of interest about it. For instance, a transistor has different models depending on the amplitude and frequency of the signal applied to it. Usually, we choose a model for the system under consideration which is simple and, at the same time, provides good insight into its behavior.

When all the components of a system have been modeled, a mathematical representation of the system can be derived by the application of physical laws governing the system. For example, an electric circuit whose components have been modeled as resistors, capacitors and inductors can be mathematically characterized by the application of the Kirchhoff's laws.

We will demonstrate the above procedure by the example of an armature- controlled d.c. motor given below.

1. Example. Consider a separately excited d.c. motor [4.1]. Assume that the field winding (usually the stator) is connected to a constant voltage source, but the armature winding (usually the rotor) is connected to a variable voltage source v(t). Thus the field current i_f can be considered a constant. The voltage $e_a(t)$ can be varied with time to change the angular velocity $\omega(t)$ of the rotor. We are interested in the relationship between $\omega(t)$ and $e_a(t)$.

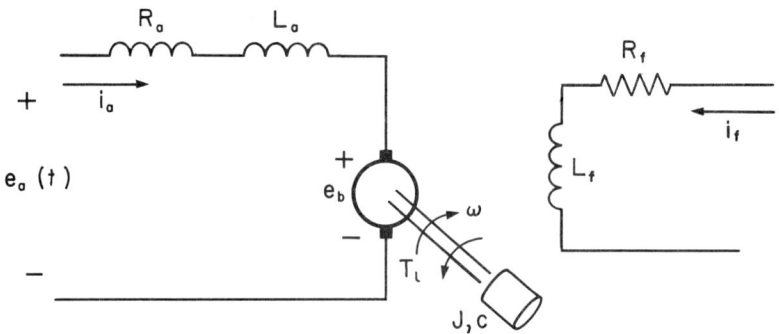

Fig. 4.2.1. Schematic diagram of a separately excited d.c. motor

Figure 1 shows a schematic diagram of the d.c. motor. In this figure, R_a and L_a represent the resistance and the inductance of interpole winding in the armature. The circle is an ideal voltage source $e_b(t)$ representing the back emf due to the rotation of the armature conductors in the magnetic field. Similarly, R_f and L_f indicate the resistance and the inductance of the field winding. Note that the nonlinearities and the dependence of the parameters on time in these windings have been neglected.

The magnetic flux ϕ_f of the field winding is a constant k_1 since i_f is assumed to be constant. The torque T_r developed on the rotor shaft is proportional to the product of ϕ_f and i_a; thus

$$T_r(t) = k_2 i_a(t) \tag{1}$$

where k_2 is a constant. Application of KVL (Kirchhoff's voltage law) to the armature circuit yields

$$e_a(t) = L_a \frac{di_a(t)}{dt} + R_a i_a(t) + e_b(t) \tag{2}$$

where $e_b(t)$ is proportional to the angular velocity of the rotor:

$$e_b(t) = k_3 \omega(t) \tag{3}$$

The rotor torque $T_r(t)$ and the angular velocity $\omega(t)$ are related by Newton's second law of dynamics:

$$T_r(t) = T_d(t) + J \frac{d\omega(t)}{dt} + C\omega(t) \quad (4)$$

where $T_d(t)$ is the load torque on the rotor shaft, C is the viscous friction constant and J is the moment of inertia of the load.

Combining equation (1) to (4) yields the following system of constant-coefficient differential equations as the mathematical representation of the armature-controlled d.c. motor:

$$\frac{di_a(t)}{dt} = -\frac{R_a}{L_a} i_a(t) - \frac{k_3}{L_a} \omega(t) + \frac{1}{L_a} e_a(t) \quad (5)$$

$$\frac{d\omega(t)}{dt} = \frac{k_2}{J} i_a(t) - \frac{C}{J}\omega(t) - \frac{1}{J}T_d(t) \quad (6)$$

An alternative mathematical representation of the above system can be obtained by combining differential equations (5) and (6) to express a relationship between $\omega(t)$ and $e_a(t)$. (See Problem 4.1.).

In system theory we study the properties of systems using their mathematical representations. We are mainly concerned with idealized system representations. That is, we assume that all system components are lumped, linear and time-invariant. This, of course, seems to be an oversimplification since hardly any systems in real-life satisfies these assumptions. Nonetheless, it is convenient in that a general theory has been developed for linear time-invariant systems. Furthermore, it still provided a great deal of insight into the general system behavior. However, one must be aware of its limitations.

4.3 THE CONCEPT OF STATE

Associated with each system are two important sets of attributes. One set, denoted by $u_1, u_2, \ldots, u_r$, indicates the *inputs* of the system, or the *control variables*. The inputs represent the stimuli which are applied to the system from external sources. The other set, denoted by $y_1, y_2, \ldots, y_m$, indicates the *outputs* of the system. The outputs of a system can be measured externally. In Example 4.2.1 $e_a(t)$ and $T_d(t)$ are inputs and $\omega(t)$ is an output of the system.

The inputs and the outputs of a system are related through the dynamic equations describing the system. In *l.t.i.* systems, these equations will be linear constant coefficient differential or difference equations. More specifically, for a SISO system (also referred to as a *scalar* system), shown in Fig. 1 the dynamic equations have the general form

$$p_n y^{(n)}(t) + p_{n-1} y^{(n-1)}(t) + \ldots + p_1 \dot{y}(t) + p_0 y(t) =$$

$$q_m u^{(m)}(t) + q_{m-1} u^{(m-1)}(t) + \ldots + q_1 \dot{u}(t) + q_0 u(t) \quad (1)$$

for continuous-time systems, and

$$p_n y(k+n)+p_{n-1}y(k+n-1)+...+p_1 y(k+1)+p_0 y(k) =$$

$$q_m u(k+m)+q_{m-1}u(k+m-1)+...+q_1 u(k+1)+q_0 u(k) \qquad (2)$$

for discrete-time systems. If $n \geq d$, the corresponding system is called *proper* and if $n > m$, it is called *strictly* proper.

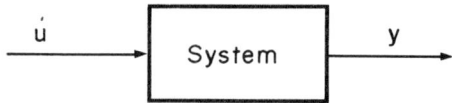

Fig. 4.3.1. A single-input single-output (scalar) system

Description of MIMO systems (also referred to as *multivariable* systems) in the above form is cumbersome because it involves a system of coupled equations of the form (1) or (2). Using the concept of *state* of a system enables us to describe scalar or multivariable systems by first-order vector differential or difference equations.

Loosely speaking, the state of a system is a collection of information which contains a history of the system; that is, the knowledge of the state and the input(s) of a system will be adequate to calculate its output(s). More precisely, we have the following definition.

1. Definition. A state of a system at time t_o is a minimum amount of information which together with the knowledge of the input(s) for $t \geq t_o$ is sufficient to uniquely determine the output(s) of the system for $t \geq t_o$.

△

The state of a finite-dimensional system can be expressed by a finite dimensional vector $\mathbf{x}(t)=[x_1, x_2, \ldots, x_n]'$ called a *state vector*. Each component x_j of $\mathbf{x}(t)$ is called a *state variable*. Note that state of a system is not necessarily unique.

Figure 2 shows a multivariable system. The vector $\mathbf{u}=[u_1, u_2, \ldots, u_r]'$ is the *input vector* and the vector $y=[y_1, y_2, \ldots, y_m]'$ is the *output vector* for this system. The input vector, the output vector and the state vector of a system each belong to a vector space of a proper dimension. That is $\mathbf{u} \epsilon U$, $\mathbf{y} \epsilon Y$ and $\mathbf{x} \epsilon X$ where the vector spaces U, Y and X, referred to as the *input space, output space* and the *state space* of the system, have dimensions r, m and n, respectively.

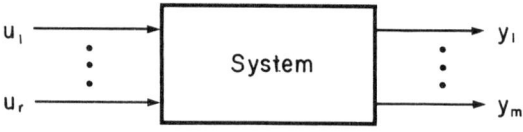

Fig. 4.3.2. A multi-input multi-output (multivariable) system

STATE-SPACE REPRESENTATION OF LINEAR SYSTEMS

2. Example. Consider Example 4.2.1. The inputs to the system are the armature voltage $e_a(t)$ and the load torque $T_d(t)$. Thus the input vector at time t can be defined as $\mathbf{u}(t)=[e_a(t),T_d(t)]'$. The only output of the system is $y(t) = \omega(t)$. Furthermore, note from (4.2.5) and (4.2.6) that the vector $\mathbf{x}(t)=[i_a(t),\omega(t)]'$ may be used to express the state of this system. (See Definition 4.3.1.)

4.4 STATE-SPACE REPRESENTATION OF LINEAR SYSTEMS

4.4.1 State and Output Equations As mentioned in the previous section, using the concept of state allows us to describe any system by a first-order vector differential or difference equation. This is very useful since the study of linear systems, independent of the order of the differential or the difference equations describing them, can be based on the study of linear *first-order* vector differential or difference equations. The following example illustrates this point.

1. Example. Once again consider Example 4.2.1. Write equations (4.2.5) and (4.2.6) in the vector matrix form to obtain

$$\frac{d}{dt}\begin{bmatrix} i_a(t) \\ \omega(t) \end{bmatrix} = \begin{bmatrix} -\frac{R_a}{L_a} & -\frac{k_3}{L_a} \\ \frac{k_2}{J} & -\frac{C}{J} \end{bmatrix} \begin{bmatrix} i_a(t) \\ \omega(t) \end{bmatrix} + \qquad (1)$$

$$\begin{bmatrix} \frac{1}{L_a} & 0 \\ 0 & -\frac{1}{J} \end{bmatrix} \begin{bmatrix} e_a(t) \\ T_d(t) \end{bmatrix}$$

or, using the input, output and state vectors defined in Example 4.3.2,

$$\dot{\mathbf{x}}(t)=\mathbf{A}\mathbf{x}(t)+\mathbf{B}\mathbf{u}(t) \qquad (2)$$

where constant matrices **A** and **B** are defined in the obvious manner. Note that these matrices depend exclusively on the known parameters of the system. Equation (2) is a linear first-order vector differential equation called the *state equation* of the system under consideration.

The output $\omega(t)$ can also be expressed in terms of the state vector $\mathbf{x}(t)$:

$$y(t)=\omega(t)=\mathbf{C}\mathbf{x}(t) \qquad (3)$$

where $\mathbf{C} = [0\ 1]$. Equation (3) is called the *output equation* of the system.

In general, if **u**, **y** and **x** are the input vector, the output vector and the state vector of a linear continuous-time system, it can be represented by the following state and output equations:

$$\dot{\mathbf{x}}(t)=\mathbf{A}(t)\mathbf{x}(t)+\mathbf{B}(t)\mathbf{u}(t) \qquad (4)$$

$$\mathbf{y}(t) = \mathbf{C}(t)\mathbf{x}(t) + \mathbf{D}(t)\mathbf{u}(t) \tag{5}$$

where for r inputs, m outputs and n state variables, matrix **A** is nxn, **B** is nxr, **C** is mxn and **D** is mxr. If the system is time-invariant, these matrices will be constant and the state and output equations will have the form

$$\dot{\mathbf{x}}(t) = \mathbf{A}\mathbf{x}(t) + \mathbf{B}\mathbf{u}(t) \tag{6}$$

$$\mathbf{y}(t) = \mathbf{C}\mathbf{x}(t) + \mathbf{D}\mathbf{u}(t) \tag{7}$$

In the case of nonlinear time-varying systems, (4) and (5) will become

$$\dot{\mathbf{x}}(t) = \mathbf{f}(\mathbf{x}(t), \mathbf{u}(t), t) \tag{8}$$

$$\mathbf{y}(t) = \mathbf{g}(\mathbf{x}(t), \mathbf{u}(t), t) \tag{9}$$

where **f** and/or **g** are nonlinear vector-valued functions. If a nonlinear system is time-invariant, the state and output equations will have no explicit dependence on t and they become

$$\dot{\mathbf{x}}(t) = \mathbf{f}(\mathbf{x}(t), \mathbf{u}(t)) \tag{10}$$

$$\mathbf{y}(t) = \mathbf{g}(\mathbf{x}(t), \mathbf{u}(t)) \tag{11}$$

For linear time-varying discrete-time systems the state and output equations have the form

$$\mathbf{x}(k+1) = \mathbf{A}(k)\mathbf{x}(k) + \mathbf{B}(k)\mathbf{u}(k) \tag{12}$$

$$\mathbf{y}(k) = \mathbf{C}(k)\mathbf{x}(k) + \mathbf{D}(k)\mathbf{u}(k) \tag{13}$$

which are analogous to (4) and (5). For *l.t.i.* discrete-time systems, matrices **A,B,C** and **D** in (12) and (13) will be constant, i.e. they will have no dependence on k.

A linear system represented by the state and output equations discussed above is referred to as the system (**A,B,C,D**). Matrix **A** in (4), (6) and (12) is known as the *system matrix*.

2. Example. Consider a multivariable system with two inputs u_1 and u_2, and two outputs y_1 and y_2 expressed by the following system of differential equations:

$$\dddot{y}_1 + 2\ddot{y}_1 + \ddot{y}_2 - 3(\dot{y}_1 + \dot{y}_2) + t(y_1 - y_2) = u_1 \tag{14}$$

STATE-SPACE REPRESENTATION OF LINEAR SYSTEMS

$$\ddot{y}_2 + 5\dot{y}_2 - 3\dot{y}_1 + 2(y_2 - y_1) = u_2 \tag{15}$$

This is a linear time-varying system. To determine a state equation for it, let $x_1 = y_1, x_2 = \dot{y}_1, x_3 = \ddot{y}_1, x_4 = y_2$ and $x_5 = \dot{y}_2$. Then (14) and (15) yield

$$\dot{x}_3 + 2x_3 + \dot{x}_5 - 3(x_2 + x_5) + t(x_1 - x_4) = u_1 \tag{16}$$

$$\dot{x}_5 + 5x_5 - 3x_2 + 2(x_4 - x_1) = u_2 \tag{17}$$

which after some algebraic manipulations result in the following state equation:

$$\frac{d}{dt}\begin{bmatrix} x_1 \\ x_2 \\ x_3 \\ x_4 \\ x_5 \end{bmatrix} = \begin{bmatrix} 0 & 1 & 0 & 0 & 0 \\ 0 & 0 & 1 & 0 & 0 \\ -(2+t) & 0 & -2 & 2+t & 8 \\ 0 & 0 & 0 & 0 & 1 \\ 2 & 3 & 0 & -2 & -5 \end{bmatrix}\begin{bmatrix} x_1 \\ x_2 \\ x_3 \\ x_4 \\ x_5 \end{bmatrix} + \begin{bmatrix} 0 & 0 \\ 0 & 0 \\ 1 & -1 \\ 0 & 0 \\ 0 & 1 \end{bmatrix}\begin{bmatrix} u_1 \\ u_2 \end{bmatrix} \tag{18}$$

Also, $x_1 = y_1$ and $x_4 = y_2$ result in the output equation

$$\begin{bmatrix} y_1 \\ y_2 \end{bmatrix} = \begin{bmatrix} 1 & 0 & 0 & 0 & 0 \\ 0 & 0 & 0 & 1 & 0 \end{bmatrix}\begin{bmatrix} x_1 \\ x_2 \\ x_3 \\ x_4 \\ x_5 \end{bmatrix} \tag{19}$$

Note that matrix **D** is equal to zero in this example; that is, the output does not directly depend on the input. (See Problem 4.3.)

3. Example. Consider a simple RLC network shown in Fig. 1. It is desired to find a state-space model for it.

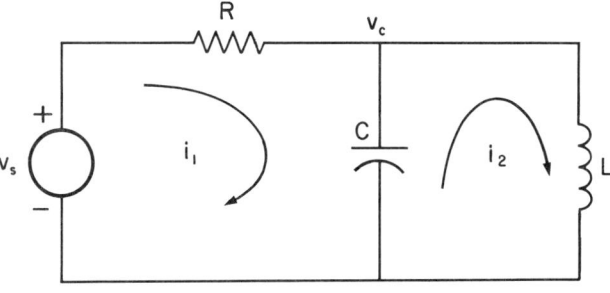

Fig. 4.4.1. An RCL network

LINEAR CONTROL SYSTEMS

As a general rule, the number of state variables is equal to the number of energy storing elements such as capacitors and inductors. The common choices of state variables in this case are voltage across the capacitor v_c and current through inductor i_2. Thus, let $x_1=v_c$, $x_2=i_2$, $u=v_s$, $y_1=i_1$, $y_2=i_2$. The necessary equations for the states can be obtained from Kirchhoff's laws i.e.

$$v_s = Ri_1 + v_c \qquad (20a)$$

$$C\dot{v}_c = i_1 - i_2 \qquad (20b)$$

$$v_c = L di_2/dt \qquad (20c)$$

Now eliminating i_1 in (20b) by using (20a) and noting the definitions of input and states we have,

$$\dot{x}_1 = -\frac{1}{RC}x_1 - \frac{1}{C}x_2 + \frac{1}{RC}u \qquad (21a)$$

$$\dot{x}_2 = \frac{1}{L}x_1 \qquad (21b)$$

with

$$y_1 = i_1 = \frac{1}{R}u - \frac{1}{R}x_1 \qquad (22a)$$

$$y_2 = i_2 = x_2 \qquad (22b)$$

In matrix form (21)-(22) can be rewritten as,

$$\begin{bmatrix}\dot{x}_1\\\dot{x}_2\end{bmatrix} = \begin{bmatrix}-1/RC & -1/C\\1/L & 0\end{bmatrix}\begin{bmatrix}x_1\\x_2\end{bmatrix} + \begin{bmatrix}1/RC\\0\end{bmatrix}u \qquad (23)$$

$$\begin{bmatrix}y_1\\y_2\end{bmatrix} = \begin{bmatrix}-1/R & 0\\0 & 1\end{bmatrix}\begin{bmatrix}x_1\\x_2\end{bmatrix} + \begin{bmatrix}1/R\\0\end{bmatrix}u \qquad (24)$$

4.4.2 Simulation Diagrams A useful aid in determining the state and output equations from differential or difference equations describing a linear system is *simulation diagrams*. They consist of different blocks each of which describes a function or an operation as shown in Fig. 2. Simulation diagrams also help in digital computer simulation of continuous-time systems e.g. Continuous System Modeling Program: CSMP [4.3] or discrete-time systems e.g. Digital Simulation Language: DSL [4.4].

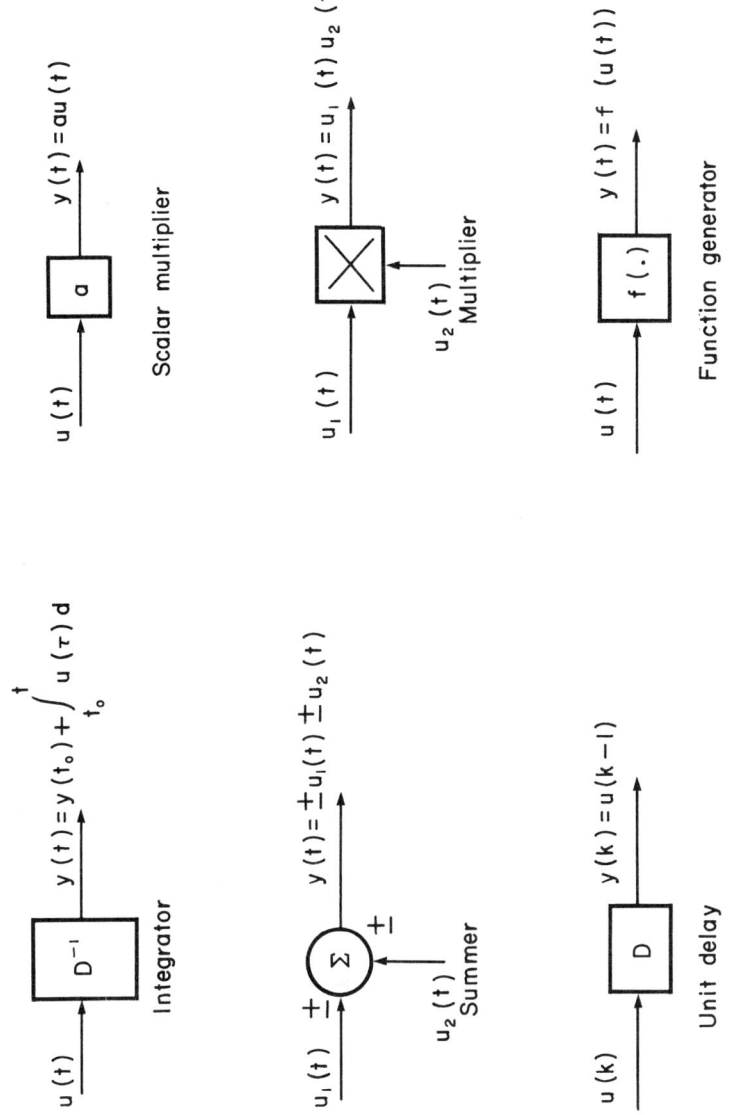

Fig. 4.4.2. Typical blocks in a Simulation diagram

The general approach to determine the simulation diagram for a linear continuous-time system is to generate the highest-order derivative(s) in the differential equation(s) describing it and to integrate successively to obtain the output(s). This is conveniently arranged in the form of a nested sequence of integrations. The following example demonstrates this.

4. Example. Consider a *l.t.i.* system described by

$$\dddot{y} + 7\ddot{y} + 5\dot{y} + 6y = \ddot{u} + 8\dot{u} - 2u \tag{25}$$

Write (25) as

$$\dddot{y} = -7\ddot{y} - 5\dot{y} - 6y + \ddot{u} + 8\dot{u} - 2u \tag{26}$$

or

$$y = D^{-1}(u - 7y + D^{-1}[8u - 5y + D^{-1}(-2u - 6y)]) \tag{27}$$

The corresponding simulation diagram is shown in Fig. 3.

△

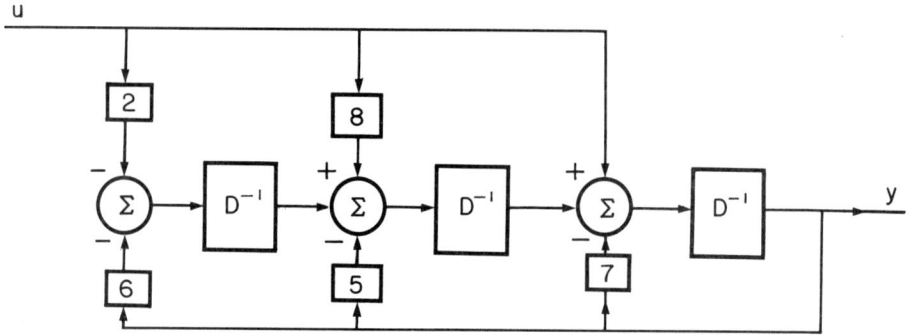

Fig. 4.4.3. Simulation diagram for Example 4.4.4

The simulation diagram for linear discrete-time systems is similar to that for linear continuous-time system except instead of integrators, unit delay blocks must be used. It can be obtained by following a procedure similar to that for continuous-time systems; that is, generate the output with the largest argument and successively pass it through unit delay blocks until the output at discrete time k is found.

5. Example. Consider a *l.t.i.* system described by the difference equation

$$y(k+3) + 7y(k+2) + 5y(k+1) + 6y(k) = u(k+2) + 8u(k+1) - 2u(k) \tag{28}$$

which is similar to (25). Solving (28) for $y(k+3)$, we obtain

$$y(k+3) = -7y(k+2) - 5y(k+1) - 6y(k) + u(k+2) + 8u(k+1) - 2u(k) \qquad (29)$$

or

$$y(k) = D\{u(k) - 7y(k) + D[8u(k) - 5y(k) + D(-2u(k) - 6y(k))]\} \qquad (30)$$

where

$$D(f(k)) \triangleq f(k-1) \qquad (31)$$

The simulation diagram can then be obtained using (30). It is the same as that shown in Fig. 3 except the integrators $(D-1)$ must be replaced with unit delay blocks (D).

Δ

Simulation diagrams for linear multivariable systems can be obtained in a similar manner. Nonlinear systems can also be simulated as discussed above where nonlinearities can be taken into account by using nonlinear blocks (such as multipliers).

4.4.3 State Equations in Standard Forms Linear time-invariant SISO systems can be described by state equations in standard forms. In this section we will discuss two standard forms known as *companion forms*. The two forms are referred to as the *controllable companion form* and the *observable companion form* for reasons which will be discussed in Chapter 8. A Third standard form for the state equations of l.t.i. systems, known as the *Jordan form*, is discussed in Problem 4.5.

4.4.3.1 The Controllable companion form. The last row of the system matrix in this form consists of the negative of the last n coefficients of the left-coefficients of the right hand side of the equation describing the system (i.e. Equation 4.3.1 or 4.3.2) divided by the first coefficient. All the other elements are zero except those on the upper diagonal which are unity. Also, all the elements of the "b" vector are zero except the last one which is unity.

To see how this form is obtained, consider the differential equation (4.3.1). For convenience, let $m = n$ and without loss of generality let $p_n = 1$. (if $m < n$, set the corresponding coefficients q_i equal to zero.) Taking the Laplace transform of both sides of (4.3.1) yields

$$Y(s) = \frac{q_n s^n + q_{n-1} s^{n-1} + \ldots + q_1 s + q_0}{s^n + p_{n-1} s^{n-1} + \ldots + p_1 s + p_0} U(s) \qquad (32)$$

write (32) as

$$Y(s) = (q_n s^n + q_{n-1} s^{n-1} + \ldots + q_1 s + q_0) Z(s) \qquad (33)$$

where

$$Z(s) = \frac{U(s)}{s^n + p_{n-1}s^{n-1} + \cdots + p_1 s + p_0} \tag{34}$$

Now define state variables as follows:

$$x_1(t) \triangleq z(t),\ x_2(t) \triangleq \dot{x}_1(t) = \dot{z}(t),\ldots,\ x_n(t) \triangleq \dot{x}_{n-1}(t) = z^{(n-1)}(t). \tag{35}$$

Therefore, using (34) we have

$$\dot{x}_n(t) = z^{(n)}(t) = -p_0 x_1(t) - p_1 x_2(t) - \cdots - p_{n-1} x_n(t) + u(t) \tag{36}$$

and the state equation is

$$\begin{bmatrix} \dot{x}_1 \\ \dot{x}_2 \\ \cdot \\ \cdot \\ \cdot \\ \dot{x}_n \end{bmatrix} = \begin{bmatrix} 0 & 1 & 0 & \cdots & 0 \\ 0 & 0 & 1 & \cdots & \\ \cdot & & & & \\ \cdot & & & & 0 \\ 0 & & & 0 & 1 \\ -p_0 & -p_1 & -p_2 & \cdots & -p_{n-1} \end{bmatrix} \begin{bmatrix} x_1 \\ x_2 \\ \cdot \\ \cdot \\ \cdot \\ x_n \end{bmatrix} + \begin{bmatrix} 0 \\ 0 \\ \cdot \\ 0 \\ 0 \\ 1 \end{bmatrix} u(t) \tag{37}$$

Also from (33), (35) and (36) we have

$$y(t) = [q_0 - p_0 q_n,\ q_1 - p_1 q_n,\ \ldots,\ q_{n-1} - p_{n-1} q_n]\,\mathbf{x}(t) + q_n u(t) \tag{38}$$

Equations (37) and (38) are the state and output equations in the controllable companion form. Note that the system matrix has the form described above. Note also that for $m = n-1$, i.e. $q_n = 0$, the coefficients of the output matrix in (38) are simply $(q_0, q_1, \ldots, q_{n-1})$. The controllable companion form of discrete-time systems can be obtained in an analogous manner.

4.4.3.2 The observable companion form. The system matrix in this form is the transpose of that in the previous form. Also, the output equation has the following form

$$y(t) = [0, 0, \ldots 0, 1]\mathbf{x}(t) \tag{39}$$

The observable companion form can be obtained from the simulation diagram by defining the output of each integrator (or unit delay block) as a state variable. Then the derivative (or the unit advance) of that state variable can be obtained by writing the equation characterizing the summer preceding the integrator (or the unit delay block).

6. Example. We will find both companion forms for Examples 3 and 4 of this Section. The Laplace transform of (20) yields

$$Y(s) = (s^2 + 8s - 2) Z(s) \qquad (40)$$

where

$$Z(s) = \frac{U(s)}{s^3 + 7s^2 + 5s + 6} \qquad (41)$$

The state variables are $x_1 = z$, $x_2 = \dot{z}$ and $x_3 = \ddot{z}$. Thus from (41) we have

$$\dot{x}_3 = -6x_1 - 5x_2 - 7x_3 + u \qquad (42)$$

and the state equation in controllable companion form is

$$\begin{bmatrix} \dot{x}_1 \\ \dot{x}_2 \\ \dot{x}_3 \end{bmatrix} = \begin{bmatrix} 0 & 1 & 0 \\ 0 & 0 & 1 \\ -6 & -5 & -7 \end{bmatrix} \begin{bmatrix} x_1 \\ x_2 \\ x_3 \end{bmatrix} + \begin{bmatrix} 0 \\ 0 \\ 1 \end{bmatrix} u \qquad (43)$$

Also, using (35), the output equation is

$$y(t) = [-2 \ 8 \ 1] \, x(t) \qquad (44)$$

Equations (43) and (44) are also the controllable companion form for the discrete-time system described by (28).

To find the observable companion form for (25), define state variables x_1, x_2 and x_3 as the output of the integrators in the simulation diagram of Fig. 3. Thus $\dot{x}_1 = -2u - 6y$, $\dot{x}_2 = 8u - 5y + x_1$, $\dot{x}_3 = u - 7y + x_2$ and $y = x_3$. Therefore we have

$$\begin{bmatrix} \dot{x}_1 \\ \dot{x}_2 \\ \dot{x}_3 \end{bmatrix} = \begin{bmatrix} 0 & 0 & -6 \\ 1 & 0 & -5 \\ 0 & 1 & -7 \end{bmatrix} \begin{bmatrix} x_1 \\ x_2 \\ x_3 \end{bmatrix} + \begin{bmatrix} -2 \\ 8 \\ 1 \end{bmatrix} u \qquad (45)$$

$$y = [0 \ 0 \ 1] x \qquad (46)$$

as the state and output equations in the observable companion form. Equations (45) and (46) are also the observable companion form for the discrete-time system described by (28).

126   LINEAR CONTROL SYSTEMS

      Program "STATEQ" determines the two companion forms from the equation describing the system. The source listing of "STATEQ" is given below followed by an example of a fourth-order system.

△

7. "STATEQ"

```
10    !    PROGRAM NAME :    "STATEQ"    PROG
20    PRINT "THIS PROGRAM DETERMINERS TWO TYPES OF STATE
      EQUATIONS IN COMPANION"
30    PRINT "FORM FOR A SYSTEM CHARACTERIZED BY A DIFFERENTIAL
      EQUATION "
40    PRINT "y(n)+p(n-1)*y(n-1)+...+p(0)*y =q(n-1)
      *u(n-1)+...+q(0)*u"
50    PRINT "where  y(i) is defined as (d^i)y/d(t^i).
      (Similarly for u(i).)"
60    PRINT "The system state equation will have the form :"
70    PRINT "   dx/dt= Ax+Bu ,       y= Cx"
80    PRINT LIN(3)
90    DIM A(50,50),B(50),C(0,50),B1(0,50),C1(50),A1(50,50),
      P(50),Q(50)
100   INPUT "Has SUB <<Vec>> been already LINKED (Y/N)?",C$
110   IF C$="Y" THEN 150
120   IF C$="N" THEN 140
130   GOTO 100
140   LINK "Vec",900,150
150   INPUT "Give your PRINTER option: For CRT press 1; for
      typewriter press 2",R
160   IF R=1 THEN PRINTER IS 16
170   IF R=2 THEN PRINTER IS 7,1
180   INPUT "Give the ORDER of the differential equation
      (<=50)",N
190   PRINT "The order of the differential equation is=";N
200   N=N-1
210   INPUT "Give the ORDER of the differential equation in
      u  (<50)",M
220   PRINT "The order of the differential equation in u is =";M
230   PRINT LIN(2)
240   REDIM A(N,N),B(N),C(0,N),B1(0,N),C1(N),A1(N,N),P(N),Q(M)
250   CALL Vec(P(*),N,"P")
260   MAT A=(0)
270   MAT B=(0)
280   MAT C=(0)
290   FOR I=0 TO N
300   A(N,I)=-P(I)
310   IF I=N THEN 330
320   A(I,I+1)=1
330   NEXT I
340   B(N)=1
350   CALL Vec(Q(*),M,"Q")
```

```
360   FOR I=0 TO M
370   C(0,I)=Q(I)
380   NEXT I
390   PRINT "STATE EQUATION IN COMPANION FORM 1:"
400   PRINT LIN(3)
410   PRINT "MATRIX A ="
420   PRINT LIN(1)
430   MAT PRINT A
440   PRINT LIN(2)
450   PRINT "MATRIX B ="
460   PRINT LIN(1)
470   MAT PRINT B
480   PRINT LIN(2)
490   PRINT "MATRIX C ="
500   PRINT LIN(1)
510   MAT PRINT C
520   PRINT LIN(3)
530   DISP "Press CONTINUE."
540   PAUSE
550   PRINT "STATE EQUATION IN COMPANION FORM 2:"
560   PRINT LIN(3)
570   FOR I=0 TO N
580   B1(0,I)=C(0,I)
590   C1(I)=B(I)
600   FOR J=0 TO N
610   A1(I,J)=A(J,I)
620   NEXT J
630   NEXT I
640   PRINT "MATRIX A ="
650   PRINT LIN(1)
660   MAT PRINT A1
670   PRINT LIN(2)
680   PRINT "MATIX B ="
690   PRINT LIN(1)
700   MAT PRINT B1
710   PRINT LIN(2)
720   PRINT "MATRIX C ="
730   PRINT LIN(1)
740   MAT PRINT C1
750   PRINTER IS 16
760   END
900   SUB Vec(A(*),N,A$)
910   FOR I=0 TO N
920   DISP A$;"(";I;")=";
930   INPUT A(I)
940   NEXT I
950   FOR I=0 TO N
960   PRINT A$;"(";I;")=";A(I)
970   NEXT I
980   PRINT LIN(1)
```

```
990  INPUT "Any Changes(Y/N)",X$
1000 IF (X$="Y") OR (X$="y") THEN 1030
1010 IF (X$="N") OR (X$="n") THEN 1070
1020 GOTO 990
1030 INPUT "Element Number to be Changed ",In
1040 DISP A$;"(";In;")=";
1050 INPUT A(In)
1060 GOTO 950
1070 SUBEND
```

8. Example.

THIS PROGRAM DETERMINERS TWO TYPES OF STATE 'EQUATIONS IN COMPANION FORM FOR A LINEAR TIME-INVARIANT SYSTEM CHARACTERIZED BY
 An ordinary DIFFERENTIAL or DIFFERENCE equation:
$y(n)+p(n-1)*y(n-1)+...+p(0)*y = q(n-1)*u(n-1)+...+q(0)*u$
where $y(i)$ is defined as $(d^i)y/d(t^i)$. (Similarly for $u(i)$).
OR:
$y(k+n)+p(n-1)*y(k+(n-1))+...+p(0)*y(k) = q(n-1)*u(k+(n-1))+.+q(0)*u(k)$
The system STATE 'EQUATION will have the form :
 $dx/dt = Ax+Bu, y = Cx$ OR $x(k+1)=Ax(k) + Bu(k), y(k)=Cx(k)$

The order of the differential equation is = 3
The order of the differential equation in u is = 2

P(0) = 6.0000E+00
P(1) = 5.0000E+00
P(2) = 7.0000E+00

C(0) =-2.0000E+00
C(1) = 8.0000E+00
C(2) = 1.0000E+00

STATE EQUATION IN CONTROLLABLE COMPANION FORM :

MATRIX A =
  0 1 0

  0 0 1

 -6 -5 -7

MATRIX B =
  0 0 1

MATRIX C =
-2 8 1

STATE EQUATION IN OBSERVABLE COMPANION FORM :

MATRIX A =
 0 0 -6

 1 0 -5

 0 1 -7

MATIX B =
-2 8 1

MATRIX C =
 0 0 1

The order of the differential equation is= 4
The order of the differential equation in u is = 2

P(0) = 1.0000E+00
P(1) =-3.0000E+00
P(2) = 5.0000E+00
P(3) = 0.0000E+00

Q(0) =-2.0000E+00
Q(1) = 3.0000E+00
Q(2) = 4.0000E+00

STATE EQUATION IN CONTROLLABLE COMPANION FORM :

MATRIX A =
 0 1 0 0

 0 0 1 0

 0 0 0 1

-1 3 -5 0

MATRIX B =
 0 0 0 1

MATRIX C =
-2 3 4 0

STATE EQUATION IN OBSERVABLE COMPANION FORM :

MATRIX A =
0 0 0 -1

1 0 0 3

0 1 0 -5

0 0 1 0

MATIX B =
-2 3 4 0

MATRIX C =
0 0 0 1

The position of the nonzero element in the **B** or **C** vector is of crucial importance. In fact if the system matrix has the form shown in (37) and

$$C = [1,0,0,...,0], \qquad (47)$$

then the resulting state and output equations are said to be in the *observable companion form* independent of the form of the **B** vector. Similarly, if the system matrix is the transpose of that in (37) and

$$B = [1,0,0,...,0]' \qquad (48)$$

then the resulting state and output equations are in the *controllable companion form* independent of the form of the **C** vector. The reason for using this terminology will become clear in Chapter 8.

4.5 STATE TRANSFORMATIONS

It was shown in the previous section that a system may have different state space representations depending on the choice of state. In this section we will discuss the relationship between different representations of a linear system based on the relationship between the state vectors.

Consider a linear time-varying nth-order system

$$\dot{x}(t) = A(t)x(t) + B(t)u(t) \qquad (1)$$

STATE TRANSFORMATIONS

$$y(t)=C(t)x(t)+D(t)u(t) \tag{2}$$

Let $T(t)$ be any nonsingular time-varying nxn matrix. Then

$$z(t)=T(t)x(t) \tag{3}$$

also qualifies as a state vector for the system because at each time t $x(t)$ is recoverable from $z(t)$:

$$x(t)=T^{-1}(t)z(t) \tag{4}$$

Matrix T(t) is referred to as a *state transformation matrix*. Suppose that the new system representation is

$$\dot{z}(t)=\hat{A}(t)z(t)+\hat{B}(t)u(t) \tag{5}$$

$$y(t)=\hat{C}(t)z(t)+\hat{D}(t)u(t) \tag{6}$$

The relationship between the matrices in the above two representations can be found in terms of the matrix T(t). From (1), (3) and (4) we have (dropping t for convenience)

$$\dot{z}=\dot{T}x+T\dot{x}=\dot{T}x+T(Ax+Bu)=(\dot{T}+TA)T^{-1}z+TBu \tag{7}$$

Also (2) and (4) yield

$$y=CT^{-1}z+Du \tag{8}$$

Comparison of (7) with (5), and (8) with (6) results in

$$\hat{A}=(\dot{T}+TA)T^{-1} \tag{9}$$

$$\hat{B}=TB \tag{10}$$

$$\hat{C}=CT^{-1} \tag{11}$$

$$\hat{D}=D \tag{12}$$

Thus an infinite number of representations may be found for the same system. Note from (12) that the matrix $D(t)$ is unaffected by the state transformation because it is a direct relation between output and input. In the case of *l.t.i.* systems, (9) becomes

$$\hat{A} = TAT^{-1} \tag{13}$$

and matrices **A** and $\hat{A}$ will be similar. (See Section 2.7.)

1. Example. Consider an nth-order *l.t.i.* scalar system represented by (**A**,**B**,**C**,**D**) with state vector $x(t)$ where **A** is an nxn matrix with linearly independent eigenvectors, **B** = $[b_1, b_2, \ldots, b_n]'$, **C**=$[c_1, c_2, \ldots, c_n]$ and *D*=0. The system block diagram is shown in Fig. 1. Let us use M^{-1}, the inverse of the modal matrix **M** of **A** (See Section 2.7.) as the state transformation matrix, i.e. define the new state vector $z(t)$ as

$$z(t) = M^{-1}x(t) \tag{14}$$

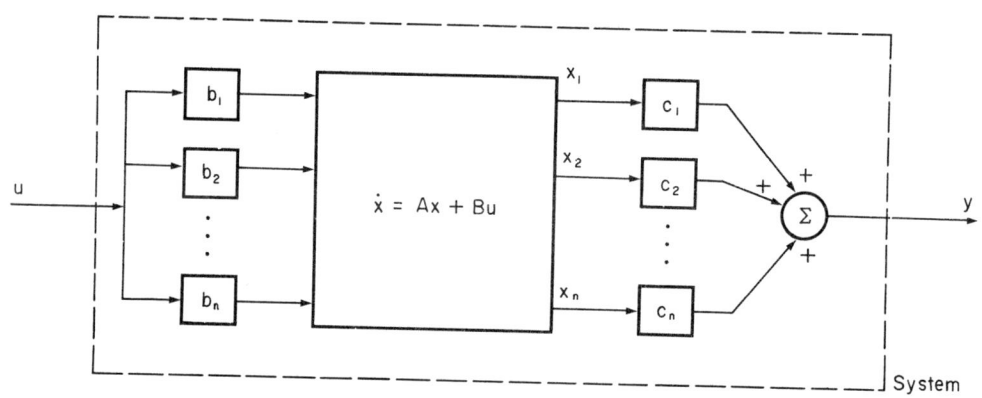

Fig. 4.5.1. Block diagram of the l.t.i. scalar system in Example 4.5.1

Then (10)-(13) imply that the new system representation ($\hat{A},\hat{B},\hat{C},\hat{D}$) with state vector $z(t)$ has the following matrices:

$$\hat{A} = M^{-1}AM = \text{diag}(\lambda_1, \lambda_2, \ldots, \lambda_n) \tag{15}$$

$$\hat{B} = M^{-1}B = [\hat{b}_1, \hat{b}_2, \ldots, \hat{b}_n]' \tag{16}$$

$$\hat{C} = CM = [\hat{c}_1, \hat{c}_2, \ldots, \hat{c}_n] \tag{17}$$

$$\hat{D} = 0 \tag{18}$$

where $\lambda_1, \lambda_2, \ldots, \lambda_n$ are eigenvalues of **A**. If $z_1, z_2, \ldots, z_n$ are components of the new state vector **z**, then (15)-(18) imply

$$\dot{z}_i = \lambda_i z_i + \hat{b}_i u_i, \ i=1,2,\ldots,n, \ y = \sum_{i=1}^{n} \hat{c}_i z_i \tag{19}$$

That is, the original nth-order system has been *decomposed* into n first-order systems as illustrated in the block diagram of Fig. 2. This is a very convenient state transformation since the analysis of a first-order system is much easier than the analysis of higher-order systems.

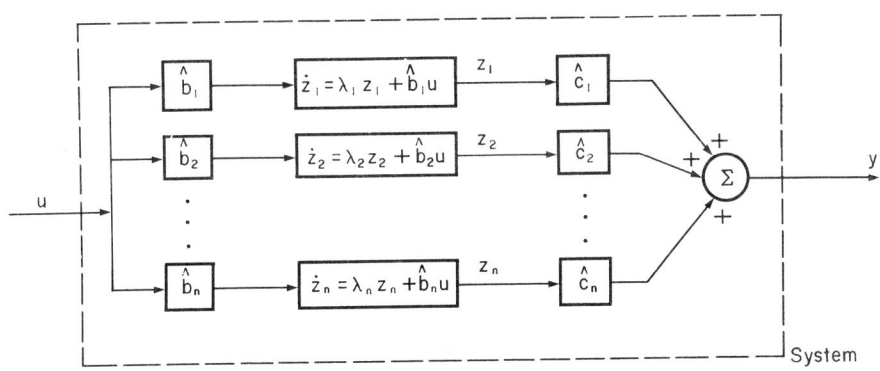

Fig. 4.5.2. Block diagram of the decomposed system in Example 4.5.1

We say that two system representations (A,B,C,D) and $(\hat{A},\hat{B},\hat{C},\hat{D})$ are *algebraically equivalent* or *strongly equivalent* if there is a one-to-one correspondence between their state variables. Thus the system described by (1) and (2) is algebraically equivalent to that described by (5) and (6) because their state vectors are related by (3) through the nonsingular state transformation matrix $T(t)$. In another form of system equivalence, known as *weak equivalence*, only the input-output relations of the two systems are the same but there is no one-to-one correspondence between the state variables. The following example illustrates the point.

2. Example. Consider the *l.t.i.* circuits shown in Fig. 3. These circuits are weakly equivalent in sinusoidal steady-state provided that $L/C = R^2$ because they both have the input-output relation

$$y(t) = R(t)u(t) \tag{19}$$

where $y(t)$ and $u(t)$ represent, respectively, the output and input phasers [4.7]. The circuits are not, however, strongly equivalent because their state spaces are of different dimensions (i.e. they are of dimensions 0 and 2).

△

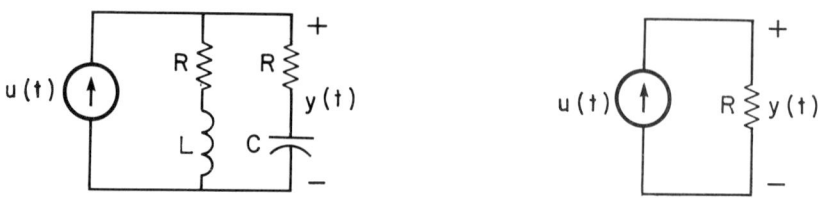

Fig. 4.5.3. Circuits in Example 4.5.2

In systems analysis we often make use of state transformations to obtain convenient representation for systems. The following algorithm provides a means for constructing a transformation matrix T which transforms any general nxn matrix into a controllable companion form [4.6]. Let

$$\hat{A} = T^{-1}AT, \hat{b} = T^{-1}b, \hat{c} = cT \qquad (20)$$

where m=r=1 and

$$T = [t^1 \ t^2 \ \cdots \ t^n] \qquad (21)$$

The columns t^i, $i=1,...n$ of the T matrix are obtained from:

$$t^n = b$$

$$t^{n-1} = Ab + a_{n-1}b$$

$$t^{n-2} = A^2b + a_{n-1}Ab + a_{n-2}b$$

$$t^1 = A^{n-1}b + a_{n-1}A^{n-2}b + ... + a_1 b \qquad (22)$$

where $a_n, a_{n-1}, \ldots, a_1, a_o$ are coefficients of the characteristic polynomial,

$$\Delta(\lambda) = a_o + a_1\lambda + a_2\lambda^2 + ... + a_{n-1}\lambda^{n-1} + \lambda^n \qquad (23)$$

The observable companion form can be found by simply evaluating $\hat{A}'$ and replacing $\hat{b}'$ and $\hat{c}'$ in place of $\hat{c}$ and $\hat{b}$, respectively.

Program "COMPAN" transforms the state equation of any *l.t.i.* scalar system into either a controllable or an observable companion form. The source listing for this program and an example of the third-order system are given below.

STATE TRANSFORMATIONS   135

3. "COMPAN"

```
10   !       PROGRAM NAME   "COMPAN" PROG
20   INPUT "Give your PRINT option: For CRT input 1; for LINE
     PRINTER input 2",P
30   IF (Pr=1) OR (Pr=2) THEN 50
40   GOTO 20
50   IF Pr=1 THEN PRINTER IS 16
60   IF Pr=2 THEN PRINTER IS 7,1
70   PRINT " 'COMPAN' TRANSFORMS THE STATE EQUATIONS OF ANY
     SINGLE-INPUT"
80   PRINT "SINGLE-OUTPUT LINEAR TIME-INVARIANT SYSTEM INTO
     A CONTROLLABLE AND/O OBSERVABLE COMPANION FORM",LIN(1)
90   PRINT "SYSTEM :    dx/dt=Ax+Bu ,   y=Cx+Du"
100  PRINT "COMPANION FORM :    dz/dt=(Ac)z+(Bc)u ,   y=(Cc)x+(Dc)u"
110  PRINT "where   A´=T*A*(T^-1),   B´=T*B,   C´=C*(T^-1),   D´=D
     and T is the"
120  PRINT "TRANSFORMATION matrix.";LIN(2)
130  INPUT "Have SUB's <<Mat>>, <<Cont>> & <<Obs>> already
     been LINKED (Y/N)?",C
140  IF C$="Y" THEN 200
150  IF C$="N" THEN 170
160  GOTO 130
170  LINK "Mat",1400,180
180  LINK "Cont",2500,190
190  LINK "Obs",3000,200
200  INPUT "Give ORDER of the system:",N
210  PRINT "ORDER of the system =";N;LIN(1)
220  CALL R(N)
230  PRINTER IS 16
240  END
250  SUB R(N)
260  OPTION BASE 1
270  DIM A(N,N),B(N,1),C(1,N),Qc(N,N),Qcinv(N,N),A1(N,N),
     Ac(N,N),Bc(N,1),Cc(1,N)
280  DIM Qo(N,N),Qoinv(N,N),A2(N,N),Ao(N,N),Bo(N,1),Co(1,N)
290  CALL Mat(A(*),N,N,"A")
300  CALL Mat(B(*),N,1,"B")
310  CALL Mat(C(*),1,N,"C")
320  INPUT "D=?",D
330  PRINT "D=";D;LIN(2)
340  CALL Cont(A(*),B(*),N,1,Qc(*))
350  Detc=DET(Qc)
360  IF ABS(Detc)<.0005 THEN Notcont
370  MAT Qcinv=INV(Qc)
380  FLOAT 4
390  PRINT "The TRANSFORMATION matrix for the CONTROLLABLE
     companion form is Tc=^(-1)=",LIN(1)
400  MAT PRINT Qcinv
410  MAT A1=Qcinv*A
```

```
420   MAT Ac=A1*Qc
430   MAT Bc=Qcinv*B
440   MAT Cc=C*Qc
450   Dc=D
460   PRINT "    MATRIX Ac=",LIN(1)
470   FOR I=1 TO N
480   FOR J=1 TO N
490   IF ABS(Ac(I,J))<.0005 THEN Ac(I,J)=0
500   NEXT J
510   IF ABS(Bc(I,1))<.0005 THEN Bc(I,1)=0
520   IF ABS(Cc(1,I))<.0005 THEN Cc(1,I)=0
530   NEXT I
540   MAT PRINT Ac
550   PRINT LIN(1),"    MATRIX Bc=",LIN(1)
560   MAT PRINT Bc
570   PRINT LIN(1),"    MATRIX Cc=",LIN(1)
580   MAT PRINT Cc
590   PRINT "Dc=";Dc;LIN(2)
600   GOTO 620
610   Notcont: PRINT "System is NOT CONTROLLABLE. Therefore
      it does not have a co rollable companion form."
620   CALL Obs(A(*),C(*),N,1,Qo(*))
630   Deto=DET(Qo)
640   IF ABS(Deto)<.0005 THEN Notobs
650   MAT Qoinv=INV(Qo)
660   PRINT LIN(2),"The TRANSFORMATION matrix for the
      OBSERVABLE companion form i To=Qo=",LIN(1)
670   MAT PRINT Qo
680   MAT A2=Qo*A
690   MAT Ao=A2*Qoinv
700   MAT Bo=Qo*B
710   MAT Co=C*Qoinv
720   Do=D
730   PRINT "    MATRIX Ao=",LIN(1)
740   FOR I=1 TO N
750   FOR J=1 TO N
760   IF ABS(Ao(I,J))<.0005 THEN Ao(I,J)=0
770   NEXT J
780   IF ABS(Bo(I,1))<.0005 THEN Bo(I,1)=0
790   IF ABS(Co(1,I))<.0005 THEN Co(1,I)=0
800   NEXT I
810   MAT PRINT Ao
820   PRINT LIN(1),"    MATRIX Bo=",LIN(1)
830   MAT PRINT Bo
840   PRINT LIN(1),"    MATRIX Co=",LIN(1)
850   MAT PRINT Co
860   PRINT "Do=";Do;LIN(2)
870   GOTO 890
880   Notobs: PRINT "System is NOT OBSERVABLE. Therefore it
      does not have an obse able companion form."
890   SUBEND
```

4. Example

'COMPAN' TRANSFORMS THE STATE EQUATIONS OF ANY SINGLE-INPUT SINGLE-OUTPUT LINEAR TIME-INVARIANT SYSTEM INTO A CONTROLLABLE AND OBSERVABLE COMPANION FORM. ORIGINAL SYSTEM: (A,B,C,D)

CONTR. COMPANION: (Ac,Bc,Cc,Dc) & OBSERV. COMPANION: (Ao,Bo,Co,Do). ORDER of the system = 3.0000E+00

Matrix A(3 x 3):

```
1.0000E+00   2.0000E+00   3.0000E+00

4.0000E+00   5.0000E+00   6.0000E+00

7.0000E+00   8.0000E+00   9.0000E+00
```

Column Vector B(3):

```
1.0000E+00

0.0000E+00

1.0000E+00
```

Row Vector C(3):

```
1.0000E+00   0.0000E+00   1.0000E+00
```

D= 1.0000E+00 The TRANSFORMATION matrix for the CONTROLLABLE companion form: 5.0000E-01 -1.0000E+00 5.0000E-01

-1.1250E+00 -1.2500E+00 1.1250E+00

6.9444E-02   8.3333E-02 -6.9444E-02

```
    MATRIX Ac=
0.0000E+00   1.0000E+00   0.0000E+00

0.0000E+00   0.0000E+00   1.0000E+00

0.0000E+00   1.8000E+01   1.5000E+01
```

MATRIX Bc=
1.0000E+00

0.0000E+00

0.0000E+00

MATRIX Cc=
2.0000E+00   2.0000E+01   3.2400E+02

Dc= 1.0000E+00 The TRANSFORMATION matrix for the OBSERVABLE companion form.
1.0000E+00   0.0000E+00   1.0000E+00

8.0000E+00   1.000E+01   1.2000E+01

1.3200E+02   1.6200E+02   1.9200E+02

MATRIX Ao=
0.0000E+00   0.0000E+00   0.0000E+00

1.0000E+00   0.0000E+00   1.8000E+01

0.0000E+00   1.0000E+00   1.5000E+01

MATRIX Bo=

2.0000E+00

2.0000E+01

3.2400E+02

MATRIX Co=
1.0000E+00   0.0000E+00   0.0000E+00

Do= 1.0000E+00

4.6 TRANSFER FUNCTION

One of the methods of describing a *l.t.i.* system in the input-output form is using the transfer function. The *transfer function* Laplace or the z-transform of the input. For a *l.t.i.* system with r inputs and m outputs the transfer function is an mxr matrix. We will discuss the

TRANSFER FUNCTION

transfer functions of continuous-time and discrete-time *l.t.i.* systems separately in Sections 4.6.1 and 4.6.2. These topics are followed by a discussion on the transfer functions of SISO systems within the context of conventional control in Section 4.6.3.

4.6.1 Continuous-Time Systems For continuous-time scalar systems the transfer function is defined as the ratio of the Laplace transform of the output and the Laplace transform of the input when all the initial conditions are zero. Thus for the scalar *l.t.i.* system described by differential equation (4.3.1) the transfer function is

$$H(s) \triangleq \frac{Y(s)}{U(s)} = \frac{q_m s^m + q_{m-1} s^{m-1} + \ldots + q_1 s + q_0}{p_n s^n + p_{n-1} s^{n-1} + \ldots + p_1 s + p_0} \tag{1}$$

Note that $H(s)$ in (1) is a scalar-valued rational function of s. Equation (1) can also be written as

$$H(s) = \frac{q_m (s-z_1)(s-z_2)\ldots(s-z_m)}{p_n (s-\lambda_1)(s-\lambda_2)\ldots(s-\lambda_n)} \tag{2}$$

where z_i $i=1,2,\ldots m$ are called the *zeros* and λ_i $i=1,2,\ldots n$ are called the *(closed-loop) poles** of the system. We will shortly see that the poles are identical to the eigenvalues of the system matrix. There may be common factors in the numerator and the denominator of (2). If these are cancelled, the remaining factors will then be referred to as the *poles* and *zeros of the transfer function*.

Equation (1) implies

$$Y(s) = H(s)U(s) \tag{3}$$

If the input is an impulse, i.e. if $u(t)=\delta(t)$, then $U(s)=1$ and $Y(s)=H(s)$. That is

$$y(t) \triangleq h(t) = L^{-1}[H(s)] \tag{4}$$

Here $h(t)$ is known as the *impulse response* of the system. Therefore, the transfer function of a scalar *l.t.i.* system is the Laplace transform of its impulse response.

For multivariable *l.t.i.* systems the Laplace transforms of the output and input vectors are related through the transfer function as in (3). The transfer function can be expressed in terms of the matrices in the system representation (A,B,C,D). To do this take the Laplace transform of the state and output equations (4.4.6) and (4.4.7) considering zero initial conditions to obtain

$$sX(s) = AX(s) + BU(s) \tag{5}$$

$$Y(s) = CX(s) + DU(s) \tag{6}$$

From the above equation we have

---

* For the definitions of closed-loop and open-loop poles of SISO systems see Section 4.6.3.

$$Y(s)=[C(sI-A)^{-1}B+D]U(s) \qquad (7)$$

Comparison with (3) yields

$$H(s)=C(sI-A)^{-1}B+D \qquad (8)$$

which is an mxr matrix known as the *transfer function matrix* of the system. The square matrix $(sI-A)^{-1}$ is called a *resolvent matrix* whose properties will be discussed in Chapter 6. Note that $(sI-A)^{-1}$ exists for all values of s which are not eigenvalues of **A**. Also note from (8) that the poles of the system are the eigenvalues of the system matrix **A**. (See Section 2.6.)

1. Example. Consider the *l.t.i.* system described by (4.4.20) which is also described by the state and output equations (4.4.43) and (4.4.44). From (4.4.20), the transfer function is

$$H(s) = \frac{s^2+8s-2}{s^3+7s^2+5s+6} \qquad (9)$$

Also, from (4.4.43) and (4.4.44) we have

$$H(s)=C(sI-A)^{-1}B+D=[-2\ 8\ 1]\begin{bmatrix} s & -1 & 0 \\ 0 & s & -1 \\ 6 & 5 & s+7 \end{bmatrix}^{-1}\begin{bmatrix} 0 \\ 0 \\ 1 \end{bmatrix}$$

$$= \frac{1}{s^3+7s^2+5s+6}[-2\ 8\ 1]\begin{bmatrix} s^2+7s+5 & s+7 & 1 \\ -6 & s^2+7s & s \\ -6s & -5s-6 & s^2 \end{bmatrix}\begin{bmatrix} 0 \\ 0 \\ 1 \end{bmatrix} \qquad (10)$$

which results in (9). The system representation (**A,B,C,D**) in (4.4.38) and (4.4.39) is said to be a *realization* of H(s). Note that a given transfer function has many realizations. The reader can verify, for example, that the alternative state and output equations (4.4.40) and (4.4.41) of this system is also a realization of the transfer function (9).

4.6.2 Discrete-Time Systems The development of transfer function for discrete-time systems is analogous to that for continuous-time systems. The transfer function of a *l.t.i.* discrete-time scalar system is defined as the ratio of the z-transform of the output and the z-transform of the input when all the initial conditions are zero. Thus for the system described by difference equation (4.3.2) the transfer function is

$$H(z) \triangleq \frac{Y(z)}{U(z)} = \frac{q_m z^d + q_{m-1} z^{d-1} + \cdots + q_1 z + q_0}{p_n z^n + p_{n-1} z^{n-1} + \cdots + p_1 z + p_0} \qquad (11)$$

which is a scalar-valued rational function of z. Equation (11) implies

$$Y(z)=H(z)U(z) \qquad (12)$$

TRANSFER FUNCTION

which is also the defining equation for the transfer function of *l.t.i.* discrete-time multivariable systems.

Consider the system described by

$$\mathbf{x}(k+1) = \mathbf{A}\mathbf{x}(k) + \mathbf{B}\mathbf{u}(k) \tag{13}$$

$$\mathbf{y}(k) = \mathbf{C}\mathbf{x}(k) + \mathbf{D}\mathbf{u}(k) \tag{14}$$

The z-transform of these equations yield

$$z\mathbf{X}(z) - z\mathbf{X}(0) = \mathbf{A}(z) + \mathbf{B}\mathbf{U}(z) \tag{15}$$

$$\mathbf{Y}(z) = \mathbf{C}\mathbf{X}(z) + \mathbf{D}\mathbf{U}(z) \tag{16}$$

If the system has zero initial conditions, then $\mathbf{x}(0) = 0$ and (15) and (16) yield

$$\mathbf{Y}(z) = [\mathbf{C}(z\mathbf{I} - \mathbf{A})^{-1}\mathbf{B} + \mathbf{D}]\mathbf{U}(z) \tag{17}$$

Therefore

$$\mathbf{H}(z) = \mathbf{C}(z\mathbf{I} - \mathbf{A})^{-1}\mathbf{B} + \mathbf{D} \tag{18}$$

which is similar to (8). Also, as we will see in Chapter 6,

$$\mathbf{H}(z) = Z[\mathbf{h}(k-1)] \tag{19}$$

where $\mathbf{h}(k)$ is the *pulse response matrix* of the system. The square matrix $(z\mathbf{I} - \mathbf{A})^{-1}$ is again called the resolvent matrix.

2. Example. Consider a *l.t.i.* discrete-time system represented by $(\mathbf{A}, \mathbf{B}, \mathbf{C}, \mathbf{D})$ where

$$\mathbf{A} = \begin{bmatrix} -1 & 1 & 0 \\ 0 & -1 & 0 \\ 0 & 0 & -3 \end{bmatrix}, \mathbf{B} = \begin{bmatrix} 1 \\ 2 \\ 0 \end{bmatrix}, \mathbf{C} = \begin{bmatrix} 0 & 1 & 1 \\ 0 & 1 & 0 \end{bmatrix}, \mathbf{D} = 0 \tag{20}$$

The transfer function of this system is, using (18),

$$\mathbf{H}(z) = \begin{bmatrix} 0 & 1 & 1 \\ 0 & 1 & 0 \end{bmatrix} \begin{bmatrix} z+1 & -1 & 0 \\ 0 & z+1 & 0 \\ 0 & 0 & z+3 \end{bmatrix}^{-1} \begin{bmatrix} 1 \\ 2 \\ 0 \end{bmatrix} = \begin{bmatrix} \dfrac{2}{z+1} \\ \dfrac{2}{z+1} \end{bmatrix} \tag{21}$$

142   LINEAR CONTROL SYSTEMS

Here again, we say that **(A,B,C,D)** is a realization of **H**(z). Note that there are other realizations of **H**(z) which are different from the one above.

△

     Program "TRANFN" calculates the transfer function matrix of any *l.t.i.* multivariable system represented by **(A,B,C,D)**. The source listing of "TRANFN" is given below followed by an example of a third-order system with two inputs and two outputs.

3. "TRANFN"

```
10    !    PROGRAM NAME : "TRANFN" PROG
20    INPUT "Give your PRINT option: For CRT input 1; for LINE
      PRINTER input 2",P
30    IF (Pr=1) OR (Pr=2) THEN 50
40    GOTO 20
50    IF Pr=1 THEN PRINTER IS 16
60    IF Pr=2 THEN PRINTER IS 7,1
70    PRINT " 'TRANFN' CALCULATES THE TRANSFER FUNCTION OF ANY
      MULTIVARIBLE"
80    PRINT " LINEAR TIME-INVARIANT SYSTEM   dx/dt=Ax+Bu,
      y=Cx+Du";LIN(1)
90    PRINT "Matrix dimensions are:    A:NxN,    B:NxR,    C:MxN,
      D:MxR";LIN(2)
100   OPTION BASE 1
110   INPUT "Have SUB's <<Mat>> & <<Resmat>> already been
      LINKED (Y/N)?",C$
120   IF C$="Y" THEN 180
130   IF C$="N" THEN 150
140   GOTO 110
150   LINK "Mat",1000,160
160   LINK "Resmat",2500,170
170   DIM A(9,9),B(9,9),C(9,9),D(9,9),Q(9,9,9),P(10),M(9,9)
      ,M1(9,9),M2(9,9),L(9,9
180   INPUT "Give matrix dimensions N, R, M (N,R,M<10, 0<=N-R,
      N-M<=8)",N,R,M
190   PRINT LIN(2),"Matrix dimensions : N=";N;",    R=";R;",
      M=";M;
200   PRINT LIN(2)
210   REDIM A(N,N),B(N,R),C(M,N),D(M,R),Q(N,N,N),P(N),M(N,N)
      ,M1(M,N),M2(M,R),L(M,
220   CALL Mat(A(*),N,N,"A")
230   CALL Mat(B(*),N,R,"B")
240   CALL Mat(C(*),M,N,"C")
250   CALL Mat(D(*),M,R,"D")
260   CALL Resmat(N,A(*),P(*),Q(*))
270   PRINT LIN(4),"H(s)=Q(s)/P(s)    where",LIN(2),"Q(s)=D*s^"
      ;N;"+SUMπQ(i)*(s^i)→
      P(s)=s^";N;"+SUMπP(i)*(s^i)→",LIN(2),"   i=0,1,...,";N-1
280   FOR I=1 TO N
290   FOR J=1 TO N
```

```
300   FOR K=1 TO N
310   M(J,K)=Q(N-I+1,J,K)
320   NEXT K
330   NEXT J
340   MAT M1=C*M
350   MAT M2=M1*B
360   FOR K=1 TO M
370   FOR J=1 TO R
380   L(K,J)=P(N-I+1)*D(K,J)
390   NEXT J
400   NEXT K
410   MAT M2=M2+L
420   IF (M=1) AND (R=1) THEN 460
430   PRINT LIN(3),"MATRIX Q(";N-I;"):",LIN(2)
440   MAT PRINT M2
450   GOTO 470
460   PRINT LIN(3),"COEFFICIENT   Q(";N-I;")=";M2(1,1)
470   NEXT I
480   PRINT LIN(3)
490   FOR I=1 TO N
500   PRINT "P(";N-I;")=";P(N-I+1),LIN(1)
510   NEXT I
520   PRINTER IS 16
530   END
```

4. EXAMPLE

'TRANFN' CALCULATES THE TRANSFER MATRIX OF ANY MULTIVARIBLE
LINEAR TIME-INVARIANT SYSTEM   (A,B,C,D)
Matrix dimensions are: A:nxn,   B:nxn,   C:mxn,   D:mxr

Matrix dimensions : n= 3 , r= 2 ,   m= 2

Matrix A(3 x 3):

  0.0000E+00 -2.0000E+00 -2.0000E+00

  0.0000E+00   0.0000E+00   1.0000E+00

  0.0000E+00 -3.0000E+00 -4.0000E+00

Matrix B(3 x 2):

  1.0000E+00   0.0000E+00

  0.0000E+00 -2.0000E+00

  1.0000E+00   0.0000E+00

Matrix C(2 x 3):

-2.0000E+00 0.0000E+00 3.0000E+00

 0.0000E+00 1.0000E+00 2.0000E+00

Matrix D(2 x 2):

 1.0000E+00 0.0000E+00

 0.0000E+00 1.0000E+00

$H(s)=Q(s)/P(s)$ where

$Q(s)=D*s\char`\^\ 3 + SUM\pi Q(i)*(s\char`\^ i)\to$, $P(s)=s\char`\^\ 3 + SUM\pi P(i)*(s\char`\^ i)\to$

   i=0,1,..., 2

MATRIX Q(2):

 5 0

 2 2

MATRIX Q(1):

-1 10

 1   7

MATRIX Q(0):

-2 -8

 0   0

P(2)= 4

P(1)= 3

P(0)= 0

4.6.3 Transfer Function of a SISO System

In many SISO systems, it is convenient to express the system in the frequency domain by taking the Laplace transform of the differential equation or the z transform of the difference equation representing the system. The ratio of the Laplace transform of the output $Y(s)$ to that of the input $U(s)$ is called the *overall* (or closed-loop) *transfer function* of the system. In Section 4.6.1 and 4.6.2 transfer functions of continuous-time and discrete-time systems were discussed from the modern control point of view. The object of this section is to present the conventional approach to the problem. For the sake of discussion let us concentrate on the dc machine of Example 4.2.1 whose model was described by Equations (4.2.1) to (4.2.4). Taking the Laplace transform of these relations, one obtains,

$$T_r(s) = k_2 I_a(s) \tag{22}$$

$$V(s) = L_a s I_a(s) + R_a I_a(s) + E_b(s) \tag{23}$$

$$E_a(s) = k_3 \Omega(s) \tag{24}$$

$$T_r(s) = T_d(s) + J s \Omega(s) + C \Omega(s) \tag{25}$$

where initial conditions $i_a(0)$ and $\omega(0)$ are assume to be zero. If the load torque is given by

$$T_d(s) = (J_d s + C_d) \Omega(s) \tag{26}$$

where J_d and C_d are the inertia and the viscous friction of the load. Equations (22) and (25) can be combined to obtain

$$(J_e s + C_e) \Omega(s) = k_2 I_a(s) \tag{27}$$

where $J_e = J + J_d$ and $C_e = C + C_d$ are the equivalent inertia and viscous friction of the motor-load assembly. Similarly, (23) and (24) can be combined to provide,

$$I_a(s) = (1/(R_a + L_a s)) e_a(s) - (k_3/(R_a + L_a s)) \Omega(s) \tag{28}$$

The above two equations and the fact that the motor's angular position $\theta(t)$ can be represented by

$$\dot{\theta}(t) = \omega(t), \text{ or } \Theta(s) = (1/s) \Omega(s) \tag{29}$$

can all be combined in a cascade fashion to give a *block-diagram* representation of the dc motor shown in Fig. 1.

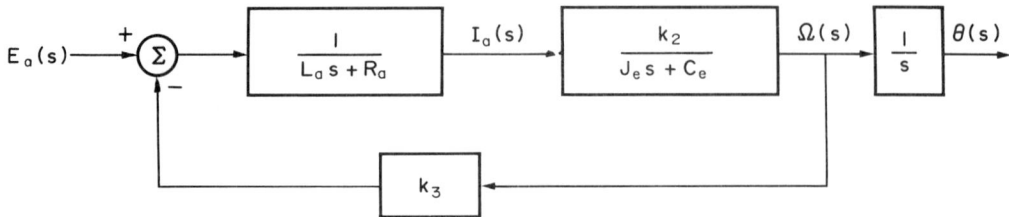

Fig. 4.6.1. A block diagram for the dc motor example

It is noted that the feedback from $\Omega(s)$ is one of inherent, not physical, type. Often one may want to close the loop by feeding the position $\Theta(s)$ or velocity $\Omega(s)$, or both, back to the forward path to achieve system regulation. One such configuration is shown in Fig. 2 for position feedback.

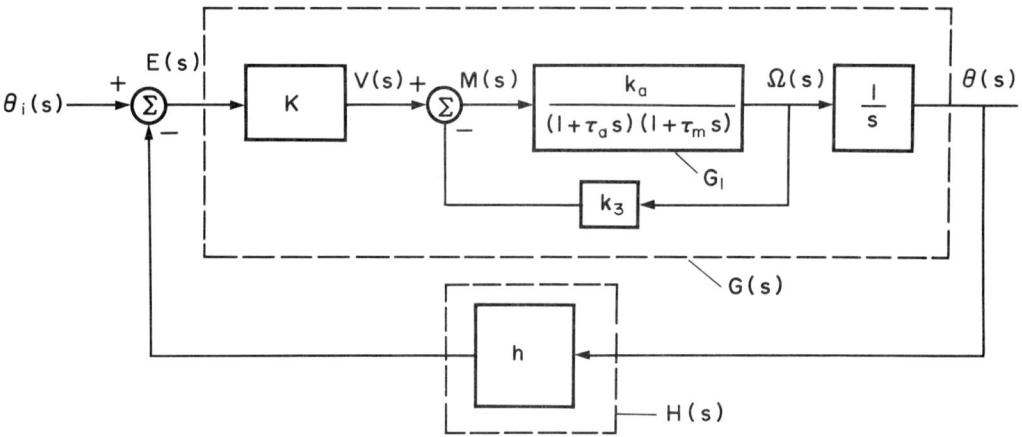

Fig. 4.6.2. Block diagram for a position feedback servo system

In this figure $k=k_2/C_e R_a$, $\tau_a=L_a/R_a$ and $\tau_m=J_e/C_e$ are the time constants of the armature circuit and equivalent load-motor, respectively. The term h in Fig. 2 is the gain of an angular position transducer such as a wirewound potentiometer. The transfer function G(s) of the dc motor of Fig. 1 is given by

$$G(s)=Kk/\{s[\tau_a\tau_m s^2+(\tau_a+\tau_m)s+1+kk_3]\} \tag{30}$$

is called the *open-loop transfer function* while H(s) is called the *feedback transfer function*. The overall or *closed-loop transfer function*, $\Theta(s)/\Theta_i(s)$ can be obtained by noting that the *error function* E(s) is given by

$$E(s)=\Theta_i(s)-H(s)\Theta(s) \tag{31}$$

where

$$\Theta(s) = G(s)E(s) \tag{32}$$

Thus eliminating $E(s)$ from the above relations yields,

$$\Theta(s) = G(s)\Theta_i(s) - G(s)H(s)\Theta(s) \tag{33}$$

or

$$\Theta(s)/\Theta_i(s) = G(s)/[1+G(s)H(s)]. \tag{34}$$

which is the desired expression.

It is worthwhile to define poles and zeros of the above transfer functions for our subsequent discussions in Chapter 7. In Section 4.6.1, zeros and closed-loop poles of a system described by a continuous-time state equation were defined. Within the context of a conventional SISO system described by (34), we have the following definition.

4. Definition Consider a SISO system with open-loop transfer function $G(s)$ and feedback transfer function $H(s)$. Let

$$G(s) \triangleq A(s)/B(s), \; 1+G(s)H(s) \triangleq C(s)/D(s). \tag{35}$$

Then the roots of $A(s)=0$, $B(s)=0$ and $C(s)=0$ are called, respectively, the *zeros, open-loop poles* and *closed-loop poles* of the system. In sequel, a program called "TRANSF" is presented to determine the transfer function of a SISO system from the polynomials representing open-loop and feedback transfer functions $G(s)$ and $H(s)$.

5. "TRANSF"

```
10    !      PROGRAM NAME : "TRANSF" PROG
20    INPUT "Give your PRINTER option: For CRT press 1; for
      typewriter press 2",
30    IF Pr=1 THEN PRINTER IS 16
40    IF Pr=2 THEN PRINTER IS 7,1
50    PRINT " 'TRANSF' RETERMINES THE CLOSED-LOOP TRANSFER
      FUNCTION OF LINEAR"
60    PRINT "SINGLE-LOOP FEEDBACK SYSTEMS.",LIN(1)
70    PRINT "Open-loop transfer function : G(s)=A(s)/B(s)"
80    PRINT "Feedback transfer function : H(s)=C(s)/D(s)"
90    PRINT "Closed-loop transfer function : R(s)=G(s)/
      [1+G(s)H(s)]=Ac(s)/Bc(s)", LIN(1)
100   PRINT "where    A(s)=SUMπA(i)*s^i→, i=0,1,...,"
```

```
110  PRINT "            B(s)=SUMπB(i)*s^i→, i=0,1,...,      ;etc.
     ";LIN(1)
120  INPUT "Have SUB's <<Vec>> & <<Polpro>> already been
     LINKED (Y/N)?",C$
130  IF C$="Y" THEN 180
140  IF C$="N" THEN 160
150  GOTO 120
160  LINK "Vec",700,170
170  LINK "Polpro",1200,180
180  INPUT "Give the order of the NUMERATOR, DENOMINATOR
     polynomials of G(s):",N
190  PRINT "Order of the NUMERATOR polynomial of G(s)=";N
200  PRINT "Order of the DENOMINATOR polynomial of G(s)="
     ;M;LIN(1)
210  INPUT "Give the order of the NUMERATOR, DENOMINATOR
     polynomials of H(s):",K
220  PRINT "Order of the NUMERATOR polynomial of H(s)=";K
230  PRINT "Order of the DENOMINATOR polynomial of H(s)="
     ;L;LIN(1)
240  N1=MAX(M+L,N+K)
250  CALL R(N,M,K,L,N1)
260  PRINTER IS 16
270  END
280  SUB R(N,M,K,L,N1)
290  N2=MIN(M+L,N+K)
300  DIM A(N),B(M),C(K),D(L),Ac(N+L),Bc(N1),R1(M+L),R2(N+K)
310  PRINT "Give coefficients of the NUMERATOR polynomial of
     G(s) in ASCENDING o er:";
320  CALL Vec(A(*),N,"A")
330  PRINT "Give coefficients of the DENOMINATOR polynomial
     of G(s) in ASCENDING rder:"
340  CALL Vec(B(*),M,"B")
350  PRINT "Give coefficients of the NUMERATOR polynomial
     of H(s) in ASCENDING o er:"
360  CALL Vec(C(*),K,"C")
370  PRINT "Give coefficients of the DENOMINATOR polynomial
     of H(s) in ASCENDING rder:"
380  CALL Vec(D(*),L,"D")
390  PRINT "Coefficients of the NUMERATOR polynomial of R(s)
     in ASCENDING order: LIN(1)
400  CALL Polpro(A(*),D(*),N,L,Ac(*))
410  FOR I=0 TO L+N
420  PRINT "Ac(";I;")=";Ac(I)
430  NEXT I
440  PRINT LIN(1),"Coefficients of the DENOMINATOR polynomial
     of R(s) in ASCENDI order:",LIN(1)
450  CALL Polpro(B(*),D(*),M,L,R1(*))
460  CALL Polpro(A(*),C(*),N,K,R2(*))
470  FOR I=0 TO N2
480  Bc(I)=R1(I)+R2(I)
```

```
490   NEXT I
500   IF N1=M+L THEN 540
510   FOR I=N2+1 TO N+K
520   Bc(I)=R2(I)
530   NEXT I
540   FOR I=N2+1 TO M+L
550   Bc(I)=R1(I)
560   NEXT I
570   FOR I=0 TO N1
580   PRINT "Bc(";I;")=";Bc(I)
590   NEXT I
```

6. EXAMPLE

'TRANSF' DETERMINES THE CLOSED-LOOP TRANSFER FUNCTION OF LINEAR SINGLE-LOOP FEEDBACK SYSTEMS.

Open-loop transfer function : $G(s)=A(s)/B(s)$
Feedback transfer function : $H(s)=C(s)/D(s)$
Closed-loop transfer function : $F(s)=G(s)/[1+G(s)H(s)]$
$=Ac(s)/Bc(s)$

where $A(s)=SUM\;A(i)*s^i$, $i=0,1,\ldots,$
       $B(s)=SUM\;B(i)*s^i$, $i=0,1,\ldots,$;etc.

Order of the NUMERATOR polynomial of $G(s)=$ 1
Order of the DENOMINATOR polynomial of $G(s)=$ 3

Order of the NUMERATOR polynomial of $H(s)=$ 0
Order of the DENOMINATOR polynomial of $H(s)=$ 1

Give coefficients of the NUMERATOR polynomial of $G(s)$ in ASCENDING order:
A(0) = 6.0000E+00
A(1) = 2.0000E+00

Give coefficients of the DENOMINATOR polynomial of $G(s)$ in ASCENDING order:
B(0) = 5.0000E+00
B(1) = 4.0000E+00
B(2) = 2.0000E+00
B(3) = 1.0000E+00

Give coefficients of the NUMERATOR polynomial of $H(s)$ in ASCENDING order:
C(0) = 2.0000E+00

Give coefficients of the DENOMINATOR polynomial of H(s) in
ASCENDING order:
D(0) = 5.0000E+00
D(1) = 1.0000E+00

Coefficients of the NUMERATOR polynomial of R(s) in ASCENDING
order:
Ac(0) = 30
Ac(1) = 16
Ac(2) = 2

Coefficients of the DENOMINATOR polynomial of R(s) in
ASCENDING order:
Bc(0) = 37
Bc(1) = 29
Bc(2) = 14
Bc(3) = 7
Bc(4) = 1

4.6.4 Signal Flow Graphs A powerful approach in obtaining the transfer function of SISO systems is through a so-called *signal flow graph*. In order to explain this presentation, let us consider the position feedback control system of Fig. 2 and draw its signal flow graph, shown in Fig. 3. As seen, each variable in the system is represented by a *node*. The system's input and output variables are denoted by an *input node* and an *output node*. A unidirectional path segment relating one incoming variable to an outgoing variable much like a block in a block diagram is called a *branch*, e.g. branch $M\Omega$. A *forward path* is a unidirectional segment originating from the input node and ending at the output one, e.g. $\theta_i EVM\Omega\theta$. A closed path terminating at the same node Vom which it originates, is called a *loop*, e.g. loops $M\Omega M$ and $EVM\Omega\theta E$. The product of the transfer functions of a given path is called the *transmittance* of that path, e.g. KG_1/s for the forward path in Fig. 3. Similar definition holds for the transmittance of a loop.

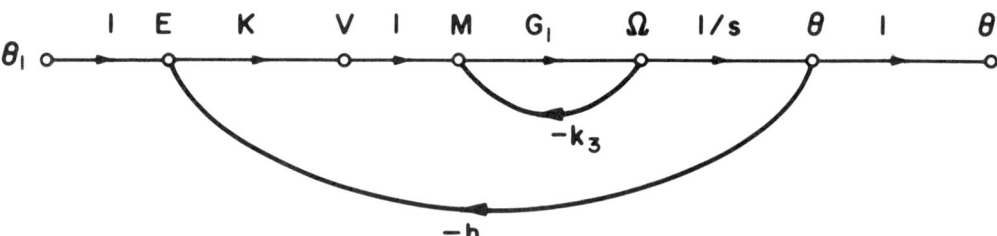

Fig. 4.6.3. A Signal Flow graph for the system of Fig. 4.6.2

TRANSFER FUNCTION

As was mentioned earlier, an important outcome of the signal flow graphs is to find the system's transfer function. This can be achieved by the Mason's Gain Formula [4.5], given below.

$$G_{c.l.} = \Sigma_n g_n \Delta_n / \Delta \tag{36}$$

where g_n is the transmittance of the nth forward path between the source and sink nodes; Δ is the *graph determinant* defined by,

$$\Delta = 1 - \Sigma_i L_{1i} + \Sigma_j L_{2j} - \Sigma_k L_{3k} + \cdots \tag{37}$$

in which

(i) L_{1i} is the loop transmittance of the ith feedback loop and $\Sigma_i L_{1i}$ is the sum of the transmittances of all feedback loops.

(ii) L_{2j} is the jth product of the loop transmittances of two non-touching loops. Loops with no common nodes are non-touching. ΣL_{2j} is the transmittance products for all possible non-touching loops taken two at a time.

(iii) L_{3k} is the kth product of the loop transmittances of three non-touching loops ΣL_{3k} is the sum of transmittances products for all possible non-touching loops taken three at a time.

and so on. Δ_n is called the *coefactor* of g_n and is obtained by evaluating the decimal of the remaining graph where all the loops touching g_n (having no nodes in common with) are disregarded. The term $\Sigma_n g_n \Delta_n$ corresponds to the summation of the products of all forward-path transmittance and their cofactors. The following example illustrates the use of Mason's gain formula

7. Example. Consider a multiloop feedforward and feedback control system shown in Fig. 4.

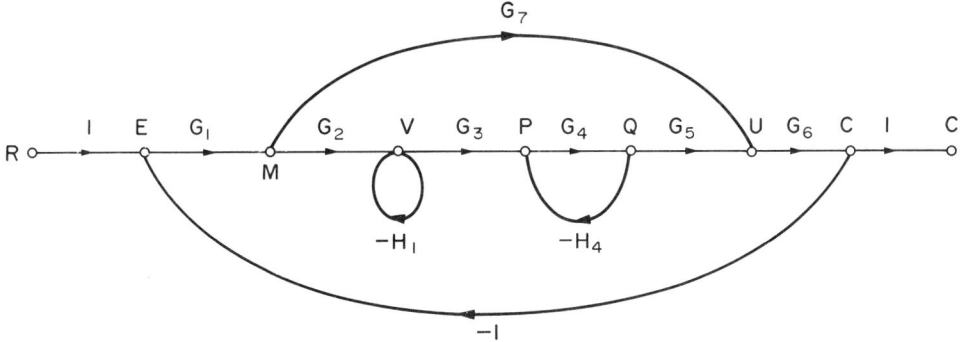

Fig. 4.6.4. Example of a signal flow graph

The system has one source node (U), one sink node (Y), six mixed nodes, two forward paths, four two-nontouching loops pairs, one three-nontouching loops group and no four nontouching loops. The transmittances of the two forward paths are

$$g_1 = G_1 G_2 G_3 G_4 G_5 G_6, \quad g_2 = G_1 G_6 G_7, \tag{38}$$

The transmittances of the 3 sets of loop groups are.

Single loops: L_{1i}

$$L_{11} = -G_1 G_2 G_3 G_4 G_5 G_6 \tag{39}$$

$$L_{12} = -G_1 G_6 G_7, \; L_{13} = -H_1, \; L_{14} = -G_4 H_4$$

Two nontouching loops: L_{2j}

$$L_{21} = G_1 G_6 G_7 H_1, L_{22} = G_1 G_4 G_6 G_7 H_4 \tag{40}$$

$$L_{23} = G_4 H_4 H_1$$

Three nontouching loops: L_k

$$L_{31} = -G_1 G_6 G_7 G_4 H_4 H_1. \tag{41}$$

The two coefactors corresponding to the two forward paths are

$$\Delta_1 = 1, \Delta_2 = 1 - (L_{13} + L_{14}) + L_{23} \tag{42}$$

which are extracted from the graph determinant

$$\Delta = 1 - (L_{11} + L_{12} + L_{13} + L_{14}) + (L_{21} + L_{22} + L_{23}) - L_{31.} \tag{43}$$

Thus, the overall transfer function of the system is

$$G_{c.l.} = (g_1 \Delta_1 + g_2 \Delta_2)/\Delta \tag{44}$$

where all terms have been defined through Equations (38)-(42).

4.7 LINEARIZATION OF NONLINEAR SYSTEMS

Most real-life systems are nonlinear time-varying systems characterized by state and output equations (4.4.8) and (4.4.9) repeated here for convenience:

LINEARIZATION OF NONLINEAR SYSTEMS

$$\dot{\mathbf{x}} = \mathbf{f}(\mathbf{x}, \mathbf{u}, t) \tag{1}$$

$$\mathbf{y} = \mathbf{g}(\mathbf{x}, \mathbf{u}, t). \tag{2}$$

No general theory for the analysis and design of nonlinear systems exists. One can deal with equations of the type (1) and (2) only by using ad hoc methods or numerical techniques. However, many nonlinear systems, particularly automatic control systems, are designed to operate in the proximity of a desired operating point. This allows the system to be linearized around its operating point. Examples are the centrifugal steam engine governor and the autopilot of an airplane. Both of these systems are highly nonlinear but they are designed to operate as close to a desired equilibrium state as possible. Another example of a nonlinear system is a transistor circuit which can be characterized using the Ebers-Moll equations [4.8]. These nonlinear equations are linearized to analyze the behavior of transistor circuits in the neighborhood of operating points.

If perturbations around a known operating point of a nonlinear system are small, the behavior of the system can be describe by a linear model. This is referred to as *small-signal analysis*. This section is concerned with the derivations of small-signal linear models around known operating points of nonlinear systems. We will cover some preliminary notions first.

4.7.1 Differentiation with Respect to a Vector Let $\mathbf{x}$ be an n-dimensional column vector with components $x_1, x_2, \ldots, x_n$. Consider a scalar-valued function of $\mathbf{x}$, $f(x_1, x_2, \ldots, x_n) \triangleq f(\mathbf{x})$. The *gradient vector* of $f(\mathbf{x})$, denoted by $\dfrac{df}{d\mathbf{x}}$, is defined as the following column vector provided that all the partial derivatives exist:

$$\frac{df}{d\mathbf{x}} = [\frac{\partial f}{\partial x_1}, \frac{\partial f}{\partial x_2}, \ldots, \frac{\partial f}{\partial x_n}]' \tag{3}$$

Now consider a vector-valued function of $\mathbf{x}$, $\mathbf{f}(\mathbf{x})$, where $\mathbf{f}$ is itself an m-dimensional column vector. Then each component $f_i(\mathbf{x})$, $i=1,2,\ldots,m$ is a scalar-valued function of $\mathbf{x}$ whose gradient is an n-dimensional column vector similar to (3). Thus the gradient of $\mathbf{f}$ is

$$\frac{d\mathbf{f}(x)}{d\mathbf{x}} = [\frac{df_1}{d\mathbf{x}}, \frac{df_2}{d\mathbf{x}}, \ldots, \frac{df_m}{d\mathbf{x}}]'$$

$$= \begin{bmatrix} \dfrac{\partial f_1}{\partial x_1} & \dfrac{\partial f_1}{\partial x_2} & \cdots & \dfrac{\partial f_1}{\partial x_n} \\ \dfrac{\partial f_2}{\partial x_1} & \dfrac{\partial f_2}{\partial x_2} & \cdots & \dfrac{\partial f_2}{\partial x_n} \\ \vdots & \vdots & \ddots & \vdots \\ \dfrac{\partial f_m}{\partial x_1} & \dfrac{\partial f_m}{\partial x_2} & \cdots & \dfrac{\partial f_m}{\partial x_n} \end{bmatrix} \tag{4}$$

which is an mxn matrix known as the *Jacobian matrix* of $\mathbf{f}(\mathbf{x})$. The Jacobian matrix plays an important role in the concept of linearization of nonlinear systems.

154 LINEAR CONTROL SYSTEMS

4.7.2 Linearization Around an Operating Point Consider a nonlinear system with n state variables, r inputs and m outputs described by (1) and (2). Suppose that for a nominal input vector $\hat{\mathbf{u}}(t)$, the resulting nominal state vector $\hat{\mathbf{x}}(t)$ and the nominal output vector $\hat{\mathbf{y}}(t)$ are known, i.e.

$$\dot{\hat{\mathbf{x}}}(t) = \mathbf{f}(\hat{\mathbf{x}}, \hat{\mathbf{u}}, t) \tag{5}$$

$$\hat{\mathbf{y}}(t) = \mathbf{g}(\hat{\mathbf{x}}, \hat{\mathbf{u}}, t) \tag{6}$$

The vectors $\hat{\mathbf{u}}, \hat{\mathbf{x}}, \hat{\mathbf{y}}$ form an operating point for the system. We would like to analyze perturbations around this operating point.

Let the input vector $\hat{\mathbf{u}}$ change to $\hat{\mathbf{u}} + \delta\mathbf{u}$ where $\delta\mathbf{u}$ is a *sufficiently small* perturbation vector. Suppose that this causes the state vector to change to $\hat{\mathbf{x}} + \delta\mathbf{x}$ and the output vector to change to $\hat{\mathbf{y}} + \delta\mathbf{y}$. We assume that $\delta\mathbf{x}$ and $\delta\mathbf{y}$ are also sufficiently small. We would like to determine $\delta\mathbf{x}$ and $\delta\mathbf{y}$ in terms of $\delta\mathbf{u}$.

Note that we must have

$$\dot{\hat{\mathbf{x}}} + \delta\dot{\mathbf{x}} = \mathbf{f}(\hat{\mathbf{x}} + \delta\mathbf{x}, \hat{\mathbf{u}} + \delta\mathbf{u}, t) \tag{7}$$

$$\hat{\mathbf{y}} + \delta\mathbf{y} = \mathbf{g}(\hat{\mathbf{x}} + \delta\hat{\mathbf{x}}, \hat{\mathbf{u}} + \delta\mathbf{u}, t) \tag{8}$$

where $\delta\dot{\mathbf{x}}$ represents the perturbation in derivative of $\mathbf{x}$ w.r.t. time. Expand (7) and (8) in Taylor series around the operating point (assuming that all the partial derivatives exist) to obtain.

$$\dot{\hat{\mathbf{x}}} + \delta\dot{\mathbf{x}} = \mathbf{f}(\hat{\mathbf{x}}, \hat{\mathbf{u}}, t) + \left(\frac{\partial \mathbf{f}}{\partial \mathbf{x}}\right)_* \delta\mathbf{x} + \left(\frac{\partial \mathbf{f}}{\partial \mathbf{u}}\right)_* \delta\mathbf{u} + higher-order\ terms \tag{9}$$

$$\hat{\mathbf{y}} + \delta\mathbf{y} = \mathbf{g}(\hat{\mathbf{x}}, \hat{\mathbf{u}}, t) + \left(\frac{\partial \mathbf{g}}{\partial \mathbf{x}}\right)_* \delta\mathbf{x} + \left(\frac{\partial \mathbf{g}}{\partial \mathbf{u}}\right)_* \delta\mathbf{u} + higher-order\ terms \tag{10}$$

where subscript * indicates evaluation at the operating point. The higher-order terms involve terms such as $(\delta x_i)^k$, $(\delta u_j)^k$, $k \geq 2$ which are negligible due to the small signal assumption. Comparison of (9) with (5), and (10) with (6) then imply

$$\delta\dot{\mathbf{x}} = \left(\frac{\partial \mathbf{f}}{\partial \mathbf{x}}\right)_* \delta\mathbf{x} + \left(\frac{\partial \mathbf{f}}{\partial \mathbf{u}}\right)_* \delta\mathbf{u} \tag{11}$$

$$\delta\mathbf{y} = \left(\frac{\partial \mathbf{g}}{\partial \mathbf{x}}\right)_* \delta\mathbf{x} + \left(\frac{\partial \mathbf{g}}{\partial \mathbf{u}}\right)_* \delta\mathbf{u} \tag{12}$$

These are the state and output equations of a linear, and in general time-varying, system $(\mathbf{A}(t), \mathbf{B}(t), \mathbf{C}(t), \mathbf{D}(t))$ with input vector $\delta\mathbf{u}$, state vector $\delta\mathbf{x}$ and output vector $\delta\mathbf{y}$ where the matrices $\mathbf{A}, \mathbf{B}, \mathbf{C}$ and $\mathbf{D}$ are the following Jacobian matrices:

LINEARIZATION OF NONLINEAR SYSTEMS

$$A(t) = (\frac{\partial f}{\partial x})_*, \quad B(t) = (\frac{\partial f}{\partial u})_*, \quad C(t) = (\frac{\partial g}{\partial x})_*, \quad D(t) = (\frac{\partial g}{\partial u})_* \tag{13}$$

Equations (11) and (12) describe the behavior of the perturbations δx and δy in terms of perturbation δu about the operating point $\hat{u}, \hat{x}, \hat{y}$. Note that the state and output equations describing the perturbations may be time-varying even if the original nonlinear system is time-invariant. Also note that the fundamental assumption for linearization to be valid is that the perturbations are small. Otherwise, the higher-order terms in (9) and (10) cannot be neglected.

1. Example.
Consider the nonlinear (time-invariant) system described by the state equation

$$\begin{bmatrix} \dot{x}_1 \\ \dot{x}_2 \\ \dot{x}_3 \end{bmatrix} = \begin{bmatrix} x_2 \\ x_3 \\ -6x_2^2 - ux_3 \end{bmatrix} \tag{14}$$

For $u = 0$ and initial conditions $x_1(0)=1, x_2(0)=0$ and $x_3(0)=0$ the solution to (14) is

$$\hat{x}_1 = 1+1/t, \quad \hat{x}_2 = -1/t^2, \quad \hat{x}_3 = 2/t^3 \tag{15}$$

The state equation for the linearized model around this operating point is

$$\delta \dot{x} = A(t)\delta x + B(t)\delta u \tag{16}$$

where

$$A(t) = (\frac{\partial f}{\partial x})_* = \begin{bmatrix} 0 & 1 & 0 \\ 0 & 0 & 1 \\ 0 & -12x_2 & -u \end{bmatrix}_* = \begin{bmatrix} 0 & 1 & 0 \\ 0 & 0 & 1 \\ 0 & 12t^{-2} & 0 \end{bmatrix} \tag{17}$$

$$B(t) = (\frac{\partial f}{\partial u})_* = \begin{bmatrix} 0 \\ 0 \\ -x_3 \end{bmatrix}_* = \begin{bmatrix} 0 \\ 0 \\ -2t^{-3} \end{bmatrix} \tag{18}$$

Note that the linearized system is time-varying.

PROBLEMS

4.1. Combine equations (4.2.5) and (4.2.6) to obtain a second-order differential equation expressing $w(t)$ as a function of $e_a(t)$. (Hint: Differentiate 4.2.6 w.r.t. time and cancel $i_a(t)$ between the two equations).

4.2. Consider the lumped linear circuit shown in Fig. P4.2 where $e_a(t)$ is the input and i_2 is the output.

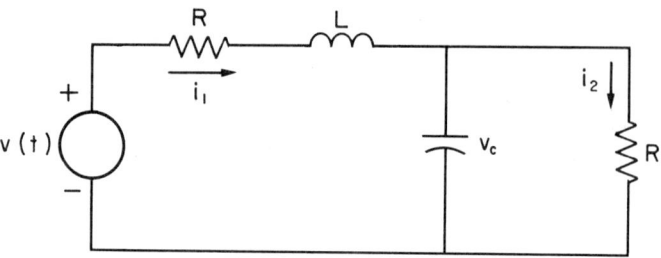

Fig. P4.2

Take i_1 and v_c as state variables and write the state and output equations for the circuit.

4.3. Show that $\mathbf{D} = \mathbf{0}$ in the output equation of a strictly proper linear system.

4.4. In Example 4.4.4 find the initial condition $x_1(0)$, $x_2(0)$ and $x_3(0)$ for both companion forms in terms of $y(0)$, $\dot{y}(0)$ and $\ddot{y}(0)$.

4.5. Consider a *l.t.i.* scalar system. Choose state variables so that the system matrix is in Jordan canonical form. (Hint: Write y(s) in partial fraction expansion form and draw the simulation diagram.)

4.6. Use "STATEQ" to determine the controllable and observable companion forms for the *l.t.i.* systems characterized by

a) $\dddot{y} - 2\ddot{y} + 5\dot{y} - 3y = 4\ddot{u} + 6\dot{u} - 8u$.

b) $y(k+4) + y(k+3) - 2y(k+1) + 5y(k) = 2u(k+3) - 6u(k+1)$.

4.7. Use "TRANFN" to determine the transfer function matrices of the following *l.t.i.* system:

a) A 4th-order scalar system

b) A 4th-order system with 3 inputs and one output.

c) A 4th-order system with one input and 3 outputs.

d) A 4th-order system with 2 inputs and 2 outputs.

4.8. Find the closed-loop transfer function of the dc machine example of Fig. 4.5.2 using Mason's gain formula.

PROBLEMS

4.9. Draw signal flow graphs for the controllable and observable companion form systems of Example 4.4.5. Hint: Take Laplace transform of individual state equations first. These graphs are sometimes called *state diagram* of the system.

4.10. Use "TRANSF" to find closed-loop transfer function of a system with open-loop transfer function $G(s) = (2x+5)/(s^4+3s^3+8s^2+5s+10)$ and $H(s) = -1$.

4.11. Consider an nth-order *l.t.i.* scalar system representation (A,B,C,D). Let

$$Q_c = [B, AB, A^2B, \ldots, A^{n-1}B]$$

If Q_c^{-1} (assuming it exists) is used as the state transformation matrix, show that the resulting state and output equations will be in the controllable companion form.

4.12. In Problem 4.11 if

$$Q_o = \begin{bmatrix} C \\ CA \\ CA^2 \\ \vdots \\ CA^{n-1} \end{bmatrix}$$

is used as the state transformation matrix, show that the resulting state and output equations will be in the observable companion form.

4.13. Use "COMPAN" to determine the state transformation matrices which transform a system representation (A,B,C,D) where

$$A = \begin{bmatrix} 0 & 4 & -2 & 0 \\ 1 & 10 & 1 & 5 \\ 2 & -1 & 8 & 4 \\ -3 & 0 & 4 & -3 \end{bmatrix}, \quad B = \begin{bmatrix} -1 \\ 0 \\ 2 \\ 5 \end{bmatrix}$$

$$C = [2\ -3\ 0\ 1],\ D = 0$$

to controllable and observable companion forms

4.14. If f is a scalar-valued function and **x** and **y** are, respectively, n- and m-dimensional vectors, show that

a) If $f(x) = x'Ay$ (a bilinear form), then $\dfrac{df}{dx} = Ay$.

b) If $f(x) = y'Ax$ (a bilinear form), then $\dfrac{df}{dx} = A'y$.

c) If $f(\mathbf{x}) = \mathbf{x}'\mathbf{A}\mathbf{x}$ (a quadratic form), then $\dfrac{df}{d\mathbf{x}} = (\mathbf{A}+\mathbf{A}')\mathbf{x}$

4.15. Consider a scalar-valued function f of an n-dimensional vector **x**. The second derivative of f w.r.t. **x**, called *Hessian matrix*, is an nxn matrix defined as

$$\mathbf{H} \triangleq \frac{d^2 f}{d\mathbf{x}^2} = \begin{bmatrix} \dfrac{\partial^2 f}{\partial x_1^2} & \dfrac{\partial^2 f}{\partial x_1 \partial x_2} & \cdots & \dfrac{\partial^2 f}{\partial x_1 \partial x_n} \\ \dfrac{\partial^2 f}{\partial x_2 \partial x_1} & \dfrac{\partial^2 f}{\partial x_2^2} & \cdots & \dfrac{\partial^2 f}{\partial x_2 \partial x_n} \\ \vdots & \vdots & & \vdots \\ \dfrac{\partial^2 f}{\partial x_n \partial x_1} & \dfrac{\partial^2 f}{\partial x_n \partial x_2} & \cdots & \dfrac{\partial^2 f}{\partial x_n^2} \end{bmatrix}$$

Find the Hessian matrix of

$$f(\mathbf{x}) = 2x_1 + x_1 x_2 + x_2 x_3^2 + x_4^3$$

4.16. Consider a nonlinear system described by

$$\ddot{y} - 2u\dot{y} + y^2 = 0.$$

Determine a state equation for this system and linearize it around the operating point obtained when $y(0) = 1$, $\dot{y}(0) = 0$ and $u = 0$.

REFERENCES

[4.1] R. Stein and W. T. Hunt, Jr., *Electric Power System Components*, Van Nostrand Reinhold, 1979.

[4.2] Y. Chu, *Digital Simulation of Continuous Systems*, McGraw-Hill, New York, 1969.

[4.3] F. H. Speckhart and W. L. Green, "A Guide to Using CSMP - the Continuous Modeling Program," Prentice Hall, Englewood Cliffs, N.J., 1976.

[4.4] W. M. Syn and M. H. Dost, *DSL/dp Digital Simulation Language Double Precision*, IBM Corporation, Gen. Prod. Div., San Jose, CA. Sept., 1980.

[4.5] S. J. Mason, Feedback Theory: Further Properties of Signal Flow Graphs, Proc. IEE, Vol 44, No. 7, July 1956, pp 920-926.

[4.6] C. T. Chen, *Introduction to Linear System Theory,* Holt, Rinehart and Winston, New York, 1970.

[4.7] E. S. Kuh and C. A. Desoer, *Basic Circuit Theory,* McGraw-Hill, New York, 1970.

[4.8] D. O. Pederson, J. J. Studer and R. J. Whinnery *Introduction to Electronics Systems, Circuits and Devices,* McGraw-Hill, New York, 1966.

CHAPTER 5

MODEL REDUCTION OF LARGE-SCALE LINEAR SYSTEMS

5.1 INTRODUCTION

In the previous chapter modeling of linear systems was considered. In any modeling task, two often conflicting factors prevail - "simplicity" and "accuracy". These factors lead to two desirable attributes for a system model - "reduced computation" and "simplified structure." These properties are of more concern for systems with high dimensions and composite interactions among different subsystems, referred to generally as *Large-Scale Systems*. The exact definition of "large-scale" has not been universally accepted. A system is sometimes said to be of large scale if the conventional techniques of system theory and control cannot be successfully applied to it. Another definition has been that a system is large in scale which can be decoupled into a number of subsystems or can be collectively controlled by many controllers. These notions lead to two popular large-scale systems control schemes - "hierarchical" and "decentralized" [5.1].

The realistic models of such systems are so high in dimension that a direct simulation or design would be neither computationally desirable nor physically possible in many cases. A multiarea large-scale power system, for example, is a very high-dimensional system which is composed of several subsystems (plants) connected by tie lines. It goes without saying that the reduction of system models is highly desirable.

There have been two primary schemes for modeling large-scale systems. These two schemes, which have been around for some time, are "perturbation" and "aggregation". They have been carried on from the economics and mathematics, respectively, to system and control theories. A perturbed model is based on ignoring certain interactions of dynamic or structural nature in the system. The basic benefits from this model are computational and structural realizations. However, these benefits cannot be at the expense of key system properties such as stability.

The other scheme for large-scale systems modeling, namely aggregation, refers to the case where the system is described by a "coarser" set of state variables than it would be otherwise. Here again the underlying factor in aggregating a system is to be able to retain the key qualitative properties such as stability. Siljak [5.2] has viewed aggregation as a natural process in the sense that the system stability which is described by several state variables is entirely represented by a single variable - the Lyapunov function. Aggregation has been applied to both time- and frequency-domain models. These will be both considered in this chapter. Many aggregation methods and also most applications of perturbation methods have been proposed for linear stationary systems with state-space representation, i.e.

$$\dot{x}(t) = Ax(t) + Bu(t), x(0) = x_o \tag{1}$$

$$y(t) = Px(t) \tag{2}$$

where x,u and y are n-, r- an m- dimensional state, control and output vectors, respectively; A,B and P are constant matrices of appropriate dimensions and x_o is the initial state vector.

The model reduction of a large-scale linear system can also be achieved if the system is described in frequency domain. The system (1)-(2) is first represented in the s-domain by an mxr dimensional transfer function matrix,

$$H(s) = P(sI-A)^{-1}B \tag{3}$$

whose elements $h_{ij}(s)$, i=1,..., m; j=1,...,r are rational functions of s. Transfer function matrix H(s) can be expressed as

$$H(s) = (D_o + D_1 s + \ldots + D_{n-1} s^{n-1})/(e_o + e_1 s + \ldots + e_n s^n) \tag{4}$$

where D_i, i=0,1,..., n-1 are constant mxr matrices, and the denominator polynomial is det(sI-A).

Based on the domain in which a large-scale system model is represented, the model reduction methods are grouped into two categories: *time domain* and *frequency domain*. The time domain methods are commonly associated with state-space models (1)-(2) while the frequency-domain methods are associated with transfer function matrix representation (3)-(4). The frequency-domain methods are, in turn, divided into two groups: SISO systems and MIMO systems. By far, the greatest effort in model reduction techniques has been for SISO systems in frequency domain, represented by the following transfer function,

$$G(s) = (d_o + d_1 s + d_2 s^2 + \ldots + d_{n-1} s^{n-1})/(e_o + e_1 s + e_2 s^2 + \ldots + e_n s^n) \tag{5}$$

The various methods of linear large-scale systems model reduction are summarized in Fig. 1. As shown, each of the techniques may consist of different versions or modifications. Clearly, the treatment of these various methods is not within the scope of this text. In this chapter perturbation and a select number of aggregation schemes will be treated. For a more detailed treatment of the subject, the reader may consult other works [5.1, 5.3].

INTRODUCTION

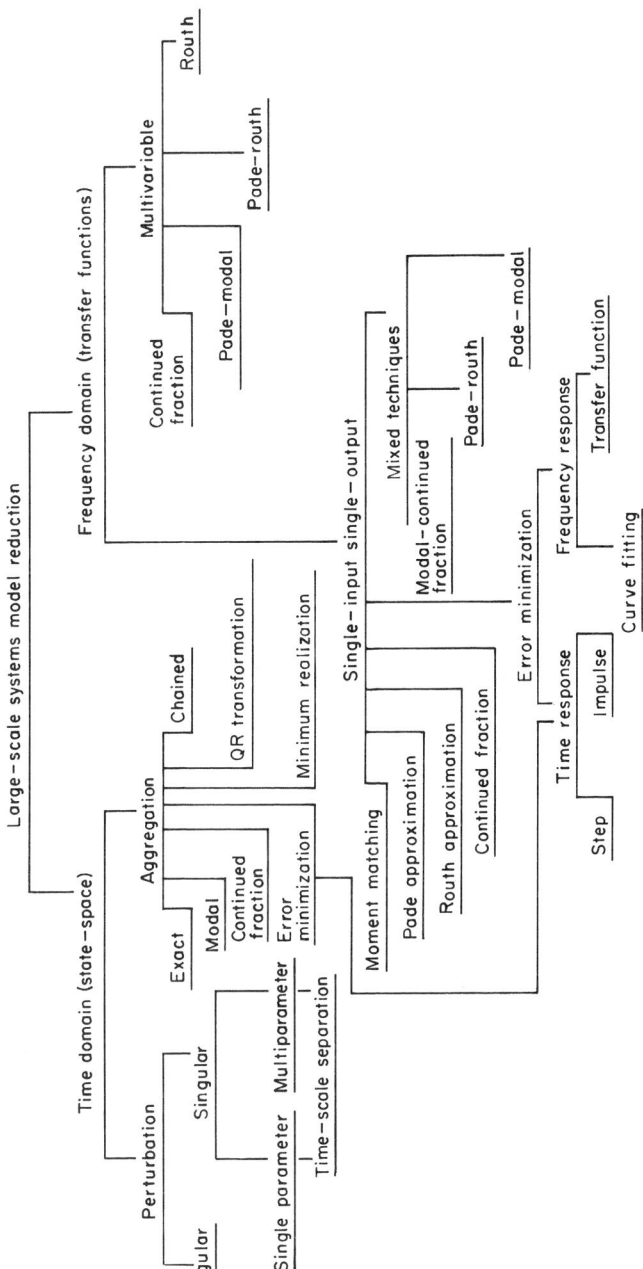

Fig. 5.1.1. A listing of large-scale systems model reduction techniques

5.2 PERTURBATION METHODS

As noted before, perturbation method is based on the approximation of system's structure through neglecting certain interactions within the model which lead to lower order. There are two basic classes of perturbations for large-scale systems modeling purposes: regularly perturbed or weakly-coupled models, and singularly perturbed or strongly-coupled models.

5.2.1 Regular Perturbation

In many industrial control systems some interactions are ignored in order to reduce computational burdens for system analysis, design or both. Examples of such practice are in chemical process control and space guidance [5.4] where different subsystems are designed for flow, pressure, and temperature control in an otherwise coupled process, or the design for each of the three axes of a 3-axis attitude control system. However, the computational advantages obtained by neglecting weakly coupled subsystems are often offset by a loss in the overall system performance. In this section the regularly perturbed models for large-scale system are considered.

Consider the following large-scale system which is split into k linear coupled subsystems,

$$
\begin{bmatrix} \dot{x}_1 \\ \dot{x}_2 \\ \cdot \\ \cdot \\ \cdot \\ \dot{x}_k \end{bmatrix} = \begin{bmatrix} A_1 & \epsilon A_{12} & \cdot & \cdot & \epsilon A_{1k} \\ \epsilon A_{21} & A_2 & \epsilon A_{23} & \cdots & \epsilon A_{2k} \\ \cdot & & & & \cdot \\ \cdot & & & & \epsilon A_{k-1,k} \\ \epsilon A_{k1} & \cdots & & \epsilon A_{k,k-1} & A_k \end{bmatrix} \begin{bmatrix} x_1 \\ x_2 \\ \cdot \\ \cdot \\ \cdot \\ x_k \end{bmatrix} \quad (1)
$$

$$
+ \begin{bmatrix} B_1 & \epsilon B_{12} & & \\ \epsilon B_{21} & \cdot & & \\ \cdot & \cdot & & \\ \cdot & \cdot & & \\ \cdot & B_k & \end{bmatrix} \begin{bmatrix} u_1 \\ u_2 \\ \cdot \\ \cdot \\ u_k \end{bmatrix}
$$

where ϵ is a small positive coupling parameter, x_i and u_i are the ith subsystem state and control vectors, respectively and all matrices are assumed to be constant. Note that all matrices A_i and B_i, i=1,2,...,k are diagonal and ϵ appears in front of all off-diagonal matrices. A special case of (1) is when k=2 which is called ϵ coupled system:

$$
\begin{bmatrix} \dot{x}_1 \\ \dot{x}_2 \end{bmatrix} = \begin{bmatrix} A_1 & \epsilon A_{12} \\ \epsilon A_{21} & A_2 \end{bmatrix} \begin{bmatrix} x_1 \\ x_2 \end{bmatrix} + \begin{bmatrix} B_1 & \epsilon B_{12} \\ \epsilon B_{21} & B_2 \end{bmatrix} \begin{bmatrix} u_1 \\ u_2 \end{bmatrix}. \quad (2)
$$

It is clear that when $\epsilon = 0$, the above system decouples into two subsystems:

PERTURBATION METHODS 165

$$\dot{\hat{x}}_1 = A_1\hat{x}_1 + B_1\hat{u}_1 \tag{3}$$

$$\dot{\hat{x}}_2 = A_2\hat{x}_2 + B_2\hat{u}_2$$

which correspond to two approximate (aggregated) models one for each subsystem. In this way, the computation associated with simulation, design, etc. will be reduced drastically, especially for large system order n and for more than two subsystems.

Consider a coupled A matrix in (2) and assume that A_1, A_{12}, A_{21} and A_2 are $n_1 \times n_1$, $n_1 \times n_2$, $n_2 \times n_1$ and $n_2 \times n_2$, respectively with $n = n_1 + n_2$ being the order of the original large-scale coupled system. Furthermore, let

$$\lambda_i\{A_1\} = \lambda_i, \ i=,2,\ldots,n_1; \lambda_j\{A_2\} = \lambda_j, \ j=n_1+1,\ldots,n; \ \lambda_k\{A\} = \lambda_k, \ k=1,2,\ldots,n \tag{4}$$

be the eigenvalues of A_1, A_2 and A, respectively. Assume that the eigenvalues of A_1 and A_2 are widely separated from each other. Without any loss of generality one can take $|\lambda_j A_2| \ll |\lambda_i A_1|$. Let the eigenvalues of A_1 be on or outside a circle with radius $R = \min |\lambda_i A_1|$, $i=1,2,\ldots,n_1$ an the eigenvalues of A_2 be on or inside a circle with radius $r = \max|\lambda_j A_2|$, $j=n_1+1, n_1+2,\ldots,n$ as show in Fig. 1. If the following conditions are satisfied, then the system is said to be *weakly coupled* [5.5, 5.6]:

$$(i) \quad (r/R) \ll 1 \tag{5}$$

$$(ii) \quad (n_1\epsilon_{12}\epsilon_{21})/R^2 \ll 1 \tag{6}$$

where $\epsilon_{12} = \max |(A_{12})_{i,j}|$ and $\epsilon_{21} = \max |(A_{21})_{k,d}|$ for $i,j = 1,2,\ldots, n_1$ and $k, d = 1,2,\ldots, n_2$. The term (r/R) is called the *separation ratio* while ϵ_{12} and ϵ_{21} represent the maximum of the moduli of the elements of A_{12} and A_{21} submatrices, respectively. Relations (5)-(6) represent conditions which approximate the roots of the following characteristic polynomial,

$$\det(\lambda I_n - A) = 0 \tag{7}$$

by the roots of the following two polynomials,

$$\det(\lambda I_{n_1} - A_1) = 0, \ \det(\lambda I_{n_2} - A_2) = 0 \tag{8}$$

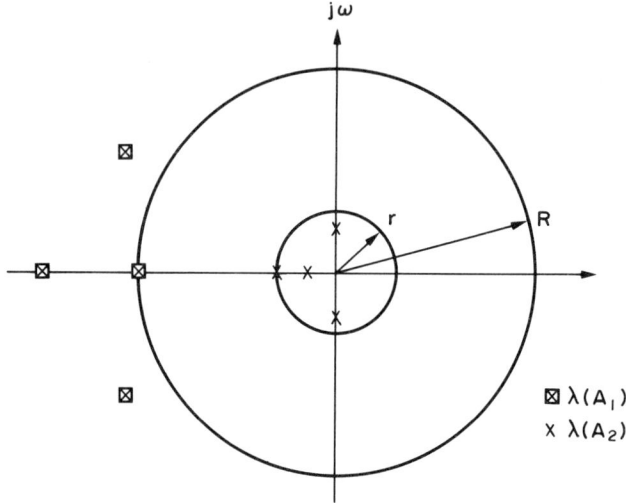

Fig. 5.2.1. Relative locations of the eigenvalues of A_1 and A_2 submatrices in weakly coupled system.

The following example illustrates a weakly-coupled system.

1. Example. Consider a 6th-order matrix with the following structure:

$$A = \begin{bmatrix} 0 & 1 & 0 & 0 & 0 & 0 \\ 0 & 0 & 1 & 0.2 & 0.05 & 0 \\ -400 & -170 & -2.3 & 0.05 & -0.02 & 0 \\ 0 & -0.01 & -0.02 & 0 & 1 & 0 \\ 0 & 0.01 & -0.015 & 0 & 0 & 1 \\ 0 & 0 & 0 & -0.04 & -0.53 & -1 \end{bmatrix}. \quad (9)$$

The eigenvalues of A are -9.94, -8.08, -4.97, -0.9816, -0.7924 and -0.52 indicating that the first three are much farther away from the $j\omega$-axis than the last three. For a two 3×3 partitioning, it can be seen that submatrices A_1 and A_2 are both in companion forms with the following eigenvalues:

$$\lambda_i\{A_1\} = \{-5, -10, -8\}; \quad \lambda_j\{A_2\} = \{-0.1, -0.8, -0.5\} \quad (10)$$

implying that $r=0.8$, $R=5$ and $(r/R)=0.16$ which is much smaller than 1; hence condition (5) holds. The values of ϵ_{12} and ϵ_{21} are 0.2 and 0.02, respectively and

$$(n_1\epsilon_{12}\epsilon_{21})/R^2 = 0.00048 \ll 1.$$

Therefore, it is seen that system matrix (9) is weakly coupled. The remainder of this section is devoted to singularly-perturbed systems.

5.2.2 Singular Peturbation

Singularly perturbed or strongly-couple systems are those whose variables have widely distinct rates of change with time. Singular perturbation differs from regular perturbation in that the perturbation is to the left-hand side of the system's state equation, i.e. a small parameter is multiplied by the time derivative of the state vector. In practice many systems, most of them large in dimension, posses fast changing variables displaying a singularly perturbed characteristic. A notable example is a power system in which the frequency and voltage transients vary from a few seconds in generator regulators, shaft stored energy and speed governor motion, to several minutes in prime mover motion, stored thermal energy and load voltage regulators [5.7]. In fact some of the *order reduction* techniques which are discussed in Sections 5.2 and 5.3 can be explained in terms of singular perturbation [5.8].

Consider a linear singularly perturbed system

$$\dot{x}(t) = A_1 x(t) + A_{12} z(t) + B_1 u(t), \; x(t_o) = x_o \tag{11}$$

$$\epsilon \dot{z}(t) = A_{21} x(t) + A_2 z(t) + B_2 u(t), \; z(t_o) = z_o \tag{12}$$

where $\epsilon > 0$ is a small scalar parameter, x and z are k- and p- dimensional vectors, respectively and u is the mx1 control vector. When $\epsilon = 0$, the n=k+p dimensional system (11)-(12) is reduced to a k-dimensional system since the differential equation (12) reduces to an algebraic equation. If A_2 is nonsingular, (11) and (12) become

$$\dot{\hat{x}}(t) = (A_1 - A_{12} A_2^{-1} A_{21}) \hat{x} + (B_1 - A_{12} A_2 B_2) \hat{u} \tag{13}$$

$$\hat{z}(t) = -A_2^{-1} A_{21} \hat{x} - A_2^{-1} B_2 \hat{u}. \tag{14}$$

Equation (13) is an approximate aggregated model for (11)-(12) which in effect means that the n eigenvalues of the original system are approximated by the k eigenvalues of $A_1 - A_{12} A_2^{-1} A_{21}$ in (13). This observation follows the same line of argument when discussing conditions for weakly-couple systems considered my Milne [5.5] and Aoki [5.6, 5.9], Bailey and Ramapriyan [5.10].

5.2.2.1 Boundary Layer Correction

It is noted that in singular perturbation going from (11)-(12) to (13) the initial condition z_o of $z(t)$ is lost and the values of $\hat{z}(t_o)$ and $z(t_o) = z_o$ are, in general, different; the difference is termed a leftside *boundary layer* which corresponds to the fast transients of (11)-(12). To investigate this phenomenon which in effect explains under what condition $\hat{x}$ and $\hat{z}$ approximate x and z, let u be zero in (11)-(12) and let

$$\eta(t) = z(t) - \hat{z}(t) = x(t) + A_2^{-1} A_{21} \hat{x}(t) \tag{15}$$

denote the error between z and $\hat{z}$. Also choose a matrix $E_1(\epsilon)$ so that when

$$\eta(t) = \mathbf{x}(t) + \mathbf{A}_2^{-1}\mathbf{A}_{21}\mathbf{x}(t) + \epsilon \mathbf{E}_1(\epsilon)\mathbf{x}(t) \tag{16}$$

is substituted in (11) and (12) with $\mathbf{u}=0$, η will be separated from $\mathbf{x}$ as

$$\dot{\mathbf{x}}(t) = (\mathbf{A}_1 - \mathbf{A}_{12}\mathbf{A}_2^{-1}\mathbf{A}_{21} + \epsilon \mathbf{E}_2)\mathbf{x}(t) + \mathbf{A}_{12}\eta(t) \tag{17}$$

$$\epsilon \dot{\eta}(t) = (\mathbf{A}_2 + \epsilon \mathbf{E}_3)\eta(t). \tag{18}$$

It can be shown that there exists an ϵ^* such that $\mathbf{E}_i = \mathbf{E}_i(\epsilon)$, i=1,2,3 are bounded over [0, ϵ^*] [5.8]. As $\epsilon \to 0$, the eigenvalues of (18) would tend to infinity as $\lambda\{\mathbf{A}_2/\epsilon\}$ would. Now a new time variable τ called *stretched time-scale*, is defined:

$$\tau = (t - t_o)/\epsilon \tag{19}$$

where $\tau = 0$ at $t = t_o$ and $dt = \epsilon d\tau$. For a change from t to τ, system (18) becomes,

$$d\eta(\tau)/d\tau = (\mathbf{A}_2 + \epsilon \mathbf{E}_3)\eta(\tau) \tag{20}$$

which continuously depends on ϵ and at $\epsilon = 0$ it becomes

$$d\eta(\tau)/d\tau = \mathbf{A}_2 \eta(\tau), \eta(0) = \hat{\mathbf{z}}(t_o) - \mathbf{z}(t_o). \tag{21}$$

Equation (21) constitutes the so-called boundary layer correction for $\mathbf{x}(t) = \hat{\mathbf{x}}(t) + \eta((t-t_o)/\epsilon)$. From the above formulation it can be shown that the *slow* and *fast* states $\mathbf{x}(t)$ and $\mathbf{z}(t)$ are,

$$\mathbf{x}(t) = \hat{\mathbf{x}}(t) + 0(\epsilon) \tag{22}$$

$$\mathbf{z}(t) = \hat{\mathbf{z}}(t) + \eta(\tau) + 0(\epsilon) \tag{23}$$

where $0(\epsilon)$ is the "large-0"-order of ϵ and whose norm is less than $d\epsilon$ with d being a constant. It is noted that the boundary layer correction is only significant for the first few seconds away from t_o and reduces to zero after some time $t = t_1$ as an exponential decay in $\tau = (t - t_o)/\epsilon$.

5.2.2.2 Time-Scale Separation

Systems which posses multi-time scales often possess distinct clusters of eigenvalues. It is useful to show that linked with this system property there is a distinct possibility of decoupling the system into subsystems: *slow* and *fast*. One usual scheme to show this decoupling characteristic has been decomposition. Here an iterative scheme based on a successive weakening of the coupling between slow and fast subsystems will be presented. The method has been developed by Kokotovic et.al. [5.11].

Consider a unforced linear singularly perturbed system

$$\dot{x} = Ax + Ez, \quad x(t_o) = x_o \tag{24}$$

$$\epsilon \dot{z} = Cx + Fz, \quad z(t_o) = z_o \tag{25}$$

where matrix F is assumed to be nonsingular. It was deduced already that when $\epsilon = 0$, corresponding to a *quasi-steady-state*, (qss) and actual x and z differ from qss values x and z mainly in their fast portions of the response. We will follow this notion in accordance with [5.7] to develop an iterative procedure for the separation of time scales. Let

$$\eta_1(t) = x - \hat{x} = x + F^{-1}Cx \tag{26}$$

which transforms (24)-(25) into the following equations, similar to (17)-(18):

$$\dot{x} = A_1 x + E\eta_1 \tag{27a}$$

$$\epsilon \dot{\eta}_1 = C_1 x + F_1 \eta_1 \tag{27b}$$

where

$$A_1 \triangleq A - EF^{-1}C, \quad C_1 \triangleq \epsilon F^{-1}CA_1 \triangleq \epsilon B_1, \quad F_1 \triangleq F + \epsilon F^{-1}CE. \tag{28}$$

The Equations (27)-(28) are in a linear singularly perturbed form as in the unforced case of (11)-(12). However, the important difference is that in the latter case, as evident from (28), slow state x has weaker presence in (27b).

In a similar fashion, after ith step (27)-(28) become

$$\dot{x} = A_i x + E\eta_i \tag{29}$$

$$\epsilon \dot{\eta}_i = C_i x + F_i \eta_i \tag{30}$$

where, similar to (28),

$$A_i \triangleq A_{i-1} - EF_{i-1}^{-1}C_{i-1}, \quad A_o = A \tag{31a}$$

$$C_i \triangleq \epsilon F_{i-1}^{-1}C_{i-1}A_i \triangleq \epsilon^i B_i, \quad C_o = C \tag{31b}$$

$$F_i = F_{i-1} + \epsilon F_{i-1}^{-1}C_{i-1}E, \quad F_o = F. \tag{31c}$$

where C_i is reduced to an order $0(\epsilon^i)$. A combination of (26) similar terms for $\eta_2, \ldots,$ and

$$\eta_i = \eta_{i-1} + F_{i-1}^{-1}C_{i-1}x \tag{32}$$

reveals that

$$\sum_{k=1}^{i}(\eta_k - \eta_{k-1}) = \eta_i - x = \sum_{k=1}^{i}(F_{k-1}^{-1}C_{k-1})x \tag{33}$$

indicating that the slow state x remains the same while the new fast state η_i has identical meaning with x. As the iteration i approaches infinity, we will have $A_\infty = A - EF^{-1}C + 0(\epsilon)$ and $F_\infty = F + 0(\epsilon)$. It is noted that even after the ith iteration the fast state η_i still influences x as shown by (29). Solving for η_i and substitution in (29) yields

$$\dot{x} - \epsilon EF_i^{-1}\dot{\eta}_i = (A_i - EF_i^{-1}C_i)x \triangleq A_{i+1}x \tag{34}$$

which suggests that

$$\zeta_1 = x - \epsilon EF_i^{-1}\eta_i \tag{35}$$

is the slow part of x. Following this observation, the slow subsystem (29) becomes

$$\dot{\zeta}_1 = A_{i1}\zeta_1 + \epsilon A_{i1}EF_i^{-1}\eta_i \triangleq A_{i1}\zeta_1 + E_{i1}\eta_i. \tag{36}$$

Note that E_{i1} is $0(\epsilon)$, indicating that the influence of the fast state η_i in the slow system has been reduced. In general

$$\dot{\zeta}_j = A_{ij}\zeta_j + E_{ij}\eta_i, \quad \zeta_j(0) = \zeta_j^o \tag{37}$$

$$\epsilon\dot{\eta}_i = C_i \dot{\zeta}_j + F_{ij}\eta_i, \quad \eta_i(0) = \eta_i^o \tag{38}$$

where

$$A_{ij+1} = A_{ij} - E_{ij}F_{ij}^{-1}C_i, \quad A_{io} = A_i \tag{39a}$$

$$E_{ij+1} = \epsilon A_{ij+1}E_{ij}F_{ij}^{-1}, \quad E_{io} = E \tag{39b}$$

$$F_{ij+1} = F_{ij} + \epsilon C_i E_{ij}F_{ij}^{-1}, \quad F_{io} = F_i. \tag{39c}$$

This completes the iterative separation of slow and fast modes since in slow subsystem (37) its interaction matrix E_{ij+1} is $0(\epsilon^j)$ while in fast subsystem (38) its interaction matrix C_i is $0(\epsilon^i)$. Using previous discussions on weakly-couple systems conditions and the above development, the following steps summarizes the separation of time-scales.

1. Algorithm

1. Set $i=j=0$ and start with $A_i=A$, $C_i=C$, E and $F_i=F$ in (24)-(25)

2. Evaluate A_{i+1}, C_{i+1} and F_{i+1} from (31). Set $i=i+1$;

3. Use A_{ij}, E_{ij}, F_{ij} in (39) to compute A_{ij+1}, E_{ij+1} and F_{ij+1}, Set $j=j+1$

4. Check for conditions for weakly-coupled systems outline by (5)-(6),

$$(a) f_1 = (r/R) \ll 1, \quad (b) f_2 = n_f \delta_{sf}\delta_{fs}/R^2 \ll 1 \tag{40}$$

where

$$r = \max_k |\lambda_k\{A_{ij}\}| \quad \delta_{sf} = \max_{d,k} |\{E_{ij}\}_{d,k}| \tag{41}$$

$$R = \min_k |\lambda_k\{F_{ij}\}| \quad \delta_{fs} = \max_{d,k} |\{C_i/\epsilon\}_{d,k}|.$$

The terms δ_{sf} and δ_{fs} are the maximum moduli of elements in the interaction matrices between slow-fast and fast-slow subsystems, respectively and n_f is the order of the fast subsystem.

The above algorithm was programmed in extended BASIC and called "SEPTIM" whose source listing is shown below, followed by a sample run for a 7th-order system.

2. "SEPTIM"

```
10      !           PROGRAM NAME : "SEPTIM"   PROG
20      OPTION BASE 1
30      PRINT " THIS PROGRAM PERFORMS THE ITERATIVE SEParation
        of TIMe-scale"
40      DIM A(8,8),E(8,9),C(9,8),F(9,9),Finv(9,9),Fc(9,8),
        Efc(8,8)
50      DIM Ca(9,8),Fca(9,8),Ce(9,9),Fce(9,9),A10(8,8),
        E10(9,8),F10(9,9)
60      DIM Ef(8,9),Cef(9,9)
70      INPUT "Order of 1st & 2nd subsystems n1 & n2 ?",N1,N2
80      PRINT "Order of 1st subsystem n1 =";N1;" Order of 2nd
        subsytem N2=";N2
90      INPUT "Number of time-scale separation iterations ?",Iter
100     PRINT "Number of time-scale separation iterations =";Iter
110     REDIM A(N1,N1),E(N1,N2),C(N2,N1),F(N2,N2),Finv(N2,N2),
        Fc(N2,N1)
120     REDIM Efc(N1,N1),Ca(N2,N1),Fca(N2,N1),Ce(N2,N2),
        Fce(N2,N2),A10(N1,N1)
130     REDIM E10(N1,N2),F10(N2,N2),Ef(N1,N2),Cef(N2,N2)
140     INPUT "Have You LINKED SUB <<Mat>>?",S$
150     IF (S$="Y") OR (S$="y") THEN 190
160     IF (S$="N") OR (S$="n") THEN 180
170     GOTO 140
180     LINK "Mat",1000,190
190     CALL Mat(A(*),N1,N1,"A")
200     CALL Mat(E(*),N1,N2,"E")
210     CALL Mat(C(*),N2,N1,"C")
220     CALL Mat(F(*),N2,N2,"F")
230     ! Iterations i begins here
240     FOR I=1 TO Iter
250     MAT Finv=INV(F)
260     MAT Fc=Finv*C
270     MAT Efc=E*Fc
280     MAT A=A-Efc
290     MAT Ca=C*A
300     MAT Fca=Finv*Ca
310     MAT Ce=C*E
320     MAT Fce=Finv*Ce
330     MAT F=F+Fce
340     MAT C=Fca
350     ! ITERATION j BEGINS HERE
360     IF I<>1 THEN 400
370     MAT A10=A
380     MAT E10=E
390     MAT F10=F
400     MAT Finv=INV(F10)
410     MAT Fc=Finv*C
420     MAT Efc=E10*Fc
```

```
430    MAT A10=A10-Efc
440    MAT Ef=E10*Finv
450    MAT E10=A10*Ef
460    MAT Cef=C*Ef
470    MAT F10=F10+Cef
480    NEXT I
490    PRINT "At iteration i=j=";I-1;"Aij:"
500    MAT PRINT A10
510    PRINT "At iteration i=j=";I-1;"Eij:"
520    MAT PRINT E10
530    PRINT "At iteration i=j=";I-1;"Cij:"
540    MAT PRINT C
550    PRINT "At iteration i=j=";I-1;"Fij:"
560    MAT PRINT F10
570    END
```

THIS PROGRAM PERFORMS THE ITERATIVE SEParation of TIMe-scale
Order of 1st subsystem n1 = 2   Order of 2nd subsytem N2= 5
Number of time-scale separation iterations = 3
Matrix (2 x 2):

-5.8000E-01	0.0000E+00
0.0000E+00	-1.0000E+00

Matrix (2 x 5):

0.0000E+00	-2.6900E-01	0.0000E+00	2.0000E-01	0.0000E+00
0.0000E+00	0.0000E+00	0.0000E+00	1.0000E+00	0.0000E+00

Matrix C(5 x 2):

0.0000E+00	0.0000E+00
0.0000E+00	0.0000E+00
-1.4100E-01	0.0000E+00
0.0000E+00	0.0000E+00
-1.7300E+02	6.6700E+01

Matrix F(5 x 5):

-5.0000E+00	2.1200E+00	0.0000E+00	0.0000E+00	0.0000E+00
0.0000E+00	0.0000E+00	3.7700E+02	0.0000E+00	0.0000E+00
1.4100E-01	-2.0000E-01	-2.8000E-01	0.0000E+00	0.0000E+00
0.0000E+00	0.0000E+00	0.0000E+00	8.3800E-02	2.0000E+00
-1.1600E+02	4.0900E+01	0.0000E+00	-6.6700E+01	-1.6700E+01

At iteration i=j= 3 Aij:
-.75412000819]   .20241889 32 25

-2.31480388696   3.56241442057E-02

At iteration i=j= 3 Eij:
3.37570265248E-03   -2.77565959093E-04   2.66952701053E-02   2.97446676845E-05
-2.43909802021E-06

1.04501195604E-02   -6.42235191231E-04   .103862464336   9.74220818910E-05
-1.46338545938E-05

At iteration i=j= 3 Cij:
6.84027965182E-03   3.67480223982E-04

1.44933970933E-03   4.60871100590E-05

-5.66455427621E-05 -2.23862135111E-06

-9.09365908051E-03 -3.08013688201E-03

-3.25103410984E-02   1.83830537062E-02

At iteration i=j= 3 Fij:
-5.01002310268   2.00467252904   -.611447639342   8.30506125378E-02
-2.72943926652E-04

-2.61532900775E-02 -.271724634453   375.535769831   .19594978261
-6.23915817390E-04

.140966034145   -.199996330028   -.280300189128   9.98007041151E-07
2.63867278672E-07

.231653201b2   -.670227453174   1.56417152069   -.495900374726
1.99186642442

-116.007379597   40.9299440046   -6.65381633889E-02   -66.6760870292
-16.699755835

3. Example. In the above sample example for "SEPTIM", the system in its original form had coupling factors of $f_1 = 0.2686$ and $f_2 = 62.4$ indicating that the slow and fast subsystems are highly coupled. The iterative time-scale separation of "SEPTIM" reduced f_1 to 0.16825 and f_2 to 0.00108 as shown in Fig. 2. The resulting slow and fast subsystems are:

$$\text{Slow: } \dot{\zeta}_3 = \begin{bmatrix} -0.754 & 0.202 \\ -2.315 & 0.035 \end{bmatrix} \zeta_3 \qquad (42)$$

$$\text{Fast: } \epsilon\dot{\eta}_3 = \begin{bmatrix} -5.01 & 2.005 & -0.611 & 0.083 & 0 \\ 0.026 & -0.272 & 375.53 & 0.196 & 0 \\ 0.141 & -0.199 & -0.28 & 0 & 0 \\ 0.231 & -0.67 & 1.564 & 0.496 & 1.992 \\ -116 & 40.93 & 0.066 & -66.67 & -16.69 \end{bmatrix} \eta_3. \qquad (43)$$

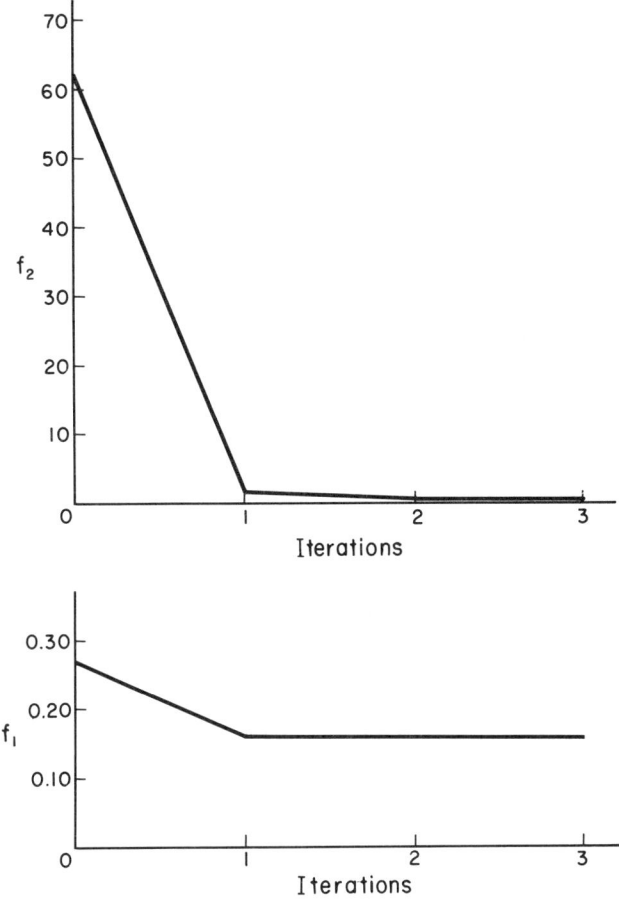

Fig. 5.2.2. Coupling factors f_i, i=1,2 versus iteration number of time-scale of separation

Note that the iterative time-scales separation has been applied for **A,E,C** and **F**. This implies that the actual value of parameter ϵ does not have to be explicitly specified for the method to converge.

5.3 AGGREGATION METHODS–TIME DOMAIN

The next two sections are concerned with reduction of large-scale systems by aggregation. The intuitive notion behind an aggregated model is to combine certain system variables which, in effect, involves weighted averaging of the state vectors to find an approximate model for a large-scale system [5.6, 5.9, 5.12]. Consider the following *l.t.i.* system,

$$\dot{\mathbf{x}}(t) = \mathbf{A}\mathbf{x}(t) + \mathbf{B}\mathbf{u}(t), \quad \mathbf{x}(0) = \mathbf{x}_o \tag{1}$$

and assume that its order n is large. Moreover, let an aggregate *l.t.i.* model of the same system be

$$\dot{z}(t) = Fz(t) + Gu(t), \quad z(0) = z_o \tag{2}$$

where $z(t)$ is the k-dimensional (k<n) aggregated state vector, F and G are $k \times k$ and $k \times m$ constant matrices, respectively. The aggregation problem is to find an aggregated model (2) which could adequately represent the system(1). The adequacy of the aggregated model depends on the precise objective of aggregation be it response characteristics, optimality, stability, etc. In sequel, two aggregation methods will be discussed.

5.3.1 Exact Aggregation

Assume that x is observed directly and (5.3.1) is controllable. The so-called "dynamic exactness" or "aggregatability" condition is

$$z(t) = Cx(t), \quad z_o = Cx_o \tag{3}$$

where C is a $k \times n$ constant matrix, known as "aggregation matrix", whose rank is assumed to be k. A closer look at (1)-(3) indicates that dynamic exactness is

$$FC = CA, \quad G = CB \tag{4}$$

using the solvability condition of Penrose [5.12] for matrix C, (4) implies that

$$F = CAC'(CC')^{-1}. \tag{5}$$

Thus once the aggregation matrix C is known, the aggregated pair (F,G) can be obtained from (4) and (5). Aoki [5.6] has shown that (4) leads to the aggregated state $z=Cx$ which is a linear combination of certain modes of x. The eigenvalues of F are contained in the set of all the eigenvalues of A, corresponding to the modes of unaggregated state vector x retained in the aggregation state vector z [5.13].

It is noted that the concept of linear system aggregation is in fact a generalization of the "dominant poles" assumption, a common procedure for classical linear control system design. It should be indicated that preservation of dynamic exactness through (4) in an effort to find matrix C requires knowledge of all eigenvalues of A making it impractical for large systems. An alternative approach which does not require the eigenvalues of A has also been proposed by Aoki [5.12]. Considering the controllability matrices of (1)-(2).*

$$W_A \triangleq [B, AB, \ldots, A^{n-1}B] \tag{6}$$

---

* For a discussion on the controllability of linear systems the reader may consult Chapter 8.

and

$$W_F \triangleq [G, FG, \ldots, F^{n-1}G] \qquad (7)$$

it can be seen from (4) that $W_F = CW_A$. Thus, C can be obtained by the following pseudo-inverse

$$C = W_F W_A^+ = W_F W_A^T (W_A W_A^T)^{-1}. \qquad (8)$$

Therefore, if F is specified, e.g. $F = \text{diag}(\lambda_1, \lambda_2, \ldots, \lambda_k)$ and choosing G to make the pair (F,G) controllable, i.e. rank $W_F = k$, C is obtained by (8).

Thus far, the discussion on the aggregation method has insisted on the dynamic exactness, i.e. when $x(t)$ corresponds to an exact representation of the response of certain linear combinations of $x_j(t)$, $j=1,2,\ldots k$ need not correspond to physical variables in general. Thus, one can consider $z(t)$ as an approximation to actual physical variables, i.e. the aggregated state vector $x(t)$ must nearly satisfy $z \approx Lx$ in addition to (2) where L is a $k \times n$ matrix chosen to pick those components of x which are to be approximate. The nature of the approximation (3) will, of course, be influenced by the choice of the aggregation matrix C. Although the choice of C for physical exactness is very restricted [5.13], a few model aggregation techniques have been proposed by a number of authors which can be used to find an approximate aggregation matrix C. It must be noted that the dynamic exactness or perfect aggregation FC=CA of condition (4) is not in general satisfied. In fact, as discussed by Jamshidi [5.11], it is a minimization of the square of the norm FC=CA unless a consistency condition $CAC^+C = CA$ is satisfied. Here C^+ is the *generalized inverse* of C.

The exact aggregation has been programmed in BASIC and called, "EXAGR" whose source program is given below, followed by a sample example.

4. "EXAGR"

```
10    !       PROGRAM NAME : "EXAGR" PROB
20    PRINT "THIS PROGRAM PERFORMS SYSTEM AGGREGATION VIA
      EXACT METHOD"
30    OPTION BASE 1
40    DIM A(9,9),B(9,2),C(5,9),F(5,5),G(5,2),Ct(9,5)
50    DIM Cct(5,5),Cctin(5,5),Ff(9,5),Wa(9,9),Wat(9,9),V(5,5)
60    DIM Wa2t(9,9),Wa2in(9,9),Wprod(9,9),Wf(5,9),Gf(5,2),
      W(5,5)
70    INPUT "Original System Order n",N
80    INPUT "Number of System Inputs m",M
90    INPUT "Reduced System Order l",L
100   PRINT "n=";N;"m=";M;"l=";L
110   Nm=N*M
120   REDIM A(N,N),Wa2t(N,N),Wa2in(N,N),Wprod(Nm,N),Gf(L,M),
      W(L,L)
130   REDIM V(L,L),B(N,M),C(L,N),F(L,L),G(L,M),Ct(N,L)
140   REDIM Cct(L,L),Cctin(L,L),Ff(N,L)
150   REDIM Wa(N,Nm),Wf(L,Nm),Wat(Nm,N)
```

```
160    INPUT "Has SUB <<Cont>> & <<Mat>> been LINKED ?",S$
170    IF (S$="Y") OR (S$="y") THEN 220
180    IF (S$="N") OR (S$="n") THEN 200
190    GOTO 160
200    LINK "Cont",1000,210
210    LINK "Mat",2000,220
220    CALL Mat(A(*),N,N,"A")
230    CALL Mat(B(*),N,M,"B")
240    Kk=0
250    INPUT "GIVE AGGREGATION CODE... 1 FOR KNOWN C, 2 FOR
       KNOWN F & G",Icode
260    ON Icode GOTO 270,410
270    PRINT "Case # 1 : KNOWN AGGREGATION MATRIX C"
280    CALL Mat(C(*),L,N,"C")
290    MAT Ct=TRN(C)
300    MAT Cct=C*Ct
310    MAT Cctin=INV(Cct)
320    MAT Ff=Ct*Cctin
330    MAT Ct=A*Ff
340    MAT F=C*Ct
350    MAT G=C*B
360    PRINT "MATRIX F:"
370    MAT PRINT F;
380    PRINT "MATRIX G:"
390    MAT PRINT G;
400    GOTO 750
410    ! EVALUATE CONTROLLABILITY MATRIX Wa
420    PRINT "Case # 2 : KNOWN AGGREGATED MATRICES F & G"
430    CALL Cont(A(*),B(*),N,M,Wa(*))
440    PRINT "CONTROLLABILITY MATRIX Wa:"
450    MAT PRINT Wa;
460    PRINT "INPUT MATRIX G"
470    CALL Mat(G(*),L,M,"G")
480    PRINT "INPUT AGGREGATED MATRIX F"
490    CALL Mat(F(*),L,L,"F")
500    ! EVALUATE CONTROLLABILITY MATRIX Wf
510    MAT Cct=IDN
520    FOR Iv=1 TO N
530    IF Iv<>1 THEN 560
540    MAT Gf=G
550    GOTO 580
560    Kk=Kk+M
570    MAT Gf=Cct*G
580    FOR I=1 TO L
590    FOR J=1 TO M
600    Wf(I,J+Kk)=Gf(I,J)
610    NEXT J
620    NEXT I
630    MAT V=F*Cct
640    MAT Cct=V
```

```
650    NEXT Iv
660    PRINT "CONTROLLABILITY MATRIX Wf:"
670    MAT PRINT Wf;
680    MAT Wat=TRN(Wa)
690    MAT Wa2t=Wa*Wat
700    MAT Wa2in=INV(Wa2t)
710    MAT Wprod=Wat*Wa2in
720    MAT C=Wf*Wprod
730    PRINT "AGGREGATION MATRIX C:"
740    MAT PRINT C;
750    INPUT "DO YOU LIKE TO REPEAT FOR A SECOND CASE (Y/N)",R$
760    IF (R$="Y") OR (R$="y") THEN 240
770    END
```

5. Example

```
THIS PROGRAM PERFORMS SYSTEM AGGREGATION VIA EXACT METHOD
n= 3 m= 1 l= 2

Matrix A( 3 x 3 ):

-2.0000E-01   1.0000E+00   1.0000E+00

 1.0000E+00  -4.0000E+00   0.0000E+00

 1.0000E+00   0.0000E+00  -6.0000E+00


Column Vector B( 3 ):

 1.0000E+00

 1.0000E+00

 1.0000E+00
Case # 1 : KNOWN AGGREGATION MATRIX C

Matrix C( 2 x 3 ):

 1.0000E+00   0.0000E+00   0.0000E+00

 0.0000E+00   5.0000E-01   5.0000E-01

MATRIX F:
-.2   2

 1  -5
```

MATRIX G:
 1

 1

Case # 2 : KNOWN AGGREGATED MATRICES F & G
CONTROLLABILITY MATRIX wa:
 1   1.8 -8.36

 1 -3   13.8

 1 -5   31.8

INPUT MATRIX G

Column Vector G(2):

 1.0000E+00

 1.0000E+00

INPUT AGGREGATED MATRIX F

Matrix F(2 x 2):

-2.0000E-01   2.0000E+00

 1.0000E+00 -4.0000E-01

CONTROLLABILITY MATRIX wf:
 1   1.8   .84

 1   .6 1.56

AGGREGATION MATRIX C:
 1.4372623576 -1.48669201362   1.04942965716

  .95817490501 -.457794675902   .499619771534

Case # 2 : KNOWN AGGREGATED MATRICES F & G
CONTROLLABILITY MATRIX wa:
 1   1.8 -8.36

 1 -3   13.8

 1 -5   31.8

INPUT MATRIX G

Column Vector G(2):

1.0000E+00

1.0000E+00

INPUT AGGREGATED MATRIX F

Matrix F(2 x 2):

-2.0000E-01   0.0000E+00

0.0000E+00   -4.0000E+00

CONTROLLABILITY MATRIX Wf:
1 -.2   .04

1 -4   16

AGGREGATION MATRIX C:
 .543726235654   .551330798578   -9.50570342798E-02

-.32319391625   1.59885931592   -.27566539882

5.3.2 Modal Aggregation

One of the more natural schemes for aggregating a large-scale system is to retain the modes of (1) and (2) in the same proportion that components of $x(t)$ are represented by output $y(t)$, i.e. aggregation process must retain all the dominant modes of the Original system. One of the first attempts along this line is due to Davison [5.14, 5.15] whose initial effort is considered as a special case of models built by aggregation [5.12]. In fact, it is shown that modal aggregation is a special case of exact aggregation in which dynamic exactness condition **FC=CA** is satisfied.

Consider the system (1) and let the reduced order model be

$$\dot{x}_r(t) = A_r x_r(t) + B_r u_r(t) \tag{9}$$

when $x_r(t)$ is a vector of states with dominant modes to be retained, A_r and B_r are $k \times k$ and $k \times m$ matrices given by:

$$A_r = M_r \Lambda_r M_r^{-1} \tag{10}$$

$$B_r = M [M^{-1} B]_k \tag{11}$$

where **M** is the modal matrix of **A**, $[\mathbf{M}^{-1}\mathbf{B}]_k$ is a $k \times m$ matrix consisting of the first k rows of $\mathbf{M}^{-1}\mathbf{B}$ and $\mathbf{M}_r$ is a $k \times k$ matrix (assumed to be nonsingular) which includes the upper k elements of the k dominant eigenvectors of **A** corresponding to the retained states. Thus **M** can be represented by

$$\mathbf{M} = \begin{bmatrix} \mathbf{M}_r & \mathbf{M}_{12} \\ \mathbf{M}_{21} & \mathbf{M}_{22} \end{bmatrix}. \qquad (12)$$

The matrix $\Lambda_r = diag(\lambda_1,...\lambda_k)$ consists of the k dominant eigenvalues of **A**.

An alternative representation of matrices of the aggregated model (9) is [5.16]

$$\mathbf{A}_r = \mathbf{M}_r \mathbf{P} \Lambda \mathbf{P}' \mathbf{M}_r^{-1}, \quad \mathbf{B}_r = \mathbf{M}_r \mathbf{P} \mathbf{M}^{-1} \mathbf{B} \qquad (13)$$

where **P** is a $k \times n$ transformation matrix,

$$\mathbf{P} = [\mathbf{I}_k \mid \mathbf{0}] \qquad (14)$$

and the aggregation matrix **C** in $\mathbf{x}_r = \mathbf{C}\mathbf{x}$ is

$$\mathbf{C} = \mathbf{M}_r \mathbf{P} \mathbf{M}^{-1}. \qquad (15)$$

We will now show that the modal aggregation of Davison [5.14, 5.15] is a special case of the exact aggregation of Aoki [5.6, 5.9, 5.12]. Let ξ be the right eigenvector of **A** corresponding to eigenvalues λ, i.e. $\mathbf{A}\xi = \lambda \xi$. Premultiplying both sides of this equality by **C** leads to $\mathbf{CA}\xi = \lambda \mathbf{C}\xi$. Now denoting $\gamma = \mathbf{C}\xi$ and remembering the condition (4), this relation can be rewritten as $\mathbf{FC}\xi = \mathbf{F}\gamma = \lambda \gamma$ which indicates that γ is a right eigenvector of matrix **F** under perfect aggregation condition, provided that $\mathbf{C}\xi \neq 0$. Thus **F** inherits a set of eigenvalues $\{\lambda_1, \lambda_2, \ldots, \lambda_k\}$ of **A**, corresponding to those eigenvectors of **A**, i.e. the set of all row vectors γ_j such that $\gamma_j \mathbf{A} = \lambda_j \gamma_j$ be the rows of the aggregation matrix, i.e.

$$\mathbf{C} = \begin{bmatrix} \gamma_1 \\ \gamma_2 \\ \cdot \\ \gamma_k \end{bmatrix} \qquad (16)$$

then the aggregated system (2) reduces to

$$\dot{\mathbf{z}} = \Lambda_k \mathbf{z} + \mathbf{G}\mathbf{u} \qquad (17)$$

where $\Lambda_k = diag(\lambda_1, \lambda_2, \ldots, \lambda_k)$ and $\mathbf{G} = \mathbf{CB}$. Moreover, it is easy to see that this aggregation does satisfy the dynamic exactness (perfect aggregation) condition (4). Thus, this development indicates that if the aggregation matrix $\mathbf{C}$ is properly chosen, exact aggregation can be used to retain the dominant modes of the system. (See Problem 5-10).

The source program and a sample example for modal aggregation are given below.

6. "MODAGR"

```
10    ! PROGRAM NAME : "MODAGR" PROG
20    PRINT "`MODAGR` PERFORMS MODAL AGGREGATION"
30    OPTION BASE 1
40    DIM A(9,9),B(9,2),Mod(9,9),M11(6,6),M12(6,3),M21(6,3),M22(3,3),Modinv(9,9)
50    DIM F(6,6),G(6,2),R(2,2),Q(9,9),B1(6,2)
60    DIM Lambda(9,9),Gama(9,2),Am(9,9)
70    DIM A11(6,6),A12(6,3),Lamb2(3,3),Gama2(3,2)
80    DIM Lamb2in(3,3),D(3,2),M11in(6,6),H(3,6),Hm12(3,3),Li(3,2),Q22h(3,6)
90    INPUT "System Order n,Input no. m,Aggregated System Order k",N,M,K
100   PRINT "Order n =";N;"Input no. m=";M;"Aggregated System Order k=";K
110   L=N-K
120   REDIM A(N,N),B(N,M),Mod(N,N),M11(K,K),M22(L,L),M12(K,L),M21(L,K)
130   REDIM F(K,K),G(K,M),R(M,M),Q(N,N),B1(K,M),Modinv(N,N),Gama(N,M)
140   REDIM A11(K,K),A12(K,L),Lamb2(L,L),Gama2(L,M),Lambda(N,N),Am(N,N)
150   REDIM Lamb2in(L,L),D(L,M),M11in(K,K),H(L,K),Hm12(L,L),Li(L,M)
160   INPUT "Is <<SUB>> Mat LINKED ?",S$
170   IF (S$="Y") OR (S$="y") THEN 210
180   IF (S$="N") OR (S$="n") THEN 200
190   GOTO 160
200   LINK "Mat",1000,210
210   CALL Mat(A(*),N,N,"A")
220   CALL Mat(B(*),N,M,"B")
230   PRINT "Give LSS Modal matrix in order of DOMINANCY of eigenvalues"
240   CALL Mat(Mod(*),N,N,"Modal")
250   MAT Modinv=INV(Mod)
260   MAT Am=A*Mod
270   MAT Lambda=Modinv*Am
280   MAT Gama=Modinv*B
290   FOR I=1 TO N
300   FOR J=1 TO N
310   I1=I-K
320   J1=J-K
330   IF (I<=K) AND (J<=K) THEN GOTO 390
340   IF (I<=K) AND (J>K) THEN GOTO 420
350   IF (I>K) AND (J<=K) THEN GOTO 450
360   M22(I1,J1)=Mod(I,J)
370   Lamb2(I1,J1)=Lambda(I,J)
380   GOTO 460
390   A11(I,J)=A(I,J)
400   M11(I,J)=Mod(I,J)
410   GOTO 460
420   A12(I,J1)=A(I,J)
430   M12(I,J1)=Mod(I,J)
440   GOTO 460
450   M21(I1,J)=Mod(I,J)
460   NEXT J
470   NEXT I
480   MAT Lamb2in=INV(Lamb2)
490   FOR J=1 TO M
500   FOR I=1 TO L
510   Gama2(I,J)=Gama(K+I,J)
520   NEXT I
530   FOR I=1 TO K
540   B1(I,J)=B(I,J)
550   NEXT I
560   NEXT J
```

```
570  MAT D=Lamb2in*Gama2
580  MAT D=(-1)*D
590  MAT M11in=INV(M11)
600  MAT H=M21*M11in
610  MAT Hm12=H*M12
620  MAT M22=M22-Hm12
630  MAT Li=M22*D
640  MAT F=A12*H
650  MAT F=A11+F
660  MAT G=A12*Li
670  MAT G=B1+G
680  PRINT " RESULTING AGGREGATED MATRICES:"
690  PRINT "    MATRIX F    "
700  MAT PRINT F;
710  PRINT "    MATRIX G    "
720  MAT PRINT G;
730  END
```

7. Example

```
'MODAGR' PERFORMS MODAL AGGREGATION
Order n = 3 Input no. m= 1 Aggregated System Order k= 2
```

Matrix A(3 x 3):

 5.0000E-01   5.0000E-01   0.0000E+00

 0.0000E+00   1.0000E+00   0.0000E+00

 8.3333E-01 -2.1667E+00 -3.3330E-01

Column Vector B(3):

 1.0000E+00

 1.0000E+00

 2.0000E+00

Give LSS Modal matrix in order of DOMINANCY of eigenvalues

Matrix Modal(3 x 3):

 0.0000E+00   1.0000E+00   1.0000E+00

 0.0000E+00   0.0000E+00   1.0000E+00

 1.0000E+00   1.0000E+00 -1.0000E+00

AGGREGATION METHODS—FREQUENCY DOMAIN

RESULTING AGGREGATED MATRICES:
 MATRIX F
.5 .5

0 1

 MATRIX G
1

1

5.4 AGGREGATION METHODS — FREQUENCY-DOMAIN

A lot of effort has been devoted to the reduction of frequency-domain models. By far, the largest effort has been on the reduction of SISO systems. However, MIMO systems modeled by (5.1.3)-(5.1.4) are also discussed here.

5.4.1 Moment Matching

This method is based on determining a set of time functions of the full model and matching them for the reduced model by choosing a number of appropriate parameters without having to obtain full model's time or frequency responses [5.17]. The technique is essentially a match of the time-moments of the full model's impulse response to those of the reduced model. Consider an nth-order large-scale SISO system,

$$H(s) = \frac{a_{21} + a_{22}s + \ldots a_{2,n} s^{n-1}}{1 + a_{12}s + a_{13}s^2 + \ldots + a_{1,n+1}s^n} \tag{1}$$

with impulse response $h(t)$.

$$H(s) = \int_0^\infty h(t) e^{-st} dt. \tag{2}$$

A power series expansion of e^{-st} about $s=0$, results in,

$$H(s) = \int_0^\infty h(t)\{1 - st + \frac{(st)^2}{2!} + \cdots\} dt$$

$$= \int_0^\infty h(t) dt - s \int_0^\infty t h(t) dt + s^2 \int_0^\infty \frac{t^2 h(t)}{2!} dt - \ldots \tag{3a}$$

or

$$H(s) = c_0 + c_1 s + c_2 s^2 + \cdots \tag{3b}$$

where

$$c_i = \frac{(-1)^i}{i!} \int_0^\infty t^i h(t) dt = \frac{(-1)^i}{i!} M_i . \tag{3c}$$

and M_i is the ith-time moment of h(t). Direct division of (1) yields

$$H(s) = a_{21} - a_{31} s + a_{41} s^2 - a_{51} s^3 + \cdots \tag{4}$$

In (4), a_{21} is the zeroth-term coefficient of the numerator in (1) and the remaining coefficients are obtained from the following recursion [5.17]:

$$a_{k,d} = a_{k-1,1} \cdot a_{1,d+1} - a_{k-1,d+1}. \tag{5}$$

The jth moment can be obtained by dividing $a_{j+1,1}$ by $j!$ for $j=0,1,2,...$ Let the full model be represented by (1), then by using (3)-(5), the following $2n$th order vector equation results:

$$\begin{bmatrix} c_0 \\ c_1 \\ c_2 \\ \cdot \\ \cdot \\ c_{n-1} \\ \hline c_n \\ c_{n+1} \\ \cdot \\ \cdot \\ \cdot \\ c_{2n-1} \end{bmatrix} = \begin{bmatrix} 0 & 0 & & & & 0 & & & & \\ -c_1 & 0 & & & & & & & & \\ -c_2 & -c_0 & & & 0 & & & 0 & & \\ \cdot & \cdot & & & & & & & & \\ -c_{n-2} & -c_{n-3} & -c_0 & \cdot & 0 & & & & & \\ \hline -c_{n-1} & -c_{n-2} & -c_1 & -c_0 & \cdot & -c_0 & & & 0 & 0 \\ -c_n & -c_{n-1} & \cdot & -c_1 & \cdot & & & & & 0 \\ \cdot & \cdot & & & & & & & & \\ \cdot & \cdot & & & & & & & & \\ -c_{2n-2} & -c_{2n-3} & \cdot & & & \cdot & & -c_1 & -c_0 & 0 \end{bmatrix}$$

$$\begin{bmatrix} a_{12} \\ a_{13} \\ a_{14} \\ \cdot \\ \cdot \\ \cdot \\ a_{1,n} \\ \hline 0 \\ 0 \\ \cdot \\ \cdot \\ \cdot \\ 0 \end{bmatrix} + \begin{bmatrix} a_{21} \\ a_{22} \\ a_{23} \\ \cdot \\ \cdot \\ \cdot \\ a_{2,n} \\ \hline 0 \\ 0 \\ \cdot \\ \cdot \\ \cdot \\ 0 \end{bmatrix} \tag{6}$$

which can be rewritten by

$$\begin{bmatrix} \hat{c}_1 \\ \hat{c}_2 \end{bmatrix} = \begin{bmatrix} C_{11} & C_{12} \\ C_{21} & C_{22} \end{bmatrix} \begin{bmatrix} \hat{a}_1 \\ 0 \end{bmatrix} + \begin{bmatrix} \hat{a}_2 \\ 0 \end{bmatrix} \tag{7}$$

where C_{11}, $C_{12}=0$ and C_{22} are $n \times n$ matrices, and $\hat{c}_i, \hat{a}_i$, $i = 1,2$ are $n \times 1$ vectors. Partitioning the set of two equations and solving for reduced model denominator, the numerator coefficients vectors $\hat{a}_1$ and $\hat{a}_2$ are given as

$$\hat{a}_1 = C_{21}^{-1}\hat{c}_2, \quad \hat{a}_2 = \hat{c}_1 - C_{11}C_{21}^{-1}\hat{c}_2 = \hat{c}_1 - C_{11}\hat{a}_1. \tag{8}$$

Note that once a Routhian array (standard Routh table) is formed from (5) the moment coefficients c_j, $j = 0,1,2,...$ can be obtained by $c_j = (-1)^j a_{j+2,1}$, $j=0,1,...$. Moreover, when the moments c_j, $j=0,1,2,...$ are obtained, the $2n \times 2n$ matrix C of (6) will be defined and the coefficients $\hat{a}_1$ and $\hat{a}_2$ will be immediate from (8).

Submatrix C_{21} is normally nonsingular and its singularity implies that the given set of moments can be matched by a simpler model. The computer program for moment matching is combined with Pade' approximation discussed next.

5.4.2 Pade' Approximation

The second model reduction method stems from the theory of Pade' approximation introduced by Pade' [5.18], extended by Wall [5.19] and has been used for model reduction by others [5.19-5.20]. Consider a function

$$f(x) = c_o + c_1 x + c_2 x^2 + ... \tag{9}$$

and a rational function $[U_m(x)/V_n(x)]$ where $U_m(x)$ and $V_n(x)$ are mth-and nth-order polynomials in x, respectively and $m \leq n$. The rational function $[U_m(.)/V_n(.)]$ is said to be a "Pade' approximant" of f(x), if and only if the first (m+n) terms of power series expansion of f(x) and $[U_m(x)/V_n(x)]$ are identical.

For the function f(x) in (9) to be approximated, let the following Pade' approximant, be defined,

$$\frac{U_m(s)}{V_n(x)} = \frac{a_o + a_1 x + a_2 x^2 + \cdots + a_m x^m}{b_o + b_1 x + b_2 x^2 + \cdots + b_n x^n} \tag{10}$$

For the first (m+n) terms of (9) and (10) to be equivalent, the following set of relations must hold:

$$a_o = b_o c_o$$

$$a_1 = b_o c_1 + b_1 c_o$$

$$a_2 = b_o c_2 + b_1 c_1 + b_2 c_o$$

$$a_m = b_o c_m + b_1 c_{m-1} + \ldots + b_m c_o$$

$$0 = b_o c_{m+1} + b_1 c_m + \ldots + b_{m+1} c_o$$

$$0 = b_o c_{m+n} + b_1 c_{m+n-1} + \ldots + b_{n-1} c_{m+1} + c_m \tag{11}$$

Once the coefficients c_i, i=0,1,2,... are determined, (11) can be written in vector form as

$$\begin{bmatrix} c_{m+1} & c_m & \cdot & \cdots & \cdot & c_1 \\ c_{m+2} & c_{m+1} & c_m & \cdots & \cdot & c_2 \\ c_{m+3} & c_{m+2} & c_{m+1} & \cdots & \cdot & c_3 \\ \cdot & \cdot & \cdot & \cdots & \cdot & \cdot \\ \cdot & \cdot & \cdot & \cdots & \cdot & c_m \\ c_{m+n} & c_{m+n-1} & \cdot & \cdots & c_{m+2} & c_{m+1} \end{bmatrix} \begin{bmatrix} b_o \\ b_1 \\ b_2 \\ \cdot \\ \cdot \\ b_{n-1} \end{bmatrix} = \begin{bmatrix} -c_o \\ -c_1 \\ -c_2 \\ \cdot \\ \cdot \\ -c_m \end{bmatrix} \tag{12a}$$

$$\begin{bmatrix} a_0 \\ a_1 \\ a_2 \\ \cdot \\ \cdot \\ \cdot \\ a_m \end{bmatrix} = \begin{bmatrix} c_0 & 0 & \cdot & \cdot & \cdot & \cdot & 0_0 \\ c_1 & c_0 & 0 & \cdot & \cdot & \cdot & 0_0 \\ c_2 & c_1 & c_0 & 0 & \cdot & \cdot & 0 \\ \cdot & \cdot & \cdot & \cdot & 0 & \cdot & \cdot \\ \cdot & \cdot & \cdot & \cdot & \cdot & \cdot & \cdot \\ \cdot & \cdot & \cdot & \cdot & \cdot & \cdot & \cdot \\ c_m & c_{m-1} & \cdot & \cdot & \cdot & c_1 & c_0 \end{bmatrix} \begin{bmatrix} b_0 \\ b_1 \\ b_2 \\ \cdot \\ \cdot \\ \cdot \\ b_{n-1} \end{bmatrix} \quad (12b)$$

It is noted that in this reformulation of (11), $b_n = b_{m+1} = 1$. Moreover, in a compact form (12) is given by

$$\mathbf{C}_1 \hat{\mathbf{b}} = -\hat{\mathbf{c}}, \quad \hat{\mathbf{a}} = \mathbf{C}_2 \hat{\mathbf{b}} \qquad (13)$$

where $\mathbf{C}_1$ and $\mathbf{C}_2$ are $(n \times n)$ and $(m+1)$, respectively, $\hat{\mathbf{a}}, \hat{\mathbf{b}}$ and $\hat{\mathbf{c}}$ are $(m+1)$-, n- and $(m+1)$-dimensional column vectors, defined by (12).

The source program and sample example for moment matching and Pade' approximation are given below:

7. "MOMPAD"

```
10   !    PROGRAM NAME : "MOMPAD" PROG
20   PRINT " 'MOMPAD' AGGREGATES A LINEAR SYSTEM THROUGH :"
30   PRINT "1. Moment Matching (Code=1) or 2. Pade Approximation (Code=2)"
40   DIM Acoef(9),Bcoef(9),Ab(9,9),Ba(9,9),Bainv(9,9),Brhs(9)
50   DIM C1(9,9),C21(9,9),C21inv(9,9),Com(9),Cmpn(9),C1121(9,9)
60   DIM Aden(9),Anum(9),W1(9),C(18),Ncoef(15),Dcoef(16)
70   INPUT "System Order n ?",Nn
80   PRINT "System Order n =";Nn
90   Nmom=2*Nn
100  REDIM C(Nmom)
110  MAT Dcoef=ZER
120  MAT Ncoef=ZER
130  PRINT "Give Numerator Coefficients in Ascending Order, 0,1,2,..."
140  FOR I=0 TO Nn-1
150    DISP "Ncoef(";I;")";
160    INPUT Ncoef(I)
170    PRINT "Ncoef(";I;")=";Ncoef(I)
180  NEXT I
190  PRINT "Give Denominator Coefficients in Ascending Order, 0,1,2,..."
200  FOR I=0 TO Nn
210    DISP "Dcoef(";I;")";
220    INPUT Dcoef(I)
230    PRINT "Dcoef(";I;")=";Dcoef(I)
240  NEXT I
250  CALL Moment(Nn,Nmom,Ncoef(*),Dcoef(*),C(*))
260  INPUT "Order of REDUCED model k?",K
270  IF K>Nn THEN 260
280  Iv=N=K-1
290  REDIM C1(N,N),C21(N,N),C21inv(N,N),Com(N),Cmpn(N),C1121(N,N)
300  REDIM Ab(N,N),Ba(N,N),Acoef(N),Bcoef(N),Bainv(N,N),Brhs(N)
310  REDIM Aden(N),Anum(N),W1(N)
320  MAT C1=ZER
```

```
330    MAT Ab=ZER
340    FOR I=0 TO N
350    Brhs(I)=-C(I)
360    Com(I)=C(I)
370    Cmpn(I)=C(I+Nn-1)
380    FOR J=0 TO N
390    C21(I,J)=-C(Iv)
400    Ba(I,J)=C(Iv+1)
410    Iv=Iv-1
420    NEXT J
430    Iv=N+I+1
440    FOR J=0 TO I
450    Ab(I,J)=C(I-J)
460    NEXT J
470    IF I=0 THEN 510
480    FOR J=0 TO I-1
490    Cl(I,J)=-C(I-J-1)
500    NEXT J
510    NEXT I
520    INPUT "Give program CODE (1or2)?",Code
530    IF (Code<>1) AND (Code<>2) THEN 610
540    ON Code GOTO 550,700
550    MAT C21inv=INV(C21)
560    MAT Aden=C21inv*Cmpn
570    MAT Cl121=Cl*C21inv
580    MAT W1=Cl121*Cmpn
590    MAT Anum=Com-W1
600    PRINT "Case # 1 Moment Matching : REDUCED MODEL Numerator Coefficients:"
610    MAT PRINT Anum;
620    PRINT "REDUCED MODEL Denominator Coefficients in Ascending Order:"
630    REDIM Anum(N+1)
640    FOR I=1 TO N+1
650    Anum(I)=Aden(I-1)
660    NEXT I
670    Anum(0)=1
680    MAT PRINT Anum;
690    GOTO End
700    MAT Bainv=INV(Ba)
710    MAT Bcoef=Bainv*Brhs
720    MAT Acoef=Ab*Bcoef
730    PRINT "Case # 2 Pade Approximation : REDUCED MODEL Numerator Coefficients:"
740    PRINT "coefficients: ai, i=0,1,2,..."
750    MAT PRINT Acoef;
760    PRINT "REDUCED MODEL Denominator Coefficients in Ascending Order:"
770    REDIM Bcoef(N+1)
780    Bcoef(N+1)=1
790    MAT PRINT Bcoef;
800    End: INPUT "Do you like to try the other method (Y/N)",M$
810    IF (M$="Y") OR (M$="y") THEN 520
820    END
830    SUB Moment(Num,Nmom,N(*),D(*),C(*))
840    C(0)=N(0)/D(0)
850    FOR I=1 TO Nmom-1
860    Ecsum=0
870    FOR J=1 TO I
880    Ecsum=Ecsum+D(J)*C(I-J)
890    NEXT J
900    C(I)=(N(I)-Ecsum)/D(0)
910    NEXT I
920    PRINT "Moments in Ascending Order:"
930    REDIM C(Nmom-1)
940    MAT PRINT C;
950    SUBEND
```

´MOMPAD´ AGGREGATES A LINEAR SYSTEM THROUGH :
1. Moment Matching (Code=1) or 2. Pade Approximation (Code=2)
System Order n = 4
Give Numerator Coefficients in Ascending Order, 0,1,2,...
Ncoef(0)= 400
Ncoef(1)= 200

```
Ncoef( 2 )= 50
Ncoef( 3 )= 10
Give Denominator Coefficients in Ascending Order, 0,1,2,...
Dcoef( 0 )= 400
Dcoef( 1 )= 660
Dcoef( 2 )= 292
Dcoef( 3 )= 33
Dcoef( 4 )= 1
Moments in Ascending Order:
 1  -1.15   1.2925  -1.350625   1.3773b125  -1.3904790625   1.39699745313
-1.40025347267

Case # 1 Moment Matching : REDUCED MODEL Numerator Coefficients:
 1                  5.3003125004



REDUCED MODEL Denominator Coefficients in Ascending Order:
 1                  6.4503125004              6.4503125004

Case # 2 Pade Approximation : REDUCED MODEL Numerator Coefficients:
coefficients: ai, i=0,1,2,...
.2556727389   .86289549376


REDUCED MODEL Denominator Coefficients in Ascending Order:
.2556727389   1.1569191435    1
```

8. Example. The poles of the original model were at -1, -2, -10 and -20 while the two 2nd-order reduced models provided poles at -0.8156, -0.1914 and -0.8582, -0.2978, respectively for moment matching and Pade' approximation.

5.4.3 Mixed Methods and Multi-Variable Systems

One of the main issues in most model reduction methods is the unpredictable results of finding a stable reduced model from an unstable full model and vice versa. The latter is due to the excitation of unretained modes through feedback, a characteristic similar to that of some aggregation methods discussed in Section 5.3. These problems can be dealt with in different ways: e.g. (i) combination of "Pade'" and "Modal" and (ii) Combination of "Pade'" and "Routh" Here, mixed methods and matrix continued fraction method will be described and applied to multivariable systems.

5.4.3.1 Pade' -Modal

The experience in model reduction using Pade' approximation, as it was mentioned before, shows that this technique may very well provide unstable reduced models. The present combined method is simply to retain the dominant poles of the system, say s_i i=1,2,...,k and find the remaining $(d-k)$ poles by Pade' approximation where $d < n$ and n is the order of the full model.

Specifically it implies that the last k rows of matrix C_1 in (12a) and (13) will be replaced by the following set of k-equations satisfying the dominant poles s_i, i=1,2,...,k:

$$\begin{bmatrix} 1 & s_1 & s_1^2 & s_1^{n-1} \\ 1 & s_2 & s_2^2 & s_2^{n-1} \\ \vdots & \vdots & \vdots & \vdots \\ 1 & s_k & s_k^2 & s_k^{n-1} \end{bmatrix} \tag{14}$$

and the right-hand side of (12a)

$$[-s_1^n - s_2^n \ldots -s_k^n]'. \tag{15}$$

Depending on the choice of the dominant poles, a stable reduced-order model can be found by this mixed method.

The extension of Pade'-Model scheme to MIMO systems is now investigated. Consider the transfer function matrix H(s) of (5.1.4) rewritten in the following form:

$$\mathbf{H}(s) = \sum_{j=1}^{n} \mathbf{A}_{2j} s^{j-1} \Big/ \sum_{j=1}^{n+1} a_j s^{j-1} \tag{16}$$

and expand it into a power series,

$$H(s) = \sum_{i=0}^{\infty} C_i s^i \tag{17}$$

where C_i, $i=1,2,\ldots$ are $m \times r$ constant matrices directly proportional to the ith moment of the system. Then the reduced-order model is given by

$$M_k(s) = \sum_{j=1}^{k} \mathbf{E}_j s^j / \Delta_k(s) \tag{18}$$

where the reduced-order characteristic polynomial $\Delta_k(s)$ is defined by

$$\Delta_k(s) = (s-\lambda_1)(s-\lambda_2)\ldots(s-\lambda_k) = \sum_{j=1}^{k} f_j s^{j-1}, \quad f_{k+1}=1 \tag{19}$$

and coefficient matrices $\mathbf{E}_j$, $j=1,\ldots,k$ are obtained by

$$E_1 = f_1 C_0$$

$$E_2 = f_1 C_1 + f_2 C_0$$

$$E_3 = f_1 C_2 + f_2 C_1 + f_3 C_0$$

$$E_k = f_1 C_{k-1} + f_2 C_{k-2} + \ldots + f_{k-1} C_1 + f_k C_0 \tag{20}$$

This is the matrix version of the scalar numerator coefficients evaluated by Pade' approximation defined by (11). Before this scheme is applied to a numerical example, the matrix continued fraction method will be considered.

5.4.3.2 Matrix Continued Fraction

Chen [5.21], Shieh and Gaudiano [5.22] have extended the continued fraction method to the case of MIMO systems. The transfer function matrix of a MIMO system with r inputs and m outputs introduced by (5.1.4) is rewritten by.

$$H(s) = \left[\sum_j A_{2j} s^{j-1} \right] \left[\sum_i A_{1i} s^i \right]^{-1} \tag{21}$$

where the matrices D_d and the coefficients $e_d, d=0,1,\ldots$ in (5.1.4) have been replaced by matrices $A_{2,d+1}$ and $A_{1,d+1} = e_d I_m$ (I_m is an mxm identity matrix). It is noted that the matrix continued fraction described here is restricted to systems with equal number of inputs and outputs, i.e. m-r; thus all the matrices on the right side of (21) and $H(s)$ on the left are $m \times m$. The matrix continued fraction of Cauer second form applied to (21) results in:

$$H(s) = [H_1 + [H_2 \frac{1}{s} + [H_3 + [H_4 \frac{1}{s} + [\ldots]^{-1}]^{-1}]^{-1}]^{-1}]^{-1} \tag{22}$$

where H_i, $i=1,2,\ldots,2k$, $k \leq n$ are matrix quotients and are evaluated through the following matrix Routh algorithm [5.23],

$$\begin{array}{l} H_1 = A_{11} A_{21}^{-1} \\ H_2 = A_{21} A_{31}^{-1} \\ H_3 = A_{31} A_{41}^{-1} \end{array} \quad \begin{array}{l} A_{11} \quad A_{12} \quad A_{13} \ldots \\ A_{21} \quad A_{22} \quad A_{23} \ldots \\ A_{31} \quad A_{32} \quad \ldots \ldots \\ A_{41} \ldots \end{array} \tag{23}$$

where

$$A_{ij} = A_{i-2,j+1} - H_{i-2} A_{i-1,j+1}, \quad i=3,4,\ldots,j=1,2,\ldots$$

$$H_i = A_{i,1} (A_{i+1,1})^{-1}, \quad i=1,2,\ldots,2j, \quad j \leq n; \text{and}$$

$$\det(A_{i+1,1}) \neq 0. \tag{24}$$

The reduced model by the matrix continued fraction is obtained by truncating after the kth quotient matrix H_k. For example, for a second order reduction one has

$$M_2(s) = \left[H_1 + [\frac{1}{s} H_2]^{-1} \right]^{-1} = H_2 \left[sI + H_1 H_2 \right]^{-1}. \tag{25}$$

The following example explains the two multi variable model reduction methods.

1. Example. Consider a two-input, two-output system [5.1]

$$H(s) = \frac{A_{21} + A_{22}s + A_{23}s^2 + A_{24}s^3}{a_1 + a_2 s + a_3 s^2 + a_4 s^3 + a_5 s^4} \tag{26}$$

where

$$a_1 = 240, a_2 = 360, a_3 = 204, a_4 = 36, a_5 = 2,$$

$$A_{21} = \begin{bmatrix} 2400 & 2000 \\ 4320 & 3200 \end{bmatrix}, A_{22} = \begin{bmatrix} 1800 & 900 \\ 1440 & 1220 \end{bmatrix}$$

$$A_{23} = \begin{bmatrix} 496 & 488 \\ 528 & 396 \end{bmatrix} \text{ and } A_{24} = \begin{bmatrix} 28 & 30 \\ 12 & 24 \end{bmatrix} \tag{27}$$

The system's closed-loop poles are $-1.197 \pm j0.693$ and $-7.803 \pm j1.358$ and it is desired to keep two of these poles.

Pade'-Modal: The expansion of the system (26) and (27) in a power series yields

$$H(s) = C_o + C_1 s + C_2 s^2 + ...$$

Then

$$C_o = \begin{bmatrix} 10 & 8.34 \\ 18 & 13.34 \end{bmatrix}, C_1 = \begin{bmatrix} -7.5 & -8.75 \\ -21 & -14.91 \end{bmatrix}. \tag{28}$$

The reduced-order model is given by

$$M_2(s) = \frac{E_1 + sE_2}{\Delta_2(s)} = \frac{E_1 + sE_2}{s^2 + 2.4s + 1.91} \tag{29}$$

where E_1 and E_2 are obtained by (20) using (28):

$$E_1 = f_1 C_o = 1.91 C_o = \begin{bmatrix} 19.1 & 15.96 \\ 34.44 & 25.53 \end{bmatrix} \tag{30}$$

$$E_2 = f_1 C_1 + f_2 C_o = 1.9 C_1 + 2.4 C_o = \begin{bmatrix} 9.58 & 3.22 \\ 2.906 & 3.39 \end{bmatrix} \tag{31}$$

and $M_2(s)$ becomes

$$M_2(s) = \frac{\begin{bmatrix} 19.1 + 9.58s & 15.96 + 3.22s \\ 34.44 + 2.906s & 25.53 + 3.39s \end{bmatrix}}{s^2 + 2.4s + 1.91} \tag{32}$$

Matrix Continued Fraction: After defining the diagonal matrices, $A_{1i} = a_i I_2$, $i = 1, 2 \ldots 5$ the matrix quotients H_1 and H_2, defined by (23) and (24) turn out to be

$$H_1 = \begin{bmatrix} -0.8 & 0.5 \\ 1.08 & -0.6 \end{bmatrix}, \quad H_2 = \begin{bmatrix} 9.85 & 7.64 \\ 16 & 12 \end{bmatrix}. \tag{33}$$

Then the second-order reduced model can be found through (25):

$$M_2(s) = H_2 \left[sI + H_1 H_2 \right]^{-1} = \begin{bmatrix} 9.85 & 7.64 \\ 16 & 12 \end{bmatrix} \cdot \begin{bmatrix} s+.116 & -.106 \\ 1.04 & s+1.04 \end{bmatrix}^{-1}$$

$$M_2(s) = \frac{\begin{bmatrix} 9.8s + 2.3 & 7.6s + 1.9 \\ 16 + 4.2 & 12s + 3.1 \end{bmatrix}}{s^2 + 1.156s + 0.241} \tag{34}$$

which provides a pair of poles at -0.273 and -0.883.

It must be noted that the matrix continued fraction method has several limitations, namely the number of outputs and inputs must be equal and, as demonstrate by Calfe and Healey [5.24], it is possible for the reduce model characteristic polynomial to be of higher order than the full model's. Furthermore, the reduction of the MIMO systems by this and modal continued fraction, are a reduction in *output* and not the *state* which is an *aggregation* of system variables

PROBLEMS

5.1 Is the system

$$(\mathbf{A},\mathbf{B},\mathbf{C}) = \left(\begin{bmatrix} 0 & 1 & 0 \\ 0 & 0 & 1 \\ -0.5 & -5.6 & -6.1 \end{bmatrix}, \begin{bmatrix} 0 \\ 0 \\ 1 \end{bmatrix}, [1 \ 0 \ 1] \right)$$

weakly coupled?

5.2 For a system

$$(\mathbf{A},\mathbf{B},\mathbf{C}) = \left(\begin{bmatrix} 0 & 1 & 0 \\ 0 & 0 & 1 \\ -1 & -a & -b \end{bmatrix}, \begin{bmatrix} 1 \\ 0 \\ 1 \end{bmatrix}, [1 \ 0 \ 1] \right)$$

find the region(s) in b-a plane such that the system is weakly coupled.

5.3 Consider an eight order unforced system

$$\dot{\mathbf{x}} = \begin{bmatrix} -0.5 & 0 & 0 & -0.2 & 0.1 & 0.2 & 0 & 0.1 \\ 0 & -1 & 0 & .1 & 0 & 1 & 0 & 0 \\ 0.1 & 0 & -2 & 0 & 0 & 0 & 4 & 0 \\ 0.05 & 0 & 0 & -10 & -1 & 0 & 0 & 0 \\ -0.15 & 1 & 0 & 10 & 30 & 1 & 0 & 0 \\ 0 & 0 & 0.15 & -0.2 & -0.3 & -1 & 0 & 0 \\ -15 & 0 & 0 & -1 & 0 & 0 & 0.08 & 2 \\ -80 & 60 & -10 & 0 & -50 & 0 & 6 & -15 \end{bmatrix} \mathbf{x}$$

use Program "SEPTIM" to see whether one can separate the time scales.

5.4 Use an appropriate computer program to check whether the system

$$\dot{\mathbf{x}} = \begin{bmatrix} -0.21054 & -0.10526 & -0.007378 & 0 & 0.0706 & 0 \\ 1 & -0.03537 & -0.000118 & 0 & 0.0004 & 0 \\ 0 & 0 & 0 & 1 & 0 & 0 \\ 0 & 0 & -605.16 & -4.92 & 0 & 0 \\ 0 & 0 & 0 & 0 & 0 & 1 \\ 0 & 0 & 0 & 0 & -3906.25 & -12.5 \end{bmatrix} \mathbf{x}$$

is weakly coupled.

5.5 A singularly perturbed system is described by

$$\dot{x} = -x+z, \; x(0) = -1$$

$$\epsilon\dot{z} = -x-2z, \; z(0) = 1.$$

Find a boundary layer for this system.

5.6 For a system

$$(A,B) = \left(\begin{bmatrix} -1 & 0.5 & 0.5 \\ 2 & -10 & 0 \\ 1 & 0 & -20 \end{bmatrix}, \begin{bmatrix} 1 \\ 1 \\ 1 \end{bmatrix} \right)$$

find a second order aggregated model using the two exact aggregation methods. Are these dynamically exact?

5.7 A third-order system is described by

$$\dot{x} = \begin{bmatrix} -1.25 & 0.02 & 0 \\ -0.5 & -0.1 & -1 \\ -1 & -0.5 & -0.08 \end{bmatrix} x + \begin{bmatrix} 1 \\ 0 \\ 1 \end{bmatrix} u.$$

Find an aggregated model using the modal scheme described in Section 5.3.2.

5.8 A system is described by

$$(A,B) = \left(\begin{bmatrix} -2 & -1 & -0.5 & -0.1 & -0.05 \\ 0.5 & -1 & 0.2 & -4 & 0.8 \\ 1 & -0.1 & -0.25 & 0 & -1 \\ 0 & 1 & -0.1 & -0.1 & 0 \\ 1 & -1 & -1 & 0.5 & -5 \end{bmatrix}, \begin{bmatrix} 1 \\ 0 \\ 1 \\ 1 \\ 1 \end{bmatrix} \right)$$

Using computer routine "EIGEN" (Chapter 2) for eigenvalue-eigenvector evaluations and "MODAGR" to find an aggregated model using the modal method of Section 5.3.

5.9 A second-order system is given by

$$\dot{x} = \begin{bmatrix} -1 & 0 \\ 1 & -8 \end{bmatrix} x + \begin{bmatrix} 1 \\ 1 \end{bmatrix} u.$$

Find an aggregation matrix which would provide a dynamically exact aggregation. Repeat for both the dominant and non-dominant eigenvalues of the original system.

5.10 Show that aggregation matrix C described by Equation (5.3.16) provides perfect aggregation.

5.11 Suppose that a 5×5 A matrix has $\lambda\{A\} = \{-1, a, a, b \pm jc\}$. Then it is known [5.1] that its modal matrix is $M = [\xi_1 : \xi_2 : v_1 : v_1 : w_1]$ where $\xi_{1,2}$ are right eigenvectors, v_1 is a generalized eigenvector and $u_{1,3} = v_1 \pm jw_1$ is a pair of complex conjugate eigenvectors. Using this, extend the modal aggregation of Davison [5.14], Section 5.3.2, for repeated and complex eigenvalues.

5.12 Use modal aggregation method to find a second-order reduced model for

$$\dot{x} = \begin{bmatrix} 0 & 1 & 0 & 0 \\ 0 & 0 & 1 & 0 \\ 0 & 0 & 0 & 1 \\ -400 & -460 & -262 & -32 \end{bmatrix} x + \begin{bmatrix} 1 \\ 0 \\ 1 \\ 1 \end{bmatrix} u$$

5.12 For a SISO system with a transfer function,

$$G(s) = \frac{1+2s-3s^2}{1+s+3s^2+s^3}$$

find a reduced order model using moment matching method.

5.13 Use Pade" approximation to reduce the following 4th order system to 2nd and 3rd order:

$$G(s) = (4s^3+9s^2+18s+24)/(s^4+s^3+2s^2+12)$$

Is stability of the full order preserved?

5.14 A system with a transfer function

$$G(s) = (s^3+6s^2+11s+6)/(s^4+17s^3+82s^2+130s+100)$$

has a pair of dominant poles. It is required to find a reduced order system which preserves the dominant poles. Use Pade' approximation to find it.

5.15 For the system,

$$G(s) = (s^2+5s+5)/(s^3+12s^2+25s+50)$$

Use Pade'-Modal to find a second-order reduced model which would preserve the system's dominant poles.

5.16 Consider a two-input two-output system with the following transfer function matrix:

$$\mathbf{H}(s) = (\mathbf{A}_{21}+\mathbf{A}_{22}s+\mathbf{A}_{23}s^2)/(a_1+a_2s+a_3s^2+a_4s^3)$$

where $a_1 = 120$, $a_2 = 320$, $a_3 = 180$, $a_4 = 2$,

$$\mathbf{A}_{21} = \begin{bmatrix} 240 & 200 \\ 420 & 300 \end{bmatrix} \quad \mathbf{A}_{22} = \begin{bmatrix} 180 & 80 \\ 100 & 120 \end{bmatrix} \quad \mathbf{A}_{23} = \begin{bmatrix} 50 & 40 \\ 50 & 30 \end{bmatrix}$$

Find a second-order reduced model using matrix continued fraction.

5.17 Repeat Problem 5.16 using Pade' Modal method.

REFERENCES

[5.1] M. Jamshidi, *Large-Scale Systems-Modeling and Control,* Elsevier-North Holland Co., New York, 1983

[5.2] D. D. Siljak, *Large-Scale Dynamic Systems,* Elsevier North Holland, New York 1978

[5.3] M. Jamshidi, "An Overview on Reduction of Large-Scale Systems Time-and-Frequency-Domain Models", submitted for publication, 1983

[5.4] P. V. Kokotovic, *Feedback Design of Large Linear Systems,* J. B. Cruz, Jr., Ed., New York:McGraw Hill Co., pp. 99-137, 1972

[5.5] R. D. Milne, "The Analysis of Weakly Coupled Dynamic Systems", Int. J. Control, Vol. 2, pp. 171-199, 1965

[5.6] M. Aoki, *In Optimization Methods for Large-Scale Systems... with applications,* D. A. Wismer, Ed., New York, McGraw Hill Book Co., Chapter 5, pp. 192-232, 1971

[5.7] P. V. Kokotovic, *Overview of Multimodeling by Singular Perturbations. Systems Engineering for Power: Organizational forms for Large-Scale Systems,* L. H. Fink and T. A. Trygar, Eds., U. S. DOE, Washington, D. C., pp. 1.3-1.4, 1979

[5.8] M. Jamshidi, 1974. "Three-stage near-optimum design of nonlinear control processes", Proc IEE, Vol. 121, pp. 886-892, 1974

[5.9] M. Aoki, "Some approximation methods and control of large-scale systems", IEEE Trans. Auto. Cont., Vol. AC-13, 1978, pp. 173-182

[5.10] F. N. Bailey and H. K. Ramapriyan, "Bounds on suboptimality in the control of linear dynamic systems", IEEE Trans. Aut. Contr. 1973

[5.11] P. V. Kokotovic, J. J. Allemongo, J. R. Winkelman, and J. H. Chow, "Singular perturbation and iterative separation of time scales", IFAC J. Automatic. Vol 16, 1980, pp. 23-33

[5.12] M. Aoki, "Control of large-scale dynamic systems by aggregation", IEEE Trans. Auto. Contr., Vol. Ac-13, 1968, pp. 246-253

[5.13] N. R. Sandell, Jr., P. Varaiya, M. Athans, and M. G. Safonov, "Survey of decentralized control methods for large-scale systems", IEEE Trans. AG (Special Issue on Large-Scale Systems), 1978, Vol. AC-23, pp. 108-128

[5.14] E. J. Davison, "A method for simplifying linear dynamic systems", IEE Trans. Auto. Contr., 1966, Vol. AC-12, pp. 119-121

[5.15] E. J. Davison, "A new method for simplifying linear dynamic systems", IEEE Trans. Auto. Contr., Vol. AC-13, 1968. pp. 214-215

[5.16] M. R. Chidamabara, "Two simple techniques for simplifying large dynamic systems", Proc. JACC Univ Colorado, Boulder, CO., 1969

[5.17] M. Lal and R. Mitra, "Simplification of large system dynamics using a moment evaluation algorithm", IEE Trans Auto. Contr., Vol. AC-9, 1974, pp. 602-603

[5.18] H. Pade', "Surla represantation approachee d'unc function par des fractions rationelles", Annales scientifiques de l'Ecole Normale upieure, Vol. 9, Ser. 3, 1892, pp. 1-93

[5.19] H. S. Wall, *Analytic Theory of Continued Fractions*, New York, Van Nostrand, 1948

[5.20] Y. Shamash, "Multivariable system reduction via modal methods and Pade's approximation", IEEE Trans. Auto. Contr., Vol. AC-20, 1975, pp. 815-817

[5.21] C. F. Chen, "Model reduction of multivariable control systems by means of continued fractions", Int. J. Control, Vol. 20, 1974, pp. 225-238

[5.22] L. S. Shieh and F. F. Gaudiano, "Matrix continued fraction expansion and inversion by the generalized matrix algorithm", Int. J. Control, Vol 20, 1974, pp. 727-737

[5.23] C. F. Chen and L. S. Shieh, "A novel approach to linear model simplification", Int. J. Control, Vol. 8, l968, pp. 561-570

[5.24] M. R. Calfe and M. Healey, "Continued-fraction model-reduction technique for multi-variable systems", Proc. ILE , Vol. 121, 1974, pp. 393-395

CHAPTER 6
SOLUTION OF THE STATE EQUATION

6.1 INTRODUCTION

In Chapter 4 we showed that the behavior of any linear time-varying system can be described by the state and output equations (4.4.4) and (4.4.5) for continuous-time systems, and by (4.4.12) and (4.4.13) for discrete-time systems. For the sake of discussion let us first concentrate on continuous-time systems, i.e. equations (4.4.4) and (4.4.5) which are repeated here for convenience:

$$\dot{\mathbf{x}}(t) = \mathbf{A}(t)\mathbf{x}(t) + \mathbf{B}(t)\mathbf{u}(t) \tag{1}$$

$$\mathbf{y}(t) = \mathbf{C}(t)\mathbf{x}(t) + \mathbf{D}(t)\mathbf{u}(t) \tag{2}$$

In order to analyze the system we must solve equations (1) and (2). That is, given any initial state $\mathbf{x}(t_o)$ and any input $\mathbf{u}(t)$ for $t \geqslant t_o$, we must determine the state vector $\mathbf{x}(t)$ and the output vector $\mathbf{y}(t)$ for $t \geqslant t_o$.

If solution $\mathbf{x}(t)$ to (1) is obtained, then $\mathbf{y}(t)$ can be determined from (2) by the substitution of $\mathbf{x}(t)$ and $\mathbf{u}(t)$ and by simple matrix multiplication and addition. Therefore, the main task is that of solving the state equation (1). Similarly, in the discrete-time case the main task is to solve the state equation (4.4.12) repeated here:

$$\mathbf{x}(k+1) = \mathbf{A}(k)\mathbf{x}(k) + \mathbf{B}(k)\mathbf{u}(k) , \tag{3}$$

i.e. determining $\mathbf{x}(k)$ for $k \geqslant k_o$ where the input $\mathbf{u}(k)$, $k \geqslant k_o$, and the initial state $\mathbf{x}(k_o)$ are known.

This chapter is concerned with the solution of equations (1) and (3). We will first deal with the case where the input $\mathbf{u}$ is identically zero and will solve the state equation for both continuous-time and discrete-time systems. Then we will provide the complete solution in both cases. Adjoint and dual systems will also be introduced in this chapter.

6.2 THE HOMOGENEOUS CASE - TRANSITION MATRIX

When the input is identically zero, the state equation becomes

$$\dot{x}(t) = A(t)x(t) \tag{1}$$

in the continuous-time case, and

$$x(k+1) = A(k)x(k) \tag{2}$$

in the discrete-time case. Equations (1) and (2) are referred to as the *homogeneous* or the *unforced* state equations.

6.2.1 Continuous-Time Systems

Let $x \in R^n$ and assume that at the initial time t_o, the initial state is

$$x(t_o) \triangleq x_o \tag{3}$$

We are interested in the solution $x(t)$ of (1) for $t \geq t_o$ subject to initial state (3). The first question that arises is that of *existence* and *uniqueness* of solutions. It has been shown [6.1, 6.2] that if $A(t)$ is real-valued and piecewise continuous, then a unique solution to (1) always exists for any initial state (3).*

1. Definition. An $n \times n$ matrix $\Phi(t)$ is called a *fundamental matrix* of (1) if it satisfies the following two conditions:

$$a)\ \dot{\Phi}(t) = A(t)\Phi(t)\ ,\ b)\ det\,(\Phi(t)) \neq 0, \quad \underline{for\ all}\ t \geq t_o \qquad \triangle \tag{4}$$

From the above definition note that if $\Phi(t)$ is a fundamental matrix of (1), then $\Phi(t)C$ is also a fundamental matrix of (1) where C is any constant $n \times n$ nonsingular matrix (Problem 6.1).

2. Lemma. If $\Phi(t)$ is a solution of (1) and for a specific time t_o $det\,(\Phi(t_o)) \neq 0$, then $det\,(\Phi(t)) \neq 0$ for all t and, therefore $\Phi(t)$ is a fundamental matrix of (1). (See Problem 6.2.)

Let $\Phi(t) = [\Phi_1(t), \Phi_2(t), \ldots, \Phi_n(t)]$. Using the above lemma one can find a fundamental matrix of (1) by assuming n linearly independent initial states $\Phi_1(t_o), \Phi_2(t_o), \ldots$ and $\Phi_n(t_o)$ and solving (1) in each case to obtain $\Phi_1(t), \Phi_2(t), \ldots$ and $\Phi_n(t)$. One convenient choice for $\Phi(t_o)$ is the identity matrix.

3. Example. Consider

$$\dot{x}(t) = \begin{bmatrix} 0 & -1 \\ 1 & 0 \end{bmatrix} x(t)\ ,\ t \geq 0 \tag{5}$$

---

* For the nonlinear state equation $\dot{x} = f(x,u,t)$, a *sufficient* condition for the existence of a unique solution is that the differential equation satisfies a *Lipschitz* condition, i.e. a positive constant k exists such that $||f(x_1,u,t) - f(x_2,u,t)|| \leq k\,||x_1 - x_2||$ for all x_1, x_2 and t [6.2, 6.9].

Let $\Phi(0) = [\Phi_1(0), \Phi_2(0)] = I$. Then it is easy to verify that $\Phi_1(t) = [\cos t \ \sin t]'$ and $\Phi_2(t) = [-\sin t \ \cos t]'$ and a fundamental matrix of (5) is

$$\Phi(t) = [\Phi_1(t), \Phi_2(t)] = \begin{bmatrix} \cos t & -\sin t \\ \sin t & \cos t \end{bmatrix} \tag{6}$$

Another fundamental matrix $\hat{\Phi}(t)$ can be found, for example as follows:

$$\hat{\Phi}(t) = \Phi(t)C = \begin{bmatrix} \cos t & -\sin t \\ \sin t & \cos t \end{bmatrix} \begin{bmatrix} 1 & 2 \\ 0 & -1 \end{bmatrix} = \begin{bmatrix} \cos t & 2\cos t + \sin t \\ \sin t & 2\sin t - \cos t \end{bmatrix} \tag{7}$$

Note that $\det(\Phi(t)) = 1$ and $\det(\hat{\Phi}(t)) = -1$. $\triangle$

If $A(t)$ in (1) is a constant matrix, then $\exp(At)$ will be a fundamental matrix because

$$\exp(At) \triangleq I + At + \frac{A^2 t^2}{2!} + \ldots + \frac{A^k t^k}{k!} + \ldots \tag{8}$$

implies that

$$\frac{d}{dt} \exp(At) = A + A^2 t + \ldots + \frac{A^k t^{k-1}}{(k-1)!} + \ldots = A \exp(At). \tag{9}$$

Also $\det(\exp(At)) \neq 0$ for all t since $\exp(A0) = I$. (See Lemma 2.)

4. Definition. An $n \times n$ matrix $\Phi(t, t_o)$ is called the state *transition matrix* or, simply, the *transition matrix* of (1) if it satisfies the following two conditions:

a) $\dfrac{\partial}{\partial t} \Phi(t, t_o) = A(t) \Phi(t, t_o)$, b) $\Phi(t_o, t_o) = I$, <u>for all</u> t_o <u>for all</u> $t \geq t_o$ $\triangle$ (10)

Note from the above definition that the solution of (1) can be written as

$$x(t) = \Phi(t, t_o) x(t_o). \tag{11}$$

6.2.2 Discrete-Time Systems

Now consider the discrete homogeneous difference equation (2). Let $x \in R^n$ and assume that at the initial discrete time k_o the initial state is

$$x(k_o) = x_o \tag{12}$$

We would like to find the solution $x(k)$ to (2) for $k \geq k_o$ subject to initial state (12). The developments in this case are parallel to those in the continuous-time case.

5. Definition An $n \times n$ matrix $\Phi(k)$ is called a fundamental matrix of (2) if it satisfies the following two conditions:

$$a) \; \Phi(k+1) = A(k)\Phi(k) \;,\; b) \; det(\Phi(k)) \neq 0, \quad k \geq k_o \qquad (13)$$

6. Example. Consider

$$x(k+1) = \begin{bmatrix} 1 & 0 \\ k & 1 \end{bmatrix} x(k), \; k \geq 0 \qquad (14)$$

to find a fundamental matrix of (14) let $\Phi(0) = [\Phi_1(0), \Phi_2(0)] = I$. Then (see Problem 6.4),

$$\Phi(k) = [\Phi_1(k), \Phi_2(k)] = \begin{bmatrix} 1 & 0 \\ \frac{k(k-1)}{2} & 1 \end{bmatrix} \qquad (15)$$

Another fundamental matrix $\hat{\Phi}(k)$ is

$$\hat{\Phi}(k) = \Phi(k)C = \begin{bmatrix} 1 & 0 \\ \frac{k(k-1)}{2} & 1 \end{bmatrix} \begin{bmatrix} -1 & 0 \\ 1 & 2 \end{bmatrix} = \begin{bmatrix} -1 & 0 \\ 1 - \frac{k(k-1)}{2} & 2 \end{bmatrix} \qquad (16)$$

Note that $det(\Phi(k)) = 1$ and $det(\hat{\Phi}(k)) = -2$. $\quad \Delta$

If $A(k)$ in (2) is a constant matrix, then A^k will be a fundamental matrix provided that A is nonsingular.

7. Definition. An $n \times n$ matrix $\Phi(k,k_o)$ is called the *(state) transition matrix* of (2) if it satisfies the following two conditions: written as

$$x(k) = \Phi(k,k_o)x(k_o). \qquad (18)$$

Furthermore,

$$\Phi(k,k_o) = \prod_{p=k_o}^{k-1} A(p) \quad \text{for} \quad k > k_o. \qquad (19)$$

6.3 PROPERTIES OF THE TRANSITION MATRIX

The transition matrix $\Phi(t,t_o)$ of system (6.2.1) has the following properties:

a) Separation property: for any fundamental matrix $\Phi(t)$ of (6.2.1),

$$\Phi(t,t_o) = \Phi(t)\Phi^{-1}(t_o). \qquad (1)$$

b) Uniqueness property: there is a unique $n \times n$ matrix $\Phi(t, t_o)$ which is the transition matrix of (6.2.1).

c) Transition property:

$$\Phi(t,t_o) = \Phi(t,t_1)\Phi(t_1,t_o), \quad \text{for all } t, t_o, t_1 \tag{2}$$

d) Inversion property:

$$\Phi(t,t_o) = \Phi^{-1}(t_o,t). \tag{3}$$

e) Determinant property:

$$\det(\Phi(t,t_o)) = \exp\left[\int_{t_o}^{t} tr(A(\tau))d\tau\right] \tag{4}$$

Property (a) follows from the definition of the transition matrix. Property (b) is a direct result of property (a) (Problem 6.5). Properties (c) and (d) also follow from property (a).

The properties of the transition matrix $\Phi(k,k_o)$ of system (6.2.2) are similar to those of the continuous-time transition matrix $\Phi(t,t_o)$. These are:

a) Separation property: for any fundamental matrix $\Phi(k)$ of (6.2.2),

$$\Phi(k,k_o) = \Phi(k)\Phi^{-1}(k_o). \tag{5}$$

b) Uniqueness property: the transition matrix $\Phi(k,k_o)$ of (6.2.2) is unique.

c) Transition property:

$$\Phi(k,k_o) = \Phi(k,k_1)\Phi(k_1,k_o), \quad \text{for all } k, k_o, k_1 \tag{6}$$

d) Inversion property:

$$\Phi(k,k_o) = \Phi^{-1}(k_o,k). \tag{7}$$

e) Determinant property:

$$\det(\Phi(k,k_o)) = [\det(A(k-1))][\det(A(k-2))]...[\det(A(k_o))], \quad \text{for all } k > k_o \tag{8}$$

The major difference between the transition matrices of continuous-time and discrete-time systems is that $\Phi(t,t_o)$ always has an inverse, however, $\Phi(k,k_o)$ may not always have an inverse. This occurs in some degenerate cases where the system matrix $A(k)$ is singular for some k. Thus relations (5) and (7) hold only when $A(k)$ is nonsingular for all k.

6.4 CALCULATION OF THE TRANSITION MATRIX

If the transition matrix of a system is known, the solution to the homogeneous state equation can be determined from (6.2.11) and (6.2.18). In this section we describe methods of calculating the transition matrix for continuous-time and discrete-time systems. Both time-invariant and time-varying cases will be considered.

6.4.1 Time-Invariant Continuous Case

Consider the homogeneous state equation

$$\dot{\mathbf{x}}(t) = \mathbf{A}\mathbf{x}(t) \tag{1}$$

where $\mathbf{A}$ is a constant $n \times n$ matrix. We showed in Section 6.2 that $\Phi(t) = e^{\mathbf{A}t}$ is a fundamental matrix of (1). The transition matrix of (1) is

$$\Phi(t, t_o) = e^{\mathbf{A}(t-t_o)} \tag{2}$$

because it satisfies conditions (6.2.10). Since (1) is time-invariant, we can conveniently assume $t_o = 0$ with no loss of generality: if $t_o \neq 0$, replace t by $t - t_o$ in the solution. Thus the main task in determining the transition matrix of (1) is to calculate $e^{\mathbf{A}t}$.

A number of methods are available for the calculation of $e^{\mathbf{A}t}$ [6.3]. The three most common methods are the *eigenvalue method*, the *Cayley-Hamilton Method* and the *Laplace Transform Method* which will be described below. We have already encountered two of these methods in Sections 2.8 and 2.10.

6.4.1.1 *The Eigenvalue Method*

If $\mathbf{M}$ is the modal matrix of $\mathbf{A}$, then from Theorem 2.9.3 we have

$$e^{\mathbf{A}t} = \mathbf{M}e^{\mathbf{J}t}\mathbf{M}^{-1} \tag{3}$$

where $\mathbf{J}$ is the Jordan canonical form of $\mathbf{A}$. If $\mathbf{A}$ has distinct eigenvalues $\lambda_1, \lambda_2, ..., \lambda_n$, then

$$\mathbf{J} = \Lambda = diag(\lambda_1, \lambda_2, \ldots, \lambda_n) \tag{4}$$

and

$$e^{\mathbf{J}t} = diag(e^{\lambda_1 t}, e^{\lambda_2 t}, \ldots, e^{\lambda_n t}). \tag{5}$$

If $\mathbf{A}$ does not have distinct eigenvalues, then

$$\mathbf{J} = diag(\mathbf{J}_1, \mathbf{J}_2, \ldots, \mathbf{J}_p). \tag{6}$$

Matrices $\mathbf{J}_i$, $i=1, 2, ..., p$ in (6) are $m_i \times m_i$ where m_i is the multiplicity of eigenvalue λ_i. (See Section 2.8.) Thus (Problem 6.8),

CALCULATION OF THE TRANSITION MATRIX

$$e^{\mathbf{J}t} = diag(e^{\mathbf{J}_1 t}, e^{\mathbf{J}_2 t}, \ldots, e^{\mathbf{J}_r t}). \tag{7}$$

The exponential of each Jordan block $\mathbf{J}_i$ can be calculated easily. (See Problem 6.9.)

6.4.1.2 The Cayley-Hamilton Method

In Section 2.10 we showed that

$$e^{\mathbf{A}t} = \sum_{i=0}^{n-1} c_i(t) \mathbf{A}^i \tag{8}$$

where $c_i(t)$, $i=0, 1, \ldots, n-1$ are linearly independent functions of time. Since $\mathbf{J} = \mathbf{M}^{-1}\mathbf{A}\mathbf{M}$, (8) implies that

$$e^{\mathbf{J}t} = \sum_{i=0}^{n-1} c_i(t) \mathbf{J}^i \tag{9}$$

Equation (9) can be used to evaluate $c_i(t)$, $i=0, 1, \ldots, n-1$ which can then be substituted in (8) to obtain $e^{\mathbf{A}t}$. For example, if $\mathbf{A}$ has distinct eigenvalues $\lambda_1, \lambda_2, \ldots, \lambda_n$, then $\mathbf{J} = \Lambda$ and (9) reduces to

$$e^{\lambda_k t} = \sum_{i=0}^{n-1} c_i(t) \lambda_k^i, \ k = 1, 2, \ldots, n \tag{10}$$

which is a system of n algebraic equations with n unknowns. It can be written as

$$\begin{bmatrix} c_0(t) \\ c_1(t) \\ \cdot \\ \cdot \\ \cdot \\ c_{n-1}(t) \end{bmatrix} = \mathbf{V}^{-1} \begin{bmatrix} e^{\lambda_1 t} \\ e^{\lambda_2 t} \\ \cdot \\ \cdot \\ \cdot \\ e^{\lambda_n t} \end{bmatrix} \tag{11}$$

where $\mathbf{V}$ is a Vandermonde matrix formed by the eigenvalues $\lambda_1, \lambda_2, \ldots, \lambda_n$. (See Equation (2.6.12).)

An extension of the above technique can be used for the evaluation of $c_i(t)$ in (9) if $\mathbf{A}$ has multiple eigenvalues. If eigenvalue λ_j has multiplicity m_j, then $e^{\lambda_j t} = \sum_{i=0}^{n-1} c_i(t) \lambda_j^i$ and its $m_j - 1$ derivatives w.r.t. λ_j will be used to obtain m_j equations corresponding to λ_j. This then results in n equations similar to (10) from which $c_0(t), c_1(t), \ldots, c_n(t)$ can be evaluated.

6.4.1.3 The Laplace Transform Method

The Laplace transform of (1) yields

$$s\mathbf{X}(s) - \mathbf{x}_o = \mathbf{A}\mathbf{X}(s). \tag{12}$$

Thus

$$X(s) = (sI-A)^{-1}x_o. \qquad (13)$$

Here, $X(s)$ is the Laplace transform of the vector solution $x(t)$ to (1) with initial state x_o. Comparison of (13) with the Laplace transform of (6.2.11) implies that

$$L[\Phi(t, t_o)] = (sI-A)^{-1} \qquad (14)$$

That is, the Laplace transform of the transition matrix in the time-invariant continuous case is the *resolvent matrix*. Setting $t_o = 0$ in (2) and (14) yields

$$e^{At} = L^{-1}[(sI-A)^{-1}]. \qquad (15)$$

For an alternative method of showing (15) see Problem 6.10.

Computation of $(sI-A)^{-1}$ is difficult especially for high-order systems. An algorithm for the calculation of $(sI-A)^{-1}$ will be given in Section 6.5.

1. Example. Consider the matrix of Example 2.8.2:

$$A = \begin{bmatrix} 2 & 1 & 0 \\ 1 & 2 & 1 \\ 0 & -1 & 2 \end{bmatrix} \qquad (16)$$

with eigenvalue $\lambda=2$ of multiplicity 3. We will calculate e^{At} using the three methods discussed above. The modal matrix M and the Jordan canonical form J of A are given in Example 2.8.6. Using the eigenvalue method, (3) implies that

$$e^{At} = \begin{bmatrix} 1 & 1 & 1 \\ 0 & 1 & 1 \\ -1 & -1 & 0 \end{bmatrix} \begin{bmatrix} e^{2t} & te^{2t} & \frac{t^2}{2}e^{2t} \\ 0 & e^{2t} & te^{2t} \\ 0 & 0 & e^{2t} \end{bmatrix} \begin{bmatrix} 1 & -1 & 0 \\ -1 & 1 & -1 \\ 1 & 0 & 1 \end{bmatrix}$$

$$= \begin{bmatrix} (1+\frac{t^2}{2})e^{2t} & te^{2t} & \frac{t^2}{2}e^{2t} \\ te^{2t} & e^{2t} & te^{2t} \\ -\frac{t^2}{2}e^{2t} & -te^{2t} & (1-\frac{t^2}{2})e^{2t} \end{bmatrix} \qquad (17)$$

where e^{Jt} has been obtained by using the results of Problem 6.9. In the Cayley–Hamilton method we have to solve the following three equations to obtain c_o, c_1 and c_2:

CALCULATION OF THE TRANSITION MATRIX

$$e^{\lambda t} = \sum_{i=0}^{2} c_i(t)\lambda^i = c_0(t) + c_1(t)\lambda + c_2(t)\lambda^2 \tag{18}$$

$$\frac{d}{d\lambda}(e^{\lambda t}) = te^{\lambda t} = c_1(t) + 2c_2(t)\lambda \tag{19}$$

$$\frac{d^2}{d\lambda^2}(e^{\lambda t}) = t^2 e^{\lambda t} = 2c_2(t) \tag{20}$$

For $\lambda = 2$ the solution is

$$c_0(t) = (1-2t+2t^2)e^{2t}, \quad c_1(t) = (t-2t^2)e^{2t}, \quad c_2(t) = \frac{t^2}{2}e^{2t} \tag{21}$$

and from (8) we have

$$e^{At} = c_0(t)\mathbf{I} + c_1(t)\mathbf{A} + c_2(t)\mathbf{A}^2 \tag{22}$$

resulting in (17). Using Laplace transform method, we have

$$(s\mathbf{I}-\mathbf{A})^{-1} = \begin{bmatrix} s-2 & -1 & 0 \\ -1 & s-2 & -1 \\ 0 & 1 & s-2 \end{bmatrix}^{-1} = \frac{1}{(s-2)^3} \begin{bmatrix} (s-2)^2+1 & s-2 & 1 \\ s-2 & (s-2)^2 & s-2 \\ -1 & -s+2 & (s-2)^2-1 \end{bmatrix} \tag{23}$$

Using (15) and calculating the inverse Laplace transform of each element of the above matrix again yields e^{At} as in (17).

Note that if $\mathbf{A}$ has complex (conjugate) eigenvalues, say $\lambda = \sigma + j\omega$, then e^{At} will have terms such as $e^{\sigma t}\sin\omega t$ and $e^{\sigma t}\cos\omega t$. (See Problem 6.14.) If eigenvalues are multiple, then such terms will also be multiplied by powers of t. From the above discussion we observe that the eigenvalues of matrix $\mathbf{A}$ are important factors in determining the behavior of System (1). This fact will be further illustrated in Chapter 7.

The Cayley-Hamilton method is used in Program "EXPATA" to analytically determine e^{At} for any real square matrix $\mathbf{A}$. The source listing for this program is given below. Two third-order examples, one with real but multiple eigenvalues and one with distinct but complex-conjugate eigenvalues, follow the source listing.

2. "EXPATA"

```
10    !     PROGRAM NAME :  "EXPATA" PROG
20    INPUT "Give your PRINT option: For CRT input 1; for
      LINE PRINTER input 2",P
30    IF (Pr=1) OR (Pr=2) THEN 50
40    GOTO 20
50    IF Pr=1 THEN PRINTER IS 16
60    IF Pr=2 THEN PRINTER IS 7,1
70    PRINT " ´EXPATA´ DETERMINES AN ANALYTIC EXPRESSION FOR
      THE EXPONENTIAL"
80    PRINT "OF A REAL SQUARE MATRIX.",LIN(1)
90    PRINT "    Matrix :   A ,      Its exponential : EXP(A*t)
      ";LIN(2)
100   DATA SUBROUTINES,Mat,Expat,Lambda,Vanmat,Detinv,Coemat,
      Prtmat,Eigen
110   V$="Have the "
120   FOR I=1 TO 9
130   READ F$
140   A$=" "
150   IF I>1 THEN V$="Has "
160   DISP V$;F$;" already been LINKED (Y/N)?";
170   INPUT A$
180   A$=UPC$(A$)
190   IF A$<>"Y" THEN 220
200   IF I=1 THEN 330
210   GOTO 320
220   IF A$<>"N" THEN 140
230   ON I GOTO 320,240,250,260,270,280,290,300,310
240   LINK F$,1000,320
250   LINK F$,2000,320
260   LINK F$,2500,320
270   LINK F$,3000,320
280   LINK F$,4000,320
290   LINK F$,5000,320
300   LINK F$,5500,320
310   LINK F$,6000,320
320   NEXT I
330   INPUT "Give ORDER of the matrix:",N
340   PRINT "Order of the matrix=";N;LIN(1)
350   IF (N<1) OR (N<>INT(N)) THEN 330
360   CALL F(N)
370   PRINTER IS 16
380   END
390   SUB F(N)
400   OPTION BASE 1
410   DIM A(N,N),Evr(N),Evi(N),M(N),Cr(N,N,N),Ci(N,N,N),
      Gr(N,N),Gi(N,N)
420   CALL Mat(A(*),N,N,"A")
430   CALL Expat(N,A(*),P,Evr(*),Evi(*),M(*),Cr(*),Ci(*))
```

```
440   PRINT LIN(2),"EXP(A*t) = SUM π C(i,j) * [ (t^(j-1))
      * EXP(Eigenvalue(i)*t) ",LIN(2),"i = 1 TO";P;"
      ( NUMBER OF DISTINCT EIGENVALUES )"
450   PRINT LIN(1),"j = 1 TO M(i) ( MULTIPLICITY OF
      EIGENVALUE(i) )"
460   J1=0
470   FOR I=1 TO P
480   PRINT LIN(2),"Eigenvalue(";I;") = ";
490   IF (Evr(I)<>0) OR (Evi(I)<>0) THEN 520
500   PRINT "0"
510   GOTO 570
520   IF Evr(I)<>0 THEN PRINT Evr(I);"   ";
530   IF Evi(I)=0 THEN PRINT LIN(1)
540   IF Evi(I)>0 THEN PRINT "+ j";Evi(I)
550   IF Evi(I)<0 THEN PRINT "- j";-Evi(I)
560   PRINT "M(";I;") = ";M(I),LIN(1)
570   FOR J=1 TO M(I)
580   J1=J1+1
590   FOR K=1 TO N
600   FOR L=1 TO N
610   Gr(K,L)=Cr(J1,K,L)
620   Gi(K,L)=Ci(J1,K,L)
630   IF ABS(Gr(K,L))<.0005 THEN Gr(K,L)=0
640   IF ABS(Gi(K,L))<.0005 THEN Gi(K,L)=0
650   NEXT L
660   NEXT K
670   PRINT LIN(1),"MATRIX C(";I;",";J;"):",LIN(2),
      "REAL PART:",LIN(1)
680   CALL Prtmat(Gr(*),N,N)
690   PRINT "IMAGINARY PART:",LIN(1)
700   CALL Prtmat(Gi(*),N,N)
710   NEXT J
720   NEXT I
730   SUBEND
```

3. Example

'EXPATA' DETERMINES AN ANALYTIC EXPRESSION FOR THE EXPONENTIAL
OF A REAL SQUARE MATRIX.

    Matrix : A ,   Its exponential : EXP(A*t)

Order of the matrix= 3

Matrix A(3 x 3):

-1.0000E+00   0.0000E+00   0.0000E+00

0.0000E+00 -4.0000E+00   4.0000E+00

0.0000E+00 -1.0000E+00   0.0000E+00

EXP(A*t) = SUM π C(i,j) * [(t^(j-1)) * EXP(Eigenvalue(i)*t)]

i = 1 TO 2 (NUMBER OF DISTINCT EIGENVALUES)

j = 1 TO M(i) (MULTIPLICITY OF EIGENVALUE(i))

Eigenvalue(1) = -1

M(1) = 1

MATRIX C(1 , 1):

REAL PART:

1.0000E+00   0.0000E+00   0.0000E+00

0.0000E+00   0.0000E+00   0.0000E+00

0.0000E+00   0.0000E+00   0.0000E+00

IMAGINARY PART:

0.0000E+00   0.0000E+00   0.0000E+00

0.0000E+00   0.0000E+00   0.0000E+00

0.0000E+00   0.0000E+00   0.0000E+00

Eigenvalue(2) = -2

M(2) = 2

MATRIX C(2 , 1):

REAL PART:

0.0000E+00   0.0000E+00   0.0000E+00

```
 0.0000E+00  1.0000E+00  0.0000E+00

 0.0000E+00  0.0000E+00  1.0000E+00
```

IMAGINARY PART:

```
 0.0000E+00  0.0000E+00  0.0000E+00

 0.0000E+00  0.0000E+00  0.0000E+00

 0.0000E+00  0.0000E+00  0.0000E+00
```

MATRIX C(2 , 2):

REAL PART:

```
 0.0000E+00  0.0000E+00  0.0000E+00

 0.0000E+00 -2.0000E+00  4.0000E+00

 0.0000E+00 -1.0000E+00  2.0000E+00
```

IMAGINARY PART:

```
 0.0000E+00  0.0000E+00  0.0000E+00

 0.0000E+00  0.0000E+00  0.0000E+00

 0.0000E+00  0.0000E+00  0.0000E+00
```

4. Example

Order of the matrix= 3

Matrix A(3 x 3):

```
 1.0000E+00  2.0000E+00  0.0000E+00

-1.0000E+00  3.0000E+00  0.0000E+00

 2.0000E+00 -4.0000E+00  3.0000E+00
```

EXP(A*t) = SUM π C(i,j) * [(t^(j-1)) * EXP(Eigenvalue(i)*t)]

i = 1 TO 3 (NUMBER OF DISTINCT EIGENVALUES)

j = 1 TO M(i) (MULTIPLICITY OF EIGENVALUE(i))

Eigenvalue(1) =   1.99999999996   + j .99999999995
M(1) = 1

MATRIX C(1 , 1):

REAL PART:

  5.0000E-01   0.0000E+00   0.0000E+00

  0.0000E+00   5.0000E-01   0.0000E+00

 -1.0000E+00   1.0000E+00   0.0000E+00

IMAGINARY PART:

  5.0000E-01 -1.0000E+00   0.0000E+00

  5.0000E-01 -5.0000E-01   0.0000E+00

  0.0000E+00   1.0000E+00   0.0000E+00

Eigenvalue(2) =   1.99999999996   - j .99999999995
M(2) = 1

MATRIX C(2 , 1):

REAL PART:

  5.0000E-01   0.0000E+00   0.0000E+00

  0.0000E+00   5.0000E-01   0.0000E+00

 -1.0000E+00   1.0000E+00   0.0000E+00

IMAGINARY PART:

 -5.0000E-01   1.0000E+00   0.0000E+00

```
-5.000 0E-01    5.0000E-01    0.0000E+00

 0.0000E+00   -1.000 0E+00   0.0000E+00
```

```
Eigenvalue( 3 ) = 3

M( 3 ) = 1

MATRIX C( 3 , 1 ):

REAL PART:

 0.0000E+00    0.0000E+00    0.0000E+00

 0.0000E+00    0.0000E+00    0.0000E+00

 2.0000E+00   -2.000 0E+00   1.0000E+00


IMAGINARY PART:

 0.0000E+00    0.000 0E+00   0.0000E+00

 0.0000E+00    0.0000E+00    0.0000E+00

 0.0000E+00    0.0000E+00    0.0000E+00
```

6.4.2 Time-Invariant Discrete Case

Consider the homogeneous state equation

$$x(k+1) = Ax(k) \qquad (24)$$

where A is a constant $n \times n$ matrix. In Section 6.2 we showed that if A is nonsingular, then A^k is a fundamental matrix of (24). The transition matrix of (24) is

$$\Phi(k, k_o) = A^{k-k_o} \qquad (25)$$

because it satisfies both conditions (6.2.17). Here again we can assume $k_o = 0$ with no loss of generality: if $k_o \neq 0$, replace k by $k - k_o$ in the solution. Thus the main task in determining the transition matrix of (24) is to calculate A^k.

Methods of calculating A^k are analogous to those for calculating e^{At}, i.e., the *eigenvalue method*, the *Cayley-Hamilton method* and the *z transform method*. These will be discussed below.

6.4.2.1 *The Eigenvalue Method*

If **M** is the modal matrix of **A**, then from Theorem 2.9.3 we have

$$\mathbf{A}^k = \mathbf{M}\mathbf{J}^k\mathbf{M}^{-1} \tag{26}$$

where **J** is the Jordan canonical form of **A**. If **A** has distinct eigenvalues $\lambda_1, \lambda_2, ..., \lambda_n$, then (4) holds and

$$\mathbf{J}^k = diag(\lambda_1^k, \lambda_2^k, \ldots, \lambda_n^k). \tag{27}$$

If **A** does not have distinct eigenvalues, then (6) holds and

$$\mathbf{J}^k = diag(\mathbf{J}_1^k, \mathbf{J}_2^k, \ldots, \mathbf{J}_p^k). \tag{28}$$

The *kth* power of each Jordan block $\mathbf{J}_i$ can be calculated easily. (See Problem 6.9.)

6.4.2.2 *The Cayley-Hamilton Method*

In Section 2.10 we showed that

$$\mathbf{A}^k = \sum_{i=0}^{n-1} c_i(k)\mathbf{A}^i \text{ for } k \geq n \tag{29}$$

where c_i, $i=0, 1, ..., n-1$ are functions of k. Since $\mathbf{J} = \mathbf{M}^{-1}\mathbf{A}\mathbf{M}$, (29) implies that

$$\mathbf{J}^k = \sum_{i=0}^{n-1} c_i(k)\mathbf{J}^i \tag{30}$$

The above equation can be used to evaluate $c_i(k)$, $i=0, 1, ..., n-1$ which can then be substituted in (29) to obtain $\mathbf{A}^k$. This procedure is similar to that explained in Subsection 6.4.1.2 for the calculation of $e^{\mathbf{A}t}$.

6.4.2.3 *The z-Transform Method*

The z transform of (24) yields

$$z\mathbf{X}(z) - z\mathbf{x}_o = \mathbf{A}\mathbf{X}(z). \tag{31}$$

Thus

$$\mathbf{X}(z) = (z\mathbf{I}-\mathbf{A})^{-1}z\mathbf{x}_o. \tag{32}$$

Here $\mathbf{X}(z)$ is the z transform of the vector solution $\mathbf{x}(k)$ to (24) with initial state $\mathbf{x}_o$. Comparison of (32) with the z transform of (6.2.18) implies that

ERRATA

JAMSHIDI & MALEK-ZAVAREI: Linear Control Systems

	For	Read
Page 15, definition 5	WCV	W=V
Page 24, equation 7	$(A-\lambda-I)$	$(A-\lambda I)$
Page 31, line 19	$(A-\lambda I)u_2$	$(A-\lambda I)^2 u_2$
Page 116, line 3	n≥d	n≥m
Page 133, equation 19	$R(t)u(t)$	$Ru(t)$
Page 140, equation 11	$z^d + q_{m-1}z^{d-1}$	$z^m + q_{m-1}z^{m-1}$
Page 155, line 9	For u=0 and initial conditions $x_1(0)=1$, $x_2(0)=0$ and $x_3(0)=0$ the solution to (14) is	For u = 0 a nominal solution to (14) is
Page 156, problem 4.6a	$\ddot{y} - 2y$	$\dddot{y} - 2y$
Page 158, problem 4.16	$y(0) = 1$, $y(0) = 0$	$\hat{x}_1 = -6/t^2$, $\hat{x}_2 = 12/t^3$
Page 217, equation 34	z^{-1}	z^{-1}
Page 282, fig. caption	Fig. 7.7.2. Polar plot for an RC-filter	The unit circle in the z-plane
Page 311, Fig. 7.6.1	K = 25	K = 25/4
Page 311, line 14	realaxis	real axis
Page 312, line 22	ss = -5/2	s = -5/2
Page 366, equation 8	delete "Then" and run on equation	
Page 382, line 19	an	and
Page 391, equation 7	d	dt
Page 398, line 11	B, AB, ^{2}B...	B, AB, $A^2 B$...
Page 489, equation 12	$(p_{n-1}-k_{n-1})^{n-1}+\ldots+(p_1-k_1)+(p_0-k_0)$	$(p_{n-1}+k_{n-1})^{n-1}+\ldots+(p_1+k_1)+(p_0+k_0)$
Page 515, equation 23	$\dfrac{K_1^2+K_2^2+2K_2^2-2K_1 K_2}{K_1(K_2-K_1)}$	$\dfrac{(1-K_1)[1+K_2-K_1)^2]+2(2-K_1)(1+K_2-K_1)}{K_1(K_1-2)}$
Page 538, equation 7	$-x'Qx$	$-1/2 x'Qx$
Page 544, problem 11.5	+4	+1
Page 544, problem 11.9	K>0	0<K 10

CALCULATION OF THE TRANSITION MATRIX

$$\mathbf{Z}[\Phi(k,k_o)] = (z\mathbf{I} - \mathbf{A})^{-1}z. \tag{33}$$

Setting $k_0 = 0$ in (25) and (33) yields

$$\mathbf{A}^k = z^{-1}[(z\mathbf{I} - \mathbf{A})^{-1}z] \tag{34}$$

Here again, as in the continuous case, the main difficulty is the computation of $(z\mathbf{I} - \mathbf{A})^{-1}$. This will be taken up in Section 6.5.

5. Example. Consider the matrix

$$\mathbf{A} = \begin{bmatrix} 2 & 0 \\ 1 & 2 \end{bmatrix} \tag{35}$$

with eigenvalue $\lambda = 2$ of multiplicity 2. We will calculate $\mathbf{A}^k$ using the three methods discussed above. The modal matrix $\mathbf{M}$ and the Jordan canonical form $\mathbf{J}$ of $\mathbf{A}$ are easily found to be

$$\mathbf{M} = \begin{bmatrix} 0 & 1 \\ 1 & 0 \end{bmatrix}, \quad \mathbf{J} = \mathbf{M}^{-1}\mathbf{A}\mathbf{M} = \begin{bmatrix} 2 & 1 \\ 0 & 2 \end{bmatrix} \tag{36}$$

Using the eigenvalue method, (26) implies that

$$\mathbf{A}^k = \begin{bmatrix} 0 & 1 \\ 1 & 0 \end{bmatrix} \begin{bmatrix} 2^k & k2^{k-1} \\ 0 & 2^k \end{bmatrix} \begin{bmatrix} 0 & 1 \\ 1 & 0 \end{bmatrix} = \begin{bmatrix} 2^k & 0 \\ k2^{k-1} & 2^k \end{bmatrix} \tag{37}$$

where $\mathbf{J}^k$ has been obtained by using the result of Problem 6.9. In the Cayley-Hamilton method we have to solve the following two equations to obtain c_o and c_1:

$$\lambda^k = c_o(k) + c_1(k)\lambda$$

$$\frac{d}{d\lambda}(\lambda^k) = k\lambda^{k-1} = c_1(k) \tag{38}$$

For $\lambda = 2$ the solution is

$$c_o(k) = (1-k)2^k, \quad c_1(k) = k2^{k-1} \tag{39}$$

and from (29) we have

$$\mathbf{A}^k = c_o(k)\mathbf{I} + c_1(k)\mathbf{A} \tag{40}$$

which yields (37). Finally, using the z transform method, we have

$$(zI-A)^{-1}z = \begin{bmatrix} z-2 & 0 \\ -1 & z-2 \end{bmatrix}^{-1} z = \frac{1}{(z-2)^2}\begin{bmatrix} z(z-2) & 0 \\ z & z(z-2) \end{bmatrix} \tag{41}$$

Using (34) and calculating the inverse z transform of each element of the above matrix again yields A^k as in (37). △

Here again note that, similar to the continuous case, the eigenvalues of A determine the behavior of system (24). This will be further demonstrated in Chapter 7.

Program "APOWRK" uses the z transform method to determine A^k for any real square matrix A. The source listing for this program is given below followed by two examples.

6. "APOWRK"

```
10      ! PROGRAM NAME : "APOWRK" PROG
20      PRINT " 'APOWRK' DETERMINES ANALYTIC EXPRESSION FOR THE
        TRANSITION MATRIX"
30      PRINT "  OF A LINEAR TIME-INVARIANT DISCRETE-TIME
        SYSTEM, i.e. (A^k)"
40      DATA SUBROUTINES,Inpmat,Expat,Lambda,Vanmat,Detinv,
        Coemat,Prtmat,Eigen
50      V$="Have the "
60      FOR I=1 TO 9
70      READ F$
80      A$=" "
90      IF I>1 THEN V$="Has "
100     DISP V$;F$;" already been LINKED ? (Y/N)";
110     INPUT A$
120     A$=UPC$(A$)
130     IF (A$<>"Y") AND (A$<>"YES") THEN 160
140     IF I=1 THEN 270
150     GOTO 260
160     IF (A$<>"N") AND (A$<>"NO") THEN 80
170     ON I GOTO 260,180,190,200,210,220,230,240,250
180     LINK F$,1000,260
190     LINK F$,2000,260
200     LINK F$,3000,260
210     LINK F$,4000,260
220     LINK F$,5000,260
230     LINK F$,6000,260
240     LINK F$,7000,260
250     LINK F$,8000,260
260     NEXT I
270     INPUT "Order of SQUARE matrix A =?",N
280     IF (N<1) OR (N<>INT(N)) THEN 270
290     CALL F(N)
300     END
310     SUB F(N)
320     OPTION BASE 1
```

```
330   DIM A(N,N),Evr(N),Evi(N),M(N),Cr(N,N,N),Ci(N,N,N),
      Gr(N,N),Gi(N,N)
340   CALL Inpmat(N,N,"A",A(*))
350   CALL Expat(N,A(*),P,Evr(*),Evi(*),M(*),Cr(*),Ci(*))
360   PRINT "A ^ k = SUM OF [ C(i,j) * P(k) * ( EIGENVALUE(i)
      ^ (k+j-1) ) ]"
370   PRINT LIN(1),"j = 1 TO M(i) ( MULTIPLICITY OF
      EIGENVALUE(i) )"
380   PRINT LIN(1),"i = 1 TO";P;"( NUMBER OF DISTINCT
      EIGENVALUES )"
390   PRINT LIN(1),"IF j = 1 THEN P(k) = 1"
400   PRINT LIN(1),"IF j > 1 THEN P(k) = k * (k-1) * ... *
      (k-j+2)"
410   J1=0
420   FOR I=1 TO P
430   PRINT LIN(5),"EIGENVALUE(";I;") = ";
440   IF (Evr(I)<>0) OR (Evi(I)<>0) THEN 470
450   PRINT "0"
460   GOTO 520
470   IF Evr(I)<>0 THEN PRINT Evr(I);" ";
480   IF Evi(I)=0 THEN PRINT LIN(1)
490   IF Evi(I)>0 THEN PRINT "+   j ";Evi(I)
500   IF Evi(I)<0 THEN PRINT "-   j ";-Evi(I)
510   PRINT "M(";I;") = ";M(I),LIN(2)
520   FOR J=1 TO M(I)
530   J1=J1+1
540   FOR K=1 TO N
550   FOR L=1 TO N
560   Gr(K,L)=Cr(J1,K,L)
570   Gi(K,L)=Ci(J1,K,L)
580   NEXT L
590   NEXT K
600   PRINT LIN(1),"MATRIX C(";I;",";J;"):",LIN(2),
      "REAL PART",LIN(1)
610   CALL Prtmat(Gr(*),N,N)
620   PRINT LIN(2),"IMAGINARY PART",LIN(1)
630   CALL Prtmat(Gi(*),N,N)
640   NEXT J
650   NEXT I
660   SUBEND
```

7. Example

```
'APOWRK' DETERMINES AN ANALYTIC EXPRESSION FOR THE TRANSITION
  MATRIX OF A LINEAR TIME-INVARIANT DISCRETE-TIME
  SYSTEM, i.e. (A^k)
DIMENSION OF MATRIX A = 3 BY 3
```

LINEAR CONTROL SYSTEMS

MATRIX A:

```
 1.00000E+00 -1.00000E+00  1.00000E+00
 0.00000E+00  1.00000E+00  1.00000E+00
 0.00000E+00  0.00000E+00  1.00000E+00
```

A ^ k = SUM OF [C(i,j) * P(k) * (EIGENVALUE(i) ^ (k+j-1))]

j = 1 TO M(i) (MULTIPLICITY OF EIGENVALUE(i))

i = 1 TO 1 (NUMBER OF DISTINCT EIGENVALUES)

IF j = 1 THEN P(k) = 1

IF j > 1 THEN P(k) = k * (k-1) * ... * (k-j+2)

EIGENVALUE(1) =   .99999999999

M(1) = 3

MATRIX C(1 , 1):

REAL PART

```
 1.0000E+00  0.0000E+00  0.0000E+00
 0.0000E+00  1.0000E+00  0.0000E+00
 0.0000E+00  0.0000E+00  1.0000E+00
```

IMAGINARY PART

```
 0.0000E+00  0.0000E+00  0.0000E+00
 0.0000E+00  0.0000E+00  0.0000E+00
 0.0000E+00  0.0000E+00  0.0000E+00
```

MATRIX C(1 , 2):

REAL PART

```
 0.0000E+00 -1.0000E+00  1.0000E+00
```

CALCULATION OF THE TRANSITION MATRIX

```
0.0000E+00   0.0000E+00   1.0000E+00

0.0000E+00   0.0000E+00   0.0000E+00
```

IMAGINARY PART

```
0.0000E+00   0.0000E+00   0.0000E+00

0.0000E+00   0.0000E+00   0.0000E+00

0.0000E+00   0.0000E+00   0.0000E+00
```

MATRIX C(1 , 3):

REAL PART

```
0.0000E+00   0.0000E+00  -5.0000E-01

0.0000E+00   0.0000E+00   0.0000E+00

0.0000E+00   0.0000E+00   0.0000E+00
```

IMAGINARY PART

```
0.0000E+00   0.0000E+00   0.0000E+00

0.0000E+00   0.0000E+00   0.0000E+00

0.0000E+00   0.0000E+00   0.0000E+00
```

Example. Matrix A in this example is the same as that in Example 5.

```
'APOWRK' DETERMINES ANALYTIC EXPRESSION FOR THE TRANSITION MATRIX
OF A LINEAR TIME-INVARIANT DISCRETE-TIME SYSTEM, i.e. (A^k)
2.00000E+00   0.00000E+00

1.00000E+00   2.00000E+00


   ^ k = SUM OF [ C(i,j) * P(k) * ( EIGENVALUE(i) ^ (k+j-1) ) ]

   = 1 TO M(i) ( MULTIPLICITY OF EIGENVALUE(i) )

   = 1 TO 1 ( NUMBER OF DISTINCT EIGENVALUES )
```

IF j = 1 THEN P(k) = 1

IF j > 1 THEN P(k) = k * (k-1) * ... * (k-j+2)

EIGENVALUE(1) = 2

M(1) = 2

MATRIX C(1 , 1):

REAL PART

  1.0000E+00   0.0000E+00
  0.0000E+00   1.0000E+00

IMAGINARY PART

  0.0000E+00   0.0000E+00
  0.0000E+00   0.0000E+00

MATRIX C(1 , 2):

REAL PART

  0.0000E+00   0.0000E+00
  1.0000E+00   0.0000E+00

IMAGINARY PART

  0.0000E+00   0.0000E+00
  0.0000E+00   0.0000E+00

6.4.3 Transition Matrix in the Time-Varying Case.

Unlike the *l.t.i.* case, there is no general method for calculation of the transition matrix of linear time-varying systems. The available methods are either very tedious or apply to very special cases. These methods can be categorized as follows:

a) Using separation property of the transition matrix. Sometimes the separation property (6.3.1) or (6.3.5) can be used to obtain the transition matrix of linear time-varying systems. This, of course, requires the knowledge of a fundamental matrix of the system. (See Problem 6.15.)

b) Direct solution of the state equation. If the homogeneous state equation can be solved directly, the solution can be put in the form $\mathbf{x}(t) = \Phi(t,t_o)\mathbf{x}(t_o)$ in the continuous case, or $\mathbf{x}(k) = \Phi(k,k_o)\mathbf{x}(k_o)$ in the discrete case to obtain the transition matrix. This method is illustrated in Problem 6.16.

c) The following theorem provides a method for evaluation of the transition matrix of a special class of linear time-varying systems.

9. Theorem. If $\mathbf{A}(t)$ and $\mathbf{A}(\tau)$ commute for all t, τ, i.e. if

$$\mathbf{A}(t)\mathbf{A}(\tau) = \mathbf{A}(\tau)\mathbf{A}(t), \quad \text{for all } t, \text{ for all } \tau \tag{42}$$

then the transition matrix of the system $\dot{\mathbf{x}}(t) = \mathbf{A}(t)\mathbf{x}(t)$ is

$$\Phi(t,t_o) = \exp[\int_{t_o}^{t} \mathbf{A}(\tau)d\tau] \tag{43}$$

Proof. First note that (2) implies $\Phi(t_o,t_o) = \exp[0] = \mathbf{I}$. Also from (2.9.7) we have

$$\exp[\int_{t_o}^{t} \mathbf{A}(\tau)d\tau] = \mathbf{I} + \int_{t_o}^{t} \mathbf{A}(\tau)d\tau + \frac{1}{2!} \int_{t_o}^{t} \mathbf{A}(\tau)d\tau \int_{t_o}^{t} \mathbf{A}(\eta)d\eta + \ldots \tag{44}$$

Taking the derivative of (44) w.r.t t yields

$$\frac{\partial}{\partial t}\exp[\int_{t_o}^{t} \mathbf{A}(\tau)d\tau] = \mathbf{A}(t) + \frac{1}{2!}[\mathbf{A}(t)\int_{t_o}^{t} \mathbf{A}(\eta)d\eta + \int_{t_o}^{t} \mathbf{A}(\tau)d\tau\mathbf{A}(t)] + \ldots \tag{45}$$

Also from (44) we have

$$\mathbf{A}(t)\exp[\int_{t_o}^{t} \mathbf{A}(\tau)d\tau] = \mathbf{A}(t) + \mathbf{A}(t)\int_{t_o}^{t} \mathbf{A}(\tau)d\tau + \ldots \tag{46}$$

It can be easily verified that (45) and (46) are equal if and only if

$$\mathbf{A}(t) \int_{t_o}^{t} \mathbf{A}(\tau) d\tau = \int_{t_o}^{t} \mathbf{A}(\tau) d\tau \mathbf{A}(t). \tag{47}$$

Differentiating (47) w.r.t. τ and multiplying both sides by -1 yields condition (42). q.e.d. △

Note that condition (42) is satisfied when

a) $\mathbf{A}(t)$ is a constant matrix.

b) $\mathbf{A}(t)$ is diagonal

c) $\mathbf{A}(t) = f(t)\mathbf{C}$ where $f(t)$ is any scalar function and $\mathbf{C}$ is any constant matrix.

10. Example. Consider $\dot{\mathbf{x}}(t) = \mathbf{A}(t)\mathbf{x}(t)$ where

$$A(t) = \begin{bmatrix} 2t & 0 \\ 0 & 1 \end{bmatrix} \tag{48}$$

Since $A(t)$ and $A(\tau)$ commute for all t, τ, we have $\Phi(t,t_o) = e^{\mathbf{B}}$ where

$$\mathbf{B} = \int_{t_o}^{t} \mathbf{A}(\tau) d\tau = \begin{bmatrix} t^2 - t_o^2 & 0 \\ 0 & t - t_o \end{bmatrix} \tag{49}$$

Thus

$$\Phi(t,t_o) = \begin{bmatrix} e^{t^2 - t_o^2} & 0 \\ 0 & e^{t - t_o} \end{bmatrix} \quad △ \tag{50}$$

Note from the above example that in linear time-varying systems the eigenvalues of the system matrix $\mathbf{A}(t)$ do not necessarily determine the dynamic behavior of the system as they do in the l.t.i. case. (Also see Problem 6.17.) We will further demonstrate this fact in Chapter 7.

6.5 RESOLVENT MATRIX

As discussed in the previous section the resolvent matrix $(s\mathbf{I}-\mathbf{A})^{-1}$ or $(z\mathbf{I}-\mathbf{A})^{-1}$ plays an important role in the analysis of l.t.i. systems. In continuous-time systems $(s\mathbf{I}-\mathbf{A})^{-1}$ is the Laplace transform of $e^{\mathbf{A}t}$, and in discrete-time systems $(z\mathbf{I}-\mathbf{A})^{-1}$ is the z transform of the matrix sequence $\mathbf{0}, \mathbf{I}, \mathbf{A}, \mathbf{A}^2, \ldots$ For low-order systems, especially when matrix $\mathbf{A}$ has many zero elements, the resolvent matrix can be computed by direct calculation of the inverse of the matrix $s\mathbf{I}-\mathbf{A}$ (or $z\mathbf{I}-\mathbf{A}$). This direct method, however, is not efficient for high-order systems. In this section an algorithm for the efficient evaluation of the resolvent matrix will be presented and some of its important properties will be pointed out. For convenience the term $(s\mathbf{I}-\mathbf{A})^{-1}$ will be used for the resolvent matrix throughout this section. However, the properties, of course, hold also in the discrete-time case where s is replaced by z.

1. Lemma. The resolvent matrix is a strictly proper rational function of s. More precisely, if the system matrix **A** is $n \times n$, we can always write

$$(s\mathbf{I}-\mathbf{A})^{-1} = \frac{Q(s)}{p(s)} = \frac{\mathbf{Q}_{n-1}s^{n-1} + \mathbf{Q}_{n-2}s^{n-2} + ... + \mathbf{Q}_1 s + \mathbf{Q}_0}{s^n + p_{n-1}s^{n-1} + ... + p_1 s + p_0} \qquad (1)$$

where $\mathbf{Q}_i$, $i=0, 1, 2, ..., n-1$, are constant $n \times n$ matrices.

Proof. We can write

$$(s\mathbf{I}-\mathbf{A})^{-1} = \frac{adj(s\mathbf{I}-\mathbf{A})}{\det(s\mathbf{I}-\mathbf{A})} \qquad (2)$$

where $adj(s\mathbf{I}-\mathbf{A})$ indicates the adjoint matrix of $s\mathbf{I}-\mathbf{A}$, i.e. the transpose of the $n \times n$ matrix of cofactors of $s\mathbf{I}-\mathbf{A}$. The elements of $adj(s\mathbf{I}-\mathbf{A})$ are determinants of $(n-1) \times (n-1)$ submatrices of $s\mathbf{I}-\mathbf{A}$ formed by deleting row-column pairs [6.4]. Therefore they are polynomials in s of order at most $n-1$. Thus it is established that the numerator of $(s\mathbf{I}-\mathbf{A})^{-1}$ has the form shown in (1). Also we have

$$\det(s\mathbf{I}-\mathbf{A}) = (-1)^n \det(\mathbf{A}-s\mathbf{I}) = (-1)^n \Delta(s) \qquad (3)$$

where $\Delta(s)$ is the characteristic polynomial of **A** which is an nth-order polynomial in s. (See Section 2.6.) Therefore the denominator of $(s\mathbf{I}-\mathbf{A})^{-1}$ has the form shown in (1). q.e.d.

2. Example. Let

$$A = \begin{bmatrix} 0 & -1 & -1 \\ 1 & -2 & -1 \\ 0 & 0 & -2 \end{bmatrix} \qquad (4)$$

Thus

$$(s\mathbf{I}-\mathbf{A})^{-1} = \begin{bmatrix} s & 1 & 1 \\ -1 & s+2 & 1 \\ 0 & 0 & s+2 \end{bmatrix}^{-1} = \frac{1}{\det(s\mathbf{I}-\mathbf{A})} \begin{bmatrix} (s+2)^2 & -(s+2) & -s-1 \\ s+2 & s(s+2) & -s-1 \\ 0 & 0 & s(s+2)+1 \end{bmatrix}$$

$$= \frac{\mathbf{Q}_2 s^2 + \mathbf{Q}_1 s + \mathbf{Q}_0}{s^3 + 4s^2 + 5s + 2} \qquad (5)$$

where

$$\mathbf{Q}_2 = \mathbf{I}, \quad \mathbf{Q}_1 = \begin{bmatrix} 4 & -1 & -1 \\ 1 & 2 & -1 \\ 0 & 0 & 2 \end{bmatrix}, \quad \mathbf{Q}_0 = \begin{bmatrix} 4 & -2 & -1 \\ 2 & 0 & -1 \\ 0 & 0 & 1 \end{bmatrix} \qquad \triangle$$

The following theorem referred to as *Leverrier's algorithm* [6.5] or *Fadeeva's method* [6.6] provides a procedure for evaluation of the resolvent matrix.

3. Theorem. The matrix coefficients Q_i and the scalar coefficients p_i, $i=0, 1, 2, ..., n-1$ in (1) can be determined as follows:

$$Q_{i-1} = Q_i A + p_i I , \quad i=1,2,...,n-1 \tag{6}$$

$$Q_{n-1} = 1 \tag{7}$$

$$p_i = -\frac{1}{n-i} tr(Q_i A) , \quad i=0, 1, 2, ..., n-1 \tag{8}$$

Furthermore, matrices A and Q_i, $i=0, 1, 2, ..., n-1$ commute and we have

$$0 = Q_o A + p_o I \tag{9}$$

Proof. Rearrange (1) as

$$(Q_{n-1}s^{n-1} + Q_{n-2}s^{n-2} +...+ Q_1 s + Q_o)(sI-A)$$

$$= (s^n + p_{n-1}s^{n-1} +...+ p_1 s + p_o)I \tag{10}$$

Equating the like powers of s in (10) yields

$$Q_{n-1} = I \tag{11}$$

$$-AQ_i + Q_{i-1} = p_i I, \quad i=n-1, n-2, ..., 2, 1 \tag{12}$$

$$-Q_o A = p_o I \tag{13}$$

which are the desired relations. If (1) is rearranged as

$$(sI-A)(Q_{n-1}s^{n-1} + Q_{n-2}s^{n-2} +...+ Q_1 s + Q_o)$$

$$= (s^n + p_{n-1}s^{n-1} +...+ p_1 s + p_o)I \tag{14}$$

then equating the like powers of s will yield

$$Q_{n-1} = I \tag{15}$$

$$-AQ_i + Q_{i-1} = p_i I, \; i = n-1, \, n-2, \, ..., \, 2, \, 1 \tag{16}$$

$$-AQ_o = p_o I \tag{17}$$

Comparison of (12) with (16) and (13) with (17) indicates that the matrices A and Q_i, $i = 0, 1, 2, ..., n-1$ commute. The proof of relation (8) is more involved an will not be presented here. It can be found in any one of the References [6.5 - 6.8]. q.e.d. Δ

Equations (6), (7) and (8) can be used to evaluate the matrix coefficients Q_i and scalar coefficients p_i successively for $i = n-1, n-2, ..., 1, 0$. Equation (9) can be used to check the accuracy of the computations.

4. Example. Consider matrix A in Example 2. Using the algorithm given in Theorem 3 we find

$$Q_2 = I \; , \; p_2 = -tr(Q_2 A) = -tr(A) = 4,$$

$$Q_1 = Q_2 A + p_2 I = A + 4I = \begin{bmatrix} 4 & -1 & -1 \\ 1 & 2 & -1 \\ 0 & 0 & 2 \end{bmatrix}$$

$$p_1 = -\frac{1}{2} tr(Q_1 A) = -\frac{1}{2}(-1-5-4) = 5$$

$$Q_o = Q_1 A + p_1 I = \begin{bmatrix} 4 & -2 & -1 \\ 2 & 0 & -1 \\ 0 & 0 & 1 \end{bmatrix}$$

$$p_o = -\frac{1}{3} tr(Q_o A) = -\frac{1}{3}(-2-2-2) = 2$$

which are the coefficients found in Example 2. We also have

$$Q_o A + p_o I = 0$$

which verifies the accuracy of the preceding computations. Δ

Program "RESMAT" uses the algorithm presented in Theorem 3 to evaluate the resolvent matrix $(sI-A)^{-1}$ of any real square matrix A. The source listing of "RESMAT" followed by a fifth-order example is given below.

5. "RESMAT"

```
10    !         PROGRAM NAME : "RESMAT" PROG
20    INPUT "Give your PRINT option: For CRT input 1; for LINE
      PRINTER input 2",P
30    IF (Pr=1) OR (Pr=2) THEN 50
40    GOTO 20
50    IF Pr=1 THEN PRINTER IS 16
60    IF Pr=2 THEN PRINTER IS 7,1
70    PRINT " 'RESMAT' FINDS THE ANALYTIC EXPRESSION FOR THE "
80    PRINT " RESOLVENT MATRIX [sI-A]^-1 = Q(s)/P(s) WHERE :"
90    PRINT "    Q(s) = SUM πQ(i)*(s^i)→ & P(s) = s^N + SUM
      πP(i)*(s^i)→,"
100   PRINT "     i = 0,1,...,N-1. N is the order of square
      matrix A.",LIN(2)
110   DATA subroutines,Mat,Resmat,Prtmat
120   V$="Have the "
130   FOR I=1 TO 4
140   READ F$
150   A$=" "
160   IF I>1 THEN V$="Has "
170   DISP V$;F$;" already been linked ? (Y/N)";
180   INPUT A$
190   A$=UPC$(A$)
200   IF (A$<>"Y") AND (A$<>"y") THEN 230
210   IF I=1 THEN 290
220   GOTO 280
230   IF (A$<>"N") AND (A$<>"n") THEN 150
240   ON I GOTO 280,250,260,270
250   LINK F$,600,280
260   LINK F$,1600,280
270   LINK F$,2600,280
280   NEXT I
290   INPUT "Order of matrix A ?",N
300   IF (N<1) OR (N<>INT(N)) THEN 290
310   CALL R(N)
320   PRINTER IS 16
330   END
340   SUB R(N)
350   OPTION BASE 1
360   DIM A(N,N),P(N),Q(N,N,N),M(N,N)
370   CALL Mat(A(*),N,N,"A")
380   CALL Resmat(N,A(*),P(*),Q(*))
390   PRINT LIN(4),"[sI-A]^-1 = Q(s)/P(s), where",LIN(2),"Q(s)
      = SUM πQ(i)*(s^i), P(s) = s^";N;" + SUM πP(i)*(s^i)",
      LIN(2),"i = 0,1,...,";N-1
400   FOR I=1 TO N
410   PRINT LIN(3),"MATRIX Q(";N-I;"):",LIN(2)
420   FOR J=1 TO N
430   FOR K=1 TO N
```

```
440   M(J,K)=Q(N-I+1,J,K)
450   NEXT K
460   NEXT J
470   CALL Prtmat(M(*),N,N)
480   NEXT I
490   PRINT LIN(3)
500   FOR I=1 TO N
510   PRINT "P(";N-I;") = ";P(N-I+1),LIN(1)
520   NEXT I
530   SUBEND
```

6. Example

```
'RESMAT' FINDS THE ANALYTIC EXPRESSION FOR THE
RESOLVENT MATRIX [sI-A]^-1 = Q(s)/P(s) WHERE :
   Q(s) = SUM πQ(i)*(s^i)→ & P(s) = s^N + SUM πP(i)*(s^i)→,
   i = 0,1,...,N-1. N is the order of square matrix A.


Matrix A( 5 x 5 ):

  1.0000E+00   0.0000E+00   2.0000E+00   4.0000E+00  -3.0000E+00

  0.0000E+00   4.0000E+00   0.0000E+00   1.0000E+00   4.0000E+00

  1.0000E+00   0.0000E+00   5.0000E+00   2.0000E+00   3.0000E+00

 -2.0000E+00  -1.0000E+00   0.0000E+00   1.0000E+00   4.0000E+00

  7.0000E+00   1.0000E+00   2.0000E+00   3.0000E+00   1.0000E+00


[sI-A]^-1 = Q(s)/P(s), where

Q(s) = SUM πQ(i)*(s^i),        P(s) = s^ 5 + SUM πP(i)*(s^i)
i = 0,1,..., 4


MATRIX Q( 4 ):

  1.0000E+00   0.0000E+00   0.0000E+00   0.0000E+00   0.0000E+00

  0.0000E+00   1.0000E+00   0.0000E+00   0.0000E+00   0.0000E+00

  0.0000E+00   0.0000E+00   1.0000E+00   0.0000E+00   0.0000E+00
```

| 0.0000E+00 | 0.0000E+00 | 0.0000E+00 | 1.0000E+00 | 0.0000E+00 |
| 0.0000E+00 | 0.0000E+00 | 0.0000E+00 | 0.0000E+00 | 1.0000E+00 |

MATRIX Q(3):

-1.1000E+01	0.0000E+00	2.0000E+00	4.0000E+00	-3.0000E+00
0.0000E+00	-8.0000E+00	0.0000E+00	1.0000E+00	4.0000E+00
1.0000E+00	0.0000E+00	-7.0000E+00	2.0000E+00	3.0000E+00
-2.0000E+00	-1.0000E+00	0.0000E+00	-1.1000E+01	4.0000E+00
7.0000E+00	1.0000E+00	2.0000E+00	3.0000E+00	-1.1000E+01

MATRIX Q(2):

1.8000E+01	-7.0000E+00	-1.8000E+01	-4.5000E+01	5.2000E+01
2.6000E+01	2.7000E+01	8.0000E+00	5.0000E+00	-2.4000E+01
1.1000E+01	1.0000E+00	2.9000E+01	1.0000E+00	-1.3000E+01
4.8000E+01	1.1000E+01	4.0000E+00	4.8000E+01	-3.8000E+01
-7.4000E+01	-1.0000E+01	2.0000E+00	3.0000E+00	4.6000E+01

MATRIX Q(1):

9.9000E+01	6.9000E+01	5.0000E+01	1.4000E+02	-2.6400E+02
-1.4400E+02	-2.7600E+02	4.4000E+01	8.0000E+01	5.0000E+01
-5.3000E+01	-1.0000E+01	-2.1400E+02	6.5000E+01	4.9000E+01

```
-3.1000E+02  -4.2000E+01   4.0000E+01  -2.1000E+02   6.6000E+01

 2.4400E+02   3.0000E+00  -4.6000E+01  -1.6100E+02  -1.0900E+02
```

MATRIX Q(0):

```
-2.2500E+02  -1.2800E+02  -8.0000E+01  -8.7000E+01   4.2500E+02

 9.0000E+01   6.2000E+02   3.2000E+01  -5.3400E+02  -1.7000E+02

-5.4000E+01  -5.6000E+01   6.7600E+02  -4.3800E+02  -2.1400E+02

 6.1200E+02   1.0800E+02  -2.8800E+02   5.4000E+02   1.0800E+02

-2.4300E+02   6.4000E+01   4.0000E+01   3.9900E+02   1.4300E+02
```

```
P( 4 ) =    -12
P( 3 ) =     56
P( 2 ) =   -355
P( 1 ) =   1754
P( 0 ) =  -2844
```

7. Example. Matrix A in this example is the same as that in Example 2.

```
'RESMAT' FINDS THE ANALYTIC EXPRESSION FOR THE
RESOLVENT MATRIX [sI-A]^-1 = Q(s)/P(s) WHERE :
   Q(s) = SUM πQ(i)*(s^i)→ & P(s) = s^N + SUM πP(i)*(s^i)→,
   i = 0,1,...,N-1. N is the order of square matrix A.
```

Matrix A(3 x 3):

```
 0.0000E+00  -1.0000E+00  -1.0000E+00
```

1.0000E+00 -2.0000E+00 -1.0000E+00

0.0000E+00   0.0000E+00 -2.0000E+00

$[sI-A]^{-1} = Q(s)/P(s)$, where

$Q(s) = \text{SUM } \pi Q(i)*(s^i)$, $\quad P(s) = s^3 + \text{SUM } \pi P(i)*(s^i)$
$i = 0,1,\ldots, 2$

MATRIX Q(2):

1.0000E+00   0.0000E+00   0.0000E+00

0.0000E+00   1.0000E+00   0.0000E+00

0.0000E+00   0.0000E+00   1.0000E+00

MATRIX Q(1):

4.0000E+00 -1.0000E+00 -1.0000E+00

1.0000E+00   2.0000E+00 -1.0000E+00

0.0000E+00   0.0000E+00   2.0000E+00

MATRIX Q(0):

4.0000E+00 -2.0000E+00 -1.0000E+00

2.0000E+00   0.0000E+00 -1.0000E+00

0.0000E+00   0.0000E+00   1.0000E+00

P(2) = 4

P(1) = 5

P(0) = 2

8. Corollary to Theorem 3.
The resolvent matrix can be written as

$$(s\mathbf{I}-\mathbf{A})^{-1} = \sum_{i=0}^{n-1} \mathbf{A}^i F_i(s) \tag{18}$$

where

$$F_i(s) = \frac{s^{n-i-1} + p_{n-1}s^{n-i-2} + \ldots + p_{i+2}s + p_{i+1}}{s^n + p_{n-1}s^{n-1} + \ldots + p_1 s + p_0}, \quad i=0, 1, 2, \ldots, n-1 \tag{19}$$

Proof. From (6) and (7) by successive substitution we have

$$\mathbf{Q}_{n-1} = \mathbf{I} \tag{20a}$$

$$\mathbf{Q}_{n-2} = \mathbf{A} + p_{n-1}\mathbf{I} \tag{20b}$$

$$\mathbf{Q}_{n-3} = (\mathbf{A} + p_{n-1}\mathbf{I})\mathbf{A} + p_{n-2}\mathbf{I} = \mathbf{A}^2 + p_{n-1}\mathbf{A} + p_{n-2}\mathbf{I} \tag{20c}$$

$$\mathbf{Q}_1 = \mathbf{A}^{n-2} + p_{n-1}\mathbf{A}^{n-3} + \ldots + p_3\mathbf{A} + p_2\mathbf{I} \tag{20d}$$

$$\mathbf{Q}_0 = \mathbf{A}^{n-1} + p_{n-1}\mathbf{A}^{n-2} + \ldots + p_3\mathbf{A}^2 + p_2\mathbf{A} + p_1\mathbf{I} \tag{20e}$$

Substitution of the above relations for $\mathbf{Q}_i$ in (1) and collecting the coefficients for powers 0 to $n-1$ of A yields (18) and (19). q.e.d. △

Note that post-multiplication of (20) by A and then using (9) results in the Cayley-Hamilton theorem (Theorem 2.9.4). Equation (18) provides an alternative method for the calculation of $e^{\mathbf{A}t}$:

$$e^{\mathbf{A}t} = \sum_{i=0}^{n-1} \mathbf{A}^i f_i(t) \tag{21}$$

where

$$f_i(t) = \mathbf{L}^{-1}[F_i(s)], \quad i=0, 1, 2, \ldots, n-1 \tag{22}$$

The resolvent matrix can also be expanded into matrix partial fraction form. That is, we can write

$$(s\mathbf{I}-\mathbf{A})^{-1} = \sum_{i=1}^{p} \sum_{j=1}^{m_i} \frac{\mathbf{K}_{ij}}{(s-\lambda_i)^j} \tag{23}$$

where m_i is the multiplicity of eigenvalue λ_i, $i=1, 2, ..., p$ of $\mathbf{A}$. Equation (23) is similar to (3.4.3) except here $\mathbf{K}_{ij}$ are matrix residues. Relation (3.4.4) may be used to evaluate these matrix residues. The inverse Laplace transform of (23) results in $e^{\mathbf{A}t}$:

$$e^{\mathbf{A}t} = \mathbf{L}^{-1}[(s\mathbf{I}-\mathbf{A})^{-1}] = \sum_{i=1}^{p} \sum_{j=1}^{m_i} \mathbf{K}_{ij} \frac{t^j e^{\lambda_i t}}{(j-1)!} \tag{24}$$

See Equation (3.4.5). Equation (24) is referred to as the *spectral representation* of $e^{\mathbf{A}t}$.

9. Example. Again consider matrix $\mathbf{A}$ in Example 2. We can write (5) as

$$(s\mathbf{I}-\mathbf{A})^{-1} = \frac{\mathbf{Q}_2 s^2 + \mathbf{Q}_1 s + \mathbf{Q}_0}{(s+1)^2(s+2)} = \frac{\mathbf{K}_{11}}{s+1} + \frac{\mathbf{K}_{12}}{(s+1)^2} + \frac{\mathbf{K}_{21}}{s+2} \tag{25}$$

where the matrix residues can be found as follows:

$$\mathbf{K}_{11} = \frac{d}{ds}[(s+1)^2(s\mathbf{I}-\mathbf{A})^{-1}]_{s=-1} = \begin{bmatrix} -1 & 0 & 1 \\ 0 & -1 & 1 \\ 0 & 0 & 0 \end{bmatrix}$$

$$\mathbf{K}_{12} = [(s+1)^2(s\mathbf{I}-\mathbf{A})^{-1}]_{s=-1} = \begin{bmatrix} -1 & 1 & 0 \\ -1 & 1 & 0 \\ 0 & 0 & 0 \end{bmatrix}$$

$$\mathbf{K}_{21} = [(s+2)(s\mathbf{I}-\mathbf{A})^{-1}]_{s=-2} = \begin{bmatrix} 0 & 0 & -1 \\ 0 & 0 & -1 \\ 0 & 0 & -1 \end{bmatrix}$$

Thus

$$e^{\mathbf{A}t} = \mathbf{L}^{-1}[(s\mathbf{I}-\mathbf{A})^{-1}] = \mathbf{K}_{11}e^{-t} + \mathbf{K}_{12}te^{-t} + \mathbf{K}_{21}e^{-2t}$$

$$= \begin{bmatrix} (t+1)e^{-t} & -te^{-t} & -e^{-t}+e^{-2t} \\ te^{-t} & (1-t)e^{-t} & -e^{-t}+e^{-2t} \\ 0 & 0 & e^{-2t} \end{bmatrix} \tag{26}$$

6.6 THE FORCED CASE - COMPLETE SOLUTION

In this section we will solve the linear state equation with input, i.e. equation (6.1.1) in the continuous-time case, and equation (6.1.3) in the discrete-time case. We will see that the knowledge of the transition matrix is essential in obtaining the solution in both cases.

6.6.1 Continuous-Time Systems

The state equation to be solved in the continuous-time case is

THE FORCED CASE—COMPLETE SOLUTION

$$\dot{x}(t) = A(t)x(t) + B(t)u(t), \quad t \geq t_o \tag{1a}$$

$$x(t_o) = x_o \tag{1b}$$

1. Theorem. The solution to (1) is

$$x(t) = \Phi(t,t_o)x_o + \int_{t_o}^{t} \Phi(t,\tau)B(\tau)u(\tau)d\tau, \quad t \geq t_o \tag{2}$$

where $\Phi(t,\tau)$ is the transition matrix of $\dot{x} = A(t)x$.

Proof. We will solve (1) using the method of variation of constants, as in the case of scalar differential equations [6.9]. We can write the solution to (1) as

$$x(t) = \Phi(t,t_o)k(t) \tag{3}$$

which is similar to the solution $\Phi(t,t_o)x_o$ for the homogeneous case except $k(t)$ in (3) is a vector function of time to be determined such that (3) satisfies (1). From (3) we have

$$\dot{x}(t) = [\frac{\partial}{\partial t}\Phi(t,t_o)]k(t) + \Phi(t,t_o)\dot{k}(t)$$

$$= A(t)\Phi(t,t_o)k(t) + \Phi(t,t_o)\dot{k}(t) \tag{4}$$

where the defining relationship (6.2.13) for the transition matrix has been used. Substitution of (4) into (1) while using (3) yields

$$A(t)\Phi(t,t_o)k(t) + \Phi(t,t_o)\dot{k}(t) = A(t)\Phi(t,t_o)k(t) + B(t)u(t)$$

or, upon cancellation of similar terms and multiplying both sides by $\Phi^{-1}(t,t_o) = \Phi(t_o,t)$,

$$\dot{k}(t) = \Phi(t_o,t)B(t)u(t) \tag{5}$$

Integrating (5) from t_o to t yields

$$k(t) = k(t_o) + \int_{t_o}^{t} \Phi(t_o,\tau)B(\tau)u(\tau)d\tau \tag{6}$$

where $k(t_o)$ can be found from (3) by setting $t = t_o$ as $k(t_o) = x(t_o)$. Substitution of $k(t)$ in (3) and using the transition property of the transition matrix yields (2). q.e.d. △

In the *l.t.i.* case where **A** and **B** are constant matrices and $\Phi(t,t_o) = e^{A(t-t_o)}$, (2) becomes

$$\mathbf{x}(t) = e^{A(t-t_o)}\mathbf{x}_o + \int_{t_o}^{t} e^{A(t-\tau)}\mathbf{B}\mathbf{u}(\tau)d\tau \qquad (7)$$

The system output, or the *complete response*, $\mathbf{y}(t)$ can be found by substituting (2) in (6.1.2):

$$\mathbf{y}(t) = \mathbf{C}(t)\Phi(t,t_o)\mathbf{x}_o + \mathbf{C}(t)\int_{t_o}^{t}\Phi(t,\tau)\mathbf{B}(\tau)\mathbf{u}(\tau)d\tau + \mathbf{D}(t)\mathbf{u}(t), \quad t \geq t_o \qquad (8)$$

If inputs $\mathbf{u}(t)$ is set equal to zero, (8) becomes

$$\mathbf{y}_{zi}(t) = \mathbf{C}(t)\Phi(t,t_o)\mathbf{x}_o, \quad t \geq t_o \qquad (9)$$

which is called the *zero-input response* of the system. Note that the zero-input response depends only on the initial state $\mathbf{x}_o$ and not on the input. If the initial state is set equal to zero in (8), it becomes.

$$\mathbf{y}_{zs}(t) = \mathbf{C}(t)\int_{t_o}^{t}\Phi(t,\tau)\mathbf{B}(\tau)\mathbf{u}(\tau)d\tau + \mathbf{D}(t)\mathbf{u}(t), \quad t \geq t_o \qquad (10)$$

which is called the *zero-state response* of the system. Note that the zero-state response depends only on the input $\mathbf{u}(t)$. Thus equation (9) expresses the following important property of linear systems.

Decomposition Property of Linear Systems. In a linear system the complete response (or the output) is the sum of its zero-state response and its zero-input response. Both the zero-state response and the zero-input response are linear functions. More precisely, we have the following two theorems.

2. Theorem. In a linear system the zero-state (z.s.) response is a linear function of the input; that is,

$$z.s. \text{ response due input } \alpha\mathbf{u}_1 + \beta\mathbf{u}_2$$

$$= \alpha(z.s. \text{ response due } \mathbf{u}_1) + \beta(z.s. \text{ response due } \mathbf{u}_2)$$

for *all* constants α and β. (11)

This property is referred to as the *superposition property* of linear systems.

THE FORCED CASE—COMPLETE SOLUTION

3. Theorem In a linear system the zero-input (z.i) response is a linear function of the initial state; that is,

$$\text{z.i. response due initial state } \alpha\, \mathbf{x}_{10} + \beta \mathbf{x}_{20}$$

$$= \alpha \text{ (z.i. response due initial state } \mathbf{x}_{10})$$

$$+ \beta \text{ (z.i. response due initial state } \mathbf{x}_{20})$$

$$\text{for all constants } \alpha \text{ and } \beta. \tag{12}$$

The proofs of the above theorems are straightforward and will not be detailed here. (See Problem (6.22).)

In the *l.t.i.* case, the complete response (8) can be written as

$$\mathbf{y}(t) = \mathbf{C}e^{\mathbf{A}(t-t_o)}\mathbf{x}_o + \mathbf{C}\int_{t_o}^{t} e^{\mathbf{A}(t-\tau)}\mathbf{B}\mathbf{u}(\tau)d\tau + \mathbf{D}\mathbf{u}(t), \quad t \geqslant t_o \tag{13}$$

6.6.2 Impulse Response Matrix

For a SISO linear system the *impulse response* is defined as the zero-state response when the input is a unit impulse. It is denoted by $h(t,\tau)$ where t is the time of observation and τ is the time at which the impulse is applied. We assume that

$$h(t,\tau) = 0 \text{ for } t < \tau. \tag{14}$$

This is referred to as *causality*. It expresses that the system output at time t does not depend upon its input(s) $u(\tau)$ for $\tau > t$ (nonanticipative system) [6.8]. See Problem 6.23. Note that we have previously encountered the impulse response in the case of SISO *l.t.i.* systems in Section 4.5. There, we defined impulse response as the inverse Laplace transform of the transfer function. We will shortly see that these two characterizations of the impulse response are consistent.

For multivariable systems with r inputs and m outputs the *impulse response matrix* $\mathbf{h}(t,\tau)$ is an $m \times r$ matrix. The element $h_{ij}(t,\tau)$ of this matrix is the *i*th component of the zero-state response at time t when the *j*th component of the input is a unit impulse applied at time τ and all the other input components are zero. Put another way, the *k*th column of $\mathbf{h}(t,\tau)$ is the zero-state response at time t when a unit impulse is applied at the *k*th input at time τ and all the other inputs are zero.

To express the impulse response matrix in terms of system matrices **A, B, C** and **D**, let

$$u(t) = \begin{bmatrix} 0 \\ \vdots \\ \delta(t-\tau) \\ 0 \\ \vdots \\ 0 \end{bmatrix} \leftarrow kth \ position. \qquad (15)$$

Then, using (10) and the above definition of impulse response matrix, the kth column of $\mathbf{h}(t,\tau)$ becomes.

$$\mathbf{h}_k(t,\tau) = \mathbf{C}(t)\int_{t_o}^{t} \Phi(t,t_1)\mathbf{B}(t_1) \begin{bmatrix} 0 \\ \vdots \\ 0 \\ \delta(t_1-\tau) \\ 0 \\ \vdots \\ 0 \end{bmatrix} dt_1 + \mathbf{D}(t) \begin{bmatrix} 0 \\ \vdots \\ 0 \\ \delta(t-\tau) \\ 0 \\ \vdots \\ 0 \end{bmatrix}, \ t \geqslant t_o \qquad (16)$$

Thus, the impulse response matrix is

$$\mathbf{h}(t,\tau) = \mathbf{C}(t)\int_{t_o}^{t} \Phi(t,t_1)\mathbf{B}(t_1)\delta(t_1-\tau)dt_1 + \mathbf{D}(t)\delta(t-\tau), \ t \geqslant t_o \qquad (17)$$

where

$$\delta(t-\tau) = diag[\delta(t-\tau)] \qquad (18)$$

Using the shifting property of the unit impulse [6.8]:

$$\int_{t_o}^{t} f(t_1)\delta(t_1-\tau)dt_1 = f(\tau) \ \text{for any function} \ f, \ \text{where} \ t_o \leqslant \tau \leqslant t, \qquad (19)$$

(17) becomes

$$\mathbf{h}(t,\tau) = \mathbf{C}(t)\Phi(t,\tau)\mathbf{B}(\tau) + \mathbf{B}(t)\delta(t-\tau), \ t \geqslant \tau \qquad (20)$$

Again we assume

$$\mathbf{h}(t,\tau) = \mathbf{0} \ \text{for} \ t < \tau. \qquad (21)$$

THE FORCED CASE—COMPLETE SOLUTION

In the *l.t.i.* case (20) becomes

$$\mathbf{h}(t,\tau) = Ce^{A(t-\tau)}B + D\delta(t-\tau), \quad t \geq \tau \tag{22}$$

Thus $\mathbf{h}(t,\tau)$ can be written as $\mathbf{h}(t-\tau)$ and we have

$$\mathbf{h}(t) = Ce^{At}B + D\delta(t), \quad t \geq 0 \tag{23}$$

where $\mathbf{h}(t) = 0$ for $t < 0$. Taking the Laplace transform of both sides, we have

$$L[\mathbf{h}(t)] = H(s) = C(sI-A)^{-1}B + D \tag{24}$$

$H(s)$ is called the *transfer function matrix* of the continuous-time system. Equation (24) is the same as (4.5.8). Note that the transfer function matrix is defined only for *l.t.i.* systems.

The zero-state response (10) can also be written as

$$\mathbf{y}_{zs}(t) = \int_{t_o}^{t} [C(t)\Phi(t,\tau)B(\tau) + D(t)\delta(t-\tau)]\mathbf{u}(\tau)d\tau, \quad t \geq t_o \tag{25}$$

where the translation property of the unit impulse had been used. Comparison of (25) with (20) yields

$$\mathbf{y}_{zs}(t) = \int_{t_o}^{t} \mathbf{h}(t,\tau)\mathbf{u}(\tau)d\tau, \quad t \geq t_o \tag{26}$$

Convolution integral (26) could also be found by using the linearity of the zero-state response. (See Problem 6.22.)

In the *l.t.i.* case (26) becomes

$$\mathbf{y}_{zs}(t) = \int_{t_o}^{t} \mathbf{h}(t-\tau)\mathbf{u}(\tau)d\tau, \quad t \geq t_o \tag{27}$$

Taking the Laplace transform of both sides of (27) we obtain

$$\mathbf{y}_{zs}(s) = H(s)U(s) \tag{28}$$

where the convolution property of the Laplace transform has been employed. (See Section 3.3.) Equation (28) is consistent with (4.5.3).

4. Example. Consider the *l.t.i:* multivariable system characterized by system matrices

$$A = \begin{bmatrix} 2 & 1 & 0 \\ 1 & 2 & 1 \\ 0 & -1 & 2 \end{bmatrix}, \quad B = \begin{bmatrix} 0 \\ 0 \\ 1 \end{bmatrix}, \quad C = \begin{bmatrix} 1 & 0 & 0 \\ 0 & 1 & 1 \end{bmatrix}, \quad D = [0] \tag{29}$$

with initial state

$$x_o = \begin{bmatrix} 0 \\ 1 \\ 0 \end{bmatrix} \tag{30}$$

and unit step input:

$$u(t) = \begin{cases} 1 & t > 0 \\ 0 & t \leqslant 0 \end{cases} \tag{31}$$

Note that the system matrix A is the same as that used in Example 6.4.1. From (13) with $t_o = 0$, the complete response is

$$y(t) = Ce^{At}x_o + C\int_0^t e^{A(t-\tau)}Bu(\tau)d\tau \tag{32}$$

Using e^{At} given in (6.4.17) we obtain

$$y(t) = \begin{bmatrix} t \\ 1-t \end{bmatrix} e^{2t} + \begin{bmatrix} t^2 - t + 1/2 \\ -t^2 + 3t + 1/2 \end{bmatrix} \frac{e^{2t}}{4} - \frac{1}{8}\begin{bmatrix} 1 \\ 1 \end{bmatrix}, \quad t \geqslant 0 \tag{33}$$

(The zero-state response portion of y(t) is called the *step response* of the system.) Also, using the resolvent matrix calculated in (6.4.23), we obtain the transfer function matrix as

$$H(s) = C(sI-A)^{-1}B + D = \frac{1}{(s-2)^3}\begin{bmatrix} 1 \\ s^2 - 3s + 1 \end{bmatrix} \tag{34}$$

(See Problem 6.25.) △

Program "COMSCS" determines the complete response of any linear system for specified inputs and initial conditions. It uses the 4*th* order Runge-Kutta method to integrate the state equation. The program can accommodate up to six inputs, eight state variables and five outputs. The inputs can be step, ramp, parabolic or sinusoidal. Although the program is written for *l.t.i:* systems, it can be easily extended to accommodate time varying matrices A(t), B(t), C(t), D(t). The source listing of "COMSCS" followed by an example is given below. The discrete-time version of this program, "COMSDS", will be given in Section 6.6.3 where matrices can be time varying.

THE FORCED CASE—COMPLETE SOLUTION

5. "COMSCS"

```
10    ! PROGRAM NAME : COMSCS "PROG"
20    PRINT " Program <<COMSCS>> gives a complete solution
      for a continuous-time"
30    PRINT " MIMO linear system in state form (A,B,C,D) "
40    OPTION BASE 1
50    COM Incode,To,Dt,Ns,Ms,Rs,Icode(6),Coef(5,5),A(8,8)
60    COM B(8,6),C(5,8),D(5,5),Ustor(6,102),Ystor(5,102),
      Xstor(8,102)
70    DIM X(101)
80    Iplot=0
90    INPUT "Have you LINKed SUB's <<Mat>>, <<Kutta>> and
      <<Plot>> (Y/N)?",M$
100   IF (M$="Y") OR (M$="y") THEN 160
110   IF (M$="N") OR (M$="n") THEN 130
120   GOTO 90
130   LINK "Mat",3000,140
140   LINK "Kutta",5500,150
150   LINK "Plot",7000,160
160   INPUT "ORDER of the system n(<= 8) ?",Ns
170   PRINT "ORDER of the system n =";Ns
180   INPUT "No. of system INPUTS m(<= 5) ?",Ms
190   PRINT "No. of system INPUTS m =";Ms
200   INPUT "No. of system OUTPUTS r(<= 6) ?",Rs
210   PRINT "No. of system OUTPUTS r =";Rs
220   IF Ns<=0 THEN 160
230   INPUT "Initial time to?",To
240   PRINT "Initial time to=";To
250   INPUT "Final time tf?",Tf
260   PRINT "Final time tf=";Tf
270   INPUT "Step size dt?",Dt
280   PRINT "Step size dt=";Dt
290   IF Dt<=0 THEN 270
300   Nb=INT((Tf-To)/Dt)+1
310   IF Nb<=0 THEN 230
320   REDIM X(Nb),Xstor(Ns,Nb+1),A(Ns,Ns),B(Ns,Ms),
      Ustor(Ms,Nb+1)
330   REDIM C(Rs,Ns),D(Rs,Ms),Icode(Ms),Ystor(Rs,Nb+1)
340   CALL Kutta1(Tf,Nb)
350   CALL Prtplt(Tf,Nb)
360   END
370   SUB Kutta1(Tf,Nb)
380   OPTION BASE 1
390   COM Incode,To,Dt,Ns,Ms,Rs,Icode(*),Coef(*),As(*)
400   COM Bs(*),Cs(*),Ds(*),Ustor(*),Ystor(*),Xstor(*)
410   DIM Ynt(Ns,1)
420   CALL Mat(As(*),Ns,Ns,"A")
430   CALL Mat(Bs(*),Ns,Ms,"B")
440   CALL Mat(Cs(*),Rs,Ns,"C")
```

```
450   CALL Mat(Ds(*),Rs,Ms,"D")
460   PRINT "INITIAL VALUES:",LIN(1)
470   FOR I=1 TO Ns
480      DISP "Y(";I;")";
490      INPUT Ynt(I,1)
500      PRINT USING 510;I,Ynt(I,1)
510      IMAGE 3X,"Y(",DD,")=",MZ.6DE
520   NEXT I
530   MAT Coef=ZER
540   PRINT "INPUT SIGNAL CODES ARE:"
550   FOR Iu=1 TO Ms
560   Ustor(Iu,1)=0
570   NEXT Iu
580   PRINT " STEP INPUT .......... 0"
590   PRINT " RAMP INPUT .......... 1"
600   PRINT " PARABOLIC INPUT ..... 2"
610   PRINT " SINUSOIDAL INPUT .... 3"
620   FOR Iu=1 TO Ms
630   PRINT "For INPUT no.";Iu;" Give INPUT SIGNAL code (0-3)"
640   INPUT "Code (0-3)",Incode
650   IF (Incode<0) OR (Incode>3) THEN 630
660   ON Incode+1 GOTO 670,710,740,770
670   INPUT "For a STEP input : r(t)=au(t); Give value of
      a",Apar
680   Coef(Iu,1)=Apar
690   Ustor(Iu,1)=Apar
700   GOTO 810
710   INPUT "For a RAMP input : r(t)=bt.u(t); Give value
      of b",Bpar
720   Coef(Iu,2)=Bpar
730   GOTO 810
740   INPUT "For a PARABOLIC input : r(t)=ct.t.u(t); Give
      value of c",Cpar
750   Coef(Iu,3)=Cpar
760   GOTO 810
770   PRINT "For a SINUSOIDAL input : r(t)=xSin(wt) :"
780   INPUT "   Give values of x & w",Xpar,Wpar
790   Coef(Iu,4)=Xpar
800   Coef(Iu,5)=Wpar
810   Icode(Iu)=Incode
820   NEXT Iu
830   CALL Kutta(Ns,To,Dt,Tf,Nb,Ynt(*),Xstor(*))
840   SUBEND
850   SUB Func(Ysv(*),X,Idm,F(*))
860   OPTION BASE 1
870   COM Incode,To,Dt,Ns,Ms,Rs,Icode(*),Coef(*),A(*)
880   COM B(*),C(*),D(*),Ustor(*),Ystor(*),Xstor(*)
890   DIM Fl(Idm,1),Rt(Ms,1)
900   It=INT((X-To)/Dt)+2
910   FOR Iu=1 TO Ms
```

THE FORCED CASE—COMPLETE SOLUTION

```
920   Ic=Icode(Iu)+1
930   ON Ic GOTO 940,960,980,1000
940   Rt(Iu,1)=Coef(Iu,Ic)
950   GOTO 1010
960   Rt(Iu,1)=Coef(Iu,Ic)*X
970   GOTO 1010
980   Rt(Iu,1)=Coef(Iu,Ic)*X*X
990   GOTO 1010
1000  Rt(Iu,1)=Coef(Iu,Ic)*SIN(Coef(Iu,Ic+1)*X)
1010  Ustor(Iu,It)=Rt(Iu,1)
1020  NEXT Iu
1030  MAT F=A*Ysv
1040  MAT F1=B*Rt
1050  MAT F=F+F1
1060  SUBEND
1070  SUB Prtplt(Tf,Nb)
1080  OPTION BASE 1
1090  COM Incode,To,Dt,Ns,Ms,Rs,Icode(*),Coef(*),A(*)
1100  COM B(*),C(*),D(*),Ustor(*),Ystor(*),Xstor(*)
1110  DIM Er(Ns,1),Yplot(Nb),X(Nb),Rt(Ms,1),Du(Rs,1),
      Yout(Rs,1),Cx(Rs,1)
1120  FOR I=1 TO Nb
1130  X(I)=To+(I-1)*Dt
1140  FOR J=1 TO Ns
1150  Er(J,1)=Xstor(J,I)
1160  NEXT J
1170  MAT Cx=C*Er
1180  FOR Iu=1 TO Ms
1190  Rt(Iu,1)=Ustor(Iu,I)
1200  NEXT Iu
1210  MAT Du=D*Rt
1220  MAT Yout=Cx+Du
1230  FOR Iy=1 TO Rs
1240  Ystor(Iy,I)=Yout(Iy,1)
1250  NEXT Iy
1260  NEXT I
1270  INPUT "Do you like to PRINT STATE VARIABLES ? ",C$
1280  IF (C$="Y") OR (C$="y") THEN 1310
1290  IF (C$="N") OR (C$="n") THEN 1380
1300  GOTO 1170
1310  FOR I=1 TO Nb
1320  FOR J=1 TO Ns
1330  Er(J,1)=Xstor(J,I)
1340  NEXT J
1350  PRINT "At t=";X(I);"x(t) Is:"
1360  MAT PRINT Er;
1370  NEXT I
1380  INPUT "Do you like to PRINT INPUT VARIABLES ? ",I$
1390  IF (I$="Y") OR (I$="y") THEN 1420
1400  IF (I$="N") OR (I$="n") THEN 1490
```

```
1410 GOTO 1380
1420 FOR I=1 TO Nb
1430 FOR Iu=1 TO Ms
1440 Rt(Iu,1)=Ustor(Iu,I)
1450 NEXT Iu
1460 PRINT "At t=";X(I);"u(t) Is:"
1470 MAT PRINT Rt;
1480 NEXT I
1490 INPUT "Do you like to PRINT OUTPUT VARIABLES ? ",O$
1500 IF (O$="Y") OR (O$="y") THEN 1530
1510 IF (O$="N") OR (O$="n") THEN 1600
1520 GOTO 1490
1530 FOR I=1 TO Nb
1540 FOR Iy=1 TO Rs
1550 Yout(Iy,1)=Ystor(Iy,I)
1560 NEXT Iy
1570 PRINT "At t=";X(I);"y(t) Is:"
1580 MAT PRINT Yout;
1590 NEXT I
1600 INPUT "Do you like to PLOT STATE VARIABLES ? ",C$
1610 IF (C$="Y") OR (C$="y") THEN 1640
1620 IF (C$="N") OR (C$="n") THEN 1760
1630 GOTO 1470
1640 FOR Ipl=1 TO Ns
1650 FOR I=1 TO Nb
1660 Yplot(I)=Xstor(Ipl,I)
1670 X(I)=To+(I-1)*Dt
1680 NEXT I
1690 Iplot=0
1700 CALL Plot(X(*),Yplot(*),Nb,Iplot,To,Tf)
1710 INPUT "Do you like to STOP PLOTTING this set of
      variables?",C$
1720 IF (C$="Y") OR (C$="y") THEN 1760
1730 IF (C$="N") OR (C$="n") THEN 1750
1740 GOTO 1710
1750 NEXT Ipl
1760 INPUT "Do you like to PLOT INPUT VARIABLES ? ",C$
1770 IF (C$="Y") OR (C$="y") THEN 1800
1780 IF (C$="N") OR (C$="n") THEN 1920
1790 GOTO 1760
1800 FOR Ipl=1 TO Ms
1810 FOR I=1 TO Nb
1820 Yplot(I)=Ustor(Ipl,I)
1830 X(I)=To+(I-1)*Dt
1840 NEXT I
1850 Iplot=0
1860 CALL Plot(X(*),Yplot(*),Nb,Iplot,To,Tf)
1870 INPUT "Do you like to STOP PLOTTING this set of
      variables?",C$
1880 IF (C$="Y") OR (C$="y") THEN 1920
```

THE FORCED CASE—COMPLETE SOLUTION

```
1890 IF (C$="N") OR (C$="n") THEN 1910
1900 GOTO 1870
1910 NEXT Ipl
1920 INPUT "Do you like to PLOT OUTPUT VARIABLES ? ",C$
1930 IF (C$="Y") OR (C$="y") THEN 1960
1940 IF (C$="N") OR (C$="n") THEN 2080
1950 GOTO 1920
1960 FOR Ipl=1 TO Rs
1970 FOR I=1 TO Nb
1980 Yplot(I)=Ystor(Ipl,I)
1990 X(I)=To+(I-1)*Dt
2000 NEXT I
2010 Iplot=0
2020 CALL Plot(X(*),Yplot(*),Nb,Iplot,To,Tf)
2030 INPUT "Do you like to STOP PLOTTING this set of
     variables?",C$
2040 IF (C$="Y") OR (C$="y") THEN 2080
2050 IF (C$="N") OR (C$="n") THEN 2070
2060 GOTO 2030
2070 NEXT Ipl
2080 SUBEND
```

6. Example. Figures 1 through 3 show time responses of states $x_i(t)$, $i=1, 2$; inputs $u_j(t)$, $j=1, 2$ and outputs $y_k(t)$, $k=1, 2, 3$ obtained from "COMSCS".

```
Program <<COMSCS>> gives a complete solution for a
continucus-time
MIMO linear system in state form (A,B,C,D)
ORDER of the system n = 2
No. of system INPUTS m = 2
No. of system OUTPUTS r = 3
Initial time to = 0
Final time tf = 10
Step size dt = .1

Matrix A( 2 x 2 ):

-2.0000E+00            -5.0000E+00

 1.0000E+00             0.0000E+00

Matrix B( 2 x 2 ):

 2.0000E+00             0.0000E+00

 0.0000E+00            -1.0000E+00
```

Matrix C(3 x 2):

 1.0000E+00   1.0000E+00

 2.0000E+00   1.0000E+00

 -1.0000E+00   1.0000E+00

Matrix D(3 x 2):

 1.0000E+00   1.0000E+00

 0.0000E+00   1.0000E+00

 1.0000E+00   2.0000E+00

```
INITIAL VALUES:
   Y( 1)= 0.000000E+00
   Y( 2)= 0.000000E+00
INPUT SIGNAL CODES ARE:
  STEP INPUT .......... 0
  RAMP INPUT .......... 1
  PARABOLIC INPUT ..... 2
  SINUSOIDAL INPUT .... 3
For INPUT no. 1  Give INPUT SIGNAL code (0-3)
For INPUT no. 2  Give INPUT SIGNAL code (0-3)

At t= 0 x(t) Is:
 0

 0


At t= .1 x(t) Is:
 .180558333333

 4.34583333333E-03

  At t= .2 x(t) Is:
   .324827190868

   1.48987059548E-02

  At t= .3 x(t) Is:
   .437386662201

   2.825209721
```

…

At t= 0 u(t) Is:
 1

 0

At t= .1 u(t) Is:
 1

 .05

At t= .2 u(t) Is:
 1

 .15

…

At t= 0 y(t) Is:
 1

 0

 1

At t= .1 y(t) Is:
 1.23490416667

 .41546249 99 99

 .9237875

…

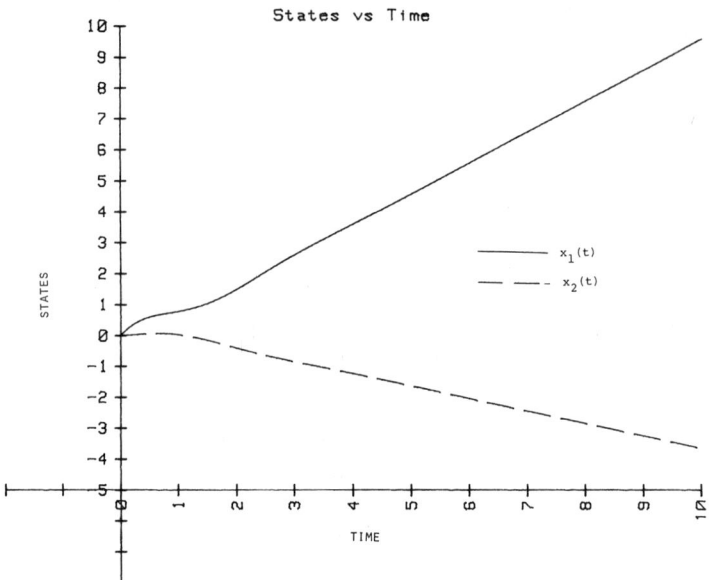

Fig. 6.6.1. State variables trajectories of Example 6.6.6

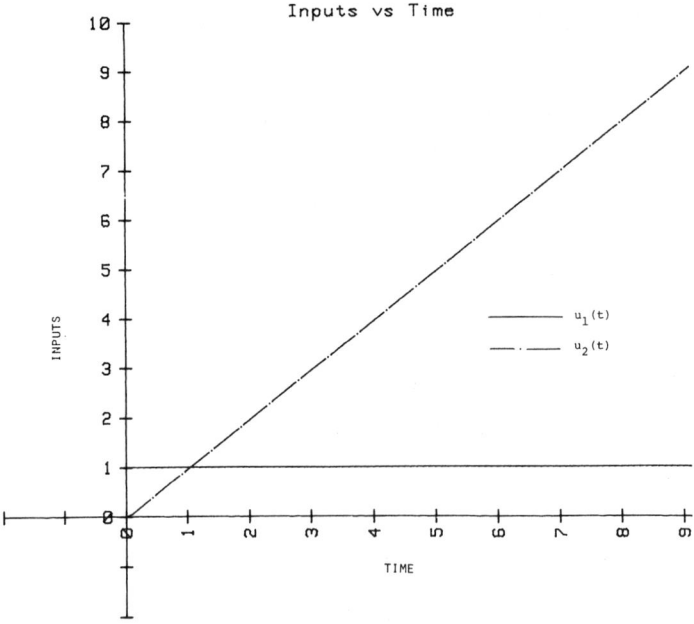

Fig. 6.6.2. Input variables trajectories of Example 6.6.6

THE FORCED CASE—COMPLETE SOLUTION

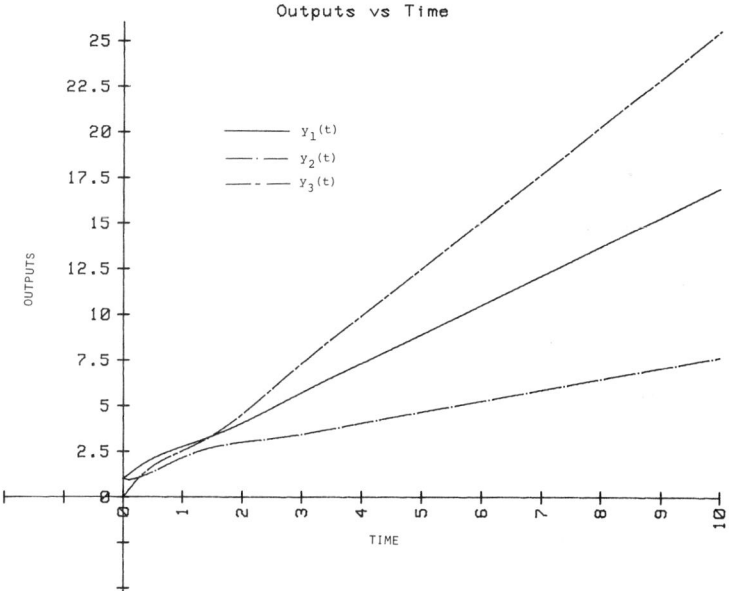

Fig. 6.6.3. Output variables trajectories of Example 6.6.6

6.6.3 Discrete-Time Systems

The state equation to be solved in the discrete-time case is

$$x(k+1) = A(k)x(k) + B(k)u(k), \quad k \geqslant k_o$$

$$x(k_o) = x_o \tag{35}$$

7. Theorem. The solution to (35) is

$$x(k) = \Phi(k,k_o)x_o + \sum_{j=k_o}^{k-1} \Phi(k,j+1)B(j)u(j), \quad k \geqslant k_o \tag{36}$$

where $\Phi(k,k_o)$ is the transition matrix of $x(k+1) = A(k)x(k)$.

Proof. From (35) we have

$$x(k_o+1) = A(k_o)x_o + B(k_o)u(k_o) \tag{37}$$

$$x(k_o+2) = A(k_o+1)x(k_o+1) + B(k_o+1)u(k_o+1)$$

$$= A(k_o+1)[A(k_o)x + B(k_o)u(k_o)] + B(k_o+1)u(k_o+1) \qquad (38)$$

and so on. It can be easily verified that in general we have

$$x(k) = \prod_{p=k_o}^{k-1} A(p)x_o + \sum_{j=k_o}^{k-1} [\prod_{p=j+1}^{k-1} A(p)]B(j)u(j), \quad k=k_o+1, k_o+2, \cdots \qquad (39)$$

Using (6.2.22) and the fact that $\Phi(k,k) = I$, (39) yields (36). △

Equation (36) is analogous to the corresponding equation (2) for the continuous case. The intergal in (2) has been replaced by a sum in (36). In the *l.t.i.* case where A and B are constant matrices and $\Phi(k,k_o) = A^{k-k_o}$, (36) becomes

$$x(k) = A^{k-k_o}x_o + \sum_{j=k_o}^{k-1} A^{k-j-1}Bu(j) \quad k \geqslant k_o \qquad (40)$$

The system output or *the complete response* $y(k)$ can be found from (4.4.13) repeated here:

$$y(k) = C(k)x(k) + D(k)u(k) \qquad (41)$$

Substituting (37) in (41) yields

$$y(k) = C(k)\Phi(k,k_o)x_o + C(k)\sum_{j=k_o}^{k-1}\Phi(k,j+1)B(j)u(j) + D(k)u(k), \quad k \geqslant k_o \qquad (42)$$

Here again the decomposition property holds, i.e. similar to the continuous case, the complete response $y(k)$ is the sum of the zero-input response $y_{zi}(k)$ and the zero-state response $y_{zs}(k)$ where

$$y_{zi}(k) = C(k)\Phi(k,k_o)x_o, \quad k \geqslant k_o \qquad (43)$$

$$y_{zs}(k) = C(k)\sum_{k=k_o}^{k-1}\Phi(k,j+1)B(j)u(j) + D(k)u(k), \quad k \geqslant k_o \qquad (44)$$

Also, the zero-state response is a linear function of the input and the zero-input response is a linear function of the initial state. (See Theorems 6.6.2 and 6.6.3.)

In the *l.t.i.* case, the complete response (42) can be written as

$$y(k) = CA^{k-k_o}x_o + C\sum_{j=k_o}^{k-1} A^{k-j-1}Bu(j) + Du(k), \quad k \geqslant k_o \qquad (45)$$

which is analogous to (13).

THE FORCED CASE—COMPLETE SOLUTION

The above solution for linear discrete-time systems can be used to obtain an approximate solution to the forced state equation in the continuous case. This is done by dividing the interval of interest into many equal subintervals and holding the input constant over each subinterval. Thus, the state $\mathbf{x}(t_k)=\mathbf{x}(k)$ and the output $\mathbf{y}(t_k)=\mathbf{y}(k)$ may be calculated using equations (36) and (42) for successively increasing integer values of k starting with k_o which corresponds to the initial time t_o. Figure 4 depicts such an arrangement. The input $\mathbf{u}(t)$ is sampled at regular intervals to obtain input samples $\mathbf{u}(t_k)=\mathbf{u}(k)$. These samples, which are held constant during each subinterval by a hold circuit, are applied to the continuous-time system which produces continuous output $\mathbf{y}(t)$. The output is similarly sampled to obtain samples $\mathbf{y}(k)$.

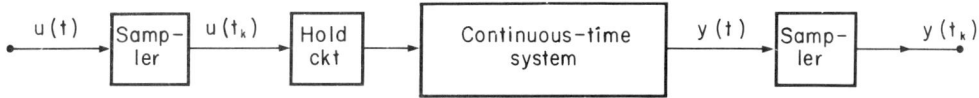

Fig. 6.6.4. Discretization of a continuous-time system

In fact, the above method is used in computer control systems where a digital computer is employed to calculate the proper input at each instant of time. Figure 5 shows the block diagram of a computer control system. The A/D block is an analog-to-digital converter (sampler) and the D/A block is a digital-to-analog converter (hold circuit). The digital computer receives output samples, compares them with a specified reference and calculates the input at each instant t_k according to a control strategy for which it is programmed. The state vector at time t_{k_o+1} can be found from (2) where the input is held constant at $\mathbf{u}(t_{k_o}) \triangleq \mathbf{u}(k_o)$:

$$\mathbf{x}(t_{k_o}+1) = \Phi(t_{k_o}+1,t_{ko})\mathbf{x}_o + [\int_{t_{k_o}}^{t_{k_o}+1} \Phi(t_{ko+1},\tau)\mathbf{B}(\tau)d\tau]\mathbf{u}(k_o) \tag{46}$$

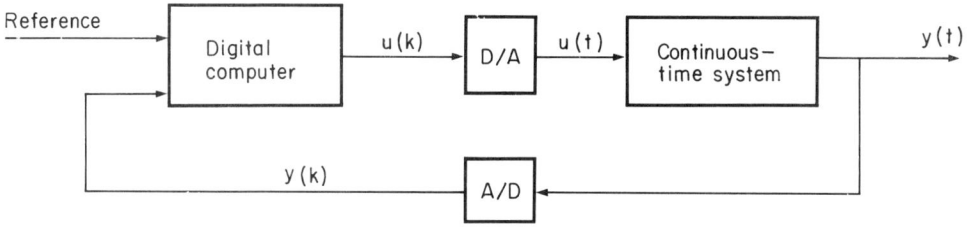

Fig. 6.6.5. Block diagram of a computer control system

Similarly, the state vector at time t_{k+1}, i.e. $\mathbf{x}(t_{k+1})\triangleq\mathbf{x}(k+1)$, can be obtained in terms of $\mathbf{x}(t_k)\triangleq\mathbf{x}(k)$ and $\mathbf{u}(t_k)\triangleq\mathbf{u}(k)$ as follows:

$$\mathbf{x}(k+1) = \Phi(t_{k+1},t_k)\mathbf{x}(k) + [\int_{t_k}^{t_{k+1}} \Phi(t_{k+1},\tau)\mathbf{B}(\tau)d\tau]\mathbf{u}(k) \tag{47}$$

Note that (47) is a discrete state equation similar to (35) where

$$A(k) = \Phi(t_{k+1}, t_k) \tag{48}$$

$$B(k) = \int_{t_k}^{t_{k+1}} \Phi(t_{k+1}, \tau) B(\tau) d\tau \tag{49}$$

Also, the output at time t_k, i.e. $y(t_k) \triangleq y(k)$ can be obtained from (6.1.2):

$$y(k) = C(t_k)x(k) + D(t_k)u(k) \tag{50}$$

which in comparison with (41) results in

$$C(t_k) = C(k) , \quad D(t_k) = D(k) \tag{51}$$

For l.t.i. systems (48) and (49) become

$$A(k) = e^{A\Delta t} , \quad B(k) = [\int_0^{\Delta t} e^{A\sigma} d\sigma] B \tag{52}$$

where $\Delta t = t_{k+1} - t_k$ is the width of each subinterval.

6.6.4 Pulse Response Matrix

We can write the zero-state response (44) as follows:

$$y_{zs}(k) = \sum_{j=k_o}^{k-1} [C(k)\Phi(k,j+1)B(j) + D(k)\Delta(k-j)] u(j) , \quad k \geq k_o \tag{53}$$

where, analogous to the continuous case,

$$\Delta(k-j) = \text{diag } [\Delta(k-j)] \tag{54}$$

and

$$\Delta(k-j) = \begin{cases} 1 & \text{for } k=j \\ 0 & \text{for } k \neq j \end{cases} \tag{55}$$

is the Kronecker delta. Note the analogy between (53) and (25). Define

$$h(k,j) = C(k)\Phi(k,j+1)B(j) + D(k)\Delta(k-j) , \quad k \geq j \tag{56}$$

THE FORCED CASE—COMPLETE SOLUTION

which is analogous to (20). Similar to the continuous case, we also assume causality here:

$$\mathbf{h}(k,j) = 0 \text{ for } k < j \tag{57}$$

Using (56), (53) can be written as

$$\mathbf{y}_{zs}(k) = \sum_{j=k_o}^{k-1} \mathbf{h}(k,j)\mathbf{u}(j), \quad k \geqslant k_o \tag{58}$$

The term $\mathbf{h}(k,j)$ in (56) is called the *pulse response matrix* * of the discrete-time system. More precisely, for a multivariable discrete-time system with r inputs and m outputs it is an $m \times r$ matrix whose ith column is the zero-state response at time t_k (indicated by argument k) when a unit pulse is applied at the *ith* input at time t_j (Fig. 3) and all the other inputs are zero. This can be easily verified by using a development parallel to that used for the impulse response matrix in the continuous case. (See Problem 6.27.)

In the *l.t.i.* case the pulse response matrix (56) becomes

$$\mathbf{h}(k,j) = \mathbf{C}\mathbf{A}^{k-j-1}\mathbf{B} + \mathbf{D}\Delta(k-j) \tag{59}$$

Thus $\mathbf{h}(k,j)$ can be written as $\mathbf{h}(k-j-1)$ and we have

$$\mathbf{h}(k) = \mathbf{C}\mathbf{A}^k\mathbf{B} + \mathbf{D}\Delta(k+1), \quad k \geqslant 0 \tag{60}$$

which is analogous to (23). Thus, assuming $k_o = 0$ with no loss of generality, the zero-state response (58) is this case becomes

$$\mathbf{y}_{zs}(k) = \sum_{j=0}^{k-1} \mathbf{h}(k-j-1)\mathbf{u}(j), \quad k \geqslant 0 \tag{61}$$

Taking the z transform of both sides of (61) and using the convolution property of the z transform (See Section 3.7.) we obtain

$$Z[\mathbf{y}_{zs}(k)] = \mathbf{Y}_{zs}(z) = \mathbf{H}(z)\mathbf{U}(z) \tag{62}$$

where

$$\mathbf{H}(z) = Z[\mathbf{h}(k-1)] = Z[\mathbf{C}\mathbf{A}^{k-1}\mathbf{B} + \mathbf{D}\Delta(k)] = \mathbf{C}(z\mathbf{I}-\mathbf{A})^{-1}\mathbf{B} + \mathbf{D} \tag{63}$$

---

* In analogy with continuous-time systems, $\mathbf{h}(k,j)$ is sometimes referred to as the *discrete impulse response* of the system.

$H(z)$ is called the *transfer function matrix* of the discrete-time system. Equation (62) is consistent with (4.5.12) and equation (53) is the same as (4.5.18). Note that the transfer function matrix is defined only for *l.t.i.* systems.

8. Example Consider the *l.t.i.* multivariable discrete-time system characterized by matrices

$$A = \begin{bmatrix} 1 & -1 & 1 \\ 0 & 1 & 1 \\ 0 & 0 & 1 \end{bmatrix}, \quad B = \begin{bmatrix} 1 & 0 \\ 0 & 1 \\ 1 & 2 \end{bmatrix}, \quad C = \begin{bmatrix} 1 & 1 & -.75 \\ 0 & 2 & 1.5 \end{bmatrix}, \quad D = \begin{bmatrix} 1 & 0 \\ 0 & 1 \end{bmatrix} \tag{64}$$

with initial condition

$$x_o = \begin{bmatrix} 1 \\ 2 \\ 2 \end{bmatrix} \tag{65}$$

and two step inputs

$$u_1(k) = \begin{cases} 2 & k>0 \\ 0 & k\leq 0 \end{cases}, \quad u_2(k) = \begin{cases} 1 & k>0 \\ 0 & k\leq 0 \end{cases} \tag{66}$$

Note that the matrix in this example is the one used in Example 6.4.7 when program "APOWRK" was discussed to evaluate the transition matrix $\{A^k\}$. The resulting transition matrix is,

$$\Phi(k,0) = A^k = \begin{bmatrix} 1 & -k & \frac{k}{2}(3-k) \\ 0 & 1 & k \\ 0 & 0 & 1 \end{bmatrix}. \tag{67}$$

Using (40) with $k_o = 0$, the state $x(k)$ is given by

$$x(k) = \begin{bmatrix} 1-k(k-1) \\ 2+2k \\ 2 \end{bmatrix} + \sum_{j=0}^{k-1} \begin{bmatrix} 2+(k-j-1)(7-2k+2j) \\ 4k-4j-3 \\ 4 \end{bmatrix} \tag{68}$$

The output vector $y(k)$ in (42) is then given by

$$y(k) = \begin{bmatrix} \frac{7}{2}+k(3-k) \\ 8+4k \end{bmatrix} + \sum_{j=0}^{k-1} \begin{bmatrix} (k-j-1)(11-2k+2j) \\ 8(k-j) \end{bmatrix} \tag{69}$$

Program "COMSDS" provides a computational means for the solution of the discrete-time system described by (35) and (41). The source listing of "COMSDS" followed by an example is given below.

THE FORCED CASE—COMPLETE SOLUTION

9. "COMSDS"

```
10    ! PROGRAM NAME : COMSDS "PROG"
20    PRINT " Program <<COMSDS>> gives a complete solution
      for a discrete-time"
30    PRINT " MIMO linear system in state form (A(k),B(k),
      C(k),D(k))"
40    OPTION BASE 1
50    COM Incode,To,Dt,Ns,Ms,Rs,Icode(6),Coef(5,5),A(8,8)
60    COM B(8,6),C(5,8),D(5,5),Ustor(6,102),Ystor(5,102),
      Xstor(8,102)
70    DIM X(101)
80    Iplot=0
90    INPUT "Have you LINKed SUB´s <<Mat>> and <<Plot>>
      (Y/N)?",M$
100   IF (M$="Y") OR (M$="y") THEN 150
110   IF (M$="N") OR (M$="n") THEN 130
120   GOTO 90
130   LINK "Mat",3000,140
140   LINK "Plot",5500,150
150   INPUT "ORDER of the system n(<= 8) ?",Ns
160   PRINT "ORDER of the system n =";Ns
170   INPUT "No. of system INPUTS m(<= 5) ?",Ms
180   PRINT "No. of system INPUTS m =";Ms
190   INPUT "No. of system OUTPUTS r(<= 6) ?",Rs
200   PRINT "No. of system OUTPUTS r =";Rs
210   IF Ns<=0 THEN 150
220   INPUT "Initial time ko?",To
230   PRINT "Initial time ko=";To
240   INPUT "Final time Nf?",Tf
250   PRINT "Final time Nf=";Tf
260   Dt=1
270   REDIM X(Tf+1),Xstor(Ns,Tf+1),A(Ns,Ns),B(Ns,Ms),
      Ustor(Ms,Tf+1)
280   REDIM C(Rs,Ns),D(Rs,Ms),Icode(Ms),Ystor(Rs,Tf+1)
290   CALL Discr1(Tf)
300   CALL Prtplt(Tf)
310   END
320   SUB Discr1(Tf)
330   OPTION BASE 1
340   COM Incode,To,Dt,Ns,Ms,Rs,Icode(*),Coef(*),As(*)
350   COM Bs(*),Cs(*),Ds(*),Ustor(*),Ystor(*),Xstor(*)
360   DIM Xnt(Ns,1)
370   INPUT "Is the system TIME-VARYING (Y/N) ?",T$
380   IF (T$="Y") OR (T$="y") THEN 470
390   IF (T$="N") OR (T$="n") THEN 410
400   GOTO 370
410   Itime=0
420   CALL Mat(As(*),Ns,Ns,"A")
430   CALL Mat(Bs(*),Ns,Ms,"B")
```

```
440    CALL Mat(Cs(*),Rs,Ns,"C")
450    CALL Mat(Ds(*),Rs,Ms,"D")
460    GOTO 530
470    PRINT "SINCE THE SYSTEM IS TIME-VARYING YOU SHOULD
       GOT TO SUBROUTINE"
480    PRINT " <<Matrix>> WHICH IS LINKED TO THE MAIN PROGRAM
       AND DEFINE ALL THE"
490    PRINT " MATRICES A(k),B(k),C(k) and D(k) BEFORE YOU GO
       ON ANY FURTHER!"
500    PRINT "If you have EDITed SUB <<Matrix>> to fit your
       problem HIT CONT."
510    Itime=1
520    PAUSE
530    PRINT "INITIAL VALUES:",LIN(1)
540    FOR I=1 TO Ns
550       DISP "x(";I;")";
560       INPUT Xnt(I,1)
570       PRINT USING 580;I,Xnt(I,1)
580       IMAGE 3X,"x(",DD,")=",MZ.6DE
590    NEXT I
600    MAT Coef=ZER
610    PRINT "INPUT SIGNAL codes ARE:"
620    FOR Iu=1 TO Ms
630    Ustor(Iu,1)=0
640    NEXT Iu
650    PRINT " STEP INPUT .......... 0"
660    PRINT " RAMP INPUT .......... 1"
670    PRINT " PARABOLIC INPUT ..... 2"
680    PRINT " SINUSOIDAL INPUT .... 3"
690    FOR Iu=1 TO Ms
700    PRINT "For INPUT no.";Iu;" Give INPUT SIGNAL code
       (0-3)"
710    INPUT "Code (0-3)",Incode
720    IF (Incode<0) OR (Incode>3) THEN 700
730    ON Incode+1 GOTO 740,780,810,840
740    INPUT "For a STEP input : r(t)=au(t); Give value of
       a",Apar
750    Coef(Iu,1)=Apar
760    Ustor(Iu,1)=Apar
770    GOTO 880
780    INPUT "For a RAMP input : r(t)=bt.u(t); Give value of
       b",Bpar
790    Coef(Iu,2)=Bpar
800    GOTO 880
810    INPUT "For a PARABOLIC input : r(t)=ct.t.u(t); Give
       value of c",Cpar
820    Coef(Iu,3)=Cpar
830    GOTO 880
840    PRINT "For a SINUSOIDAL input : r(t)=xSin(wt) :"
850    INPUT "   Give values of x & w",Xpar,Wpar
```

THE FORCED CASE—COMPLETE SOLUTION

```
860  Coef(Iu,4)=Xpar
870  Coef(Iu,5)=Wpar
880  Icode(Iu)=Incode
890  NEXT Iu
900  CALL Discr(Tf,Xnt(*),Itime)
910  SUBEND
920  SUB Discr(Tf,Xnt(*),Itime)
930  OPTION BASE 1
940  COM Incode,To,Dt,Ns,Ms,Rs,Icode(*),Coef(*),A(*)
950  COM B(*),C(*),D(*),Ustor(*),Ystor(*),Xstor(*)
960  DIM Xk(Ns,1),Axk(Ns,1),Uk(Ms,1),Buk(Ns,1),Cxk(Rs,1),
     Duk(Rs,1)
970  IF Itime=0 THEN 990
980  CALL Matrix(A(*),B(*),C(*),D(*),0)
990  MAT Xk=Xnt
1000 FOR K=1 TO Tf
1010 MAT Axk=A*Xk
1020 FOR Iu=1 TO Ms
1030 Ic=Icode(Iu)+1
1040 ON Ic GOTO 1050,1070,1090,1110
1050 Uk(Iu,1)=Coef(Iu,Ic)
1060 GOTO 1120
1070 Uk(Iu,1)=Coef(Iu,Ic)*K
1080 GOTO 1120
1090 Uk(Iu,1)=Coef(Iu,Ic)*K*K
1100 GOTO 1120
1110 Uk(Iu,1)=Coef(Iu,Ic)*SIN(Coef(Iu,Ic+1)*K)
1120 Ustor(Iu,K)=Uk(Iu,1)
1130 NEXT Iu
1140 FOR I=1 TO Ns
1150 Xstor(I,K)=Xk(I,1)
1160 NEXT I
1170 MAT Cxk=C*Xk
1180 MAT Duk=D*Uk
1190 MAT Cxk=Cxk+Duk
1200 FOR I=1 TO Rs
1210 Ystor(I,K)=Cxk(I,1)
1220 NEXT I
1230 FOR I=1 TO Ms
1240 Ustor(I,K)=Uk(I,1)
1250 NEXT I
1260 MAT Buk=B*Uk
1270 MAT Xk=Axk+Buk
1280 IF Itime=0 THEN 1300
1290 CALL Matrix(A(*),B(*),C(*),D(*),K)
1300 NEXT K
1310 SUBEND
1320 SUB Prtplt(Tf)
1330 OPTION BASE 1
1340 COM Incode,To,Dt,Ns,Ms,Rs,Icode(*),Coef(*),A(*)
```

```
1350 COM B(*),C(*),D(*),Ustor(*),Ystor(*),Xstor(*)
1360 DIM Er(Ns,1),Yplot(Tf+1),X(Tf+1),Uk(Ms,1),Du(Rs,1),
     Yout(Rs,1),Cx(Rs,1)
1370 FOR I=1 TO Tf+1
1380 X(I)=I-1
1390 NEXT I
1400 INPUT "Do you like to PRINT STATE VARIABLES ? ",C$
1410 IF (C$="Y") OR (C$="y") THEN 1440
1420 IF (C$="N") OR (C$="n") THEN 1510
1430 GOTO 1430
1440 FOR I=1 TO Tf
1450 FOR J=1 TO Ns
1460 Er(J,1)=Xstor(J,I)
1470 NEXT J
1480 PRINT "At k=";X(I);"x(k) Is:"
1490 MAT PRINT Er;
1500 NEXT I
1510 INPUT "Do you like to PRINT INPUT VARIABLES ? ",I$
1520 IF (I$="Y") OR (I$="y") THEN 1550
1530 IF (I$="N") OR (I$="n") THEN 1620
1540 GOTO 1510
1550 FOR I=1 TO Tf
1560 FOR Iu=1 TO Ms
1570 Uk(Iu,1)=Ustor(Iu,I)
1580 NEXT Iu
1590 PRINT "At k=";X(I);"u(k) Is:"
1600 MAT PRINT Uk;
1610 NEXT I
1620 INPUT "Do you like to PRINT OUTPUT VARIABLES ? ",O$
1630 IF (O$="Y") OR (O$="y") THEN 1660
1640 IF (O$="N") OR (O$="n") THEN 1730
1650 GOTO 1620
1660 FOR I=1 TO Tf
1670 FOR Iy=1 TO Rs
1680 Yout(Iy,1)=Ystor(Iy,I)
1690 NEXT Iy
1700 PRINT "At k=";X(I);"y(k) Is:"
1710 MAT PRINT Yout;
1720 NEXT I
1730 INPUT "Do you like to PLOT STATE VARIABLES ? ",C$
1740 IF (C$="Y") OR (C$="y") THEN 1770
1750 IF (C$="N") OR (C$="n") THEN 1890
1760 GOTO 1600
1770 FOR Ipl=1 TO Ns
1780 FOR I=1 TO Tf
1790 Yplot(I)=Xstor(Ipl,I)
1800 X(I)=I-1
1810 NEXT I
1820 Iplot=0
1830 CALL Plot(X(*),Yplot(*),Tf,Iplot,To,Tf)
```

THE FORCED CASE—COMPLETE SOLUTION

```
1840 INPUT "Do you like to STOP PLOTTING this set of
     variables?",C$
1850 IF (C$="Y") OR (C$="y") THEN 1890
1860 IF (C$="N") OR (C$="n") THEN 1880
1870 GOTO 1840
1880 NEXT Ipl
1890 INPUT "Do you like to PLOT INPUT VARIABLES ? ",C$
1900 IF (C$="Y") OR (C$="y") THEN 1930
1910 IF (C$="N") OR (C$="n") THEN 2050
1920 GOTO 1890
1930 FOR Ipl=1 TO Ms
1940 FOR I=1 TO Tf
1950 Yplot(I)=Ustor(Ipl,I)
1960 X(I)=I-1
1970 NEXT I
1980 Iplot=0
1990 CALL Plot(X(*),Yplot(*),Tf,Iplot,To,Tf)
2000 INPUT "Do you like to STOP PLOTTING this set of
     variables?",C$
2010 IF (C$="Y") OR (C$="y") THEN 2050
2020 IF (C$="N") OR (C$="n") THEN 2040
2030 GOTO 2000
2040 NEXT Ipl
2050 INPUT "Do you like to PLOT OUTPUT VARIABLES ? ",C$
2060 IF (C$="Y") OR (C$="y") THEN 2090
2070 IF (C$="N") OR (C$="n") THEN 2210
2080 GOTO 2050
2090 FOR Ipl=1 TO Rs
2100 FOR I=1 TO Tf
2110 Yplot(I)=Ystor(Ipl,I)
2120 X(I)=I-1
2130 NEXT I
2140 Iplot=0
2150 CALL Plot(X(*),Yplot(*),Tf,Iplot,To,Tf)
2160 INPUT "Do you like to STOP PLOTTING this set of
     variables?",C$
2170 IF (C$="Y") OR (C$="y") THEN 2210
2180 IF (C$="N") OR (C$="n") THEN 2200
2190 GOTO 2160
2200 NEXT Ipl
2210 SUBEND
2220 SUB Matrix(A(*),B(*),C(*),D(*),K)
2230 OPTION BASE 1
2240 MAT A=ZER
2250 MAT B=ZER
2260 MAT C=ZER
2270 MAT D=ZER
2280 A(1,1)=A(1,3)=A(2,3)=A(3,3)=B(1,1)=B(3,1)=C(1,1)=
     C(1,2)=D(1,1)=1
2290 B(3,2)=C(2,2)=2
```

```
2300 A(1,2)=-K/10
2310 A(2,2)=1/2
2320 B(2,2)=D(2,2)=K/100
2330 C(1,3)=-3/4
2340 C(2,3)=-1
2350 SUBEND
```

10. Example. Figures 6 and 7 show the state and output trajectories of the system, generated by "COMSDS".

```
Program <<COMSDS>> gives a complete solution for a
discrete-time
  MIMO linear system in state form (A(k),B(k),C(k),D(k))
ORDER of the system n = 3
No. of system INPUTS m = 2
No. of system OUTPUTS r = 2
Initial time ko= 0
Final time Nf= 10
SINCE THE SYSTEM IS TIME-VARYING YOU SHOULD GOT TO SUBROUTINE
  <<Matrix>> WHICH IS LINKED TO THE MAIN PROGRAM AND DEFINE
ALL THE MATRICES A(k),B(k),C(k) and D(k) BEFORE YOU GO ON ANY
FURTHER! If you have EDITed SUB <<Matrix>> to fit your
problem HIT CONT. INITIAL VALUES:

    x( 1)= 0.000000E+00
    x( 2)= 0.000000E+00
    x( 3)= 0.000000E+00
INPUT SIGNAL codes ARE:
  STEP INPUT .......... 0
  RAMP INPUT .......... 1
  PARABOLIC INPUT ..... 2
  SINUSOIDAL INPUT .... 3
For INPUT no. 1   Give INPUT SIGNAL code (0-3)
For INPUT no. 2   Give INPUT SIGNAL code (0-3)

At k= 0 x(k) Is:
 0

 0

 0


At k= 1 x(k) Is:
 1

 0

 4
```

At k= 2 x(k) Is:
6

4.015

8

At k= 0 u(k) Is:
1

1.5

At k= 1 u(k) Is:
1

1.5

At k= 0 y(k) Is:
1

0

At k= 1 y(k) Is:
-1

-3.985

At k= 2 y(k) Is:
5.015

.06

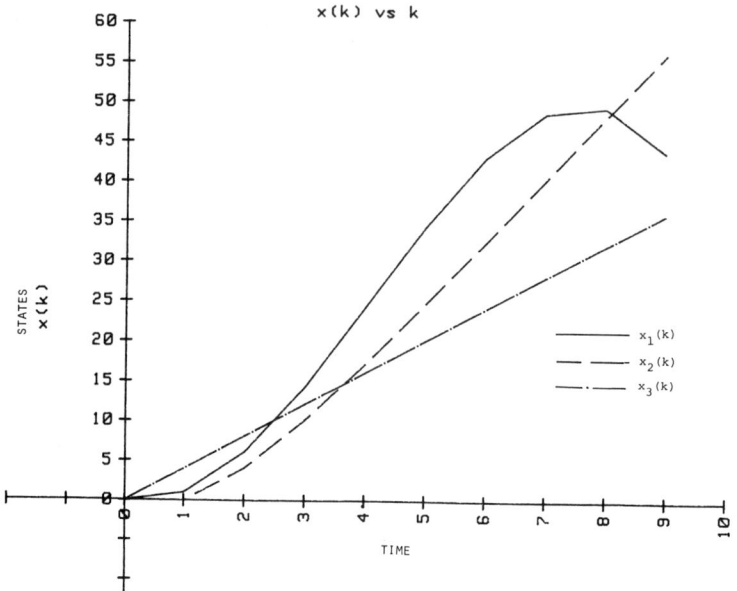

Fig. 6.6.6. State x(k) trajectories for Example 6.6.10

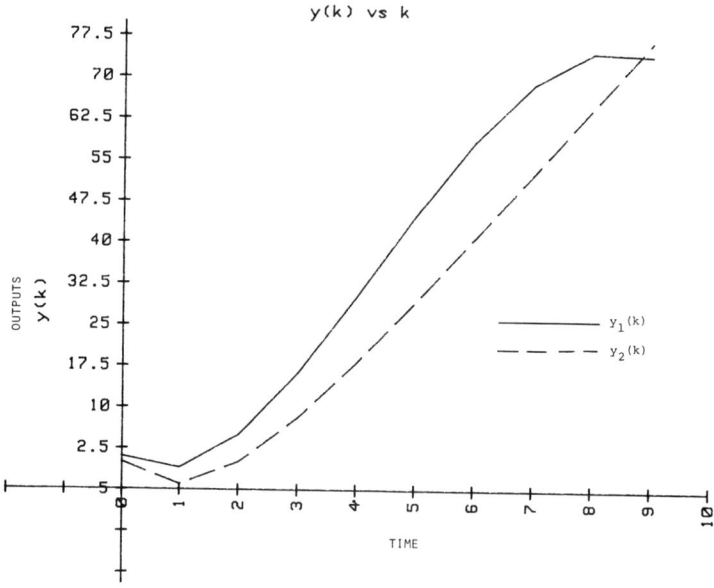

Fig. 6.6.7. Output y(k) trajectories for Example 6.6.10

6.7 ADJOINT AND DUAL SYSTEMS

In this section we will discuss the concepts of adjoint and dual systems. These concepts are very useful and have many applications particularly in optimal control systems.

1. Definition. (Adjoint state equations) The linear homogeneous differential equations

$$\dot{x}(t) = A(t)x(t) \qquad (1)$$

$$\dot{z}(t) = -A^*(t)z(t) \qquad (2)$$

are said to be *adjoints* of one another. △

Matrix $A^*(t)$ in (2) is the conjugate transpose of matrix $A(t)$. (See Section 2.5.) If matrix A is real, then A^* is simply the transpose of A. Adjoint state equations have two important properties given by the following theorems.

2. Theorem. If $x(t)$ is any solution to (1) and $z(t)$ is any solution to (2), then their inner product is constant; i.e.

$$(x(t),z(t)) = constant \quad \text{for all } t \qquad (3)$$

Proof. We will show that the derivative of the inner product (3) w.r.t. time is zero. Using (1) and (2) and the properties of the inner product we have

$$\frac{d}{dt}(x,z) = (\dot{x},z) + (x,\dot{z})$$

$$= (Ax,z) + (x,-A^*z)$$

$$= (Ax,z) + (-Ax,z) = (0,z) = 0$$

3. Example. Consider the system $\dot{x} = Ax$, $x(0) \triangleq x_o$ where

$$A = \begin{bmatrix} 0 & -4 \\ 1 & -4 \end{bmatrix} \qquad (4)$$

It can be verified that

$$e^{At} = \begin{bmatrix} (1+2t)e^{-2t} & -4te^{-2t} \\ te^{-2t} & (1-2t)e^{-2t} \end{bmatrix} \qquad (5)$$

Thus we have

$$x(t) = e^{At}x_o = \begin{bmatrix} (1+2t)x_{o1} - 4tx_{o2} \\ x_{o1} + (1-2t)x_{o2} \end{bmatrix} e^{-2t} \qquad (6)$$

The adjoint of this system is the system $\dot{z} = Bz$, $z(0) \triangleq z_o$ where

$$B = -A' = \begin{bmatrix} 0 & -1 \\ 4 & 4 \end{bmatrix} \qquad (7)$$

We find

$$z(t) = e^{Bt}z_o = \begin{bmatrix} (1-2t)e^{2t} & -te^{2t} \\ 4te^{2t} & (1+2t)e^{2t} \end{bmatrix} \begin{bmatrix} z_{o1} \\ z_{o2} \end{bmatrix}$$

$$= \begin{bmatrix} (1-2t)z_{o1} - tz_{o2} \\ 4t - z_{o1} + (1+2t)z_{o2} \end{bmatrix} e^{2t} \qquad (8)$$

It can be easily checked that $(x(t), z(t)) = x_{o1}z_{o1} + x_{o2}z_{o2} = (x_o, z_o)$ as expected.

4. Theorem. If $\Phi(t)$ is a fundamental matrix of (1), then $(\Phi^{-1}(t))^* \triangleq \Psi(t)$ is a fundamental matrix of (2).

Proof. We have

$$\Phi(t)\Phi^{-1}(t) = I$$

Thus,

$$\frac{d}{dt}(\Phi\Phi^{-1}) = \dot{\Phi}\Phi^{-1} + \Phi\frac{d}{dt}(\Phi^{-1}) = 0$$

or

$$\Phi\frac{d}{dt}(\Phi^{-1}) = -A\Phi\Phi^{-1}$$

which yields

$$\frac{d}{dt}(\Phi^{-1}) = -\Phi^{-1}A. \qquad (9)$$

Taking the conjugate transpose of both sides of (9) we obtain

$$\frac{d}{dt}(\Phi^{-1})^* = -\mathbf{A}^*(\Phi^{-1})^*. \tag{10}$$

Thus $(\Phi^{-1})^*$ is a fundamental matrix of (2).

5. Corollary. If $\Phi(t,\tau)$ and $\Psi(t,\tau)$ are the transition matrices of (1) and (2), respectively, then

$$\Psi^*(t,\tau) = \Phi(\tau,t) \tag{11}$$

Proof. From (6.3.1) we have $\Phi(t,\tau) = \Phi(t)\Phi^{-1}(\tau)$ and $\Psi(t,\tau) = \Psi(t)\Psi^{-1}(\tau)$ where $\Phi(t)$ and $\Psi(t)$ are fundamental matrices of (1) and (2), respectively. Thus

$$\Psi^*(t,\tau) = (\Psi^{-1}(\tau))^* \Psi^*(t). \tag{12}$$

From Theorem 4 we have $\Phi^{-1}(t) = \Psi^*(t)$. Therefore (12) yields.

$$\Psi^*(t,\tau) = (\Phi(\tau))\Phi^{-1}(t) = \Phi(\tau,t) \quad \text{q.e.d.} \quad \Delta$$

One of the applications of the adjoint concept is to solve a state equation when a final state instead of initial state has been specified. To illustrate this point consider the state equation (6.6.1). We can write its solution (6.6.2) as follows:

$$\mathbf{x}(t) = \Phi(t,t_o)[\mathbf{x}(t_o) + \int_{t_o}^{t} \Phi^{-1}(t,t_o)\Phi(t,\tau)\mathbf{B}(\tau)\mathbf{u}(\tau)d\tau] \tag{13}$$

Using the transition property of $\Phi(t,\tau)$ and (11), (13) yields

$$\mathbf{x}(t) = \Phi(t,t_o)[\mathbf{x}(t_o) + \int_{t_o}^{t} \Psi^*(\tau,t_o)\mathbf{B}(\tau)\mathbf{u}(\tau)d\tau] \tag{14}$$

If the state is specified at a final time t_f, (14) can be used to find the corresponding initial state $\mathbf{x}(t_o)$:

$$\mathbf{x}(t_o) = \Phi(t_o,t_f)\mathbf{x}(t_f) - \int_{t_o}^{t_f} \Psi^*(\tau,t_o)\mathbf{B}(\tau)\mathbf{u}(\tau)d\tau \tag{15}$$

This enables us to determine the state trajectory $\mathbf{x}(t)$ for $t_o < t < t_f$. (See Problem 6.31.)

The adjoint concept can also be defined in the case of discrete-time systems.

6. Definition.
The linear homogeneous discrete state equations

$$x(k+1) + A(k)x(k) \tag{16}$$

$$z(k+1) = [A^{-1}(k)]^* z(k) \tag{17}$$

are said to be adjoints of one another.

The adjoint systems in the discrete case have properties similar to the continuous case. (See Problem 6.32.) Adjoint systems have applications in optimization problems. The reader is referred to reference [6.8] (p.346) for the illustration of some of these applications.

7. Definition (Dual systems)
The linear continuous-time systems $S = (A(t),B(t),C(t),D(t))$ and $S^* = (-A^*(t),C^*(t),B^*(t),D^*(t))$ are said to be *duals* of one another. Note that

$$S : \begin{cases} \dot{x}(t)=A(t)x(t)+B(t)u(t) \\ y(t)=C(t)x(t)+D(t)u(t) \end{cases} \tag{18}$$

and

$$\bar{S} : \begin{cases} \dot{z}(t)=-\bar{A}(t)z(t)+\bar{C}(t)u(t) \\ y(t)=\bar{B}(t)z(t)+\bar{D}(t)u(t) \end{cases} \tag{19}$$

Definition of duality for the discrete-time case follows.

8. Definition.
The linear discrete-time systems

$$S = (A(k),B(k),C(k),D(k)) \text{ and } \bar{S} = (A^{-1}(k)^*,C^*(k),B^*(k),D^*(k))$$

are said to be duals of one another. △

Note that duality (similar to the adjoint concept) is reciprocal, i.e. if system S is the dual of system $\bar{S}$, then system $\bar{S}$ is the dual of system S. Also note that if system S has r inputs and m outputs, its dual system $\bar{S}$ will have m inputs and r outputs. The concept of duality of systems will be used in Chapter 8.

PROBLEMS

6.1 Show that if Φ is a fundamental matrix of $\dot{x}(t) = A(t)x(t)$, then so is $\Phi(t)C$ where C is any constant nonsingular matrix.

6.2 If matrix $\Phi(t)$ is a solution of $\dot{x}(t) = A(t)x(t)$ and for a given time to, det $(\Phi(t)) \neq 0$, show that $\Phi(t)$ is a fundamental matrix of that differential equation.

Hint: Show that det $(\Phi(t)) \neq 0$ for *all* t by contradiction or show that

$$\frac{\det(\Phi(t_2))}{\det(\Phi(t_1))} = \exp[\int_{t_1}^{t_2} tr(A(\tau))d\tau].$$

6.3 Find a fundamental matrix for $\dot{x}(t) = A(t)x(t)$ where

$$A(t) = \begin{bmatrix} 2t & 1 \\ 0 & 1 \end{bmatrix}$$

Hint: Let $\Phi(t) = [t\sigma_1(t), t\sigma_2(t)]$. Determine $t\sigma_1$ and $t\sigma_2$ by solving the state equation for initial conditions $[1, 0]'$ and $[0, 1]'$, respectively.

6.4 Find a fundamental matrix for $x(k+1) = A(k)x(k)$ where

$$A(k) = \begin{bmatrix} 1 & 0 \\ k & 1 \end{bmatrix}$$

Hint: Write separate equations for the components of $x(k)$ and use z transform where necessary.

6.5 Show that the transition matrix of $\dot{x} = A(t)x(t)$ is unique.

Hint: Use $\Phi(t,t_o) = \Phi(t)\Phi^{-1}(t_o)$ and the result of Problem 6.1.

6.6 Prove that [32]

$$\det(e^{At}) = \exp[tr(A\ t)].$$

Note that this is a special case of (6.3.4).

6.7 Prove the determinant properties of continuous-time and discrete-time transition matrices, i.e. relations (6.3.4) and (6.3.8).

6.8 If square matrix A is block diagonal, i.e. if

$$A = \begin{bmatrix} M_1 & . & 0 & . \\ . & M_2 & . & . \\ 0 & . & . & M_n \end{bmatrix}$$

where M_i, $i=1, 2, \ldots, n$ are square matrices, show that

$$A^k = Block-diag\ (M_1^k, M_2^k, \ldots, M_n^k)$$

and conclude that

$$e^{At} = Block-diag\ (e^{M_1 t}, e^{M_2 t}, \ldots, e^{M_n t})$$

6.9 Show that for the $m \times m$ Jordan block

$$J = \begin{bmatrix} \lambda & 1 & . & 0 \\ . & \lambda & 1 & . \\ . & . & . & 1 \\ 0 & . & . & \lambda \end{bmatrix}$$

we have

$$e^{Jt} = \begin{bmatrix} e^{\lambda t} & te^{\lambda t} & . & \frac{t^{m-1}}{(m-1)!}e^{\lambda t} \\ 0 & e^{\lambda t} & te^{\lambda t} & . & te^{\lambda t} \\ 0 & . & . & e^{\lambda t} \end{bmatrix}$$

and

$$J^k = \begin{bmatrix} \lambda^k & k\lambda^{k-1} & k! \frac{\lambda^{k-m+1}}{(m-1)!(k-m+1)!} \\ 0 & \lambda^k & k\lambda^{k-1} & k\lambda^{k-1} \\ 0 & . & & \lambda^k \end{bmatrix}$$

6.10 Use (2.9.7) to show that $e^{At} = L^{-1}[(sI-A)^{-1}]$.

Hint: Write $(sI-A)^{-1}$ as $\frac{1}{s}(I - \frac{A}{s})^{-1}$ and expand it.

6.11 For the matrix

$$A = \begin{bmatrix} 1 & 2 & 0 \\ -1 & 3 & 0 \\ 2 & -4 & 3 \end{bmatrix}$$

calculate e^{At} and A^k using the methods discussed in Section 6.4. Use programs "EXPATA" and "APOWRK" to verify the results.

6.12 Repeat Problem 6.11 for

$$A = \begin{bmatrix} 0 & -1 & -1 \\ 1 & -2 & -1 \\ 0 & 0 & -2 \end{bmatrix}$$

Use program "EXPATA" to show that

$$e^{At} = \begin{bmatrix} (t+1)e^{-t} & -te^{-t} & -e^{-t}+e^{-2t} \\ te^{-t} & (1-t)e^{-t} & -e^{-t}+e^{-2t} \\ 0 & 0 & e^{-2t} \end{bmatrix}$$

6.13 Repeat Problem 6.11 for

$$A = \begin{bmatrix} 1 & 0 & 0 \\ -1 & 1 & 0 \\ 2 & 4 & 1 \end{bmatrix}$$

Use program "APOWRK" to show that

$$A^k = \begin{bmatrix} 1 & 0 & 0 \\ -k & 1 & 0 \\ 2k(2-k) & 4k & 1 \end{bmatrix}.$$

6.14 Show that if $A = \begin{bmatrix} 0 & -\omega^2 \\ 1 & 0 \end{bmatrix}$, $e^{At} = \begin{bmatrix} \cos \omega t & -\omega \sin \omega t \\ \frac{1}{\omega} \sin \omega t & \cos \omega t \end{bmatrix}$

6.15 Find the transition matrices of the systems in Problems 6.3 and 6.4.

6.16 Find the transition matrix for $\dot{x} = A(t)x$ where

$$A(t) = \begin{bmatrix} t^2 & 0 \\ t & 1 \end{bmatrix}$$

Hint: Solve for components of x separately and use

$$x(t) = \Phi(t,t_o)x(t_o).$$

6.17 Determine the fundamental matrix corresponding to system matrix

$$A(t) = \begin{bmatrix} -4 & 3e^{-4t} \\ -e^{4t} & 0 \end{bmatrix},$$

6.18 Show that if

$$A(t) = \begin{bmatrix} 2 & -e^t \\ e^{-t} & 1 \end{bmatrix},$$

then the transition matrix of $\dot{x} = A(t)x$ is

$$\Phi(t,t_0) = \begin{bmatrix} e^{2(t-t_0)}\cos(t-t_0) & e^{-2t+t_0}\sin(t-t_0) \\ e^{t_0-2t}\sin(t-t_0) & e^{t_0-t}\cos(t-t_0) \end{bmatrix}$$

6.19 If in the homogeneous linear time-varying system $\dot{x} = A(t)x$ matrix $A(t)$ is periodic, i.e. if $A(t) = A(t+\omega)$, show that the transition matrix of this system can be written as

$$\Phi(t,t_0) = P(t,t_0) e^{R(t-t_0)}$$

where $P(t,t_0) = P(t+\omega,t_0)$ and R is a constant matrix.

6.20 Consider the matrix

$$A = \begin{bmatrix} 1 & 2 & -2 \\ 0 & 0 & 1 \\ 0 & -3 & -4 \end{bmatrix}$$

Determine the resolvent matrix $(sI-A)^{-1}$, a) directly, b) using the algorithm given in Theorem 6.5.3, and c) using "RESMAT".

6.21 For matrix A in Problem 6.20,

a) Use methods of Section 6.4 to evaluate e^{At}.

b) Use "RESMAT" to calculate $(sI-A)^{-1}$ and apply methods of Section 6.5, i.e. equations (6.5.21) and (6.5.24) to evaluate e^{At}.

c) Verify the results in (a) and (b) by using "EXPATA" to evaluate e^{At}.

6.22 Use equations (6.6.9) and (6.6.10) to prove Theorems 6.6.2 and 6.6.3.

6.23 Use the superposition property of linear systems to show relation 6.6.26. Use this to show that in non-anticipative systems (i.e. systems in which the output at time t does not depend on inputs at times larger then t) $h(t,\tau) = 0$ for $t < \tau$.

Hint: Divide the interval of interest $[t_0,t]$ into n subintervals. Approximate the input $u(t)$ as a series of n weighted impulses. Using the superposition property of linear systems find the zero-state response to this approximate input as a sum. Then let $n \to \infty$ to obtain the convolution intergal.

PROBLEMS

6.24 Prove that two linear time-varying algebraically equivalent systems with the same initial state produce the same complete response when excited by the same input (i.e. show that they have the same impulse response matrix and the same zero-input response).

6.25 In Example 6.6.4 use the transfer function matrix (6.6.34) to determine the impulse response matrix and use (6.6.26) to verify that the step response is as given in (6.6.33).

6.26 Consider the continuous-time system (A, B, C, D) where

$$A = \begin{bmatrix} 1 & 2 & -2 \\ 0 & 0 & 1 \\ 0 & -3 & -4 \end{bmatrix}, \; B = \begin{bmatrix} 0 & 2 \\ 1 & 0 \\ -2 & -1 \end{bmatrix}, \; C = [1, 0, -1]$$

with inputs $u_1(t)$ = unit step function and $u_2(t)$ = unit ramp function. Use "COMSCS" to find plots of each state variable and the output versus time.

6.27 Use equations (6.6.43) and (6.6.44) to prove the linearity of zero-state response and zero-input response in discrete-time systems.

6.28 Use a development similar to that given in equations (6.6.15) to (6.6.20) for the continuous case to show that the pulse response matrix of a linear discrete-time system is given by (6.6.56).

6.29 Consider the discrete-time system (A, B, C, D) where matrices A, B and C are given in Problem 6.26. Use "COMSDS" to find plots of each state variable and the output versus k.

6.30 Consider the systems $\dot{x}(t) = A(t)x(t) + B(t)u(t)$ and $\dot{z}(t) = -A^*(t)z(t)$. Show that

$$(z(t_1)k, x(t_1)) - (z(t_o), x(t_o)) = \int_{t_o}^{t_1} (z(\tau), B(\tau)u(\tau)) d\tau.$$

6.31 Consider the system

$$\dot{x} = \begin{bmatrix} -1 & 3 \\ 0 & -2 \end{bmatrix} x + \begin{bmatrix} 1 \\ -1 \end{bmatrix} u, \; y = [0 \; 1]x.$$

Use the adjoint system to find all the states $x(0)$ which result in $y(3)=1$.

6.32 Consider the discrete adjoint state equations (6.7.11) and (6.7.12).

    a) If $x(k)$ is any solution to (6.7.11) and $z(k)$ is any solution to (6.7.12), show that $(x(k), z(k))$ = constant.

    b) If $\Phi(k)$ is a fundamental matrix of (6.7.11), show that $(\Phi^{-1}(k))^*$ is a fundamental matrix of (6.7.12).

c) If $\Phi(k,k_o)$ is the transition matrix of (6.7.11), show that $(\Phi^{-1}(k,k_o))^*$ is the transition matrix of (6.7.12).

REFERENCES

[6.1] R. W. Brockett, *Finite-Dimensional Linear Systems*, Wiley, 1970

[6.2] C. A. Desoer, *Notes for a Second Course on Linear Systems*, Van Nostrand Reinholt, 1970

[6.3] C. Moler and C. Van Loan, "Nineteen Dubious Ways to Compute the Exponential of a Matrix," *SIAM Review*, Vol. 20, No. 4, Oct., pp 801-836, 1978

[6.4] A. S. Householder, *The Theory of Matrices in Numerical Analysis*," Blaisdell, New York, 1964

[6.5] D. M. Wiberg, *State Space and Linear Systems*, McGraw-Hill, 1971

[6.6] V. N. Fadeeva, *Computational Methods of Linear Algebra*, Dover, 1959

[6.7] H. H. Rosenbrock, *State-Space and Multivariable Theory*, Nelson, 1970

[6.8] L. A. Zadeh and C. A. Desoer, *Linear System Theory*, McGraw-Hill, 1963

[6.9] E. Coddington and N. Levinson, *Theory of Ordinary Differential Equations*, McGraw-Hill, 1955

CHAPTER 7

SYSTEM STABILITY

7.1 INTRODUCTION

In previous chapters we discussed methods of modeling systems and solving the differential or difference equations describing them. This chapter is concerned with the study of system stability. Generally speaking, we call a system stable if it returns to equilibrium after the occurrence of a disturbance. That is, a system is stable if its response due to a bounded initial condition or a bounded input does not grow without bound. The concept of stability is of great importance to the system designer. An unstable system is not acceptable since it results in breaking down or burning out of components during the operation.

We will consider system stability in two cases. In one case the system is free (i.e. it has zero input) but there is nonzero initial condition. This is referred to as *zero-input stability* or *internal stability*. In the other case the system is relaxed (i.e. it has zero initial condition) but it has nonzero input. This is referred to as *zero-state stability* or *external stability* [7.1]. We will formalize these concepts in the next section. For linear systems, the zero-input stability and the zero-state stability can be characterized by the properties of the transition matrix and the transfer function matrix, respectively. We will present the corresponding necessary and sufficient conditions in Section 7.3.

We will show that the stability of l.t.i. systems is completely characterized by the eigenvalues of the system matrix. Methods of this characterization include the Routh-Hurwitz criterion, the root-locus method, the Bode diagram and the Nyquist method for continuous-time systems and the Jury-Blanchard criterion for discrete-time systems. Further, the root-locus, Nyquist and Bode methods provide measures which indicate how close the system is to instability. These methods will be discussed in subsequent sections.

For nonlinear and/or time-varying systems, the stability characterization is more complicated. We will discuss the circle criterion for the stability of linear time-varying systems. Finally, Lyapunov method which is the main tool for studying the stability of nonlinear systems will be presented. Computer programs which have been developed for most of these methods will also be introduced in the corresponding sections.

7.2 STABILITY CONCEPTS AND DEFINITIONS

We will consider the concepts and definitions related to zero-input stability and zero-state stability in this section.

7.2.1 Zero-Input Stability

Let us begin by defining equilibrium states of systems.

1. Definition. A state x_e of a system is said to be an *equilibrium state* if it has the following property:

$$x(t_o) = x_e => x(t) = x_e \text{ for all } t > t_o$$

provided that no input is applied. △

Consider a (nonlinear time-varying) system described by the differential equation

$$\dot{x}(t) = f(x(t), u(t), t) \qquad (1)$$

or the difference equation

$$x(k+1) = f(x(k), u(k), k) \qquad (2)$$

Then in the continuous-time system (1) the equilibrium states are the solutions to

$$f(x_e, 0, t) = 0, \quad t \geqslant t_o \qquad (3)$$

and in the discrete system (2) they are the solutions to

$$f(x_e, 0, k) = x_e, \quad k \geqslant k_o \qquad (4)$$

Note that a system may have more than one equilibrium state. For linear systems characterized by $\dot{x}(t) = A(t)x(t)$, the state $x_e = 0$, i.e. the origin of the state space, is always an equilibrium state. For nonlinear systems we can always assume that $x_e = 0$ is an equilibrium state with no loss of generality: if in system (1) the equilibrium state under consideration is at $x_e \neq 0$, then define a new state vector $x = x - x_e$ so that $\dot{x} = f(x(t), u(t), t)$ and $f(0, 0, t) = 0$. (See Problem 7.2.) Similar argument holds for system (2).

2. Definition. The equilibrium state $x_e = 0$ is *stable in the sense of Lyapunov* (abbreviated *isL*), or simply stable, if for any given scalar $\epsilon > 0$ there exists a scalar $\delta(t_o, \epsilon) > 0$ such that $\|x_o\| < \delta$ implies $\|x(t)\| < \epsilon$ for all $t \geqslant t_o$.

STABILITY CONCEPTS AND DEFINITIONS

3. Definition. The equilibrium state $x_e = 0$ is *asymptotically stable* if 1) it is stable *isL*, and 2) for any t_o and for any initial state x_o sufficiently close to 0 the state $x(t)$ tends to the origin of the state space as t tends to infinity. △

The above definitions of stability are illustrated in Fig. 1 for a two-dimensional system. If the initial state x_o is restricted to a circle of radius δ centered at the origin, the resulting trajectory $x(t)$, often indicated by $x(t,x_o,t_o)$, remains forever inside the cylinder with base radius ϵ. Further, if the origin of the state space is asymptotically stable, the trajectory converges to the line $x_1 = x_2 = 0$ as $t \to \infty$.

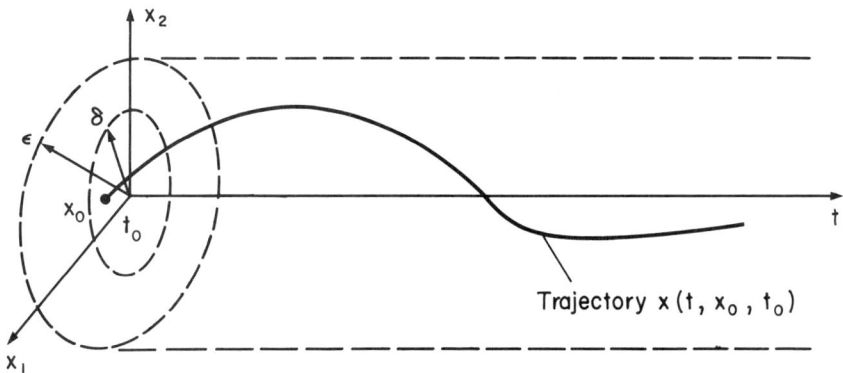

Fig. 7.2.1. Illustration of stability definitions

If in Definitions 2 and 3, δ is independent of t_o, then the stability is said to be *uniform*. Note that for time-invariant systems the stability is always uniform. Also if in Definitions 2 and 3 the initial state is not restricted to states sufficiently close to the origin of the state space, then we have *global stability* or *stability in the large*. Note that for linear systems, asymptotic stability is always global. (See Problem 7.5.) This is due to the fact that in linear systems the zero-input response is a linear function of the initial state. (See Theorem 6.6.3.) For nonlinear systems, however, an equilibrium state may be asymptotically stable without being asymptotically stable in the large. (See Problem 7.6.)

4. Example. The above definitions of stability can be illustrated by an example of a physical system consisting of a ball bearing on a surface. The cross section of such a system is shown in Fig. 2.

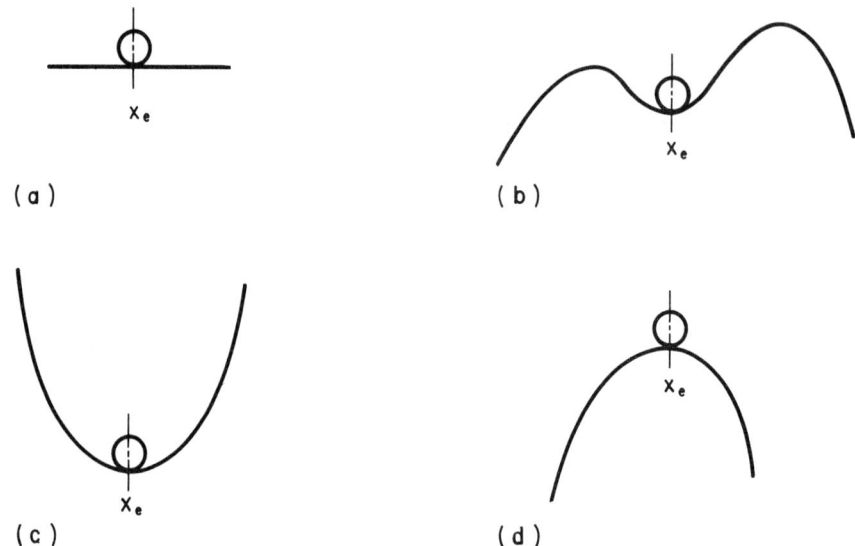

Fig. 7.2.2. Examples of equilibrium states a) neutrally stable, b) asymptotically stable (local), c) asymptotically stable (global), d) unstable

5. Definition. A linear system is said to be stable *isL* (asymptotically stable) if the equilibrium state $x_e = 0$ is stable *isL* (asymptotically stable).

7.2.2 Zero-State Stability

So far we have considered the stability of the zero-input response. We now turn our attention to the stability of relaxed systems from the input-output point of view.

6. Definition. Consider a vector space (V,F) and a norm $||.||$ on that space. We say the function $f:D \to V$, where D is the domain set of f, is bounded on D if there exists a finite positive number M such that $||f(t)|| \leq M$ for all $t \epsilon D$. △

Boundedness of matrices is defined in a similar manner, i.e. a matrix is bounded if all its columns are bounded in the sense of Definition 6.

Consider a linear system characterized by the state equation

$$\dot{x}(t) = A(t)x(t) + B(t)u(t), \quad t \geq t_o \tag{5}$$

$$x(k+1) = A(k)x(k) + B(k)u(k), \quad k \geq k_o \tag{6}$$

where **u** is an r-dimensional input vector and **A** and **B** are $n \times n$ and $n \times r$ bounded matrices respectively. Further, we assume that matrices $A(t)$ and $B(t)$ in (5) are continuous functions of time. Also consider the output equation

$$y(t) = C(t)x(t) + D(t)u(t) \tag{7}$$

or

$$y(k) = C(k)x(k) + D(k)u(k) \qquad (8)$$

The zero-state response of the system described by (5) and (7) is given by (6.6.26) repeated here for convenience:

$$y_{zs}(t) = \int_{t_o}^{t} h(t,\tau)u(\tau)d\tau, \quad t \geq t_o \qquad (9)$$

and that for the system described by (6) and (8) is given by (6.6.58):

$$y_{zs}(k) = \sum_{j=k_o}^{k-1} h(k,j)u(j), \quad k \geq k_o \qquad (10)$$

7. Definition. The system described by (5) and (7) with zero initial state ($x_o = 0$) is said to be bounded-input bounded-output stable (abbreviated *bibo* stable) if for all t_o and for all inputs $u(t)$ bounded on $[t_o,\infty)$, the output $y(t)$ is bounded on $[t_o,\infty)$. △

A similar definition applied to the discrete system described by (6) and (8).

In the next section we will show that in a system which is *bibo* stable there may be variables (i.e. state variables) which grow without bound. This, of course, indicates that the model representing the system is not adequate. We will examine this point also in Chapter 8.

7.3 STABILITY CRITERIA

7.3.1 Zero-input Stability

Consider the system characterized by

$$\dot{x}(t) = A(t)x(t), \quad t \geq t_o \qquad (1)$$

or

$$x(k+1) = A(k)x(k), \quad k \geq o \qquad (2)$$

1. Theorem. System (1) is stable *isL* if and only if there exists a constant M, which may depend on t_o, such that

$$\| \Phi(t,t_o) \| \leq M \quad \text{for all } t \geq t_o \qquad (3)$$

where $\Phi(t,t_o)$ is the transition matrix of (1).

Proof. *Sufficiency.* The solution of (1) can be written as

$$x(t) = \Phi(t,t_o)x_o \tag{4}$$

Thus,

$$\|x\| \leq \|\Phi(t,t_o)\| \, \|x_o\| \leq M \|x_o\| \text{ for all } t \geq t_o \tag{5}$$

Therefore $\|x_o\|$, $\dfrac{\epsilon}{M} = \delta$ implies that $\|x(t)\| < \epsilon$ for *all* $t \geq t_o$ which by Definition 7.2.2 implies stability *isL* of the system.

Necessity. We will prove necessity by contradiction. Suppose that the system is stable *isL* but $\Phi(t,t_o)$ is not bounded on $t \geq t_o$. Then there is at least one element of $\Phi(t,t_o)$, say $\phi_{ij}(t,t_o)$, which is not bounded. Choose the initial state as follows:

$$x_o = [0,0,...,0,1,0,....]' \tag{6}$$
$$\underset{jth\ element}{}$$

Then the *ith* component of $x(t)$ will be $\phi_{jk}(t,t_o)$ for $t \geq t_o$, which is unbounded. Therefore the system is unstable which contradicts the assumption that $\Phi(t,t_o)$ is not bounded.

2. Theorem. System (1) is asymptotically stable if and only if condition (3) holds and further,

$$\lim_{t \to \infty} \|\Phi(t,t_o)\| = 0 \text{ for all } t_o \tag{7}$$

The proof is similar to that of Theorem 1.

3. Theorem. System (1) is uniformly asymptotically stable in the large, or *exponentially stable*, if and only if positive constants M_1 and M_2 exist such that

$$\|\Phi(t,t_o)\| \leq M_1 e^{-M_2(t-t_o)} \text{ for all } t \geq t_o \text{ for all } t_o \tag{8}$$

Proof. Necessity. We have $\|x(t)\| \leq \|\Phi(t,t_o)\| \, \|x_o\|$. Therefore, if $\|x_o\| < \delta$, where δ is an arbitrary positive scalar, using (8) we obtain $\|x(t)\| < M_1 \delta e^{-M_2(t-t_o)}$ and $x(t)$ tends to 0 as $t \to \infty$. This proves uniform asymptotic stability in the large.

Sufficiency. If system (1) is uniformly asymptotically stable in the large, then for any $\delta > 0$ and any $\epsilon > 0$ there exists a time T independent of t_o such that $\|x_o\| < \delta$ implies $\|x(t)\| < \epsilon$ for *all* $t \geq t_o + T$. (See Problem 7.7.) Let $\|x_o\| = 1$ (i.e. choose $\delta > 1$) and $\epsilon = e^{-1}$. Then

$$\|x(t_o+T)\| = \|\Phi(t_o+T, t_o)x_o\| = \|\Phi(t_o+T, t_o)\| < e^{-1} \tag{9}$$

Using the transition property of the transition matrix (Section 6.3), (9) implies that for any positive integer k,

$$\|\Phi(t_o+kT, t_o)\| < e^{-k} \qquad (10)$$

Choose k such that $t_o+kT \leq t \leq t_o+(k+1)T$, $M_2 = \dfrac{1}{T}$ and $M_1 = \|\Phi(t, t_o+kT)\| e$ which is bounded due to the asymptotic stability of the system. Thus

$$\|\Phi(t,t_o)\| \leq \|\Phi(t,t_o+kT)\| \; \|\Phi(t_o+kT,t_o)\|$$

$$= M_1 e^{-1} \|\Phi(t_o+kT,T_o)\| < M_1 e^{-k-1}$$

$$= M_1 e^{-[t_o+(k+1)T-t_o]/T} \leq M_1 e^{-(t-t_o)/T} = M_1 e^{-M_2(t-t_o)} \quad q.e.d. \qquad (11)$$

For *l.t.i.* continuous-time systems, the eigenvalues of the system matrix completely characterize the stability of the system. In fact, for such systems we have

$$\Phi(t,t_o) = e^{A(t-t_o)} \qquad (12)$$

where **A** is the system matrix. (See Section 6.4.1.) The spectral representation of (12) was given in (6.5.24) as

$$\sum_{i=1}^{p} \sum_{j=1}^{m_j} \mathbf{K}_{ij} \, t^j \, \frac{e^{\lambda_j(t-t_o)}}{(j-1)!} \qquad (13)$$

where m_j is the multiplicity of eigenvalue λ_j, $i=1,2,...,p$ of **A**. Note that eigenvalues with zero real parts occur in conjugate pairs, e.g. $\lambda_2=j\omega$, $\lambda_2=-j\omega$, which result in bounded terms $Sin\,\omega t$ or $Cos\,\omega t$ in e^{At}. However, multiple eigenvalues with zero real parts result in unbounded terms $tSin\,\omega t$ or $tCos\,\omega t$. Thus we have the following corollaries of Theorems 1 and 2.

4. Corollary. A *l.t.i.* continuous system is asymptotically stable if and only if all the eigenvalues of its system matrix have negative real parts. △

The terminology often used in reference to the sign of the real parts of a system's eigenvalues is *right-half plane* (abbreviated r.h.p.) and *left-half plane* (abbreviated l.h.p.). They refer to the parts of the complex plane to the right and the left of the imaginary axis, respectively. (See Fig.1.) If the imaginary axis is included in a half plane, then the half plane is referred to as *closed*, otherwise it is said to be *open*. Thus, for example, the open l.h.p. refers to the part of the complex plane to the left of the imaginary axis but not including it.

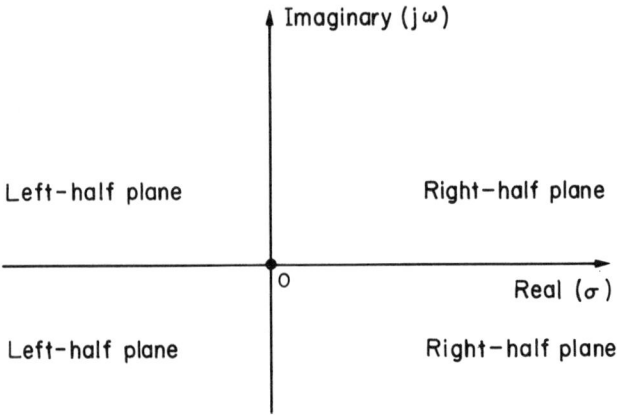

Fig. 7.3.1. The half planes in the complex plane

It is important to note that the linear time-varying systems described by (1) need not be stable even if each eigenvalue of the system matrix has a negative real part for $t \geqslant t_o$. The following example illustrates this point.

5. Example. Consider the linear time-varying system (1) where

$$A(t) = \begin{bmatrix} -4 & 3e^{4t} \\ -e^{-4t} & 0 \end{bmatrix} \qquad (14)$$

The eigenvalues of $A(t)$ are real and negative (independent of time): $\lambda_1 = -1$, $\lambda_2 = -3$. However, the system is not stable since its transition matrix

$$\Phi(t,t_o) = \frac{1}{2}\begin{bmatrix} 3e^{-\frac{5}{2}(t-t_o)} & -e^{-\frac{7}{2}(t-t_o)} & 3e^{\frac{5}{2}(t-t_o)} & -3e^{-\frac{7}{2}(t-t_o)} \\ e^{\frac{1}{2}(t-t_o)} & -e^{\frac{3}{2}(t-t_o)} & 3e^{\frac{1}{2}(t-t_o)} & -e^{\frac{3}{2}(t-t_o)} \end{bmatrix} \qquad (15)$$

is not bounded (see Problem 6.16). Thus, by Theorem 1, the system is not even stable *isL*. △

In order to insure the stability of system (1), additional conditions will be required. For example, if all the eigenvalues of $A(t)$ are in the open l.h.p. for all $t \geqslant t_o$ and if the rate of change of $A(t)$ w.r.t. time t is sufficiently small, then the system will be asymptotically stable [7.2 - 7.4].

6. Theorem. System (2) is stable *isL* if an only if there exists a constant M, which may depend on k_o, such that

$$\| \Phi(k,k_o) \| < M \quad \text{for all }, k \geqslant k_o \qquad (16)$$

where $\Phi(k,k_o)$ is the transition matrix of (2).

STABILITY CRITERIA

7. Theorem. System (2) is asymptotically stable if and only if condition (16) holds and further,

$$\lim_{k \to \infty} \| \Phi(k,k_o) \| = 0 \quad \text{for all } k_o \tag{17}$$

The proofs of the above theorems are similar to the proofs of Theorems 1 and 2. △

For *l.t.i.* discrete systems again the eigenvalues of the system matrix completely characterize system stability. In fact, for such systems we have

$$\Phi(k,k_o) = \mathbf{A}^{k-k_o} \tag{18}$$

where **A** is the system matrix. (See Section 6.4.2.). To determine the relationship between system stability and the eigenvalues λ_j of the system matrix, consider the *l.t.i.* discrete system

$$\mathbf{x}(k+1) = \mathbf{A}\mathbf{x}(k), \quad k \geq 0 \tag{19}$$

where **A** has n linearly independent eigenvectors. Use a state transformation

$$\mathbf{z}(k) = \mathbf{M}^{-1}\mathbf{x}(k) \tag{20}$$

where **M** is the modal matrix of **A**. Then the new state equation will be

$$\mathbf{z}(k+1) = \mathbf{M}^{-1}\mathbf{A}\mathbf{M}\mathbf{z}(k) = \hat{\mathbf{A}}\mathbf{z}(k) \tag{21}$$

where

$$\hat{\mathbf{A}} = diag(\lambda_1, \lambda_2, \ldots, \lambda_n) \tag{22}$$

Stability of system (21) is equivalent to the stability of system (19). However, it is more convenient to deal with system (19) in which the boundedness of the transition matrix $\hat{\mathbf{A}}^{k-k_o}$ can be related to the magnitudes of the eigenvalues of $\hat{\mathbf{A}}$ (which are the same as the eigenvalues of **A**. Note that

$$\hat{\mathbf{A}}^{k-k_o} = diag(\lambda_1^{k-k_o}, \lambda_2^{k-k_o}, \ldots, \lambda_n^{k-k_o}) \tag{23}$$

and that

$$\lim_{j \to \infty} |\lambda_i|^j \to 0 \leftrightarrow |\lambda_i| < 1, \quad i=1,2,\ldots,n. \tag{24}$$

Thus $\|\hat{\mathbf{A}}^{k-k_\bullet}\|$ is bounded and converges to 0 as $k\to\infty$ if and only if $|\lambda_i|<1$, $i=1,2,\ldots,n$. This condition is equivalent to requiring that all the eigenvalues of the system matrix be inside the unit circle in the complex plane. (See Fig. 2.) If simple eigenvalues exist on the unit circle, $\|\hat{\mathbf{A}}^{k-k_\bullet}\|$ still remains bounded but no longer converges to zero. Multiple eigenvalues on the unit circle, however, may cause $\|\hat{\mathbf{A}}^{k-k_\bullet}\|$ to explode as $k\to\infty$. (See Example 10 below.) The above argument can also be extended to the case where matrix $\mathbf{A}$ is not diagonalizable. (See Problem 7.8.) Thus we have the following corollaries to Theorems 6 and 7.

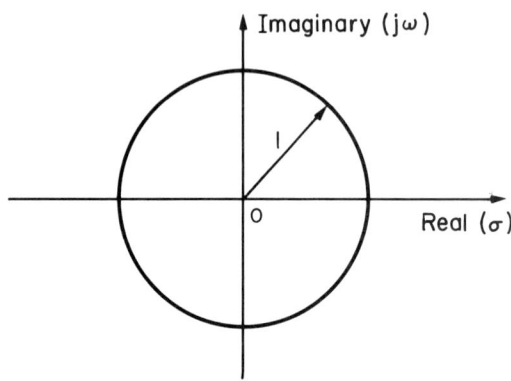

Fig. 7.7.2. Unit circle in the complex plane

8. Corollary. A l.t.i. discrete-time system is stable *isL* if and only if all the eigenvalues of its system matrix are inside the closed unit circle in the complex plane and the eigenvalues on the unit circle have Jordan blocks of orders no higher than unity.

9. Corollary. A l.t.i. discrete-time system is asymptotically stable if and only if all the eigenvalues of its system matrix are strictly inside the unit circle in the complex plane.

10. Example. Consider system (19) in two cases:

(a) $\mathbf{A} = \begin{bmatrix} 1 & 0 \\ 0 & 1 \end{bmatrix}$ with two Jordan blocks of order 1.

(b) $\mathbf{A} = \begin{bmatrix} 1 & 1 \\ 0 & 1 \end{bmatrix}$ with one Jordan block of order 2.

The eigenvalues of $\mathbf{A}$ in both cases are $\lambda_1 = \lambda_2 = 1$, i.e. in both cases $\mathbf{A}$ has multiple eigenvalues on the unit circle in the complex plane. In case (a) the system is stable *isL* since $\mathbf{x}(k+1) = \mathbf{x}(0)$ for *all* $k \geqslant 0$. In case (b), however, the system is unstable since $x_1(k+1) = x_1(0) + kx_2(0)$, which goes to ∞ as $k\to\infty$.

STABILITY CRITERIA

Program *"STABCD"* determines the stability of any *l.t.i.* continuous-time or discrete-time system. The source listing of this program followed by examples of continuous-time and discrete-time systems are given below.

11. "STABCD"

```
10   !    PROGRAM NAME : "STABCD" PROG
20   PRINT " 'STABCD' DETERMINE THE STABILITY OF THE EQUILIBRIUM
     POINT OF ANY"
30   PRINT "LINEAR CONTINUOUS-TIME OR DISCRETE-TIME SYSTEM."
     ,LIN(1)
40   PRINT "Continuous-time system:   dx/dt=Ax,"
50   PRINT "Discrete-time system:     x(k+1)=Ax(k).";LIN(2)
60   INPUT "Have SUB's <<Mat>> & <<Eigen>> already been LINKED
     (Y/N)?",A$
70   IF A$="Y" THEN 120
80   IF A$="N" THEN 100
90   GOTO 60
100  LINK "Mat",2000,110
110  LINK "Eigen",3100,120
120  INPUT "DISCRETE OR CONTINUOUS SYSTEM(D/C)?",C$
130    IF (C$="D") OR (C$="d") THEN 200! CALL STAKI
140    IF (C$="C") OR (C$="c") THEN 160! CALL STAB
150    GOTO 120
160    INPUT "Give ORDER of the system matrix:",N
170    PRINT "Order of the system matrix=";N;LIN(2)
180    CALL Stab(N)
190    GOTO 230
200    INPUT "Give ORDER of the system matrix:",N
210    PRINT "Order of the system matrix=";N;LIN(2)
220    CALL Staki(N)
230  END
240  SUB Stab(N)
250  OPTION BASE 1
260  DIM A(N,N),Vecr(N,N),Veci(N,N),Evr(N),Evi(N),Indic(N)
270  LINPUT "DO YOU WANT THE EIGENVALUES PRINTED (Y/N)?",E$
280  IF (E$="Y") OR (E$="y") THEN 310
290  IF (E$="N") OR (E$="n") THEN 330
300    GOTO 270
310    Pri=99
320    GOTO 340
330    Pri=0
340    CALL Mat(A(*),N,N,"A")
350    CALL Eigen(N,A(*),Evr(*),Evi(*),Vecr(*),Veci(*),Indic(*))
360    IF Pri=99 THEN 380
370    IF Pri=0 THEN 430
380      PRINT LIN(2),"REAL PARTS OF EIGENVALUES:",LIN(1)
390      FLOAT 4
400      MAT PRINT Evr:
```

```
410     PRINT LIN(2),"IMAGINARY PART OF EIGENVALUES;",LIN(1)
420     MAT PRINT Evi;
430     Eps=.0001
440     FOR I=1 TO N
450     IF ABS(Evr(I))<Eps THEN 470
460     GOTO 480
470     Evr(I)=0
480     NEXT I
490     FOR I=1 TO N
500     IF Evr(I)>0 THEN 740
510    IF Evr(I)=0 THEN 580
520    NEXT I
530    GOTO 720
540    FOR I=1 TO N
550    IF Evr(I)>0 THEN 740
560      FOR J=1 TO N
570      IF I=J THEN 680
580      IF Evr(I)<Evr(J)+Eps THEN 600
590      GOTO 680
600      IF Evr(I)>Evr(J)-Eps THEN 630
610      IF Evr(I)>0 THEN 740
620      GOTO 680
630      Evr(I)=Evr(J)
640      IF Evr(I)=0 THEN 660
650      GOTO 680
660      IF Evi(J)=Evi(I) THEN 740
670      GOTO 680
680 NEXT J
690    NEXT I
700     PRINT "STABILITY IN THE SENSE OF LYAPUNOV FOR THE ";
710     GOTO 750
720     PRINT "ASYMTOTICALLY STABLE";
730     GOTO 750
740     PRINT "UNSTABLE ";
750     PRINT "POINT OF EQUILIBRIUM FOR THE CONTINUOUS-TIME
        SYSTEM."
760     SUBEND
770    SUB Eigl(N)
780    OPTION BASE 1
790    DIM A(N,N),Vecr(N,N),Veci(N,N),Evr(N),Evi(N),Indic(N)
800    FIXED 0
810    CALL Mat(A(*),N,N,"A")
820    GOTO 840
830    CALL Eigen(N,A(*),Evr(*),Evi(*),Vecr(*),Veci(*),Indic(*))
840    PRINT LIN(2),"REAL COMPONENTS OF EIGENVALUES:",LIN(1)
850    FLOAT 4
860    MAT PRINT Evr;
870    PRINT LIN(2),"IMAGINARY COMPONENTS OF THE EIGENVALUES:"
       ,LIN(1)
```

```
880   MAT PRINT Evi;
890   PRINT LIN(2),"REAL COMPONENTS OF EIGENVECTORS
      [IN COLUMNS]",LIN(1)
900   MAT PRINT Vecr;
910   PRINT LIN(2),"IMAGINARY COMPONENTS OF EIGENVECTORS
      [IN COLUMNS]:",LIN(1)
920   MAT PRINT Veci;
930   SUB Staki(N)
940   OPTION BASE 1
950   DIM A(N,N),Vecr(N,N),Veci(N,N),Evr(N),Evi(N),Indic(N),
      Flag(N),R(N)
960   PRINT LIN(2),"ORDER OF MATRIX A(*)=";N
970   CALL Mat(A(*),N,N,"A")
980   LINPUT "DO YOU WANT THE EIGENVALUES PRINTED (Y/N)?",D$
990   IF (D$="Y") OR (D$="y") THEN 1020
1000  IF (D$="N") OR (D$="n") THEN 1040
1010  GOTO 980
1020  Prin=999
1030  GOTO 1050
1040  Prin=0
1050   CALL Eigen(N,A(*),Evr(*),Evi(*),Vecr(*),Veci(*),Indic(*))
1060   IF Prin=99 THEN 1080
1070   IF Prin=0 THEN 1120
1080   PRINT LIN(2),"REAL PARTS OF EIGEN VALUES:",LIN(1)
1090   MAT PRINT Evr;
1100   PRINT LIN(2),"IMAGINARY PART OF EIGEN VALUES:",LIN(1)
1110   MAT PRINT Evi;
1120   FOR I=1 TO N
1130   R(I)=SQR(Evr(I)^2+Evi(I)^2)
1140   NEXT I
1150   Eps=.001
1160   FOR I=1 TO N
1170   IF R(I)<1+Eps THEN 1190
1180    GOTO 1220
1190    IF R(I)>1-Eps THEN 1210
1200    GOTO 1230
1210    R(I)=1
1220   NEXT I
1230    FOR I=1 TO N
1240    IF R(I)>1 THEN 1530! UNSTABLE
1250    NEXT I
1260    FOR I=1 TO N
1270    IF R(I)=1 THEN 1320
1280    NEXT I
1290    FOR I=1 TO N
1300    IF R(I)<>1 THEN 1510      !AS STABILITY
1310    NEXT I
1320     FOR I=1 TO N
1330      IF R(I)<>1 THEN 1450!NEXT J
1340     FOR J=1 TO N
```

```
1350    IF I=J THEN 1440! NEXT J
1360    IF ABS(Evr(I))<ABS(Evr(J)+Eps) THEN 1380
1370    GOTO 1440 ! NEXT J
1380    IF ABS(Evr(I))>ABS(Evr(J)-Eps) THEN 1400
1390    GOTO 1440!NEXR J
1400    IF Evr(I)=Evr(J) THEN 1420
1410    GOTO 1440
1420    IF Evi(I)=Evi(J) THEN 1530 ! DOUBLE ROOT OF ABS=1
1430     GOTO 1440!NEXT J
1440     NEXT J
1450     NEXT I
1460     FOR I=1 TO N
1470      IF R(I)=1 THEN 1490! I.S.L STABILITY
1480         GOTO 1490  !PRINT AS STABILITY
1490     PRINT "STABILITY IN THE SENSE OF LYAPUNOV  FOR THE";
1500     GOTO 1540
1510     PRINT "ASYMPTOTICALLY STABLE ";
1520     GOTO 1540
1530     PRINT "UNSTABLE";
1540     PRINT " EQUILIBRIUM POINT FOR THE DISCRETE-TIME
         SYSTEM."
1550     SUBEND
```

12. Example

```
´STABCD´ DETERMINE THE STABILITY OF THE EQUILIBRIUM POINT
OF ANY
LINEAR CONTINUOUS-TIME OR DISCRETE-TIME SYSTEM.

Continuous-time system:   dx/dt=Ax,
Discrete-time system:     x(k+1)=Ax(k).


Order of the system matrix= 4


Matrix A( 4 x 4 ):

  0.0000E+00   1.0000E+00   0.0000E+00   0.0000E+00

  0.0000E+00   0.0000E+00   1.0000E+00   0.0000E+00

  0.0000E+00   0.0000E+00   0.0000E+00   1.0000E+00

 -5.0000E+00  -4.0000E+00  -3.0000E+00  -2.0000E+00
```

REAL PARTS OF EIGENVALUES:

-1.2878E+00 -1.2878E+00   2.8782E-01   2.8782E-01

IMAGINARY PART OF EIGENVALUES;

 8.5790E-01 -8.5790E-01   1.4161E+00 -1.4161E+00

UNSTABLE POINT OF EQUILIBRIUM FOR THE CONTINUOUS-TIME SYSTEM.

13. Example

'STABCD' DETERMINE THE STABILITY OF THE EQUILIBRIUM POINT OF ANY LINEAR CONTINUOUS-TIME OR DISCRETE-TIME SYSTEM.

Continuous-time system: dx/dt=Ax,
Discrete-time system: x(k+1)=Ax(k).

Order of the system matrix= 3

ORDER OF MATRIX A(*)= 3

Matrix A(3 x 3):

  5.0000E-01 -5.0000E-01   1.0000E+00

  0.0000E+00   5.0000E-01   2.0000E+00

  0.0000E+00   0.0000E+00   5.0000E-01

REAL PARTS OF EIGEN VALUES:

  .500000000001   .500000000001   .500000000001

288   LINEAR CONTROL SYSTEMS

IMAGINARY PART OF EIGEN VALUES:

0   0   0

ASYMPTOTICALLY STABLE   EQUILIBRIUM POINT FOR THE DISCRETE-TIME SYSTEM.

7.3.2 Zero-State Stability

14. Theorem. Consider the linear continuous-time system described by (7.2.5) and (7.2.7), Let $x(t_o) = 0$ and let matrices $A(t)$ and $B(t)$ be continuous and bounded on $[t_o, \infty)$. Then this system is *bibo* stable if and only if there exists a finite number M such that

$$\int_{t_o}^{t} \| \mathbf{h}(t,\tau) \|_1 d\tau < M \quad \text{for all } t_o \text{ for all } t \geqslant t_o \tag{25}$$

where $\mathbf{h}(t,\tau)$ is the impulse response matrix of the system. (See Section 4.5.1.) Note that norm $\| \cdot \|_1$ is used, i.e. $\| A \|_1 = \max_j \sum_{i=1}^{n} | a_{ij} |$. (See Problem 2.13.)

Proof. Sufficiency. Assume that (25) holds and suppose that $\mathbf{u}(t)$ is bounded, i.e. a finite number M_1 exists such that $\| \mathbf{u}(t) \| < M_1$ for all $t \geqslant t_o$. Then from (7.2.9) we have

$$\| \mathbf{y}_{zs}(t) \| \leqslant \int_{t_o}^{t} \| \mathbf{h}(t,\tau) \| \, \| \mathbf{u}(\tau) \| \, d\tau < M_1 \int_{t_o}^{t} \| \mathbf{h}(t,\tau) \| \, d\tau < M_1 M \quad \text{for all } t_o \text{ for all } t \geqslant t_o \tag{26}$$

and $\mathbf{y}_{zs}(t)$ is bounded on $[t_o, \infty)$. Therefore the system is *bibo* stable.

Necessity. We will prove necessity by contradiction. Assume that the system is *bibo* stable but (25) does not hold, i.e. for any number N there is a time T such that

$$\int_{t_o}^{T} \| \mathbf{h}(T,\tau) \|_1 d\tau = \int_{t_o}^{T} \max_j \sum_{i=1}^{n} \| h_{ij}(T,\tau) \| \, d\tau > N \tag{27}$$

This implies that for at least one element of $\mathbf{h}$, say h_{ij}, we have

$$\int_{t_o}^{T} | h_{ij}(T,\tau) | \, d\tau > N \tag{28}$$

Now define the input $\mathbf{u}(t)$ to be a vector all whose elements are zero except the jth element which is $sgn h_{ij}(T,t)$ where the signum function is defined as

$$sgn \alpha = \begin{cases} 1 & \text{if } \alpha > 0 \\ 0 & \text{if } \alpha = 0 \\ -1 & \text{if } \alpha < 0 \end{cases} \tag{29}$$

Thus $\mathbf{u}(t)$ is a bounded input. From (7.2.9) the ith component of the zero-state response to this input is

$$y_i(T) = \int_{t_o}^{T} h_{ij}(T,\tau) \operatorname{sgn} h_{ij}(T,\tau) d\tau = \int_{t_o}^{T} |h_{ij}(T,\tau)| d\tau$$

which by (28) is larger than N. Since N can be chosen arbitrarily large, the zero-state response is not bounded. That is, the system is not *bibo* stable. This is a contradiction to the original assumption; therefore, *bibo* stability implies that (25) holds. △

If $\mathbf{x}(t_o) \neq 0$, then for *bibo* stability of system (7.2.5) and (7.2.7) there will be additional requirements that the zero input system be stable *isL* and that matrix $\mathbf{C}(t)$ be bounded on (t_o, ∞).

For *l.t.i.* systems, (7.2.9) becomes

$$\mathbf{y}_{zs}(t) = \int_0^t \mathbf{h}(t-\tau) \mathbf{u}(\tau) d\tau, \quad t \geq 0 \tag{30}$$

Then it is easy to prove the following corollary of the above theorem.

15. Corollary. A *l.t.i.* continuous system is *bibo* stable if and only if there exists a finite number M such that

$$\int_0^\infty \| \mathbf{h}(\tau) \| d\tau < M \qquad \triangle \tag{31}$$

For a SISO continuous-time system with a proper rational transfer function, condition (31) is clearly equivalent to requiring that all the poles of the transfer function be in the open *l.h.p.* (Assuming that common factors in the numerator and the denominator of the transfer function have been canceled). Therefore, from Corollary 5, for such systems asymptotic stability implies *bibo* stability. This does not hold for linear time-varying systems in general. (See Problem 7.10.) However, it can be shown [7.5.6] that for linear time-varying systems *uniform* asymptotic stability and *bibo* stability are equivalent.

Note that *bibo* stability in *l.t.i.* systems does not necessarily imply asymptotic stability as demonstrated by the following example. This situation comes about when common factors exist in the numerator and the denominator of the transfer function. We will further investigate this point in Chapter 8.

16. Example. Consider the *l.t.i.* system $\dot{\mathbf{x}} = \mathbf{A}\mathbf{x} + \mathbf{B}\mathbf{u}$, $\mathbf{y} = \mathbf{C}\mathbf{x}$ where

$$\mathbf{A} = \begin{bmatrix} 1 & 1 \\ 2 & 0 \end{bmatrix}, \quad \mathbf{B} = \begin{bmatrix} 1 \\ -2 \end{bmatrix}, \quad \mathbf{C} = [1 \ \ 0]$$

The eigenvalues of **A** are $\lambda_1 = -1$, $\lambda_2 = 2$. Therefore the system is not stable. But we have

$$h(s) = \mathbf{C}(s\mathbf{I}-\mathbf{A})^{-1}\mathbf{B} = \frac{s-2}{(s+1)(s-2)} = \frac{1}{s+1} \tag{32}$$

which by the above discussion implies *bibo* stability.

For linear discrete systems described by (7.2.6) and (7.2.8) the zero-state response is given by (7.2.10). Thus we have the following Theorem which is analogous to Theorem 15.

17. Theorem. Consider the linear discrete-time system described by (7.2.6) and (7.2.8). Let $\mathbf{x}(k_o)=0$ and let matrices $\mathbf{A}(k)$ and $\mathbf{B}(k)$ be bounded for all $k \geq k_o$. Then this system is *bibo* stable if and only if there exists a finite number M such that

$$\sum_{j=k_o}^{k-1} \| \mathbf{h}(k,j) \|_1 < M \quad \text{for all } k_o \text{ for all } k \geq k_o \tag{33}$$

The proof of this theorem is similar to that of Theorem 14 and will be omitted.

For *l.t.i.* systems, (30) becomes

$$\mathbf{y}_{zs}(k) = \sum_{j=0}^{k-1} \mathbf{h}(k-j-1)\mathbf{u}(j), \quad k \geq 0 \tag{34}$$

(see (6.6.61)) and the following corollary of the above theorem characterizes *bibo* stability.

18. Corollary. A *l.t.i.* discrete system is *bibo* stable if and only if there exists a finite number M such that

$$\sum_{j=0}^{\infty} |\mathbf{h}(j)| < M \quad \triangle \tag{35}$$

For a SISO discrete system with a proper rational transfer function, condition (35) is equivalent to requiring that all the poles of the transfer function be in the open unit circle in the complex plane (assuming that common factors have been cancelled). Discussions about the relationship between asymptotic and *bibo* stability for continuous system are valid for discrete systems as well.

7.4 ROUTH-HURWITZ STABILITY CRITERION

One of the most celebrated tests for the determination of *l.t.i.* SISO system stability is the Routh-Hurwitz stability criterion [7.7, 7.8]. The criterion provides a means of checking whether the system is stable or not by predicting the number of roots of the system's characteristic equation which lie in the *r.h.p.* Consider the characteristic equation of a SISO system shown in Fig. 1:

$$p(s) = 1 + G(s) = 1 + \frac{a_o + a_1 s + \ldots + a_{n-1}s^{n-1}}{b_o + b_1 s + \ldots + b_n s^n} = 0 \tag{1}$$

or

$$p(s) = c_0 + c_1 s + \ldots + c_n s^n = 0. \tag{2}$$

The system stability is characterized by determining whether any roots of $p(s)$ in (2) is in the *r.h.p.* Let the characteristic equation (2) be rewritten in the following array form.

$$\begin{array}{c|cccc} s^n & c_n & c_{n-2} & c_{n-4} & \cdots \\ s^{n-1} & c_{n-1} & c_{n-3} & c_{n-5} & \cdots \\ s^{n-2} & d_{n-1} & d_{n-3} & d_{n-5} & \\ s^{n-3} & e_{n-1} & & & \\ & \vdots & & & \\ & \vdots & & & \\ s^0 & h_{n-1} & & & \end{array} \tag{3}$$

where

$$d_{n-1} = (-1/c_{n-1})\begin{bmatrix} c_n & c_{n-2} \\ c_{n-1} & c_{n-3} \end{bmatrix} = (c_{n-1} c_{n-2} - c_n c_{n-3})/c_{n-1} \tag{4}$$

$$d_{n-3} = (-1/c_{n-1})\begin{bmatrix} c_n & c_{n-4} \\ c_{n-1} & c_{n-5} \end{bmatrix}, \quad e_{n-1} = (-1/d_{n-1})\begin{bmatrix} c_{n-1} & c_{n-3} \\ d_{n-1} & d_{n-3} \end{bmatrix} \tag{5}$$

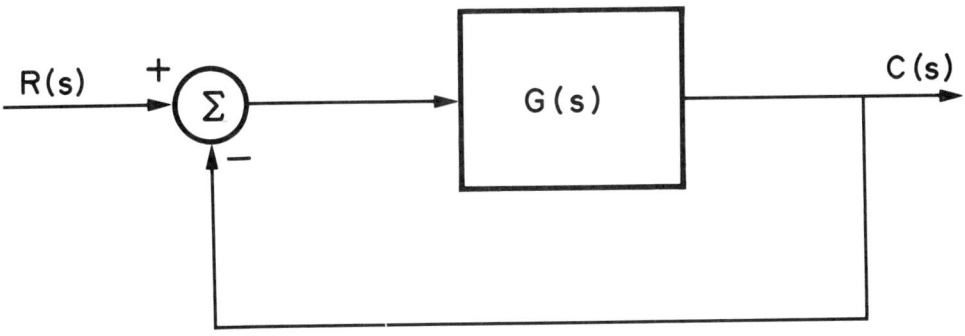

Fig. 7.4.1. Block diagram of a unity feedback control system

etc.. Then the Routh-Hurwitz criterion states that the number of closed-loop poles (roots of $p(s)=0$) which lie in the r.h.p. is equal to the number of sign changes in the first column of the array. There are three distinct cases which can be realized: (i) no zeros in column 1, (ii) one zero in column 1 and (iii) all zeros in one row. In sequel, the three cases are described by three examples followed by Program "ROUTH" treating the same numerical examples.

1. Case (i): No zeros in column 1

Consider a system with open-loop transfer function,

$$G(s) = 24/(s^4+10s^3+35s^2+50s)$$

whose characteristic equation becomes

$$s^4+10s^3+35s^2+50s+24 = 0.$$

The Routh array becomes,

s^4	1	35	24
s^3	10	50	
s^2	30	24	
s^1	42	0	
s^0	24		

with no sign changes in the first column which indicates that the system is stable.

2. Case (ii): One zero in column 1

To illustrate the second case, consider a fifth-order system,

$$G(s) = (4s^2+11s+9)/(s^5+2s^4+2s^3+2s+1)$$

whose characteristic equation is easily seen to be

$$s^5+2s^4+2s^3+4s^2+13s+10 = 0.$$

The first three rows of the Routh array are

$$\begin{array}{c|ccc} s^5 & 1 & 2 & 13 \\ s^4 & 2 & 4 & 10 \\ s^3 & 0 & 8 & \end{array}$$

which results in a zero in the first column. In order to complete the array, replace the zero element by a small positive number ϵ and proceed to complete the array and then let ϵ go to zero and check the signs of the first column entries. For this particular system, the modified Routh array becomes:

$$\begin{array}{c|ccc} s^5 & 1 & 2 & 13 \\ s^4 & 2 & 4 & 10 \\ s^3 & \epsilon & 8 & 0 \\ s^2 & 4-16/\epsilon & 10 & \\ s^1 & 8-10\epsilon^2/(4\epsilon-16) & 0 & \\ s^0 & 10 & & \end{array}$$

Note that as ϵ approaches zero the term $(4-16/\epsilon)$ approaches a large negative number while the value $8-10\epsilon^2/(4\epsilon-16)$ approaches $+8$. Therefore it is concluded that the system has 2 poles in the r.h.p. and hence it is unstable.

3. Case (iii): All zeros in one row.

For this case, consider the example, where

$$G(s) = (13+3s)/(50+24s^2+4s^3+s^4+s^5)$$

whose characteristic equation becomes

$$p(s) = s^5+s^4+4s^3+24s^2+3s+63 = 0$$

and the Routh array,

s^5	1	4	3
s^4	1	24	63
s^3	-20	-60	0
s^2	21	63	0
s^1	0	0	0
s^0			

which has resulted a row with all zeros. Under this condition, an auxiliary equation consisting of the row s^2, i.e. $21s^2+63 = 0$ or $s^2+3 = 0$ or a pair of poles on the $j\omega$-axis exists. It is easily seen that the characteristic equation can be written as

$$p(s) = (s^2+3)(s^3+s^2+s+21) = 0.$$

In general if a given row m is composed of all zeros, one can use the row immediately above it i.e. row $m+1$ and set up an auxiliary equation

$$a_{m+1}s^{m+1} + a_{m-1}s^{m-1} + \ldots \quad (6)$$

and take the derivative of it, i.e.

$$(m+1)a_{m+1}s^m + (m-1)a_{m-1}s^{m-2} + \ldots \quad (7)$$

and replace the mth row by

$$(m+1)a_{m+1} \quad (m-1)a_{m-1} \ldots \quad (8)$$

and continue on. Program "ROUTH" given below takes care of all the three cases described above.

4. "ROUTH"

```
10     ! PROGRAM NAME : "ROUTH" PROG
20     PRINT " This program determines the STABILITY of a Single-Input"
30     PRINT "  Single-Output system via ROUTH-HURWITZ CRITERION"
40     OPTION BASE 1
50     DIM A(11),B(11),X(11),R(11,11)
60     INPUT "Give degrees of Numerator(m<=10) & Denominator(n<=10)",M,N
70     PRINT "Numerator Degree m=";M;"Denominator Degree n=";N
80     IF (N>10) OR (M>10) THEN 60
90     M1=M+1
100    N1=N+1
```

```
110   IF M1>N1 THEN 140
120   S=N1
130   GOTO 150
140   S=M1
150   REDIM A(S),B(S),X(S),R(S,6)
160   ! SET EXISTING & ROUTH ARRAYS EQUAL TO ZERO
170   MAT A=ZER
180   MAT B=ZER
190   MAT X=ZER
200   MAT R=ZER
210   ! N2 = NO. OF COLUMNS IN THE ROUTH ARRAY
220   N2=(S-1)/2+1
230   PRINT "Enter NUMERATOR COEFFICIENTS in ASCENDING ORDER:"
240   FOR I1=1 TO M1
250   I=I1-1
260   DISP "A(";I;")";
270   INPUT A(I1)
280   NEXT I1
290   MAT PRINT A;
300   INPUT "Are NUMERATOR COEFFICIENTS CORRECT?",N$
310   IF (N$="N") OR (N$="n") THEN 230
320   PRINT "Enter DENOMINATOR COEFFICIENTS in ASCENDING ORDER:"
330   FOR J1=1 TO N1
340   J=J1-1
350   DISP "B(";J;")";
360   INPUT B(J1)
370   NEXT J1
380   MAT PRINT B;
390   INPUT "Are DENOMINATOR COEFFICIENTS CORRECT?",D$
400   IF (D$="N") OR (D$="n") THEN 320
410   ! FIND THE CHARACT. EQ. BY ADDING NUMERATOR & DENOMINATOR COEFFICIENTS
420   MAT X=A+B
430   ! FIND THE FIRST TWO ROWS OF THE ROUTH ARRAY
440   L=S
450   FOR J=1 TO N2
460   FOR K=1 TO 2
470   R(K,J)=X(L)
480   L=L-1
490   NEXT K
500   IF L=1 THEN 530
510   NEXT J
520   GOTO 550
530   R(1,N2)=X(1)
540   R(2,N2)=0
550   ! CALCULATE THE ROUTH MEMBERS
560   Q=0
570   Lp2=S-1
580   Lj=(S+1)/2
590   FOR I=1 TO Lp2
600   IF I=1 THEN 680
610   FOR J=2 TO Lj
620   Rp1=R(I,1)*R(I-1,J)
630   Rp2=R(I-1,1)*R(I,J)
640   Rp3=Rp1-Rp2
650   R(I+1,J-1)=Rp3/R(I,1)
660   NEXT J
670   ! IS A MEMBER OF THE FIRST COLUMN EQUAL TO ZERO?
680   W=R(I+1,1)
690   IF W<>0 THEN 860
700   ! DOES A ROW CONTAIN ALL ZEROS?
710   FOR J=2 TO Lj
720   IF (R(I+1,J)>1E-9) OR (R(I+1,J)<-1E-9) THEN 850
730   NEXT J
740   Q=Q+1
```

```
750   ! DIFFERENTIATE THE AUXILARY EQUATION
760   Lc=(S-I)/2
770   K=I-1
780   FOR J=1 TO 10
790   R(I+1,J)=R(I,J)*(Lp2-K)
800   IF Lc<1 THEN 860
810   Lc=Lc-1
820   K=K+2
830   NEXT J
840   ! REPLACE A ZERO IN THE FIRST COLUMN BY A SMALL NUMBER
850   R(I+1,1)=1E-10
860   NEXT I
870   IF Q=0 THEN 920
880   ! WARNING ON THE EXISTANCE OF AN AUXILARY EQUATION
890   PRINT "WARNING !!! Your system's CHARACTERISTIC EQUATION contains an"
900   PRINT " AUXILARY EQUATION. There could be poles on the jw-axis causing"
910   PRINT " OSSCILLATORY characteristics!!"
920   PRINT LIN(1)," ***   ROUTH-HURWITZ ARRAY   ***",LIN(1)
930   FIXED 2
940   MAT PRINT R;
950   FIXED 0
960   ! COUNT THE NUMBER OF SIGN CHANGES IN THE FIRST COLUMN
970   Q=C=0
980   IF R(1,1)<0 THEN 1000
990   C=1
1000  FOR I=2 TO S
1010  IF R(I,1)>=0 THEN 1060
1020  IF C=0 THEN 1090
1030  Q=Q+1
1040  C=0
1050  GOTO 1090
1060  IF C=1 THEN 1090
1070  Q=Q+1
1080  C=1
1090  NEXT I
1100  ! NUMBER OF POLES IN THE RIGHT-HALF-PLANE & STABILITY STATUS
1110  PRINT "   Number of ROOTS in the RIGHT-HALF-PLANE = ";Q
1120  IF Q<>0 THEN 1150
1130  PRINT LIN(1),"SYSTEM IS STABLE."
1140  GOTO 1160
1150  PRINT LIN(1),"SYSTEM IS UNSTABLE."
1160  INPUT "Do you like to RUN another program (Y/N)?",P$
1170  IF (P$="Y") OR (P$="y") THEN 60
1180  END
```

5. Example

```
 This program determines the STABILITY of a Single-Input
   Single-Output system via ROUTH-HURWITZ CRITERION
Numerator Degree m= 2 Denominator Degree n= 5
Enter NUMERATOR COEFFICIENTS in ASCENDING ORDER:
  9  11  4  0  0  0

Enter DENOMINATOR COEFFICIENTS in ASCENDING ORDER:
  1  2  0  2  2  1


   ***    ROUTH-HURWITZ ARRAY   ***

  1.00  2.00  13.00  0.00  0.00  0.00
```

ROUTH-HURWITZ STABILITY CRITERION

2.00	4.00	10.00	0.00	0.00	0.00
.00	8.00	0.00	0.00	0.00	0.00
-159999999996.00	10.00	0.00	0.00	0.00	0.00
8.00	0.00	0.00	0.00	0.00	0.00
10.00	0.00	0.00	0.00	0.00	0.00

Number of ROOTS in the RIGHT-HALF-PLANE = 2

SYSTEM IS UNSTABLE.

Numerator Degree m= 0 Denominator Degree n= 4
Enter NUMERATOR COEFFICIENTS in ASCENDING ORDER:
24 0 0 0 0

Enter DENOMINATOR COEFFICIENTS in ASCENDING ORDER:
0 50 35 10 1

   ***   ROUTH-HURWITZ ARRAY ***

1.00	35.00	24.00	0.00	0.00	0.00
10.00	50.00	0.00	0.00	0.00	0.00
30.00	24.00	0.00	0.00	0.00	0.00
42.00	0.00	0.00	0.00	0.00	0.00
24.00	0.00	0.00	0.00	0.00	0.00

Number of ROOTS in the RIGHT-HALF-PLANE = 0

SYSTEM IS STABLE.

Numerator Degree m= 1 Denominator Degree n= 5
Enter NUMERATOR COEFFICIENTS in ASCENDING ORDER:
13 3 0 0 0 0

Enter DENOMINATOR COEFFICIENTS in ASCENDING ORDER:
50 0 24 4 1 1

WARNING !!! Your system's CHARACTERISTIC EQUATION contains an AUXILARY EQUATION. There could be poles on the jw-axis causing OSSCILLATORY characteristics!!

*** ROUTH-HURWITZ ARRAY ***

1.00   4.00   3.00   0.00   0.00   0.00

1.00   24.00   63.00   0.00   0.00   0.00

-20.00 -60.00 0.00   0.00   0.00   0.00

21.00 63.00   0.00   0.00   0.00   0.00

42.00 0.00   0.00   0.00   0.00   0.00

63.00 0.00   0.00   0.00   0.00   0.00

Number of ROOTS in the RIGHT-HALF-PLANE = 2

SYSTEM IS UNSTABLE.

Numerator Degree m= 4 Denominator Degree n= 10
Enter NUMERATOR COEFFICIENTS in ASCENDING ORDER:
1 4 5 2 1 0 0 0 0 0
Enter DENOMINATOR COEFFICIENTS in ASCENDING ORDER:
10 10 5 3 4 12 9 4 5 3 1

*** ROUTH-HURWITZ ARRAY ***

1.00   5.00   9.00   5.00   10.00   11.00

3.00   4.00   12.00   5.00   14.00   0.00

3.67   5.00   3.33   5.33   11.00   0.00

-.09   9.27   .64   5.00   0.00   0.00

379.00 29.00   207.00 11.00   0.00   0.00

9.28   .69   5.00   0.00   0.00   0.00

.98   2.68   11.00   0.00   0.00   0.00

-24.67 -98.97 0.00   0.00   0.00   0.00

-1.26   11.00   0.00   0.00   0.00   0.00

-315.04 0.00   0.00   0.00   0.00   0.00

11.00   0.00   0.00   0.00   0.00   0.00

Number of ROOTS in the RIGHT-HALF-PLANE = 4

SYSTEM IS UNSTABLE.

In the terminal session shown, four examples are considered, three of which are the ones discussed above. Besides those a 10*th*-order system with 4 *r.h.p.* poles is also considered.

7.5 JURY-BLANCHARD STABILITY CRITERION

The z transform, introduced in Chapter 3, allows one to establish stability criteria for *l.t.i.* discrete-time systems similar to the Laplace transform based tests such as the Routh-Hurwitz criterion. The region of stability in the z-plane can be identified through a mapping of the right-half s plane to the z-plane. This mapping is given by (3.6.3), repeated here for convenience.

$$z = e^{sT} \tag{1}$$

where T is the sampling period. By letting $s = \sigma + j\omega$ in (1), one obtains

$$z = e^{\sigma T} e^{j\omega T} = e^{\sigma T + j\omega T} = e^{\sigma T}\cos \omega T + je^{\sigma T}\sin \omega T \tag{2}$$

which provides the corresponding mapped coordinate of the point s in the z-plane. From (2) it is clear that the $s = j\omega$ axis is mapped into a circle with unit radius centered at the origin of the z-plane (the unit circle). Furthermore, the left-half s-plane where $\sigma < 0$ is mapped into the interior of the unit circle. Figure 1 shows the corresponding regions of the two planes.

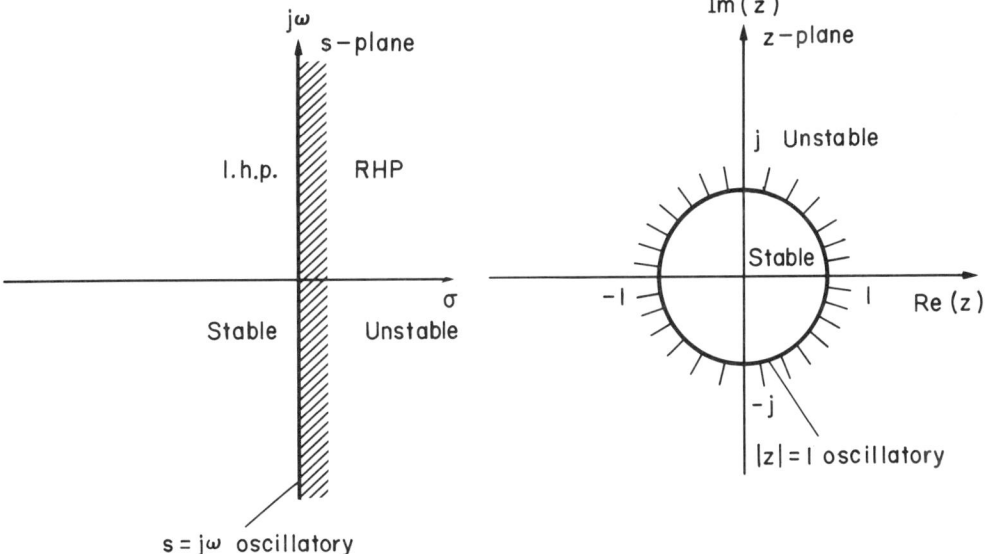

Fig. 7.5.1. Regions of stability and instability in *s*- and *z*-planes

It is therefore clear that for a discrete-time system it is necessary that the roots of the characteristic equation $1+G(z)H(z) = 0$ lie inside the unit circle (Recall Corollary 7.3.8).

One of the earlier attempts in locating the roots of the characteristic equation with respect to the unit circle is the Schur-Cohn criterion [7.9]. This criterion provides the necessary and sufficient conditions for the roots to lie inside the unit circle in terms of the signs of the "Schur-Cohn determinants." This criterion, however, is too cumbersome to be used for systems whose order is larger than 2 and is not considered here. A simpler approach, due to Jury and Blanchard [7.10] with some similarities to the Routh-Hurwitz criterion is considered instead.

Let the characteristic equation of a discrete-time system be.

$$F(z) = a_0 + a_1 z + a_2 z^2 + \ldots + a_{n-1} z^{n-1} + a_n z^n = 0 \tag{3}$$

where $a_0, a_1, \ldots, a_n$ are real coefficients with the assumption $a_n > 0$. If $a_n < 0$, it suffices to multiply both sides of (3) by -1. Then a table (see Table 1) is constructed, where the entries of the $(2k+2)$nd, $k=0,1,2,\ldots$ row are in the reverse order of those in $(2k+1)$th row. The entries of rows 3, 4,... are obtained by

$$b_k = \begin{bmatrix} a_0 & a_{n-k} \\ a_n & a_k \end{bmatrix}, \quad c_k = \begin{bmatrix} b_0 & b_{n-1-k} \\ b_{n-1} & b_k \end{bmatrix}, \quad d_k = \begin{bmatrix} c_0 & c_{n-2-k} \\ c_{n-2} & c_k \end{bmatrix}$$

$$\ldots \quad g_0 = \begin{bmatrix} p_0 & p_3 \\ p_3 & p_0 \end{bmatrix}, \quad q_2 = \begin{bmatrix} p_0 & p_1 \\ p_3 & p_2 \end{bmatrix}$$

Row	z^0	z^1	z^2	z^3		z^{n-k}		z^{n-1}	z^n
1	a_0	a_1	a_2	a_3	$\ldots$	a_{n-k}	$\ldots$	a_{n-1}	a_n
2	a_n	a_{n-1}	a_{n-2}	a_{n-3}	$\ldots$	a_k	$\ldots$	a_1	a_0
3	b_0	b_1	b_2	b_3	$\ldots$	b_{n-k}	$\ldots$	b_{n-1}	
4	b_{n-1}	b_{n-2}	b_{n-3}	b_{n-4}	$\ldots$	b_k	$\ldots$	b_0	
5	c_0	c_1	c_2	c_3	$\ldots$	c_{n-2}			
6	c_{n-2}	c_{n-3}	c_{n-4}	c_{n-5}	$\ldots$	c_0			
$\vdots$									
$2n-5$	p_0	p_1	p_2	p_3					
$2n-4$	p_3	p_2	p_1	p_0					
$2n-3$	q_0	q_1	q_2						

Then the necessary and sufficient conditions for the characteristic equation (3) to have all its roots inside the unit circle on the z-plane (i.e. asymptotic stability) are [7.9].

$$F(1)>0, \quad F(-1) \begin{cases} >0 & n \text{ even} \\ <0 & n \text{ odd} \end{cases} \tag{4a}$$

$$|a_o| < a_n$$

$$|b_o| > |b_{n-1}|$$

$$|c_o| > |c_{n-2}|$$

$$|d_o| > |d_{n-3}|$$

$$|q_o| > |q_2| \tag{4b}$$

where (4b) constitutes n-1 constraints.

Before, the above test is explained in terms of a numerical example, a special case, called "singular" is described. The singular case refers to the situation in which some or all the entries of an entire row of the Jury-Blanchard table are zero. Under this condition, the unit circle is expanded and contracted infinitesimally, i.e. moving the roots off the unit circle. The transformation to achieve this is

$$z = (1+\epsilon)z \tag{5}$$

where ϵ is a very small real number. Depending on whether ϵ is positive or negative, the unit circle is expanded or contracted, respectively. The transformation (5) can be easily applied since

$$z^n = (1+\epsilon)^n z^n = (1+n\epsilon)z^n. \tag{6}$$

Thus, one would use (6) to rewrite the characteristic equation (3) and then apply a more convenient table due to Raible [7.10], given below:

a_n	a_{n-1}	a_{n-2}	...	a_2	a_1	a_o	k_a
b_o	b_1	b_2	...	b_{n-2}	b_{n-1}		k_b
c_o	c_1	c_2	...	c_{n-2}			k_c
.	.	.	.	.	.	.	.
.	.	.	.	.	.	.	.
p_o	p_1	p_2					k_p
q_o	q_1						k_q
r_o							

where $k_a = a_0/a_n$, $k_b = b_{n-1}/b_0$, $k_c = c_{n-2}/c_0$, $\cdots$, $k_p = p_2/p_0$, $k_q = q_1/q$
$b_i = a_{n-i} - k_a a_i$, $i=0, 1, 2, \cdots, n-1$ $c_j = b_j - k_b b_{n-j-1}$, $j=0, 1, 2, \cdots, n-2$
$q_0 = p_0 - k_p p_2$, $q_1 = p_1 - k_p p_1$, $r_0 = q_0 - k_q q_1$.

Then the number of positive entries in the first column $(b_0, c_0, \ldots, p_0, q_0, r_0)$ is equal to the number of roots inside the unit circle. Similarly, the number of negative entries in the first column is equal to the number of roots outside the unit circle. Should a singularity condition appear in Raible's table, the transformation (5)-(6) can be used. By letting $\epsilon > 0$ and $\epsilon < 0$ the difference between the positive (or negative) entries in the first column of Raible's table would give the number of roots which lie on the unit circle. A more convenient form of Raible's table is given below:

a_n	a_{n-1}	a_{n-2}	$\ldots a_2$	a_1	a_0	$k_a = a_0/a_n$
$-) \ a_0 k_a$	$a_1 k_a$	$a_2 k_a$	$\ldots a_{n-2} k_a$	$a_{n-1} k_a$		
b_0	b_1	b_2	$\ldots b_{n-2}$	b_{n-1}		$k_b = b_{n-1}/b_0$
$-) \ b_{n-1} k_b$	$b_{n-2} k_b$	$b_{n-3} k_b$	$b_1 k_b$			
c_0	c_1	c_2	$\ldots c_{n-2}$			$k_c = c_{n-2}/c_0$
$-) \ \cdot$	$\cdot$	$\cdot$	$\cdot$			
p_0	p_1	p_2				$k_p = p_2/p_0$
$-) \ p_2 k_p$	$p_1 k_p$					
q_0	q_1					$k_q = q_1/q_0$
$-) \ q_1 k_q$						
r_0						

The following examples illustrate the above stability tests.

1. Example. Consider a characteristic equation

$$F(z) = 5 - 2z + 13z^2 \tag{7}$$

it is desired to check its stability. The first two conditions of the Jury-Blanchard criterion are $F(1)=16>0$ and $F(-1)=20>0$. Since $n=2$ is even, these conditions satisfy the first two stability criteria. Next, since $2n-3=1$, only one row of the table needs to be calculated.

Row	z^0	z^1	z^2
1	5	-2	13

and $|a_o| = 5 < a_2 = 13$. Hence this system is asymptotically stable. As an alternative, one can use Raible's table:

$$a_2 = 13 \qquad a_1 = -2 \qquad a_0 = 5 \quad k_a = 0.385$$
$$-) \; a_0 k_a = 1.92 \quad a_1 k_a = -0.77$$
$$b_0 = 11.08 \quad b_1 = -1.23 \qquad\qquad k_b = -0.111$$
$$-) \; b_1 k_b = 0.136$$
$$c_0 = 10.944$$

Now since both coefficients b_o and c_o are positive, the roots of the characteristic equation are inside the unit circle. In fact the roots are $z_{1,2} = 1/13 \pm j8/13$.

2. Example. Consider the equation

$$F(z) = z^4 + z^3 + z^2 + z + 1 = 0. \tag{8}$$

Since all the coefficients are equal, this case is singular. Thus the transformation (5)-(6) can be used to result

$$F((1+\epsilon)z) = (1+4\epsilon)z^4 + (1+3\epsilon)z^3 + (1+2\epsilon)z^2 + (1+\epsilon)z + 1 = 0. \tag{9}$$

The entries of the Raible's table are given below:

	$1+4\epsilon$	$1+3\epsilon$	$1+2\epsilon$	$1+\epsilon$	1	$k_a = \dfrac{1}{1+4\epsilon}$
$-)$	$\dfrac{1}{1+4\epsilon}$	$\dfrac{1+\epsilon}{1+4\epsilon}$	$\dfrac{1+2\epsilon}{1+4\epsilon}$	$\dfrac{1+3\epsilon}{1+4\epsilon}$		
$b_o =$	$\dfrac{8\epsilon}{1+4\epsilon}$	$\dfrac{6\epsilon}{1+4\epsilon}$	$\dfrac{4\epsilon}{1+4\epsilon}$	$\dfrac{2\epsilon}{1+4\epsilon}$		$k_b = 1/4$
$-)$	$\dfrac{\epsilon}{2(1+4\epsilon)}$	$\dfrac{\epsilon}{1+4\epsilon}$	$\dfrac{3\epsilon}{2(1+4\epsilon)}$			
$c_o =$	$\dfrac{15\epsilon}{2(1+4\epsilon)}$	$\dfrac{5\epsilon}{1+4\epsilon}$	$\dfrac{5\epsilon}{2(1+4\epsilon)}$			$k_c = 1/3$
$-)$	$\dfrac{5\epsilon}{6(1+4\epsilon)}$	$\dfrac{5\epsilon}{3(1+4\epsilon)}$				
$d_o =$	$\dfrac{25\epsilon}{6(1+4\epsilon)}$	$\dfrac{10\epsilon}{3(1+4\epsilon)}$				$k_d = 4/5$
$-)$	$\dfrac{8\epsilon}{3(1+4\epsilon)}$					
$e_o =$	$\dfrac{3\epsilon}{2(1+4\epsilon)}$					

It is now noted that the entries of the first column b_o, c_o, d_o and e_o are directly proportional to ϵ. These entries are all positive (or negative) when $\epsilon > 0$ (or $\epsilon < 0$). As ϵ changes from positive to negative, the three roots move from inside the unit circle to outside. Hence the system whose characteristic equation is (8) has four roots on the unit circle. In fact, the roots are $z_{1,2} = 0.39017 \pm j0.95105$ and $z_{3,4} = -0.809017 \pm j0.5877853$.

In sequel, the source listing of program "JURY" is given. This program tests the stability of discrete-time systems including the singular cases.

3. "JURY"

```
10    ! PROGRAM NAME : "JURY" PROG
20    PRINT "This program determines the STABILITY of a Discrete-Time System"
30    PRINT " via JURY-BLANCHARD (& RAIBLE) STABILITY CRITERION .",LIN(1)
40    INPUT "Give your PRINT option: for CRT input 1; for LINE PRINTER input 2",P
50    IF (P=1) OR (P=2) THEN 70
60    GOTO 40
70    IF P=1 THEN PRINTER IS 16
80    IF P=2 THEN PRINTER IS 7,1
90    PRINT "Input data format can be:"
100   PRINT "1.Characteristic Polynomial F(z)=SUM of a(i)*z^i , i=0 to n (n<=10)"
110   PRINT "OR 2.Closed-loop trasfer function evaluated as follows:"
120   PRINT "     Open loop transfer function: G(z)=A(z)/B(z)"
130   PRINT "     Feedback transfer function : H(z)=C(z)/D(z)"
140   PRINT "     where A(z)=SUM of a(i)*z^i , i=0 to n (n<=5)"
150   PRINT "           B(z)=SUM of b(i)*z^i , i=0 to n (n<=5)   ;etc.";LIN(1)
160   DIM A(10),B(10),C(10),D(10),F(20),P(20),Q(20)
170   INPUT "Have SUB's <<Vec>> & <<Polpro>> already been LINKED (Y/N)?",S$
180   IF (S$="Y") OR (S$="y") THEN 230
190   IF (S$="N") OR (S$="n") THEN 210
200   GOTO 170
210   LINK "Vec",1630,220
220   LINK "Polpro",2500,230
230   INPUT "Input data format (1 or 2) ?",Nf
240   IF (Nf=1) OR (Nf=2) THEN 260
250   GOTO 230
260   ON Nf GOTO 270,340
270   INPUT "Enter the order of the polynomial F(z):",N1
280   PRINT "Order of the polynomial : F(z)=";N1;LIN(1)
290   INPUT "ANY CHANGE (Y/N)?",Q$
300   IF (Q$="Y") OR (Q$="y") THEN 270
310   CALL Vec(F(*),N1,"A")
320   IF F(N1)>0 THEN 580
330   GOTO 520
340   INPUT "Enter the order of the polynomial A(z),B(z)",K,L
350   INPUT "Enter the order of the polynomial C(z),D(z)",M,N
360   PRINT "Order of the polynomials:";LIN(1);"A(z)=";K;"B(z)=";L;"C(z)=";M;"D(z
)=";N;LIN(2)
370   INPUT "ANY CHANGE (N/Y)?",Q$
380   IF (Q$="Y") OR (Q$="y") THEN 340
390   PRINT "Enter COEFFICIENTS of the polynomial A(Z) in ASCENDING ORDER:"
400   CALL Vec(A(*),K,"A")
410   PRINT "Enter COEFFICIENTS of the polynomial B(Z) in ASCENDING ORDER:"
420   CALL Vec(B(*),L,"B")
430   PRINT "Enter COEFFICIENTS of the polynomial C(Z) in ASCENDING ORDER:"
440   CALL Vec(C(*),M,"C")
450   PRINT "Enter COEFFICIENTS of the polynomial D(Z) in ASCENDING ORDER:"
460   CALL Vec(D(*),N,"D")
470   CALL Polpro(A(*),C(*),K,M,P(*))
480   CALL Polpro(B(*),D(*),L,N,Q(*))
490   N1=MAX(K+M,L+N)
500   MAT F=P+Q
510   IF F(N1)>0 THEN 540
520   MAT F=F*(-1)
530   PRINT "After change the signs of all coefficients of polynomial due to nega
ve A(n),"
540   PRINT "the coefficients of the polynomial F(Z) :"
550   FOR I=0 TO N1
560   PRINT "A(";I;") = ";F(I)
570   NEXT I
580   Cd=FNCfz(F(*),N1,1)
590   PRINT LIN(1);"F(1)=";Cd
600   IF Cd>0 THEN 620
610   GOTO 670 ! UNSTABLE
620   Cd=FNCfz(F(*),N1,-1)
630   PRINT "F(-1)=";Cd
640   IF (N1 MOD 2=0) EXOR (Cd>0) THEN 670   ! UNSTABLE
650   IF ABS(F(0))<=F(N1) THEN 690
660   PRINT "|A(0)|>=A(n)"
```

```
670  PRINT "This system is UNSTABLE."
680  GOTO 740
690  IF N1>2 THEN 730
700  IF ABS(F(0))=F(N1) THEN 660
710  PRINT "|A(0)|<A(n)";LIN(1);"This system is ASYMPTOTICALLY STABLE."
720  GOTO 740
730  CALL Jut(F(*),N1)
740  INPUT "Do you want to run Jury's Stability test on a new data set?(Y/N)",I$
750  IF (I$="N") OR (I$="n") THEN 780
760  PRINT LIN(3);"           *****       *****      *****        *****";LIN(3)
770  GOTO 230
780  END
790  DEF FNCfz(F(*),M,N)
800  T=F(0)
810  FOR I=1 TO M
820  T=T+F(I)*N^I
830  NEXT I
840  RETURN T
850  FNEND
860  SUB Jut(A(*),N)
870  !   "Jut" tests and prints Jury's Stability Table. It will stop testing
880  !   at the first point where violates the stable conditions.
890  DIM J(N+1,N)
900  FOR I=0 TO N
910  J(0,I)=A(I)
920  J(1,N-I)=A(I)
930  NEXT I
940  M=N
950  FOR K=2 TO 2*N-3 STEP 2
960  M=M-1
970  FOR I=0 TO M
980  J(K,I)=J(K-2,0)*J(K-2,I)-J(K-1,0)*J(K-1,I)
990  J(K+1,M-I)=J(K,I)
1000 NEXT I
1010 IF ABS(J(K,0))>ABS(J(K+1,0)) THEN 1050
1020 IF (J(K,0)=0) AND (J(K+1,0)=0) THEN 1300
1030 L=K
1040 GOTO 1090
1050 NEXT K
1060 PRINT "This system is ASYMTOTICALLY STABLE."
1070 L=N
1080 GOTO 1100
1090 PRINT "This system is UNSTABLE."
1100 INPUT "Do you want to print the Jury's Test Table ?(Y/N)",Y$
1110 IF (Y$="Y") OR (Y$="y") THEN 1140
1120 IF (Y$="N") OR (Y$="n") THEN 1430
1130 GOTO 1100
1140 PRINT LIN(2);"    ****    JURY'S STABILITY TABLE    ****";LIN(2)
1150 PRINT USING "#,K,2X";"ROW"
1160 IMAGE #,K
1170 IMAGE #,1(DD,6X)
1180 FOR I=0 TO N
1190 PRINT USING 1160;"z^"
1200 PRINT USING 1170;I
1210 NEXT I
1220 FOR I=0 TO L
1230 PRINT USING "/"
1240 PRINT USING "#,DD,2X";I
1250 FOR K=0 TO N
1260 PRINT USING "#,MD.DDE,1X";J(I,K)
1270 NEXT K
1280 NEXT I
1290 GOTO 1430
1300 CALL Raible(A(*),N,1E-9,Np)
1310 CALL Raible(A(*),N,-1E-9,Ne)
1320 Ru=ABS(Np-Ne)       !No. of roots on the unit circle
1330 Rot=MAX(Np,Ne)-Ru   !No. of roots outside the unit circle
1340 IF Rot=0 THEN 1380
1350 PRINT "No. of ROOTS OUTSIDE the UNIT CIRCLE =";Rot
1360 PRINT "This system is UNSTABLE."
1370 GOTO 1430
1380 IF Ru>0 THEN 1410
```

```
1390 PRINT "This system is ASYMPTOTICALLY STABLE."
1400 GOTO 1430
1410 PRINT "No. of ROOTS ON the UNIT CIRCLE =";Ru
1420 PRINT "This system is STABLE in the sense of Lyapunov."
1430 SUBEND
1440 SUB Raible(A(*),N,E,Neg)
1450 ! "Raible" makes the Raible's tabulation for SINGULAR CASE of Discrete-Time
1460 ! System, and count the quantities of NEGATIVE coefficients in 1st column.
1470 DIM R(N,N)
1480 Neg=0
1490 FOR I=0 TO N
1500 R(0,N-I)=(1+I*E)*A(I)
1510 NEXT I
1520 M=N
1530 FOR I=0 TO N-1
1540 Ki=R(I,M)/R(I,0)
1550 M=M-1
1560 FOR J=0 TO M
1570 R(I+1,J)=R(I,J)-R(I,M-J)*Ki
1580 NEXT J
1590 IF R(I+1,0)>0 THEN 1610
1600 Neg=Neg+1
1610 NEXT I
1620 SUBEND
```

4. Example

This program determines the STABILITY of a Linear Discrete-Time System via JURY-BLANCHARD (& RAIBLE) STABILITY CRITERION .

Input data format can be:
1.Characteristic Polynomial $F(z)$=SUM of $a(i)*z^i$, $i=0$ to n ($n<=10$) OR 2.Closed-loop transfer function evaluated as follows:

>   Open loop transfer function: $G(z)=A(z)/B(z)$
>   Feedback transfer function : $H(z)=C(z)/D(z)$
>   where $A(z)$=SUM of $a(i)*z^i$, $i=0$ to n ($n<=5$)
> $B(z)$=SUM of $b(i)*z^i$, $i=0$ to n ($n<=5$)   ;etc.

Order of the polynomial : $F(z)$= 1

Order of the polynomial : $F(z)$= 4

A(0) = 1.0000E+00
A(1) = 1.0000E+00
A(2) = 1.0000E+00
A(3) = 1.0000E+00
A(4) = 1.0000E+00

F(1)= 5
F(-1)= 1
No. of ROOTS ON the UNIT CIRCLE = 4
This system is STABLE in the sense of Lyapunov.

***** ***** ***** *****

Order of the polynomials:
A(z) = 1 B(z) = 1 C(z) = 1 D(z) = 1

Enter COEFFICIENTS of the polynomial A(Z) in ASCENDING ORDER:
A(0) = 1.0000E+00
A(1) = 3.0000E+00

Enter COEFFICIENTS of the polynomial B(Z) in ASCENDING ORDER:
B(0) = -2.0000E+00
B(1) = 2.0000E+00

Enter COEFFICIENTS of the polynomial C(Z) in ASCENDING ORDER:
C(0) = 1.0000E+00
C(1) = 3.0000E+00

Enter COEFFICIENTS of the polynomial D(Z) in ASCENDING ORDER:
D(0) = -2.0000E+00
D(1) = 2.0000E+00

the coefficients of the polynomial F(Z) :
A(0) = 5
A(1) = -2
A(2) = 13

F(1) = 16
F(-1) = 20
|A(0)| < A(n)
This system is ASYMPTOTICALLY STABLE.

***** ***** ***** *****

Order of the polynomial : F(z) = 6

A(0) = 1.0000E+00
A(1) = 5.0000E+00
A(2) = 8.0000E+00
A(3) =-6.0000E+00
A(4) = 8.0000E+00
A(5) = 3.0000E+00
A(6) = 7.0000E+00

F(1) = 26
F(-1) = 22
This system is UNSTABLE.

**** JURY'S STABILITY TABLE ****

ROW	z^0	z^1	z^2	z^3	z^4	z^5	z^6
0	1.00E+00	5.00E+00	8.00E+00	-6.00E+00	8.00E+00	3.00E+00	7.00E+00
1	7.00E+00	3.00E+00	8.00E+00	-6.00E+00	8.00E+00	5.00E+00	1.00E+00
2	-4.80E+01	-1.60E+01	-4.80E+01	3.60E+01	-4.80E+01	-3.20E+01	0.00E+00
3	-3.20E+01	-4.80E+01	3.60E+01	-4.80E+01	-1.60E+01	-4.80E+01	0.00E+00
4	1.28E+03	-7.68E+02	3.46E+03	-3.26E+03	1.79E+03	0.00E+00	0.00E+00

***** ***** ***** *****

Order of the polynomial : F(z) = 4

A(0) = 1.0000E+00
A(1) = 2.0000E+00
A(2) = 3.0000E+00
A(3) = 2.0000E+00
A(4) = 1.0000E+00

F(1) = 9
F(-1) = 1
No. of ROOTS OUTSIDE the UNIT CIRCLE = 1
This system is UNSTABLE.

***** ***** ***** *****

Order of the polynomial : F(z) = 2

A(0) = 5.0000E+00
A(1) =-2.0000E+00
A(2) = 1.3000E+01

F(1) = 16
F(-1) = 20
|A(0)|<A(n)
This system is ASYMPTOTICALLY STABLE.

7.6 THE ROOT LOCUS METHOD

The locations of SISO system's closed-loop poles have an important effect on the relative stability and the transient response of the system. The system's open-loop (or d.c.) gain is often subject to drifts and variations. Therefore, it is important for a designer to have an exact understanding of the locations of the roots of the characteristic equation as the open-loop gain varies. The *root locus method*, introduced by Evans [7.11-7.13] is the scheme to do this. This method has been used extensively throughout control engineering practice. The root locus is a graphical description of the sensitivity of system's poles with respect to a parameter, be it the open-loop gain or any other one.

Consider the characteristic equation of a SISO system:

$$1+G(s)H(s)=0 \quad or \quad G(s)H(s)=-1 \tag{1}$$

which can be satisfied if the following two magnitude and angle conditions hold:

$$|G(s)H(s)|=1, \quad \sphericalangle G(s)H(s) = (2k+1)180° \tag{2}$$

where k is an integer. The latter condition in (2) is used to plot the root locus, while the former is often used to graduate the plot in terms of the d.c. gain $0 \leqslant K \leqslant \infty$.

In this section, the conventional procedure of plotting a root locus is first introduced and then a BASIC program called "ROOTLC" which handles plots for both continuous-time and discrete-time systems is given.

7.6.1 Conventional Approach in Root Locus Plotting

We begin the discussion by drawing the root locus for a simple second-order system with open-loop transfer function.

$$G(s) = K/(s(s+5)) \tag{3a}$$

where $0 \leqslant K \leqslant \infty$. The characteristic equation of this system is

$$1+G(s) = 0 \quad or \quad s^2+5s+K = 0 \tag{3b}$$

whose roots are,

$$s_{1,2} = -5/2 \pm \sqrt{25-4K}/2. \tag{3c}$$

It is noted that for $K=0$, $s_{1,2} = 0, -5$ which correspond to the open-loop system as evident from (3a). As K is increased from zero to 25, the two roots in (3c) get closer to each other until they meet at $s_{1,2} = -5/2$. For $K > 25/4$, the roots would become complex as $s_{1,2} = -5/2 \pm j\sqrt{4K-25}/2$. until they approach $s_{1,2} = -5/2 \pm j\infty$. The locus is shown in Fig. 1.

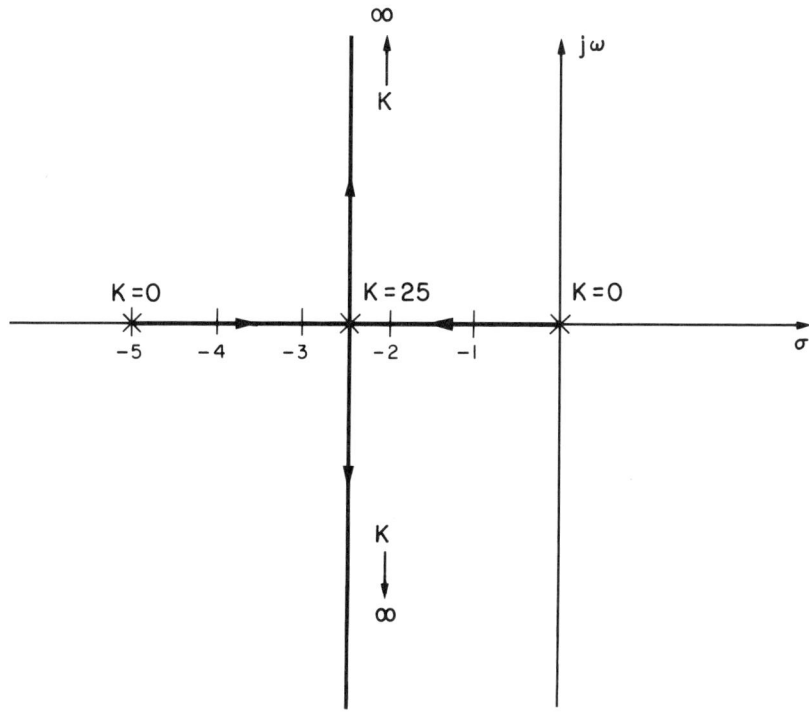

Fig. 7.6.1. The root locus of $G(s) = K/(s(s+5))$

The plotting of a root locus requires a set of conventional rules. These rules have been exhaustively discussed in literature on control system [7.14, 7.15]. Here, a brief description of these rules are presented and applied to a system with open-loop transfer function,

$$G(s) = K/(s(s^2+2s+2)) \qquad (4)$$

These seven rules are given below:

Rule 1. Branches: The *branches* or segments of a root locus, whose number is equal to the number of open-loop poles begin at the open-loop poles and end at either a system zero or go to infinity. For the system of transfer function (4), there are three branches with all three going to infinity.

Rule 2. Real-Axis Sub-branches: The portions of the root locus branches which are on the realaxis lie to the left of an odd number of poles and zeros. For the example in (4), real-axis loci exist to the left of the pole at the origin.

Rule 3. Root Locus Symmetry: Since the complex roots come in conjugate pairs, the root locus is symmetrical with respect to the real axis.

Rule 4. Asymptotes: When some branches go to infinity, they approach along asymptotes centered at

$$\sigma_a = (\sum \text{poles of } G(s) - \sum \text{zeros of } G(s))/(n_p - n_z)$$

$$= (\sum_{i=1}^{n_p} p_i - \sum_{j=1}^{n_z} z_j)/(n_p - n_z). \quad (5a)$$

with an angle

$$\phi_a = (2k+1)180° \ (n_p - n_z), \ k=0,1,...,(n_p - n_z - 1) \quad (5b)$$

where k is an integer index, n_p and n_z are number of open-loop poles and zeros. For the example in (4), $\sigma_a = -2/3$ and $\phi_a = 60°, 180°$ and $300°$.

Rule 5. Imaginary-Axis Crossing: The branches of the root locus may sometimes cross the $j\omega$-axis. This crossing can be obtained through the application of the Routh-Hurwitz criterion. For the example at hand, the characteristic equation is written as,

$$s(s^2+2s+2) + K = s^3+2s^2+2s+K = 0.$$

The Routh-Hurwitz array is

s^3	1	2
s^2	2	K
s^1	2-K/2	0
s^0	K	

which indicates that the system is stable as long as $0 < K < 4$. The auxiliary equation (row s^2) can be used to find the coordinate of the root locus and the $j\omega$-axis, i.e.

$$2s^2+K = 2s^2+4 = 0 \text{ or } s = \pm j\sqrt{2}$$

Rule 6. Break-away and Break-in Points: For the cases where two real poles or two real zeros are placed adjacent to each other, the branches of the root locus break away from or break in the negative real axis. For the system whose root locus is shown in Fig. 1, $ss = -5/2$ is a break away point. In order to find such points, the characteristic equation is rewritten as

$$K = p(s) \quad (6)$$

THE ROOT LOCUS METHOD

Then the extreme points of (6) would provide potential break-away and/or break-in points. This is achieved by finding the roots of $dK/ds=0$. As an example, consider the transfer function of a system

$$1+G(s) = 1+\frac{K(s+2)}{s(s+1)} \tag{7}$$

where a zero and two open-loop poles exist. The value of K in (6) becomes

$$K = -(s^2+s)/(s+2) \tag{8}$$

whose derivative is

$$dK/ds = (s^2+4s+2)/(s+2)^2 = 0 \tag{9}$$

resulting in s = -0.586 and -3.414. It is clear that the first point $s_1 = -0.586$ is a breakaway and $s_2 = -3.414$ is a break-in as shown in Fig. 2.

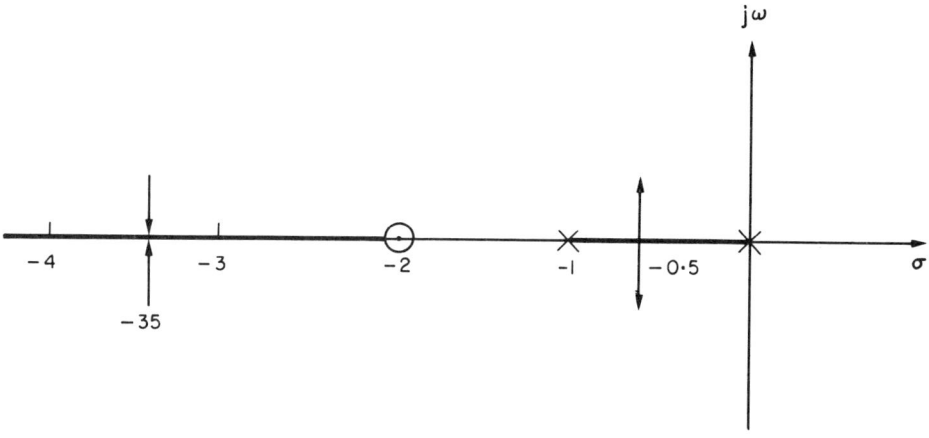

Fig. 7.6.2. Breakaway and break-in points for system of Eq. (7.6.7)

A careful look at the function K(s) indicates that it has a relative maximum at s = -0.586 and a relative minimum at s = -3.414. For the system of Eq. (4), there are no breakaway or break-in points.

Rule 7. Angle of Departure or Entry: When a system has pairs of complex conjugate open-loop poles, the corresponding branches leave these poles at angles called *departure angles*. Similarly, when there are complex conjugate zeros, the corresponding branches enter these zeros at angles called *angles of entry*. To illustrate these points, consider the poles of the system (4) represented in Fig. 3.

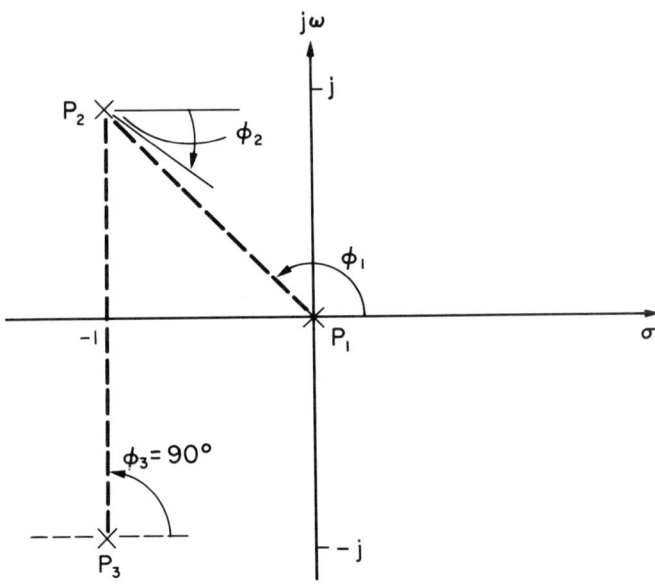

Fig. 7.6.3. Angle condition in the vicinity of a complex pole

The angle condition in (2) can now be written as,

$$\phi_1 + \phi_2 + \phi_3 = (1+2k)180° \tag{10a}$$

or

$$\phi_2 = (1+2k)180° - (\phi_1 + 90°) \tag{10b}$$

which is a value for the angle of departure from pole p_2. The value of this angle is $\phi_2 = 180 - (135° + 90°) = -45°$. The angle of departure from pole p_3 is $+45°$ by symmetry. As another example, consider the pole-zero condition of Fig. 4. Here there are two angles of entry associated with the two complex conjugate zeros. Once again, the angle condition (2) can be written as

$$\phi_1 + \phi_2 + \phi_3 + \phi_4 - \psi_1 - \psi_2 = (1+2k)180° \tag{11a}$$

or the angle of entry is,

$$\psi_1 = (\phi_1 + \phi_2 + \phi_3 + \phi_4 - 90°) = (1+2k)180°. \tag{11b}$$

THE ROOT LOCUS METHOD

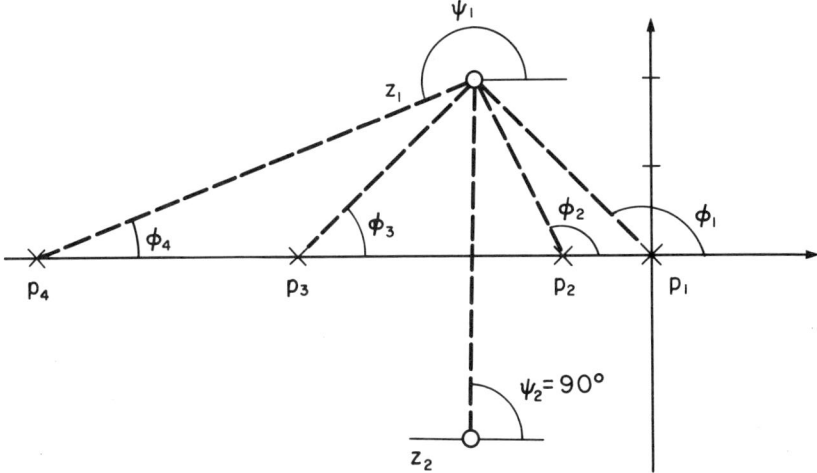

Fig. 7.6.4. Angle condition in the vicinity of a complex zero

If the result of the above seven rules, as applicable to the system of Eq. (4), are considered collectively, the root locus of Fig. 5 would result. It is noted that, in general, a number of test points are needed to complete the locus.

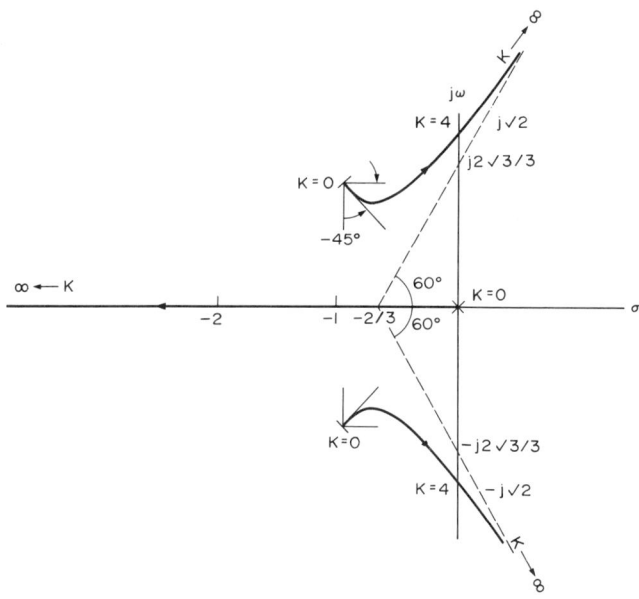

Fig. 7.6.5. A root locus for system $GH = K/(s(s^2+2s+2)$

316   LINEAR CONTROL SYSTEMS

In earlier times, the discrete line segments from each pole and zero were measured by a device called *spirule* * to locate candidate points on the locus and find its corresponding d.c. gain. In order to find the d.c. gain K for a particular point s^* on the root locus of Fig. 5 it is necessary to evaluate

$$|G(s)H(s)|_s = K/(|s|.|s+1+j|.|s+1-j|)^s = 1$$

or

$$K = |s|.|s+1+j|.|s+1-j|. \qquad (12)$$

The root locus plot will be utilized in Chapter 9 for the design of linear control systems. In this chapter the topic of interest is stability and within that context the root locus is not the fastest means for checking system stability. It does, however, provide a convenient graphical means to check the stability of the system as the d.c.gain increases. The above discussions on the conventional approach in plotting the root locus does show one point and that is the extent of detail calculations that one must perform in order to accurately plot it. It is, therefore, worth mentioning that a computer-aided approach to plot the root locus is a very useful tool. A program which can plot a root locus for continuous- or discrete-time system is described in the next section.

7.6.2 Computer-Aided Approach in Root Locus Plotting

In sequel, a program called "ROOTLC" is presented which plots the root locus of a linear continuous-time or discrete-time system.

1. Example Consider the system of equation (3a) and rewrite the characteristic equation (3b) as,

$$(1+K.0)s^2+(5+K.0)s+(0+K.1) = 0 \qquad (13)$$

for inputing to the program "ROOTLC" the terminal session and the root locus of this system are shown in the subsequent pages and Fig. 6. Note that the plot in Fig. 6 is identical to the one shown in Fig. 1. A small bug in "ROOTLC" causes occasional jumps from on closed-loop pole location to another location causing some discontinuities. However, the program has worked for many examples. Figs. 7-9, show the root loci for 3 other examples, among them is a discrete-time system shown in Fig. 9 whose slightly different version in continuous-time form was discussed in the previous section and plotted in Fig. 5. Further details on root locus are given in the problems at the end of this chapter.

2. "ROOTLC"

---

* A Spirule is a device used to sum angles and perform product and division of directed line segments [7.15].

THE ROOT LOCUS METHOD

```
10   !      PROGRAM NAME : "ROOTLC" PROG
20   !      THIS PROGRAM DRAWS THE ROOT LOCUS OF A CONTINUOUS- OR
30   !      DISCRETE-TIME SYSTEM FROM THEIR Laplace- OR Z-transform
40   PRINT "  This program draws the ROOT LOCUS of a continuous- &"
50   PRINT "   discrete-time system from their Laplace- or Z-transform"
60   INPUT "Have you LINKED SUB <<Rootfd>>(Y/N)",Q$
70   IF (Q$="Y") OR (Q$="y") THEN 110
80   IF (Q$="N") OR (Q$="n") THEN 100
90   GOTO 60
100  LINK "Rootfd",3130,110
110  DEG
120  PLOTTER IS "GRAPHICS"
130  SCALE -11,11,-11,11
140  INPUT "Order of the System?",N
150  PRINT LIN(1),"System Order=";N
160  IF N<=0 THEN 140
170  INPUT "Case # 1=CONTINUOUS & 2=DISCRETE",Icase
180   CALL Root(N,Plot$,Icase)
190  OFF KEY #0
200  BEEP
210  IF Plot$<>"9872A" THEN 240
220  PEN 0
230  GOTO 260
240  INPUT "Do you like to dump the GRAPHICS(Y/N)",Ans$
250  IF Ans$="Y" THEN DUMP GRAPHICS
260  END
270  SUB Root(N,Plot$,Icase)
280  DIM Rroot(1:N),Iroot(1:N),Rcoef(0:N),Icoef(0:N),Save(0:N)
290  DIM R(0:N),Rl(0:N),Ic(0:N),Rc(0:N),Last(0:N),Lastl(0:N)
300  DIM Arr(0:N),Arrl(0:N),Savel(0:N),Save2(0:N),Irool(0:N)
310  DIM Rrool(0:N)
320  ON KEY #0 GOTO 1450
330  Itmax=100
340  Tola=Tolf=.001
350  IF Icase=1 THEN 380
360  PRINT "Coefficients: [(Rcoef(0)+Icoef(0)*K)+(Rcoef(1)+Icoef(1)*K)*z^1+...]"
370  GOTO 390
380  PRINT "Coefficients: [(Rcoef(0)+Icoef(0)*K)+(Rcoef(1)+Icoef(1)*K)*s^1+...]"
390  PRINT LIN(1),SPA(8),"Rcoef","  Icoef",LIN(1)
400  FOR I=N TO 0 STEP -1
410      DISP "Rcoef(";I;")=";
420      INPUT Rcoef(I)
430      Save(I)=Rcoef(I)
440      DISP "Icoef(";I;")=";
450      INPUT Icoef(I)
460      Savel(I)=Icoef(I)
470      PRINT USING 490;Rcoef(I),Icoef(I)
480  NEXT I
490  IMAGE 3X,MZ.4DE,5X,MZ.4DE
500  LINPUT "CHANGES  (Y/N)",C$
510  IF (C$="N") OR (C$="n") THEN 650
520  IF (C$="Y") OR (C$="y") THEN 540
530  GOTO 500
540  INPUT "Coefficient number",I
550  IF (I<0) OR (I>N) THEN 540
560  DISP "Rcoef(";I;")=";
570  INPUT Rcoef(I)
580  Save(I)=Rcoef(I)
590  DISP "Icoef(";I;")=";
600  INPUT Icoef(I)
610  Savel(I)=Icoef(I)
620  PRINT USING 630;I,Rcoef(I),I,Icoef(I)
630  IMAGE "Rcoef(",DDD,")=",MZ.4DE,5X,"Icoef(",DDD,")=",MZ.4DE
640  GOTO 500
650  INPUT "Do you like to use the CRT or 9872A as the PLOTTER",Plot$
660  IF (Plot$="CRT") OR (Plot$="9872A") THEN 700
670  BEEP
680  PRINT "Answer: Either  'CRT' or '9872A'"
690  GOTO 650
700  IF Plot$="CRT" THEN PLOTTER IS "GRAPHICS"
```

```
710  IF Plot$="9872A" THEN PLOTTER IS "9872A"
720  Kmax=100
730  Step=.01
740  MAT R=ZER
750  IF Plot$="CRT" THEN GRAPHICS
760  Num=0
770  FOR I=N TO 0 STEP -1
780  Ic(I)=0
790  IF (Icoef(I)<>0) AND (Num1=0) THEN Num1=I
800  Rc(I)=Icoef(I)
810  NEXT I
820  IF Num1=0 THEN 840
830  CALL Rootfd(Num1,Rc(*),Ic(*),Tola,Tolf,Itmax,Rrool(*),Irool(*))
840  Count=0
850  FOR I=0 TO N
860  Last(I)=Lastl(I)=Arr(I)=Arrl(I)=0
870  NEXT I
880  Ijkl=-Step
890  Itime=-1
900  Ijkl=Ijkl+Step
910  Itime=Itime+1
920  FOR Ijkll=0 TO N
930  Rcoef(Ijkll)=Save(Ijkll)+Ijkl*Savel(Ijkll)
940  Rcoef(Ijkll)=Save(Ijkll)+Ijkl*Savel(Ijkll)
950  Icoef(Ijkll)=0
960  NEXT Ijkll
970  CALL Rootfd(N,Rcoef(*),Icoef(*),Tola,Tolf,Itmax,Proot(*),Iroot(*))
980  FOR L=1 TO N
990  Save2(L)=0
1000 NEXT L
1010 IF Ijkl=0 THEN GOSUB First
1020 FOR I=1 TO N
1030 Iroot=I
1040     IF Ijkl<>0 THEN 1100
1050     GOSUB Box
1060     IF (I<>N) OR (Plot$="CRT") THEN 1300
1070     INPUT "Which pen do you like to use to PLOT",Pen
1080     PEN Pen
1090     GOTO 1300
1100     Dif=10
1110     FOR J=1 TO N
1120     IF I=1 THEN 1160
1130     FOR L=1 TO I-1
1140     IF J=Save2(L) THEN 1200
1150     NEXT L
1160     Dif2=SQR((R(J)-Rroot(I))^2+(Rl(J)-Iroot(I))^2)
1170     IF Dif<Dif2 THEN 1200
1180     Dif=Dif2
1190     Iroot=J
1200     NEXT J
1210     Save2(I)=Iroot
1220     IF (ABS(R(Iroot))>Max) OR (ABS(Rl(Iroot))>Max) THEN 1300
1230     FOR J=1 TO Num1
1240     IF (ABS(Rroot(I)-Rrool(J))<.1) AND (ABS(Iroot(I)-Irool(J))<.1) THEN 1300
1250     NEXT J
1260     PLOT R(Iroot),Rl(Iroot),-2
1270     PLOT Rroot(I),Iroot(I),-1
1280     IF (ABS(Rroot(I))>Max) OR (ABS(Iroot(I))>Max) THEN 1300
1290     IF INT(Itime/(5*2^Count))*(5*2^Count)=Itime THEN GOSUB Arrow
1300     R(Iroot)=Rroot(I)
1310     Rl(Iroot)=Iroot(I)
1320     PENUP
1330 NEXT I
1340 FOR I=1 TO N
1350 FOR J=1 TO Num1
1360 IF (ABS(Rroot(I)-Rrool(J))<.5) AND (ABS(Iroot(I)-Irool(J))<.5) THEN 1390
1370 NEXT J
1380 IF (ABS(Rroot(I))<Max) AND (ABS(Iroot(I))<Max) THEN 1410
1390 NEXT I
1400 GOTO 1450
1410 IF INT(Itime/(5*2^Count))*(5*2^Count)=Itime THEN Count=Count+1
```

```
1420 IF Itime<>10 THEN 900
1430 Step=Step*10
1440 GOTO 890
1450 SUBEXIT
1460 Box: MOVE Rroot(I)+.02*Max,Iroot(I)+.02*Max
1470 Arr(I)=Rroot(I)
1480 Arrl(I)=Iroot(I)
1490 IPLOT -.04*Max,-.04*Max,-1
1500 IPLOT .04*Max,0,-2
1510 IPLOT -.04*Max,.04*Max,-1
1520 PENUP
1530 RETURN
1540 Arrow: DEG
1550 FOR J=1 TO N
1560 IF I=1 THEN 1600
1570 FOR L=1 TO I-1
1580 IF J=Save2(L) THEN 1610
1590 NEXT L
1600 IF SQR((Arr(J)-Rroot(I))^2+(Arrl(J)-Iroot(I))^2)<.1*Max THEN RETURN
1610 NEXT J
1620 IF ABS(Iroot(I)-Rl(Iroot))>1E-10 THEN 1650
1630 Ang=0
1640 GOTO 1660
1650 Ang=ATN((Rl(Iroot)-Iroot(I))/(R(Iroot)-Rroot(I)))
1660 IF Rroot(I)-R(Iroot)<0 THEN Ang=Ang+180
1670 PDIR Ang
1680 Arr(Iroot)=Rroot(I)
1690 Arrl(Iroot)=Iroot(I)
1700 IPLOT -.02*Max,.02*Max,-2
1710 IPLOT .02*Max,-.02*Max,-1
1720 IPLOT -.02*Max,-.02*Max
1730 RETURN
1740 First: Max=0
1750 FOR I=1 TO N
1760 IF SQR(Rroot(I)^2+Iroot(I)^2)>Max THEN Max=SQR(Rroot(I)^2+Iroot(I)^2)
1770 NEXT I
1780 FOR I=1 TO Num1
1790 IF SQR(Rrool(I)^2+Irool(I)^2)>Max THEN Max=SQR(Rrool(I)^2+Irool(I)^2)
1800 NEXT I
1810 Max=Max*4
1820 Max=INT(Max/5+.999)*5
1830 IF Max=0 THEN Max=5
1840 LOCATE 12,108,2,98
1850 SHOW -Max,Max,-Max,Max
1860 Int=INT(Max/5)
1870 IF Int=0 THEN Int=1
1880 IF Plot$="CRT" THEN 1910
1890 INPUT "Which pen do you like to use for the AXES?",Pen
1900 PEN Pen
1910 LORG 5
1920 CSIZE 3.0
1930 LINE TYPE 3.0
1940 AXES Int,Int,0,0
1950 IF Icase=1 THEN 2030
1960 DEG
1970 MOVE 1,0
1980 FOR I=0 TO 360 STEP 9
1990 X=COS(I)
2000 Y=SIN(I)
2010 PLOT X,Y,-1
2020 NEXT I
2030 IF Plot$="CRT" THEN 2060
2040 INPUT "which pen do you like to label the axes(1-4)?",Pen
2050 PEN Pen
2060 LINE TYPE 1
2070 FOR I=-INT(Max/Int)*Int TO INT(Max/Int)*Int STEP Int
2080 MOVE I,-.06*Max
2090 LABEL USING "K";I
2100 MOVE .09*Max,I
2110 IF I>=0 THEN LABEL USING "A,K";"j",I
2120 IF I<0 THEN LABEL USING "A,A,K";"-","j",ABS(I)
```

```
2130 NEXT I
2140 CSIZE 2.5
2150 LORG 3.0
2160 Place=-Max
2170 Fudge=0
2180 Num=0
2190 FOR I=0 TO N
2200 IF Save(I)=0 THEN 2220
2210 Num=Num+1
2220 IF Save1(I)=0 THEN 2240
2230 Num=Num+1
2240 NEXT I
2250 INPUT "Do you like the characteristic polynomial PRINTED OUT?(Y/N)",An$
2260 IF An$="N" THEN 2950
2270 IF Plot$="CRT" THEN 2300
2280 INPUT "Which pen do you like to use to write out the polynomial",Pen
2290 PEN Pen
2300 Place=Place+INT(Num/5)*Int+Int/2
2310 FOR I=N TO 0 STEP -1
2320 MOVE -Max+.5+Fudge*Max/2.3,Place
2330 IF I=N THEN 2640
2340 IF Fudge<4 THEN 2380
2350 Place=Place-Int
2360 Fudge=0
2370 MOVE -Max+.5+Fudge*Max/2.3,Place
2380 IF I=0 THEN 2850
2390 IF (Save(I)<>0) AND (Save1(I)<>0) THEN 2560
2400 Fudge=Fudge+1
2410 IF Save(I)=0 THEN 2490
2420 ON Icase GOTO 2430,2460
2430 LABEL USING 2440;Save(I),I
2440 IMAGE "+ ",DDD.DD,"*s^",D
2450 GOTO 2840
2460 LABEL USING 2470;Save(I),I
2470 IMAGE "+ ",DDD.DD,"*z^",D
2480 GOTO 2840
2490 ON Icase GOTO 2500,2520
2500 LABEL USING 2530;Save1(I),I
2510 GOTO 2840
2520 LABEL USING 2540;Save1(I),I
2530 IMAGE "+ ",DDD.DD,"*K*s^",D
2540 IMAGE "+ ",DDD.DD,"*K*z^",D
2550 GOTO 2840
2560 ON Icase GOTO 2570,2590
2570 LABEL USING 2610;Save(I),Save1(I),I
2580 GOTO 2600
2590 LABEL USING 2620;Save(I),Save1(I),I
2600 Fudge=Fudge+2
2610 IMAGE "+ (",DDD.DD,"+",DDD.DD,"K)*s^",D
2620 IMAGE "+ (",DDD.DD,"+",DDD.DD,"K)*z^",D
2630 GOTO 2840
2640 IF (Save(I)<>0) AND (Save1(I)<>0) THEN 2770
2650 Fudge=Fudge+1
2660 IF Save(I)=0 THEN 2740
2670 ON Icase GOTO 2680,2700
2680 LABEL USING 2710;Save(I),I
2690 GOTO 2840
2700 LABEL USING 2720;Save(I),I
2710 IMAGE DDD.DD"*s^",D
2720 IMAGE DDD.DD"*z^",D
2730 GOTO 2840
2740 LABEL USING 2750;Save1(I),I
2750 IMAGE DDD.DD"*K*S^",D
2760 GOTO 2840
2770 ON Icase GOTO 2780,2800
2780 LABEL USING 2820;Save(I),Save1(I),I
2790 GOTO 2810
2800 LABEL USING 2830;Save(I),Save1(I),I
2810 Fudge=Fudge+2
2820 IMAGE "(",DDD.DD,"+",DDD.DD,"*K)*s^",D
2830 IMAGE "(",DDD.DD,"+",DDD.DD,"*K)*z^",D
```

```
2840 NEXT I
2850 IF (Save(0)<>0) AND (Savel(0)<>0) THEN 2930
2860 IF Save(0)=0 THEN 2900
2870 LABEL USING 2880;Save(0)
2880 IMAGE " +",DDD.DD
2890 GOTO 2950
2900 LABEL USING 2910;Savel(0)
2910 IMAGE " +",DDD.DD,"*K"
2920 GOTO 2950
2930 LABEL USING 2940;Save(0),Savel(0)
2940 IMAGE " +",DDD.DD,"+",DDD.DD,"K"
2950 IF Numl=0 THEN 3080
2960 IF Plot$="CRT" THEN 2990
2970 INPUT "Which pen do you like to use for the ZEROS",Pen
2980 PEN Pen
2990 FOR I=1 TO Numl
3000 MOVE Rrool(I)+.02*Max,Irool(I)
3010 FOR J=1 TO 8
3020 Y=.02*Max*(SIN(360/8*J)-SIN(360/8*(J-1)))
3030 X=.02*Max*(COS(360/8*J)-COS(360/8*(J-1)))
3040 IPLOT X,Y
3050 NEXT J
3060 NEXT I
3070 PENUP
3080 IF Plot$="CRT" THEN 3110
3090 INPUT "Which pen do you like to use for the POLES",Pen
3100 PEN Pen
3110 RETURN
3120 SUBEND
```

3. Examples

This program draws the ROOT LOCUS of a continuous- & discrete-time system from their Laplace- or Z-transform

```
System Order= 2
Coefficients: [(Rcoef(0)+Icoef(0)*K)+(Rcoef(1)+Icoef(1)*K)
               *s^1+...]
```

Rcoef	Icoef
1.0000E+00	0.0000E+00
5.0000E+00	0.0000E+00
0.0000E+00	1.0000E+00

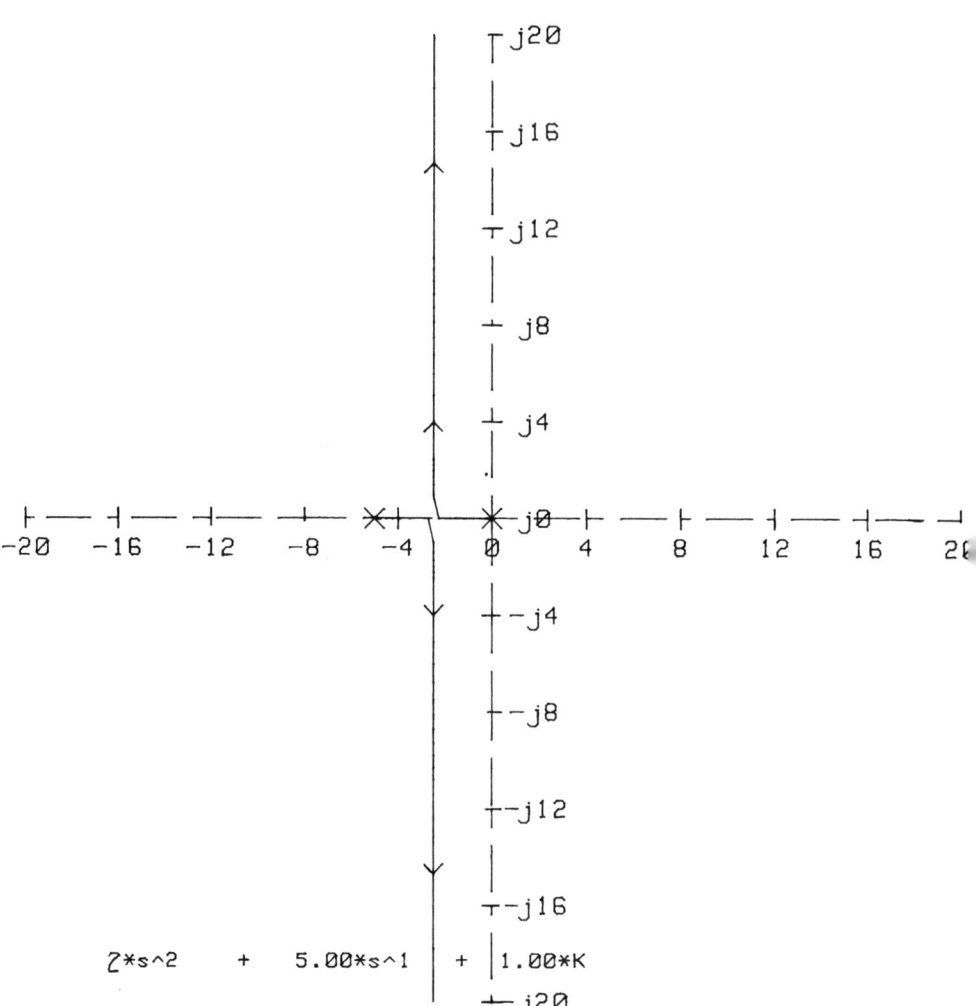

Fig. 7.6.6. Root locus for system of Example 7.6.3

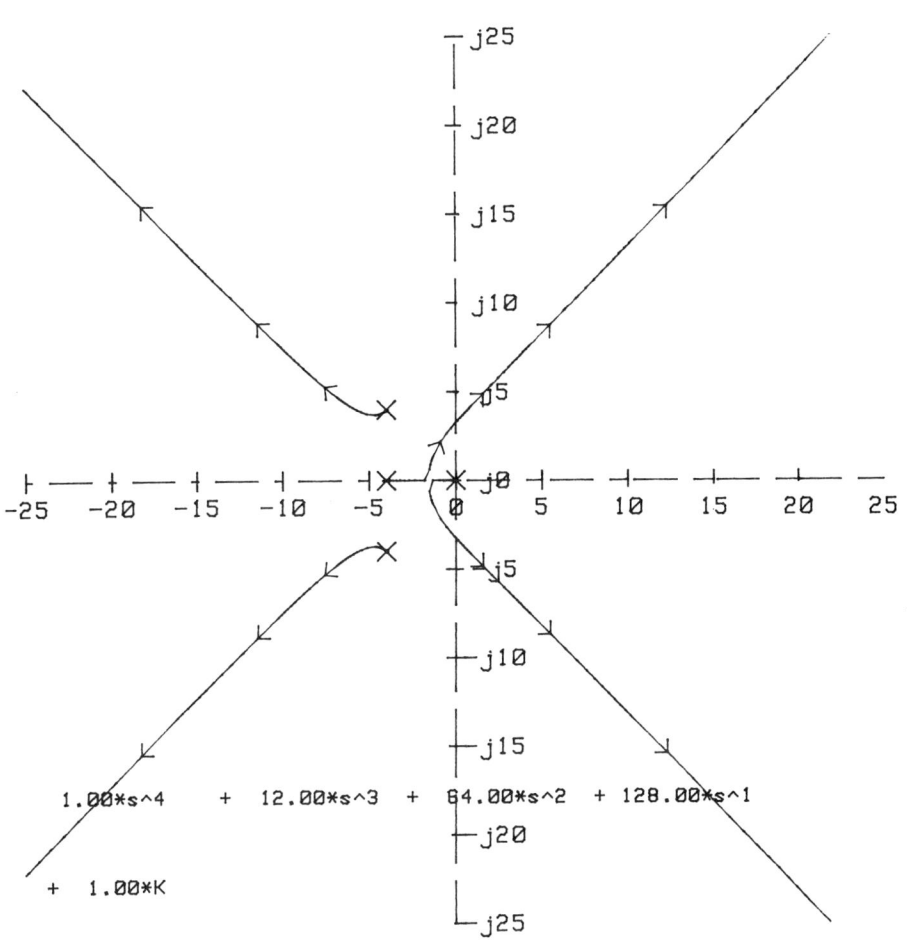

Fig. 7.6.7.

LINEAR CONTROL SYSTEMS

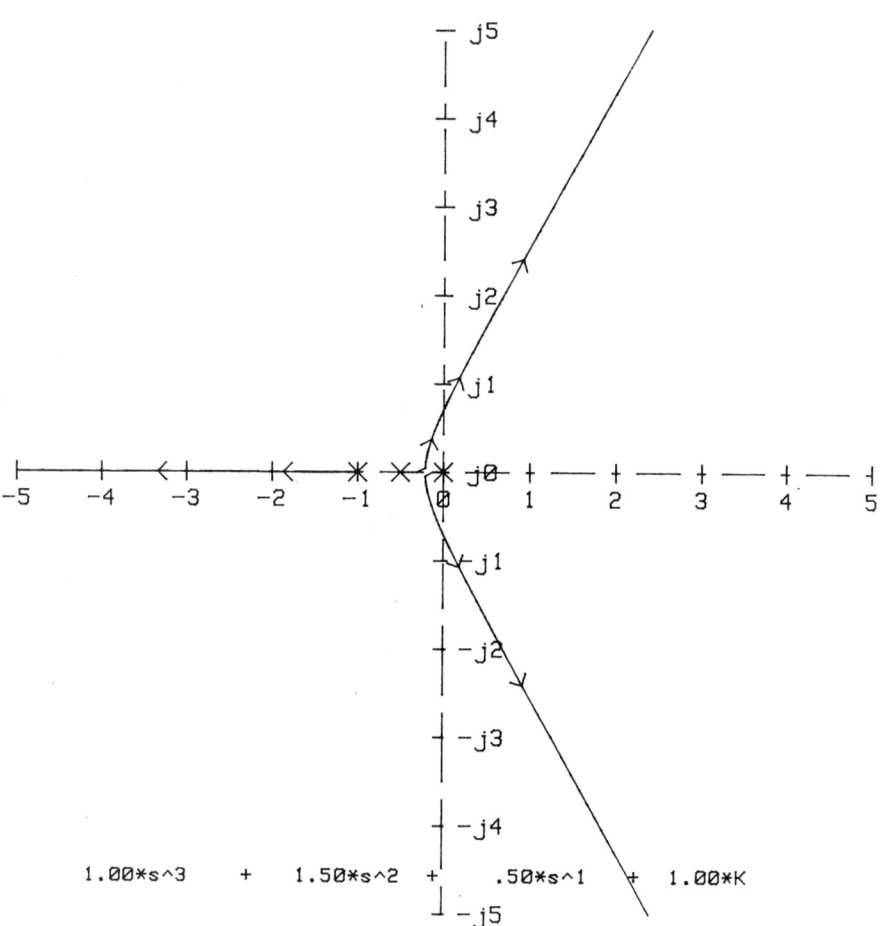

Fig. 7.6.8.

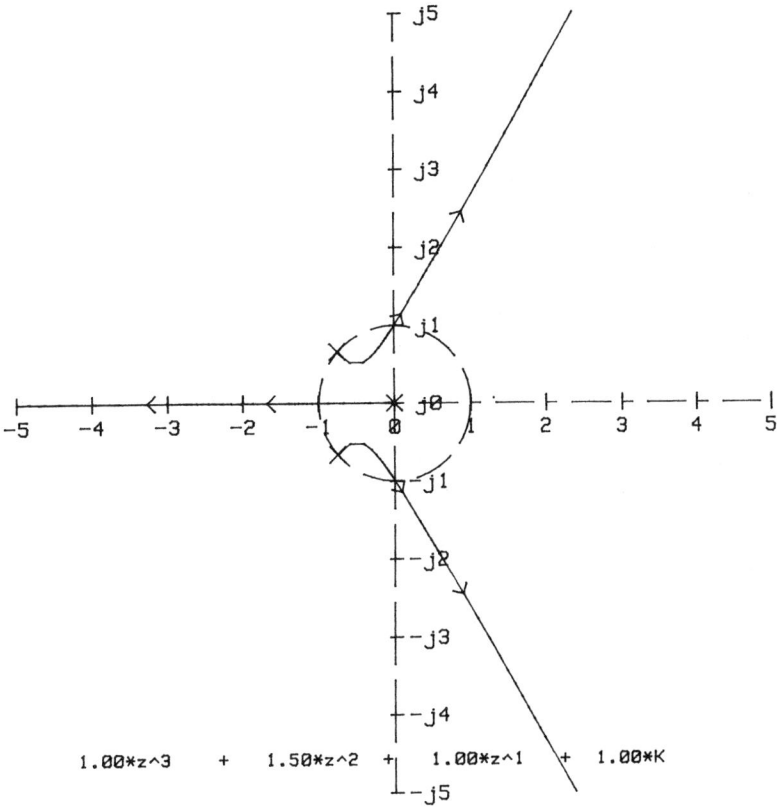

Fig. 7.6.9.

7.7 FREQUENCY-RESPONSE METHODS

In this section two other methods of checking the stability of *l.t.i.* SISO systems will be considered. These are polar (or Nyguist) diagrams [7.16] and the log-magnitude vs. phase (or Bode) diagrams [7.17]. In the first section the polar plots are first explained and then the Nyquist stability criterion is introduces. In the next section it is shown how the Bode diagrams are plotted and then by defining the system *phase margins* and *gain margins*, the stability of the system is investigated. A program, called "BNPLOT", will then be introduced which draws Bode and Nyquist plots. plots and can be considered as design specifications.

7.7.1 Nyquist Diagram

Consider the open-loop transfer function of Equation (1), represented in phasor form,

$$G(j\omega) = |G(j\omega)| \angle G(j\omega). \tag{1}$$

A plot of $\text{Im}[G(j\omega)]$ versus $\text{Re}[G(j\omega)]$ as ω increases from zero to infinity is commonly referred to as a *polar plot*. In order to illustrate how to draw a polar plot, let us consider an RC-filter network shown in Fig. 1. The transfer function of the filter is

$$G(s) = E_o(s)/E_i(s) = 1/(1+RCs) = 1/(1+\tau s) \tag{2}$$

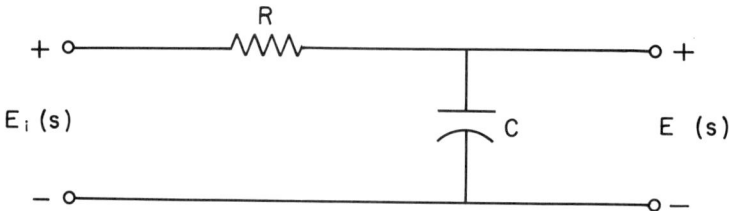

Fig. 7.7.1. An RC filter

where $\tau = RC$ is the *time constant* of the network. In phasor form, we have

$$G(j\omega) = (1+(\omega\tau)^2)^{-1/2} \angle \tau^{-1}(-\omega\tau)$$

$$= (1+(\omega\tau)^2)^{-1/2} - j(\omega\tau)(1+(\omega\tau)^2)^{-1/2} \tag{3}$$

At $\omega = 0$, it is clear that the plot starts from the point $1\angle 0° = 1 - j0$. At $\omega = 1/\tau$, the plot is at $(2)^{-1/2}\angle -45° = \sqrt{2}/2 - j\sqrt{2}/2$ and finally as $\omega \to \infty$, the plot approaches $0\angle -90° = 0 - j0$. The polar plot for the RC-filter is shown in Fig. 2. It is easy to show that the polar plot of the RC-filter is a circle with center at $(1/2, 0)$ and radius $1/2$.

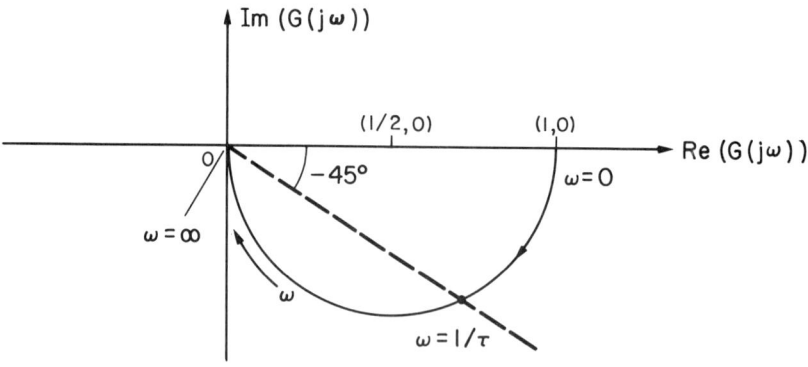

Fig. 7.7.2. Polar plot for an RC-filter

FREQUENCY-RESPONSE METHODS

1. Example. Let the open-loop transfer function of a 3rd-order system be

$$G(j\omega) = K/(j\omega(1+j\omega\tau_1)(1+j\omega\tau_2)) \qquad (4)$$

where $\tau_1 > \tau_2 > 0$. It is desired to draw the polar plot of this system. It was deduced from the *RC*-filter example that for a single open-loop pole, the polar plot approached infinity. In fact, for each zero or pole in the transfer function $G(j\omega)$, the phase angle undergoes a $+90°$ or $-90°$ shift, respectively. Hence, for the system of Equation (4), the polar plot approaches $-270°$ angle as ω approaches infinity. For low frequencies, i.e. $\omega \ll 1$, the transfer function approaches $\infty \angle -90°$. On the other hand, for high frequencies one has $0 \angle -270°$. Therefore, it becomes clear that for the polar plot to go from $\infty \angle -90°$ to $0 \angle -270°$, it must cross the negative real axis. In order to find the frequency ω_c at this intersection, one should equate $\text{Im}(G(j\omega_c))$ to zero. In this example we have

$$G(j\omega) = -K(\tau_1+\tau_2)/((1+\omega^2\tau_1^2)(1+\omega^2\tau^{2^2}))$$

$$-jK(1-\omega^2\tau_1\tau_2)/(\omega(1+\omega^2\tau_1^2)(1+\omega^2\tau_2^2)) \qquad (5)$$

which provides a frequency of $\omega_c = (\tau_1\tau_2)^{-1/2}$.

The value of the negative real-axis intersection is obtained by

$$\text{Re}(G(j\omega_c)) = -K/(1/\tau_1 + 1/\tau_2). \qquad (6)$$

Gathering the above information, the polar plot of this 3rd order system is shown in Fig. 3.

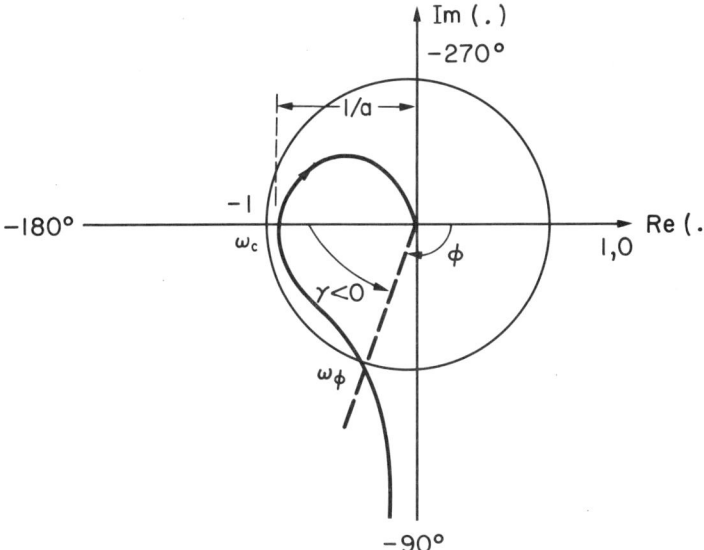

Fig. 7.7.3. Polar plot of the system $G(j\omega)=K/(j\omega(1+j\omega\tau_1).(1+j\omega\tau_2))$

The polar plot of this system can be used to illustrate the following definitions of gain and phase crossover frequencies and gain and phase margins.

2. Gain Crossover. The point at which the gain of the polar plot is unity is called the *gain crossover*. The corresponding frequency at gain crossover point is called the gain *crossover frequency* ω_ϕ (see Fig. 3).

3. Phase Margin. The angle $\gamma = 180° + \phi = 180° + \angle G(j\omega_\phi)$, where ϕ is the phase angle at the gain crossover frequency ω_ϕ, is called the *phase margin* (see angle γ in Fig. 3).

4. Phase Crossover. The point at which the polar plot crosses the $-180°$ line (negative real axis is called the *phase crossover point*. The corresponding frequency at phase crossover point is called the *phase crossover frequency*, ω_c, (see ω_c in Fig. 3).

5. Gain Margin. A factor "a" by which the gain must be changed in order to have gain and phase crossover frequencies coincide is called the *gain margin*. In polar plots (See Fig. 3) the distance on the $-180°$ line between the phase crossover point and the origin is effectively "1/a".

In the next section, the ramifications of gain and phase margins from the point of view of system stability through Bode diagrams will be discussed. In sequel, however, the stability of control systems via polar plots is best studied by the celebrated Nyquist stability criterion discussed next.

7.7.1.1 Nyquist Stability Criterion. Consider the closed-loop transfer function of a *l.t.i.* SISO system,

$$C(s)/R(s) = G(s)/(1+G(s)H(s)). \tag{7}$$

It is desired to check whether any roots of the characteristic equation $1+G(s)H(s)=0$ lie in the r.h.p. Let the characteristic equation be described by

$$A(s) = 1+G(s)H(s) = \prod_{i=1}^{n}(s-z_i)/\prod_{j=1}^{n}(s-p_j) \tag{8}$$

where z_i, i=1,...,n are roots of $A(s)=0$, i.e. the system's closed-loop poles and p_i, i=1,...,n are the open-loop poles of the system. Now, assume that a system has the pole-zero arrangements shown in Fig. 4.

FREQUENCY-RESPONSE METHODS 329

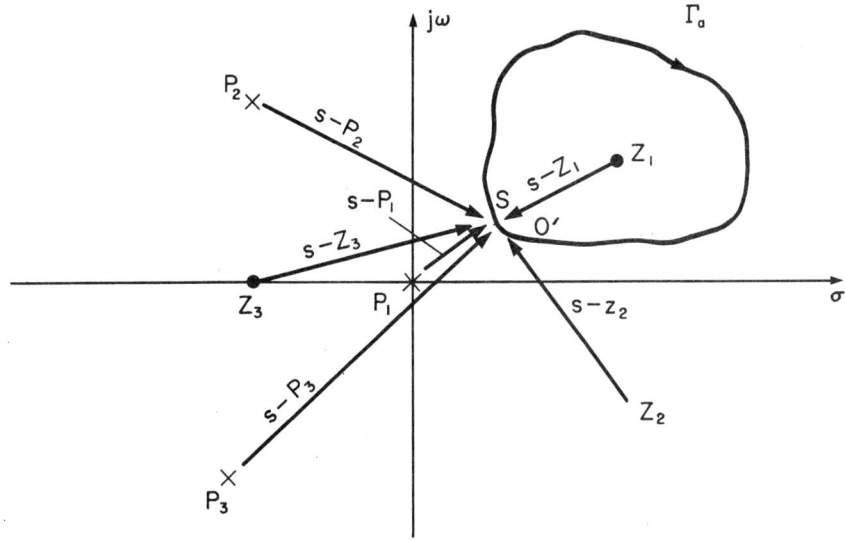

Fig. 7.7.4. Poles-zero locations for a SISO system

An arbitrary closed-loop curve Γ_a is drawn in the *r.h.p.* about zero z_1. A set of line segments are drawn from each pole and zero to a point $s = \sigma+j\omega$ on Γ_a. As point o' is moved clockwise around Γ_a, the vector $s-z_1$ rotates around point z_1 through a net angle of $-360°$. All other line segments go through a net angle of $0°$. Therefore, the $360°$ clockwise rotation (i.e. $-360°$) of the vector $s-z_1$ should be realized by the enclosure of zero z_1 of $A(s)$ by curve Γ_a. Now consider a case where a number of zeros and poles of the system exist as shown in Fig. 5.

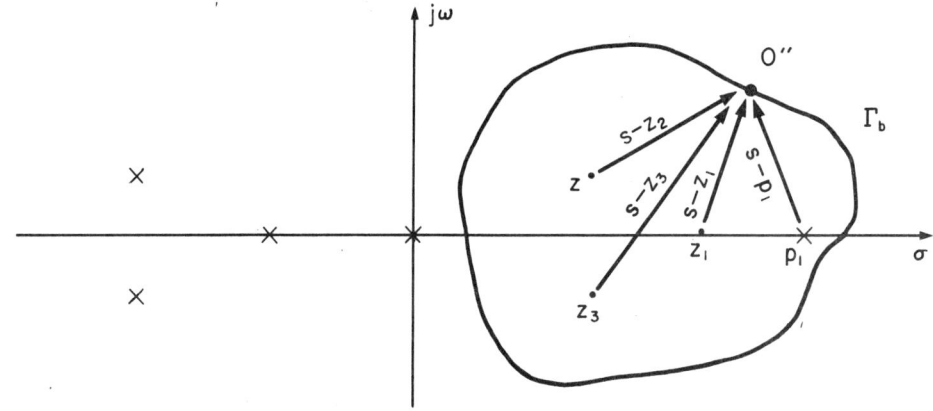

Fig. 7.7.5. Another pole-zero arrangement for a SISO system

As point O" moves once in clockwise direction around the curve Γ_b, each of the four vectors shown will undergo a net angle of $-360°$. Since the angular rotation of pole p_1 is experienced by the characteristic function in its denominator, the net angular rotation realized by $A(s)$ is equal to the net angular rotations due to pole p_1 minus the net angular rotations due to the zeros z_1, z_2 and z_3, i.e. $1+G(s)H(s)$ rotates a net angle of $1 \times 360° + (-3 \times 360°) = -720°$. Therefore, it can be stated that the net number of rotations N experienced by $1+G(s)H(s)$ by the clockwise rotation of O" on Γ_b once is

(No. poles enclosed by Γ_a) − (No. zeros enclosed by Γ_b) = 1−3 = −2

where the minus sign denotes clockwise (CW) rotation. Note that if Γ_b included only the pole p_1, $A(s)$ experiences one counterclockwise (CCW) rotation as O" is moved around Γ_b.

Now let a closed contour Γ, shown in Fig. 6, encircle all the poles and zeros of $A(s)$ with positive real part in the r.h.p. The contour Γ, shown here, is called the *Nyquist path*. At this point, based on the previous discussions, the following conclusions can be made:

(i) The total number of CW rotations of $A(s)$ due to its zeros is equal to its total number of zeros z_R in the r.h.p.

(ii) The total number of CCW rotations of $A(s)$ due to its poles is equal to its total number of poles p_R in the r.h.p.

(iii) The net number of rotations N of $A(s) = 1+G(s)H(s)$ about the origin is equal to the total number of poles p_R minus the total number of zeros z_R in the r.h.p.. N May be positive (CCW), negative (CW) or zero, i.e.

$$N = (\text{Change in phase of } 1+G(s)H(s))/360° = p_R - z_R \qquad (9)$$

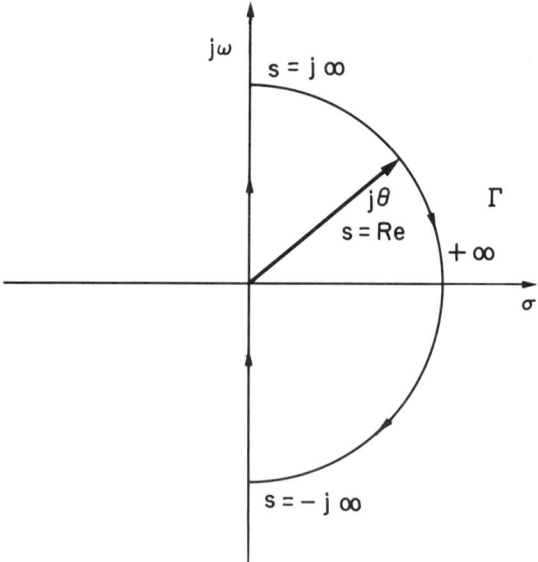

Fig. 7.7.6. A schematic of the Nyquist path

FREQUENCY-RESPONSE METHODS   331

Clearly, for a stable closed-loop system $1+G(s)H(s)$ cannot have any zeros in the r.h.p., i.e. $z_R = 0$, hence for a stable system, the net number of rotations, if any, of $A(s) = 1+G(s)H(s)$ about the origin must be $N = p_R - z_R = p_R \geqslant 0$ CCW, or equal to its r.h.p. poles p_R. In other words, if $A(s)$ goes through a net CW rotation it means that $N = p_R - z_R < 0$ and hence the system is unstable. If $N=0$, then the system may be stable if $p_R = z_R = 0$ or unstable if $p_R > 0$ and $p_R = z_R$.

It is often more convenient to discuss the rotation of $G(s)H(s)$ about the point $(-1,j0)$ instead of that of $1+G(s)H(s)$ about the origin as indicated by the two diagrams of Fig. 7. Therefore, all the discussions made regarding the number of rotations of $A(s)=1+G(s)H(s)$ about the origin can be made for the rotations of $G(s)H(s)$ about the point $(-1,j0)$. We can now state the *Nyquist Stability Criterion* as follows.

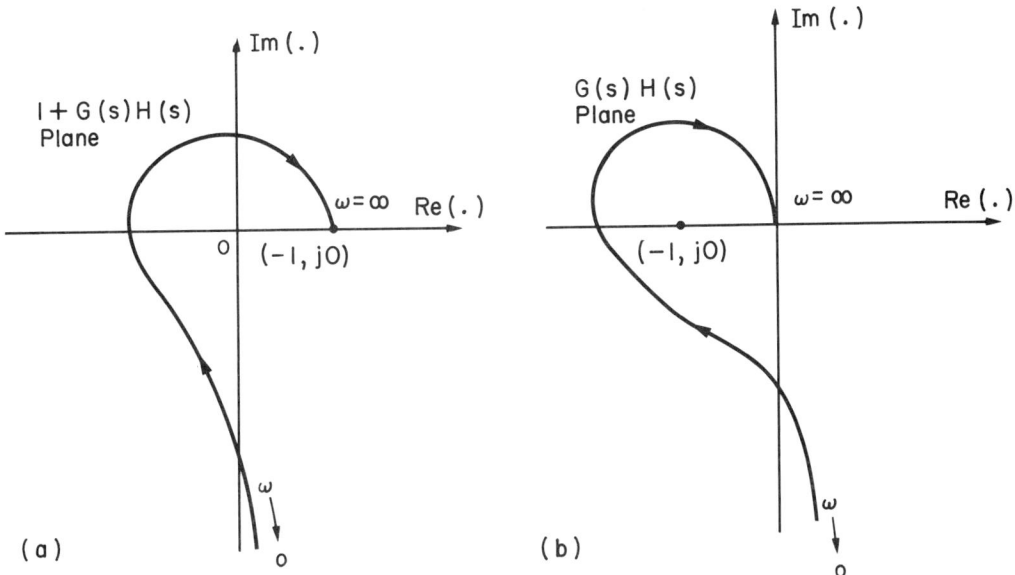

Fig. 7.7.7. The rotations of two complex functions
a) $1+G(s)H(s)$ about origin and
b) $G(s)H(s)$ about $(-1, j0)$

Consider a linear system with p_R open-loop poles in the r.h.p.. Then the closed-loop system is stable if the Nyquist diagram of $G(j\omega)H(j\omega)$ encircles the point $(-1,j0)$ exactly p_R times in the CCW direction. If either the direction or the number of encirclements is different from the criterion, then the system would be unstable.

The following example illustrates the Nyquist diagram and the stability criterion.

6. Example. Let us reconsider the system of Example 1 whose transfer function is shown in Equation (4) and its polar plot was drawn in Fig. 3. Here, the object is to determine the system's stability via the Nyquist criterion.

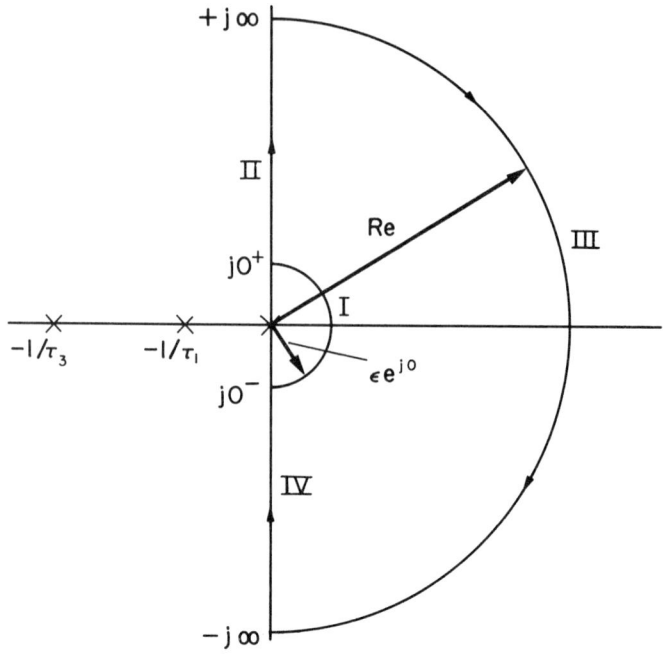

Fig. 7.7.8. Nyquist path for Example 6, $\tau_1 > \tau_2$

To begin, a Nyquist path encompassing the entire r.h.p. excluding the open-loop pole at the origin is drawn as shown in Fig. 8. The path has bypassed the open-loop pole at the origin, because we are interested in the open-loop poles in the r.h.p. only and not on the $j\omega$-axis. Moreover, the path is divided into four sections as indicated. The mapping of the s-plane to the GH-plane can thus be done in four segments. In sequel, the mapped plots for each of the four segments are given. First consider segment I. On this segment, $s = \epsilon e^{j\theta}$ with $-90° \leqslant \theta \leqslant 90°$, then

$$G(s)H(s) \cong K/(e^{j\theta}(1+\tau_1 e^{j\theta})(1+\tau_2 e^{j\theta})) = (K/\epsilon)e^{-j\theta} = (K/\epsilon)\underline{/-\theta} \qquad (10)$$

As the point s moves from $j0^-$ to $j0^+$ on segment I (see Fig. 8), the corresponding point $G(s)H(s)$ on the GH-plane moves from $+90°$ with magnitude $\lim_{\epsilon \to 0}(K/\epsilon) = \infty$ and ends at ∞ and $-90°$.

The mapping of segment II where $s = j\omega$ corresponds to the polar plot of the system as described by Example 1 and Fig. 3. and will not be repeated here. The mapping of segment IV where $s = -j\omega$ is just symmetrical to the mapping of segment II.

On the third segment, $s = Re^{j\theta}$ where R is a very large number and the angle θ moves from $+90°$ to $-90°$. Thus,

$$G(s)H(s) = K/(Re^{j\theta}(1+\tau_1 Re^{j\theta})(1+\tau_2 Re^{j\theta}))$$

$$\cong (K/R^3\tau_1\tau_2)e^{-j3\theta} = (K/R^3\tau_1\tau_2)\underline{/-3\theta} \tag{11}$$

which indicates that as s moves from $\infty + 90°$ to $\infty\angle-90°$ on segment III, the corresponding mapped point on the GH-plane moves from $0\angle-270°$ to $0\angle+270°$. When all the information about the mapping of all four segments are put together, the plot of Fig. 9 results. Now as $\epsilon \to 0$, the diagram of Fig. 9 would approach the plot shown in Fig. 10.

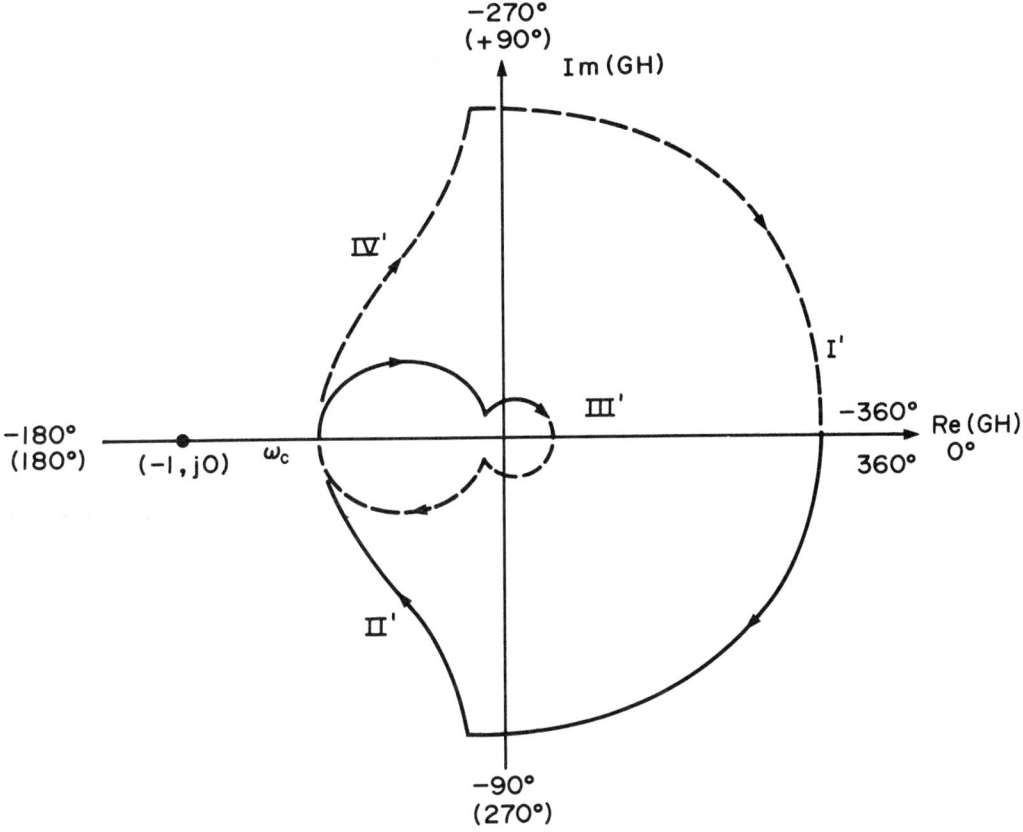

Fig. 7.7.9. A complex Nyquist diagram for system

$$GH = K/(s(1+\tau_1 s).(1+\tau_2 s))$$

334   LINEAR CONTROL SYSTEMS

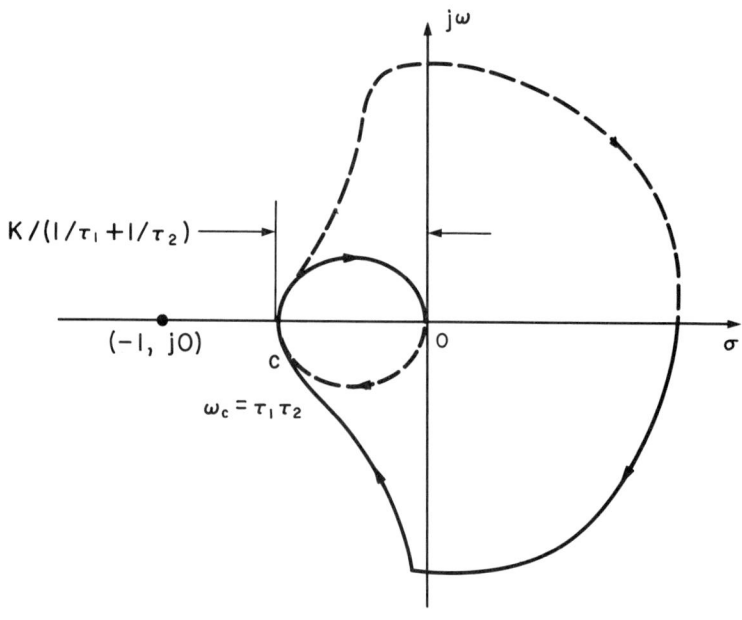

Fig. 7.7.10. Nyquist diagram of Example 7.7.6

To apply the Nyquist stability criterion, it is noted that there are no open-loop poles, i.e. $p_R = 0$. Thus for the closed-loop system to be stable, the Nyquist diagram would not encircle (-1,j0) point at all as shown in Fig. 11. Hence for the case when the negative real-axis intercept $(K/(1/\tau_1+1/\tau_2))<1$ or $K<(1/\tau_1+1/\tau_2)$, the system is stable if the dc gain K is increased to $K = 1/\tau_1+1/\tau_2$, the system would go into oscillation and the plot would pass through the point (-1,j0). For $K>(1/\tau_1+1/\tau_2)$, the plot encircles the (-1,j0) point twice in CW direction as shown in Fig. 11. Under this condition the Nyquist stability criterion predicts that the system is unstable since

$$N = p_R - z_R = 0 - z_R = -2 \tag{12}$$

or there are two closed-loop poles in the r.h.p.. A similar conclusion can be drawn using Bode plots (section 7.7.2), the Routh-Hurwitz criterion or the Root-locus (see Problem 7-32).

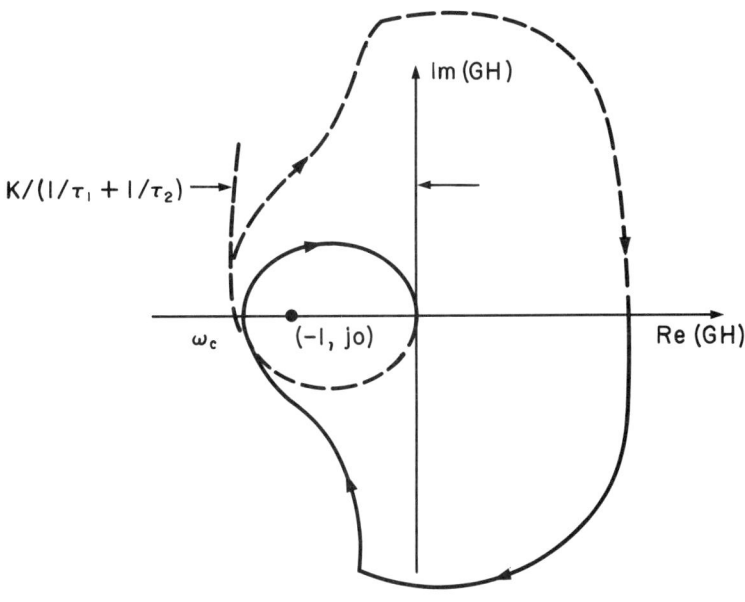

Fig. 7.7.11. A condition under which the system of
Example 7.7.6 is unstable, $K>(1/\tau_1+1/\tau_2)$

7.7.1.2 Stability Margins and Conditional Stability. In the previous section two important parameters were introduced to obtain information regarding the margins of stability- *phase and gain margins*. It is noted that the point (-1,j0) provides two conditions

$$|G(s)H(s)| = 1, \quad < G(s)H(s) = 180° \tag{13}$$

for any system. A system may be operating at a given open-loop gain K either at some "phase margin" away from the 180° or at some "gain margin" away from the value of unity magnitude, i.e. $|GH| = 1$. Below, alternative definitions and interpretations of these terms are provided. Consider a portion of a Nyquist diagram shown in Fig. 12.

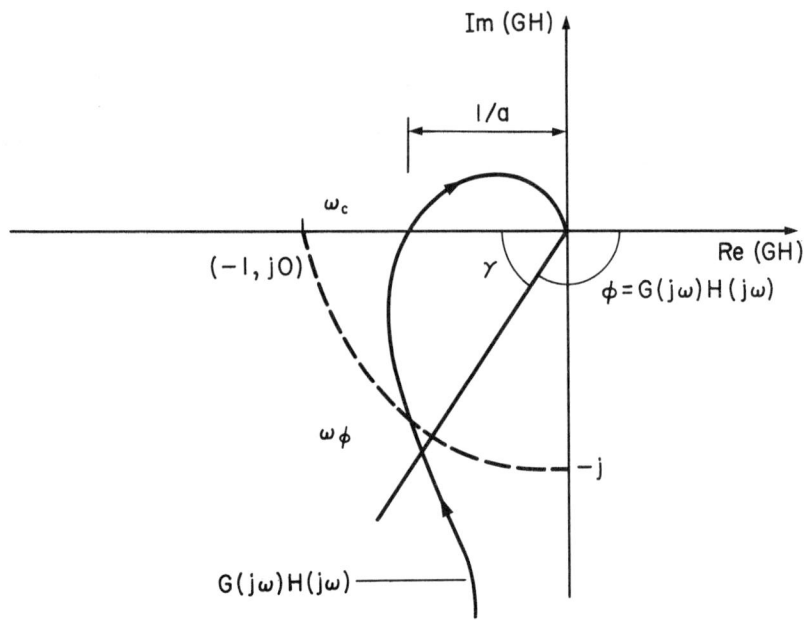

Fig. 7.7.12. **A pictorial description of phase and gain margins on a Nyquist diagram**

The angle through which the Nyquist plot would have to be rotated *CW* to pass through the (-1,j0) point is the *phase margin* γ. It is noted that γ is a direct angular measure of the (-1,j0) point avoidance and the margin of stability. Should the unit circle pass through more than one point in the 3rd quadrant, γ is measured from the closest intersection point to the 180° - line. The frequency ω_ϕ at which $|GH(j\omega)|=1$ is the *gain-crossover* frequency which was already defined by Definition 7.7.2 and in Fig. 3 for the case of Bode diagrams. Here, again the phase margin is $\gamma = 180° + \phi$.

Let 1/a denote the magnitude of a point inside the unit circle and closest to the (-1,j0) point where the polar plot crosses the negative-real axis (the $-180°$ line). Similarly, the frequency ω_c at which the plot crosses the negative-real axis is the *phase-crossover* frequency. The *gain margin* is defined as the additional factor "a" by which the point (-1,j0) can be avoided. In other words, "a" is a measure of (-1,j0) point avoidance or a ratio by which open-loop gain must increase to cause instability if all other system parameters remain constant. A good rule for stability margins is $\gamma = +30°$ and a = 3. The correlation between stability margins on root locus, Bode and Nyquist diagrams is illustrated in Problem 7-32.

The remaining concept in this section is conditional stability which is now discussed. We begin by defining what is meant by conditional stability. A system is said to be *conditionally stable* if it can become "unstable" for a "decrease" in gain. This reduction in gain is quite possible in practical cases due to, for example, aging of amplifiers or reduction of power supplies outputs. Consider the root locus of a fifth-order system with two zeros as depicted in Fig. 13.

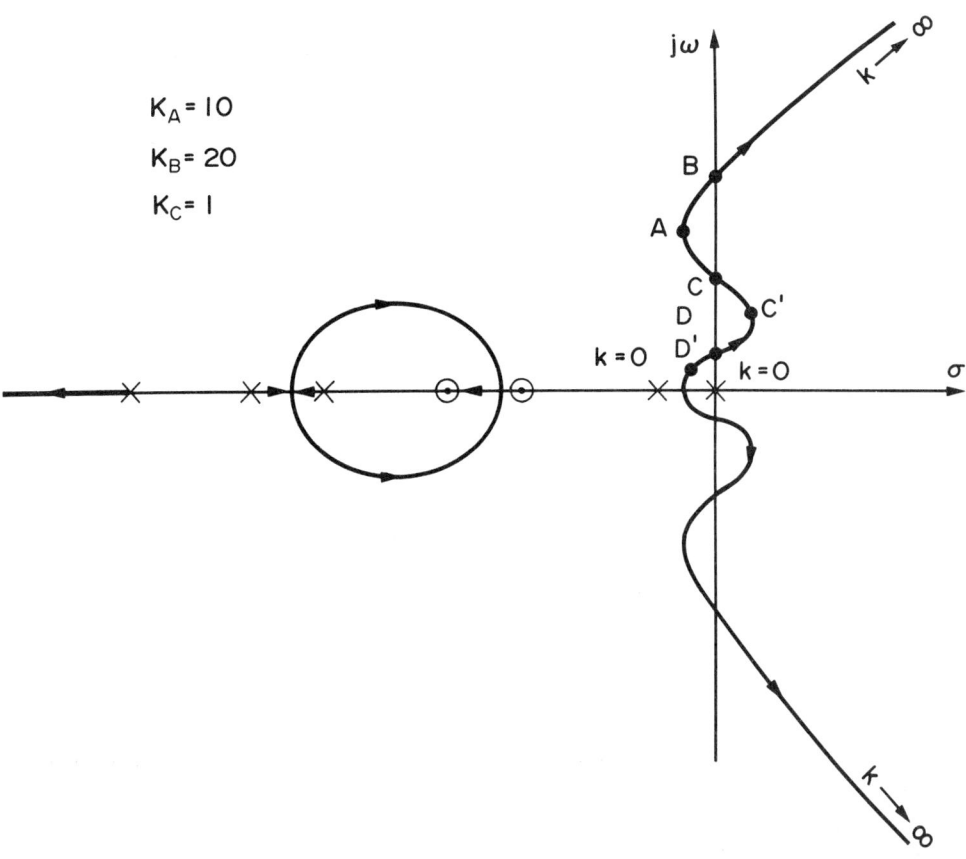

Fig. 7.7.13. Root locus of a conditionally stable system

Assume that the system is operating at point A on the root locus with a dc gain of 100. As seen, if the gain K_A is increased by a factor more than 2 or decreased by a factor more than 10, the system would be operating at points B' or C', respectively which would cause instability. Now let us consider a representative Nyquist plot of this system when it is operating at point A as is shown in Fig. 14.

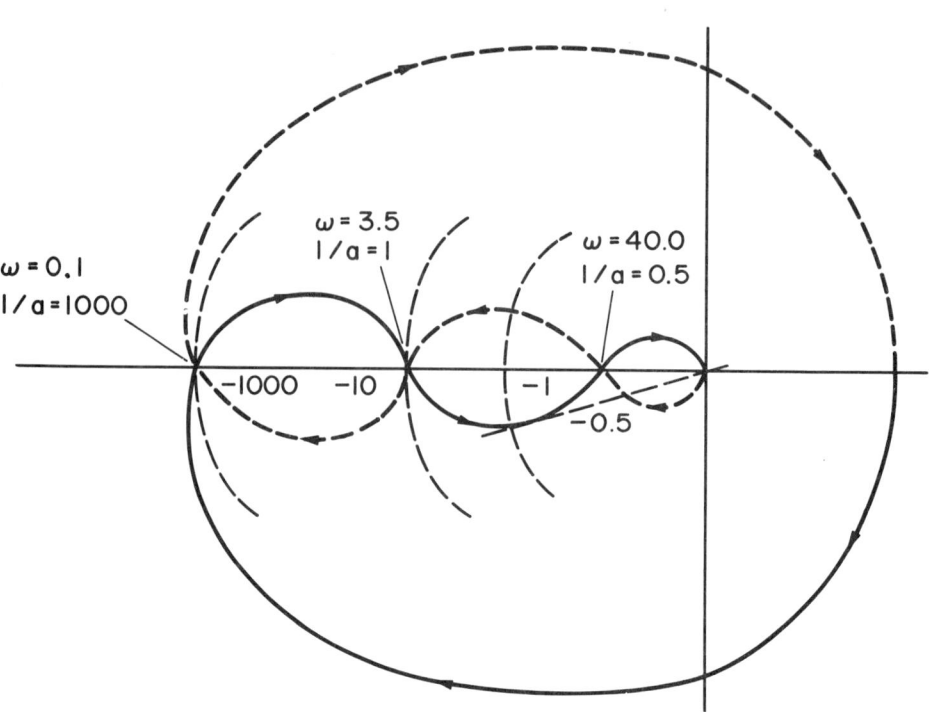

Fig. 7.7.14. Nyquist plot of a conditionally stable system (negative real axis is not to scale)

At $\omega = 40$, we have $1/a = 0.5$ or a gain margin of $a=2$. At this frequency if the gain K_A is increased from 10 to $2\times10=20$, the system oscillates (point B in Fig. 13) and from $K_A > 20$ instability results (point B'). Now at $\omega = 3.5$, $1/a=10$, the gain margin is $a=0.1$. Here if system gain K_A is decreased from 10 to $0.1\times10=1.0$, once again the system oscillates (point C in Fig. 13) and from $K_A < 1$ the system becomes unstable once again. Finally, if the gain is decreased even further to $(1/1000)\times10 = 0.01$ to correspond to $\omega = 0.1$, the system would be oscillating once more (point D in Fig. 13) and for $K_A < 0.01$, the system becomes stable again (point D'). In summary, this conditionally stable system would be stable for $0 \leqslant K < 0.01$ and $1 < K < 20$ and unstable for $K < 0$, $0.01 \leqslant K \leqslant 1$ and $K \geqslant 20$. Further comments on conditionally stable system can be found in Problem 7-33.

7.7.2 Bode Diagram

Consider the open-loop transfer function of a SISO system expressed in complex frequency domain,

$$G(s)|_{s=j\omega}=G(j\omega)=A(\omega)+jB(\omega)=|G(j\omega)|\underline{/}\,G(j\omega) \tag{14}$$

where $G(j\omega)$ is a complex variable, while $A(\omega)$ and $B(\omega)$ are real variables. Then the *frequency*

FREQUENCY-RESPONSE METHODS

response of this system is defined as the steady-state response of the system to a sinusoidal input signal. In (14), the magnitude and the phase of $G(j\omega)$ are calculated as follows:

$$|G(j\omega)| = (A^2(\omega) + B^2(\omega))^{1/2} \tag{15}$$

$$\phi(\omega) = \angle G(j\omega) = \tan^{-1}(B(\omega)/A(\omega)). \tag{16}$$

The log-magnitude and phase plots of the system are defined as the responses of $20 \log|G(j\omega)|$ and $\phi(\omega)$ versus angular frequency ω. The unit of the magnitude in logarithm base 10 is called *decibel* or *db* for short. The magnitude and phase versus frequency plots are called *Bode diagrams* in recognition of H.W. Bode who used them rather extensively in studying feedback amplifiers [7.17]. As the first example let us reconsider the *RC* filter shown in Fig. 1. The transfer function of the filter is given by (2). The logarithmic gain is

$$20 \log|G| = 20 \log(1/(1+(\omega\tau)^2)^{1/2})$$

$$= -10 \log(1+(\omega\tau)^2) \tag{17}$$

In order to plot the log-magnitude versus frequency, it is commonly done for different ranges of the frequency. For $\omega \ll 1/\tau$, (17) is approximated by

$$20 \log|G| = -10 \log(1) = 0 \; db, \; \omega \ll 1/\tau \tag{18}$$

For large frequencies, i.e. $\omega \gg 1/\tau$, the gain (17) becomes

$$20 \log|G| = -20 \log(\omega\tau), \; \omega \gg 1/\tau \tag{19}$$

while at $\omega = 1/\tau$, one has

$$20 \log|G| = -10 \log(2) = -3.01 \; db. \tag{20}$$

It is clear from (18) that the log-magnitude is constant 0 *db* for low frequencies, then it drops to -3.01 *db* at the *corner* or *break frequency* $\omega_c = 1/\tau$ and finally for high frequencies, the plot drops linearly at a slope of -20 *db* for every *decade*, i.e. a range $\omega_1 \leqslant \omega \leqslant \omega_2$ such that $\omega_2 = 10 \, \omega_1$. It is, therefore, very convenient to plot $20 \log |G|$ versus $\log \omega$. Figure 15a shows this plot.

LCS-L

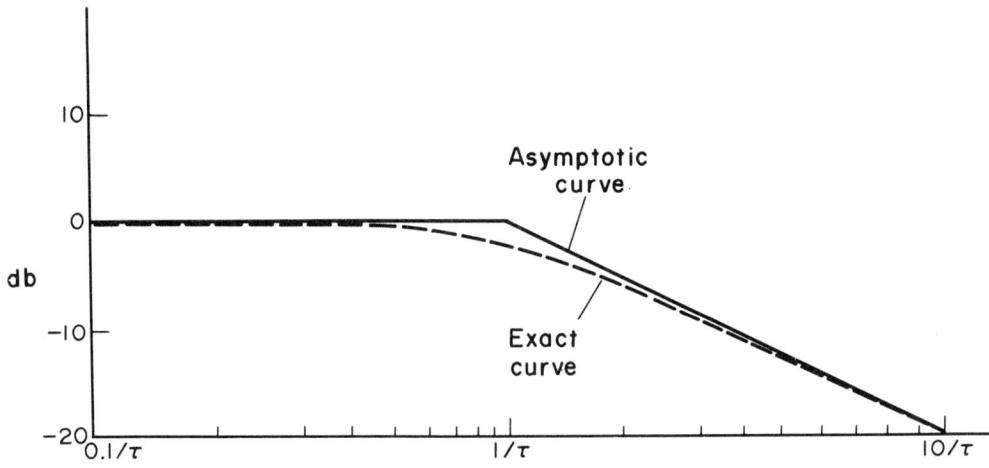

Fig. 7.7.15a. Log-magnitude plot of the Bode diagram of $(1+j\omega\tau)^{-1}$

The phase angle for the *RC* filter is given by

$$\phi(\omega) = -tan^{-1}\omega\tau \qquad (21)$$

which has the following asymptotic and exact behaviors for a wide range of frequency.

$$\phi(\omega) \to 0°, \quad \omega \ll 1/\tau$$

$$\phi(\omega) = -45°, \omega = 1/\tau$$

$$\phi(\omega) \to -90°, \omega \gg 1/\tau. \qquad (22)$$

The phase plot of the *RC* filter's Bode diagram is shown in Fig. 15b.

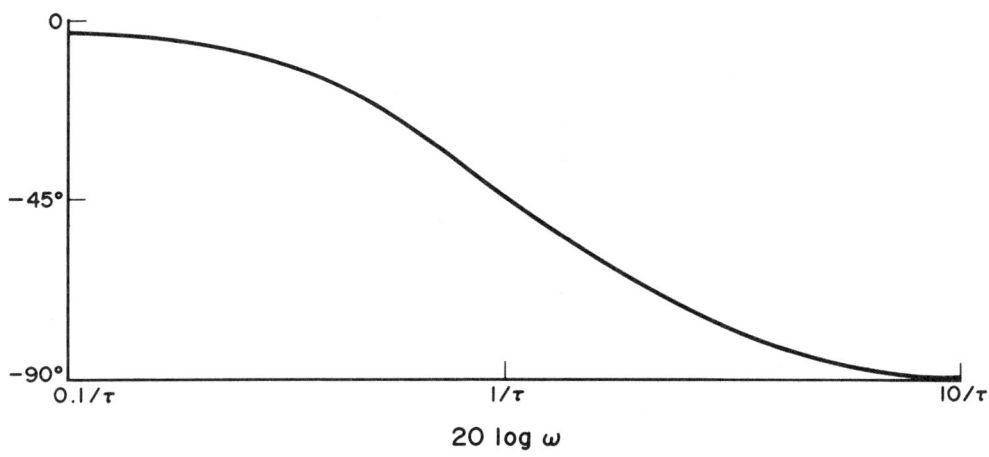

Fig. 7.7.15b. The phase angle plot of the RC filter's Bode diagram

It is a common practice to represent the transfer function in factored form in such a way that it is composed of a combination of the following terms: constant gain K, simple or multiple poles $(1+j\omega/\omega_p)^{-n}$ for $n \geq 1$, simple or multiple zeros $(1+j\omega/\omega_z)^n$, poles or zeros at the origin $(j\omega)^{\pm n}$ and quadratic terms $[1+2\xi(j\omega/\omega_c)+(j\omega/\omega_c)^2]^{\pm n}$. The log-magnitude and phase angles of each term would be plotted separately and then they are graphically added. The simple pole $(1+j\omega/\omega_1)^{-1}$ was already discussed for the RC filter example of Fig. 1. In sequel, the Bode plot of the remaining terms are discussed.

1. *Constant Gain K.* The logarithmic gain is $20 \log K$ while the angle is $0°$. The magnitude plots are horizontal lines.

2. *Poles or Zeros at the origin* $(j\omega)^{\pm 1}$. A pole or zero at the origin have a logarithmic gain as

$$20 \log |(j\omega)^{\pm 1}| = \pm 20 \log \omega \; db \tag{23}$$

which is a straight line on a semi-log paper with slope of $\pm 20 db/$ decade and a horizontal crossing at $\omega = 1$. The phase angle for this term is $\phi(\omega) = \pm 90°$.

3. *Poles or Zeros on Real Axis* $(1+j\omega/\omega_1)^{\pm 1}$. These terms have log-magnitude,

$$20 \log |(1+j\omega/\omega_1)^{\pm 1}| = \pm 10 \log(1+(\omega/\omega_1)^2) \tag{24}$$

The asymptotic behavior of this plot begins at $\pm 10 \log(\omega/\omega_1)^2 = \pm 20 \log(\omega/\omega_1)$ or a straight line with a $\pm 20db$/decade slope. The two asymptotic lines (i.e. 0 db and ± 20 db/dec.) cross each other at the point $\omega = \omega_1$ or at the corner frequency (See Fig. 15a.). However, the actual value of the logarithmic gain at $\omega = \omega_1$ is $\pm 10 \log(2) = \pm 3$ db as demonstrated for the RC filter example. The phase angle $\phi(\omega) = \pm \tan^{-1}(\omega/\omega_1)$ which begins at $0°$ for low frequencies ($\omega \ll \omega_1$), reaches $\phi(\omega_1) = \pm \tan^{-1}(1) = \pm 45°$ and approaches $\phi(\omega) = \pm \tan^{-1}(\infty) = \pm 90°$, (See Fig. 15b).

4. *Complex Conjugate Poles or Zeros.* $[1+2\xi(j\omega/\omega_2)+(j\omega/\omega_2)^2]^{\pm 1}$. Let the ratio ω/ω_2 be represented by quantity v and evaluate the logarithmic gain as,

$$\pm 20 \log|G(j\omega)| = \pm 10 \log((1-v^2)^2 + 4\xi^2 v^2) \tag{25}$$

while the phase angle is

$$\phi(\omega,\xi) = \pm \tan^{-1}(2\xi v/(1-v^2)). \tag{26}$$

Once again the asymptotic behaviors of the above plots will be investigated. When $v \ll 1$, the magnitude is,

$$\pm 10 \log(1) = 0 \ db \tag{27}$$

and the phase angle approaches $0°$. On the other end of the frequency scale, i.e. for $v \gg 1$, the magnitude is

$$\pm 10 \log(v^4) = \pm 40 \log(v) \tag{28}$$

which results in a straight line with a slope of ± 40 db/decade. The phase angle $\phi(\omega)$ approaches $0°$ asymptotically as $v \ll 1$ and reaches $\pm 180°$ as $v \gg 1$. The difference between the asymptotic and actual curves depend on the damping ratio and should be accounted for $\xi < 0.707$. Figure 16 shows the log-magnitude and phase angle for a quadratic term consisting of a pair of complex conjugate poles.

FREQUENCY-RESPONSE METHODS

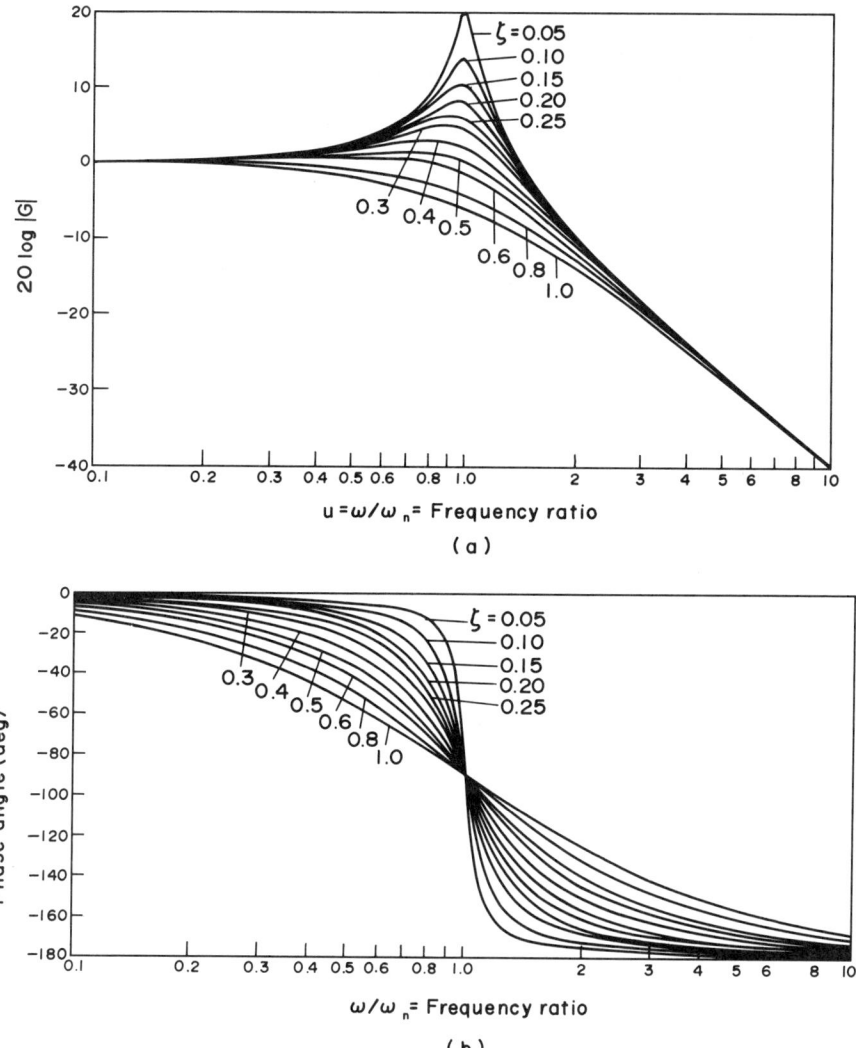

Fig. 7.7.16. Bode diagram for a pair of complex conjugate poles
(a) Magnitude, (b) Phase

The frequency ω_r at which the maximum magnitude occurs is called the *resonant frequency*. Note that as the damping ratio ξ approaches zero, the resonant frequency ω_r approaches the corner frequency ω_2. The resonant frequency is obtained by taking the derivative of the magnitude of (25) with respect to v and setting it equal to zero. The resulting equation is

$$v^2 - 1 + 2\xi^2 = 0 \tag{29}$$

or

$$\omega_r = \sqrt{1-2\xi^2)}, \xi < .0707. \quad (30)$$

The maximum value of the magnitude itself is

$$M_r = |G(\omega_r)| = (2\xi\sqrt{1-\xi^2})^{-1}, \xi < 0.707. \quad (31)$$

Table 1 summarizes the Bode plots for the four factor terms discussed above.

Table 7.7.1 Summary of the Bode results for a four-term transfer function

Term #	Factor $G(j\omega)$	ω_c	Magnitude $20 \log |G(j\omega)|$	Phase Angle $\phi(\omega)$
1	K	-	$20 \log K$	$0°$
2	$(j\omega)^{\pm n}$	-	0 db at $\omega=1$ $\pm 20n$ db/decade	$\pm 90n°$
3	$(1+j\omega/\omega_1)^{\pm n}$	ω_1	0 db $\omega \ll \omega_1$ $\pm 20n$ db/decade, $\omega \gg \omega_1$ $\pm$ 3n db correction at ω_1	$0°, \omega \ll \omega_1$ $\pm 45n°, \omega=\omega_1$ $\pm 90n°, \omega \gg \omega_1$
4	$[1+ 2\xi(j\omega/\omega_2) + (j\omega/\omega_2)^2]^{\pm n}$ $0<\xi<0.707$	ω_2	$\omega_r=\omega_2\sqrt{1-2\xi^2}$ 0 db $\omega \ll \omega_2$ ± 40 n db/dec., $\omega \gg \omega_2$ $M_r=(2\xi\sqrt{1-\xi^2})^{-1}$	$0°, \omega \ll \omega_2$ $\pm 90n°, \omega=\omega_2$ $\pm 180n°, \omega \gg \omega_2$

7. Example. Consider the following third-order system

$$G(s)H(s) = 1000(s+3)/(s(s+12)(s+50)) \quad (32)$$

In order to draw the Bode plots, (32) is rewritten in the frequency-domain by setting $s=j\omega$ as follows:

$$G(j\omega)H(j\omega) = 5(1+j\omega/3)/(j\omega(1+j\omega/12)(1+j\omega/50)) \quad (33)$$

In order to plot this transfer function, all the 5 terms in (33) are summarized in Table 2. Figure 17 shows a plot of the Bode diagrams of this example.

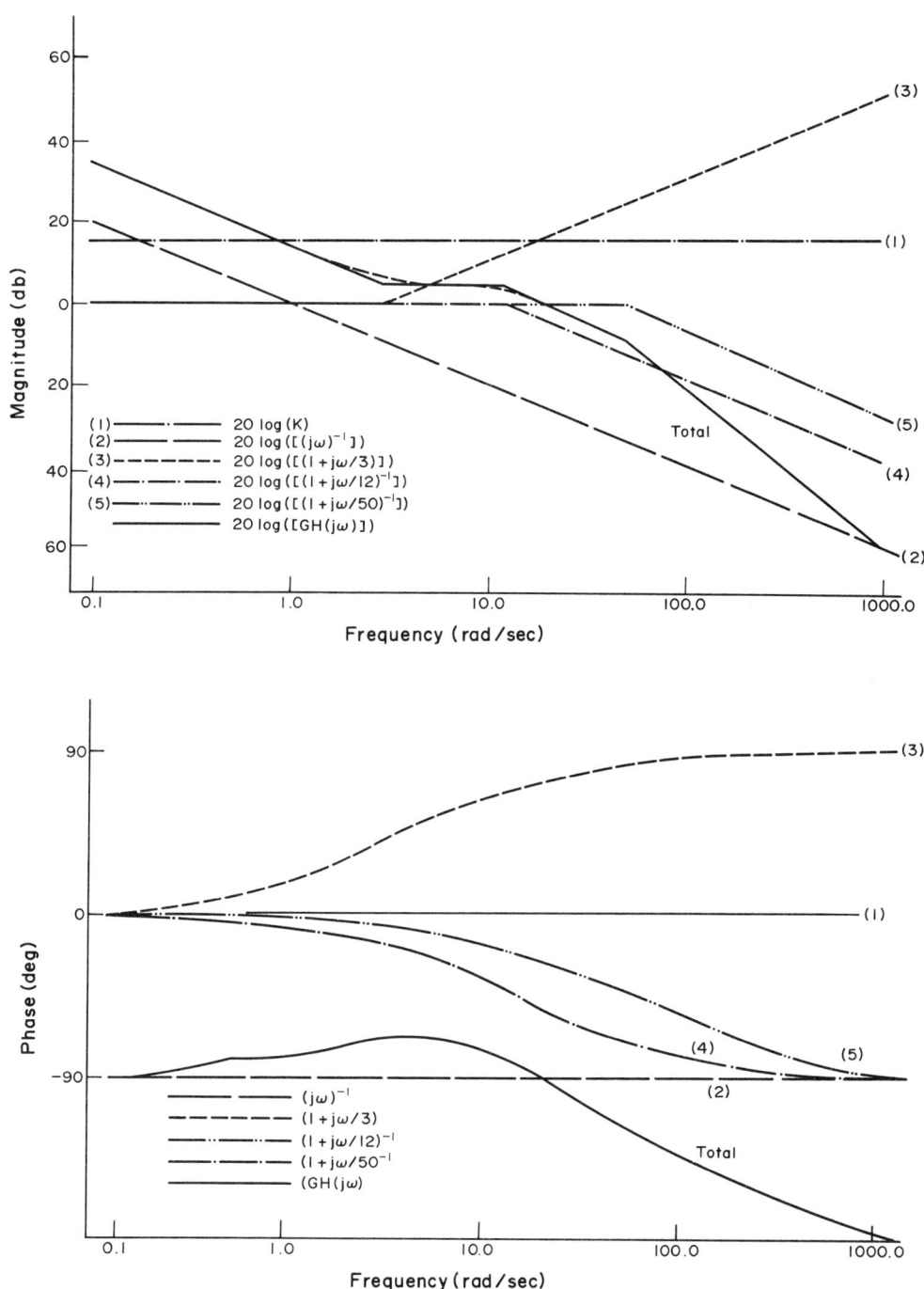

Fig. 7.7.17. Bode diagram for Example 7.7.7

Table 7.7.2 Summary of the results of Example 7.7.7

Term #	Factor $G(j\omega)$	ω_c	Logarithmic Magnitude $20 \log \|G(j\omega)\|$	Phase Angle $\phi(\omega)$
1	5	-	$20 \log 5 = 15$	$0°$
2	$(j\omega)^{-1}$	-	0db at $\omega = 1$ -20 db/decade	$-90°$
3	$1 + j\omega/3$	3	0db $\omega \ll 3$ $+20$ db/decade $\omega \gg 3$ $+3$ db correction $\omega = 3$	$0°\ \omega \ll 3$ $+45°\ \omega = 3$ $+90°\ \omega \gg 3$
4	$(1 + j\omega/12)^{-1}$	12	0db $\omega \ll 12$ -20 db/decade $\omega \gg 12$ -3db correction $\omega = 12$	$0°\ \omega \ll 12$ $-45°\ \omega = 12$ $-90°\ \omega \gg 12$
5	$(1 + j\omega/50)^{-1}$	50	0db $\omega \ll 50$ -20 db/decade $\omega \gg 50$ -3db correction $\omega = 50$	$0°\ \omega \ll 50$ $-45°\ \omega = 50$ $-90°\ \omega \gg 50$

Further Bode plots are considered in the Problems at the end of this chapter. In sequel, the gain and phase margins are redefined in terms of Bode diagrams and provide their roles in stability interpretation.

7.7.3 Phase and Gain Margins

The use of Bode diagrams in establishing the stability depend on *Gain Crossover, Phase Crossover, Phase Margin* and *Gain Margin*.

8. Gain Crossover. The point at which the gain plot crosses the unity (0db) line is called the gain crossover. The corresponding frequency at gain crossover is termed the *gain crossover frequency*, ω_ϕ (See Fig. 18).

9. Phase Margin. The angle $\gamma = 180° + \phi$ where ϕ is the phase angle at the gain crossover frequency ω_ϕ is called the *phase margin* (see Fig. 18).

10. Phase Crossover. The point at which the phase angle plot crosses the $-180°$ line is called the phase crossover. The corresponding frequency at phase crossover is termed the *phase crossover frequency* ω_c (see Fig. 18).

FREQUENCY-RESPONSE METHODS

11. Gain Margin. A factor "a" by which the gain must be changed in order to have gain and phase crossover frequencies coincide is called the *gain margin*. In terms of the transfer function at the frequency ω_c, we have

$$|G(j\omega_c)|a = 1 \tag{34}$$

which indicates that the gain margin is

$$20 \log a = -20 \log |G(j\omega_c)|. \tag{35}$$

Figure 18 shows an illustration of these definitions. It is noted that the system is stable when its phase and gain margins are positive. On the other hand, if phase and gain margins are negative the system is unstable (see Fig. 18b).

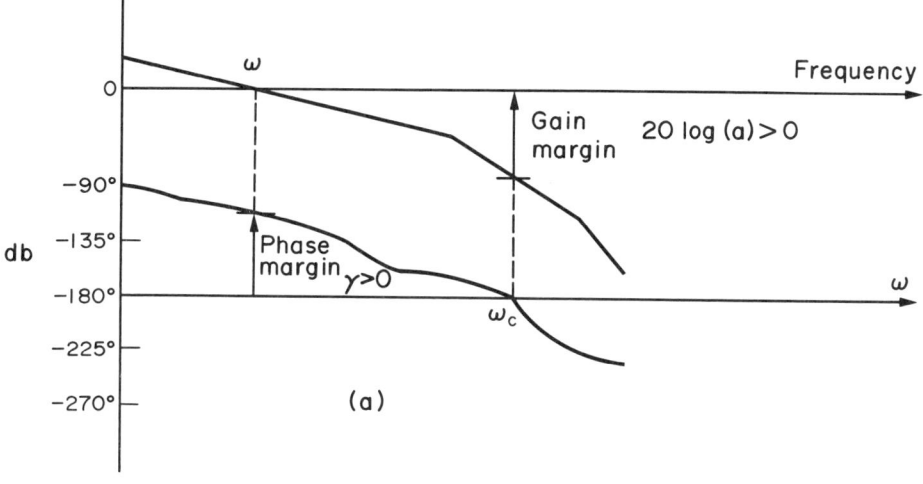

Fig. 7.7.18. Gain and Phase plots of $G(j\omega)$ showing phase and gain margins for
(a) a stable system

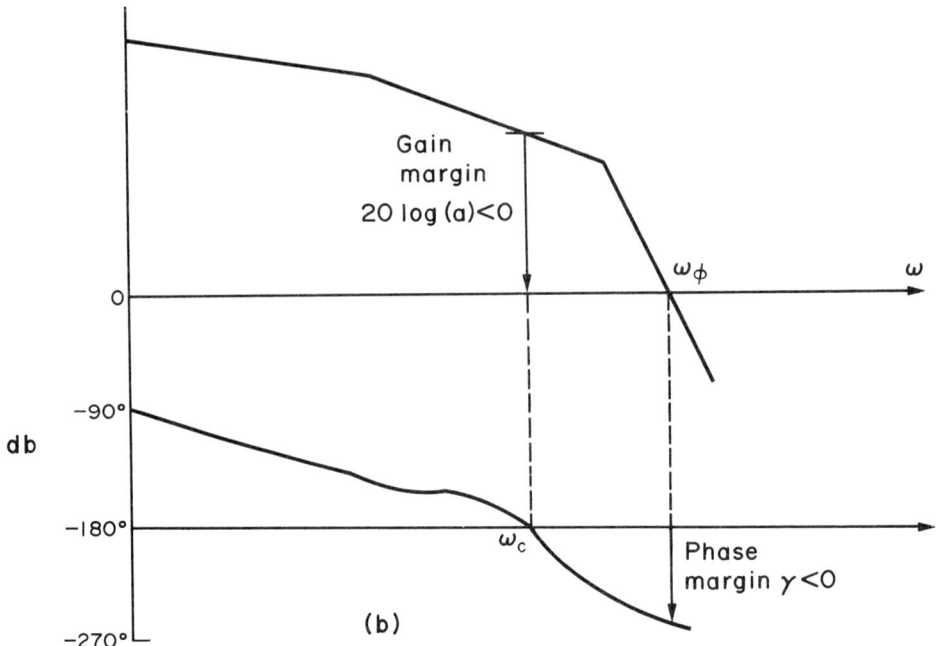

Fig. 7.7.18. Gain and Phase plots of $G(j\omega)$
showing phase and gain margins for
(b) an unstable system

A phase margin of $+45°$ to $+60°$ is commonly considered as a safe margin of stability. These margins, as will be seen in the next section and Chapter 9, can be similarly interpreted on the polar plot.

Program "BNPLOT", described in the following pages plots the Bode and Nyquist diagrams of any linear SISO continuous-time or discrete-time system. The program provides its own semi-log paper for the magnitude and phase plots of the Bode diagram and the polar coordinates for the Nyquist plot (the $j\omega$-axis portion of the mapping). The source listing of "BNPLOT" is followed by two examples:

a) $GH = 1000\,(s+3)/(s(s+12)\,(s+50))$ which corresponds to Example 7.7.7 (see also Fig. 7.7.17)

b) $GH = 1/((s+1)\,(s+2)\,(s+5))$ which corresponds to a special case of Example 7.7.1 (see also Fig. 7.7.3).

FREQUENCY-RESPONSE METHODS

12. "BNPLOT"

```
10    !     PROGRAM NAME : "BNPLOT" PROG
20    !        THIS PROGRAM DRAWS THE BODE & NYQUIST PLOTS FOR
30    !        A LINEAR CONTINUOUS- & DISCRETE-TIME SYSTEM FROM
40    !        THEIR Laplace- OR Z-transforms
50    DIM Gonogo$(3),Type$(7),Zp$(1),Pr(10),Pi(10),Zr(10),Zi(10)
60    DIM Ar(10),Ai(10),Xtitle$(30),Ytitle$(30)
70    PRINT "This program constructs BODE & NYQUIST plots for CONT. & DIS.-TIME"
80    PRINT "transfer functions of up to 10 poles & 10 zeros. Functoons must  "
90    PRINT "be of the form   K(s+Z1)(s+Z2).../(s+P1)(s+P2)  "
100   PRINT "  OR    K(z+Z1)(z+Z2).../(z+P1)(z+P2) , where K is the GAIN,"
110   PRINT " Z's are the zeros & -P's are the poles.",LIN(1)
120   PRINT " All poles & zeros should be in first-order form. Those which"
130   PRINT "are not should be factored until they are. Multiple poles or zeros"
140   PRINT "should be entered as if they were separate.",LIN(1)
150   PRINT "You will be asked to enter information after which you should hit"
160   PRINT "the CONT key to continue."
170   Iplot=0
180   INPUT "Do you wish to start a new problem (Y/N)?",Gonogo$
190   IF (Gonogo$="Y") OR (Gonogo$="y") THEN 220
200   IF (Gonogo$="N") OR (Gonogo$="n") THEN STOP
210   GOTO 170
220   INPUT "What kind of plots do you wish(BODE or NYQUIST)?",Type$
230   IF (Type$<>"BODE") AND (Type$<>"NYQUIST") THEN 220
240   PRINT PAGE,"You have asked for ";Type$;" plots"
250   INPUT "Is this correct (Y/N)?",Gonogo$
260   IF (Gonogo$="Y") OR (Gonogo$="y") THEN 290
270   IF (Gonogo$="N") OR (Gonogo$="n") THEN 220
280   GOTO 250
290   INPUT "How many ZEROS are in this function(<=10)?",Nzeros
300   INPUT "How many POLES are in this function(<=10)?",Npoles
310   PRINT LIN(1),"Your function has ";Nzeros;" zero(s) & ";Npoles;" pole(s)"
320   INPUT "Is this correct (Y/N)?",Gonogo$
330   IF (Gonogo$="Y") OR (Gonogo$="y") THEN 360
340   IF (Gonogo$="N") OR (Gonogo$="n") THEN 290
350   GOTO 320
360   INPUT "Type the GAIN of the system (K>=1)",K
370   IF K<=0 THEN 360
380   PRINT LIN(2),"The GAIN of the system is ";K
390   INPUT "Is this correct (Y/N)?",Gonogo$
400   IF (Gonogo$="Y") OR (Gonogo$="y") THEN 430
410   IF (Gonogo$="N") OR (Gonogo$="n") THEN 360
420   GOTO 390
430   IF Nzeros=0 THEN 540
440   PRINT LIN(2),"Enter the -ZEROS in complex form. If either the real or"
450   PRINT "imaginary part is 0.0, enter as such. Reminder : The -ZERO in a"
460   PRINT "term (s+3-j2) or (z+3-j2) is 3-j2.(Note:-ZERO=0=0+j0)",LIN(1)
470   Zp$="-Z"
480   Nn=Nzeros
490   GOSUB Zpread
500   FOR J=1 TO Nzeros
510   Zr(J)=Ar(J)
520   Zi(J)=Ai(J)
530   NEXT J
540   PRINT LIN(2),"Enter the -POLES in complex form. If either the real or"
550   PRINT "imaginary part is 0.0, enter as such. Reminder : The pole in a"
560   PRINT "term such as (s+3-j2) or (z+3-j2) is 3-j2.(Note:-POLE=0=0+j0)"
570   Zp$="-P"
580   Nn=Npoles
590   GOSUB Zpread
600   FOR I=1 TO Npoles
610   Pr(I)=Ar(I)
620   Pi(I)=Ai(I)
630   NEXT I
640   PRINT LIN(2),"Transfer function summary for ";Type$;" plots"
650   PRINT LIN(1),"The GAIN is ";K,LIN(1),"The -ZEROS are",LIN(1)
660   IF Nzeros=0 THEN 700
670   FOR I=1 TO Nzeros
680   PRINT "-Z";I;"= ";Zr(I);" +j";Zi(I)
```

```
690    NEXT I
700    PRINT LIN(1),"The -POLES are",LIN(2)
710    FOR I=1 TO Npoles
720    PRINT "-P";I;" = ";Pr(I);" +j";Pi(I)
730    NEXT I
740    PRINT LIN(2),"Hit the CONT key to continue"
750    PAUSE
760    IF Type$="NYQUIST" THEN 2080
770    PRINT LIN(2),"Specify a FREQUENCY RANGE you wish to plot over"
780    PRINT "Example: 0.1<=W<=1000."
790    INPUT "Input MINIMUM frequency (Must be a power of 10.)",Wmin
800    Value=FRACT(LGT(Wmin))
810    IF ABS(Value)>.000001 THEN 790
820    INPUT "Input MAXIMUM frequency (Must be a power of 10.)",Wmax
830    Value=FRACT(LGT(Wmax))
840    IF ABS(Value)>.0000001 THEN 820
850    IF Wmin>=Wmax THEN 870
860    GOTO 890
870    PRINT LIN(1),"Your MINIMUM value >= Your MAXIMUM...Try again"
880    GOTO 790
890    PRINT LIN(2),"The FREQUENCY RANGE of your plot is ";Wmin;" <=W<= ";Wmax
900    INPUT "Is this correct (Y/N)?",Gonogo$
910    IF (Gonogo$="Y") OR (Gonogo$="y") THEN 940
920    IF (Gonogo$="N") OR (Gonogo$="n") THEN 790
930    GOTO 900
940    PRINT LIN(1),"For the MAGNITUDE BODE PLOT,Specify a DECIBEL RANGE you"
950    PRINT "wish to plot over. Example: -40<=DB<=60..Must be a multiple of 20."
960    INPUT "Input the MINIMUM DB (Multiple of 20.)",Dbmin
970    Value=FRACT(Dbmin/20)
980    IF ABS(Value)>.00001 THEN 960
990    INPUT "Input the MAXIMUM DB (Multiple of 20.)",Dbmax
1000   Value=FRACT(Dbmax/20)
1010   IF ABS(Value)>.00001 THEN 990
1020   IF Dbmin>=Dbmax THEN 1020
1030   GOTO 1060
1040   PRINT LIN(1),"Your MINIMUM value >= Your MAXIMUM...Try again"
1050   GOTO 960
1060   PRINT LIN(2),"The DECIBEL RANGE of your PLOT is ";Dbmin;" <=DB<= ";Dbmax
1070   INPUT "Is this correct (Y/N)?",Gonogo$
1080   IF (Gonogo$="Y") OR (Gonogo$="y") THEN 1110
1090   IF (Gonogo$="N") OR (Gonogo$="n") THEN 960
1100   GOTO 1070
1110   PRINT LIN(1),"Always hit the CONT key to continue"
1120   PAUSE
1130   Log_interval=LGT(Wmax)-LGT(Wmin)
1140   Y_interval=(Dbmax-Dbmin)/20
1150   Ymin=Dbmin
1160   Sf=20
1170   Xtitle$="Frequency, rad/sec"
1180   Ytitle$="Magnitude, dbs"
1190   GOSUB Logplt
1200   Istart=1
1210   FOR Decade=0 TO Log_interval
1220     FOR Interval=1 TO 9
1230     Omega=Wmin*10.0^(Decade+LGT(Interval))
1240     Sumzero=0
1250     IF Nzeros=0 THEN 1290
1260     FOR I=1 TO Nzeros
1270       Sumzero=Sumzero+20*LGT(SQR(Zr(I)^2+(Omega+Zi(I))^2))
1280     NEXT I
1290     Sumzero=Sumzero+20*LGT(K)
1300     Sumpole=0
1310     FOR I=1 TO Npoles
1320       Sumpole=Sumpole+20*LGT(SQR(Pr(I)^2+(Omega+Pi(I))^2))
1330     NEXT I
1340     Db=Sumzero-Sumpole
1350     Value=(Db-Dbmin)/20
1360     GOSUB Plot_test
1370     NEXT Interval
1380   NEXT Decade
1390   PAUSE
```

FREQUENCY-RESPONSE METHODS

```
1400    EXIT GRAPHICS
1410    PRINT LIN(E(2))
1420    PRINT "For the PHASE BODE PLOT, specify a PHASE ANGLE RANGE you"
1430    PRINT "wish to plot over. Ex :-90<=PHASE<=135..Must be a multiple of 45."
1440    INPUT "Input MINIMUM PHASE ANGLE in degrees (Multiple of 45).",Phasemin
1450    Value=FRACT(Phasemin/45)
1460    IF ABS(Value)>.00001 THEN 1440
1470    INPUT "Input MAXIMUM PHASE ANGLE (Multiple of 45).",Phasemax
1480    Value=FRACT(Phasemax/45)
1490    IF ABS(Value)>.00001 THEN 1470
1500    IF Phasemin>=Phasemax THEN 1520
1510    GOTO 1540
1520    PRINT LIN(1),"Your MINIMUM value >= Your MAXIMUM...Try again"
1530    GOTO 1440
1540    PRINT LIN(1),"PHASE ANGLE RANGE of plot:";Phasemin;" <=PHASE<= ";Phasemax
1550    INPUT "Is this correct (Y/N)?",Gonogo$
1560    IF (Gonogo$="Y") OR (Gonogo$="y") THEN 1590
1570    IF (Gonogo$="N") OR (Gonogo$="n") THEN 1440
1580    GOTO 1550
1590    Y_interval=(Phasemax-Phasemin)/45
1600    Ymin=Phasemin
1610    Sf=45
1620    Xtitle$="Frequency, rad/sec"
1630    Ytitle$="Phase Angle, deg"
1640    GOSUB Logplt
1650    FOR Decade=0 TO Log_interval
1660       FOR Interval=1 TO 9
1670       Omega=wmin*10.0^(Decade+LGT(Interval))
1680       Sumzero=0
1690       Nn=Nzeros
1700       IF Nzeros=0 THEN 1770
1710       FOR I=1 TO Nn
1720       Ar(I)=Zr(I)
1730       Ai(I)=Zi(I)
1740       NEXT I
1750       GOSUB Angle_test
1760       Sumzero=Sum
1770       Nn=Npoles
1780       FOR I=1 TO Nn
1790       Ar(I)=Pr(I)
1800       Ai(I)=Pi(I)
1810       NEXT I
1820       GOSUB Angle_test
1830       Sumpole=Sum
1840       Phase=Sumzero-Sumpole
1850       Value=(Phase-Phasemin)/45
1860       Itime=0
1870       GOSUB Plot_test
1880       NEXT Interval
1890    NEXT Decade
1900    PAUSE
1910    EXIT GRAPHICS
1920    IF Iplot=1 THEN 2010
1930    INPUT "Do you like to plot on the X-Y PLOTTER (Y/N)?",P$
1940    IF (P$="Y") OR (P$="y") THEN 1970
1950    IF (P$="N") OR (P$="n") THEN 2010
1960    GOTO 1930
1970    PLOTTER IS "9872A"
1980    OUTPUT 705;"VS4;"
1990    Iplot=1
2000    GOTO 640
2010    INPUT "Do you like to draw its NYQUIST PLOT (Y/N)?",V$
2020    IF (V$="Y") OR (V$="y") THEN 2050
2030    IF (V$="N") OR (V$="n") THEN 170
2040    GOTO 2010
2050    Type$="NYQUIST"
2060    Iplot=0
2070    GOTO 640
2080    GOSUB Polarplot
2090    Istart=1
2100    FOR Decade=0 TO 5
```

```
2110    Con=10.0^(Decade-1)
2120    Del=(10.0^Decade-Con)/100
2130     FOR Interval=0 TO 99
2140     Omega=Con+Interval*Del
2150     Sumzero=1
2160     IF Nzeros=0 THEN 2200
2170     FOR I=1 TO Nzeros
2180     Sumzero=Sumzero*SQR(Zr(I)^2+(Omega+Zi(I))^2)
2190     NEXT I
2200     Sumzero=Sumzero*K
2210     Sumpole=1
2220     FOR I=1 TO Npoles
2230     Sumpole=Sumpole*SQR(Pr(I)^2+(Omega+Pi(I))^2)
2240     NEXT I
2250     Maggh=Sumzero/Sumpole
2260     IF (Maggh<Ghmin) OR (Maggh>Ghmax) THEN 2560
2270     Value=LGT(Maggh)-LGT(Ghmin)
2280     IF Maggh<1 THEN Value=ABS(LGT(Ghmin))+LGT(Maggh)
2290     Sumzero=0
2300     IF Nzeros=0 THEN 2380
2310     Nn=Nzeros
2320     FOR I=1 TO Nn
2330     Ar(I)=Zr(I)
2340     Ai(I)=Zi(I)
2350     NEXT I
2360     GOSUB Angle_test
2370     Sumzero=Sum
2380     Nn=Npoles
2390     FOR I=1 TO Nn
2400     Ar(I)=Pr(I)
2410     Ai(I)=Pi(I)
2420     NEXT I
2430     GOSUB Angle_test
2440     Sumpole=Sum
2450     Phase=Sumzero-Sumpole
2460     DEG
2470     X=Value*COS(Phase)
2480     Y=Value*SIN(Phase)
2490     RAD
2500     IF Istart=1 THEN 2530
2510     PLOT X,Y,-1
2520     GOTO 2570
2530     MOVE X,Y
2540     Istart=0
2550     GOTO 2570
2560     Istart=1
2570     NEXT Interval
2580    NEXT Decade
2590    PAUSE
2600    EXIT GRAPHICS
2610    IF Iplot=1 THEN 2700
2620    INPUT "Do you like to plot on the X-Y PLOTTER (Y/N)?",W$
2630    IF (W$="Y") OR (W$="y") THEN 2660
2640    IF (W$="N") OR (W$="n") THEN 2700
2650    GOTO 2620
2660    PLOTTER IS "9872A"
2670    OUTPUT 705;"VS4;"
2680    Iplot=1
2690    GOTO 640
2700    INPUT "Do you like to draw its BODE PLOT (Y/N)?",V$
2710    IF (V$="Y") OR (V$="y") THEN 2740
2720    IF (V$="N") OR (V$="n") THEN 170
2730    GOTO 2700
2740    Type$="BODE"
2750    Iplot=0
2760    GOTO 640
2770    STOP
2780 Plot_test:   !
2790    IF Iplot=0 THEN 2830
2800    ! INPUT "Which pen do you like for the plot (1-4)",Ipen
```

FREQUENCY-RESPONSE METHODS

```
2810 ! IF (Ipen<1) OR (Ipen>4) THEN 2912
2820 ! PEN Ipen
2830     IF Value<0 THEN 2860
2840     IF Value>Y_interval THEN 2890
2850     GOTO 2920
2860     MOVE Decade+LGT(Interval),0
2870     Istart=1
2880     GOTO 2980
2890     MOVE Decade+LGT(Interval),Y_interval
2900     Istart=1
2910     GOTO 2980
2920     IF Istart<>0 THEN 2960
2930     Istart=0
2940     DRAW Decade+LGT(Interval),Value
2950     GOTO 2980
2960     Istart=0
2970     MOVE Decade+LGT(Interval),Value
2980     RETURN
2990 Zpread: FOR I=1 TO Nn
3000   PRINT Zp$;I;" REAL ="
3010   INPUT Ar(I)
3020   PRINT Zp$;I;" IMAGINARY ="
3030   INPUT Ai(I)
3040   PRINT Zp$;I;" = ";Ar(I);" + j";Ai(I)
3050   INPUT "Is this correct (Y/N)?",Gonogo$
3060   IF (Gonogo$="Y") OR (Gonogo$="y") THEN 3090
3070   IF (Gonogo$="N") OR (Gonogo$="n") THEN 3000
3080   GOTO 3050
3090   NEXT I
3100   RETURN
3110 Logplt:  !
3120 IF Iplot=1 THEN 3140
3130 PLOTTER IS 13,"GRAPHICS"
3140 GRAPHICS
3150 LIMIT 0,184,0,140
3160 LOCATE 20,120,20,90
3170 SCALE 0,Log_interval,0,Y_interval
3180 LINE TYPE 3
3190 AXES 0,1,0,0,1,1,200
3200 LINE TYPE 1
3210 FRAME
3220 Logarithmic:  !
3230 FOR Decade=0 TO Log_interval
3240    FOR Interval=1 TO 9
3250     MOVE Decade+LGT(Interval),0
3260     DRAW Decade+LGT(Interval),150
3270    NEXT Interval
3280   MOVE Decade,0
3290   DRAW Decade,150
3300 NEXT Decade
3310 Log_labels:  !
3320   LORG 6
3330   FOR Decade=0 TO Log_interval
3340   CSIZE 4
3350   MOVE Decade,-.4
3360   Position=Wmin*10.0^Decade
3370   LABEL USING "M4D.DD";Position
3380   IF Decade=Log_interval THEN 3440
3390   CSIZE 2
3400     FOR Interval=2 TO 9
3410     MOVE Decade+LGT(Interval),-.05
3420     LABEL USING "KX";Interval
3430    NEXT Interval
3440 NEXT Decade
3450 Lin_labels:  !
3460   CSIZE 4
3470   LORG 8
3480   FOR Y_axis=0 TO Y_interval
3490   MOVE .01,Y_axis
3500   Position=Ymin+Sf*Y_axis
3510   LABEL USING "K,X";Position
```

```
3520    NEXT Y_axis
3530    Text:    !
3540    DEG
3550    LORG 3
3560    SETGU
3570    MOVE 50,3
3580    LABEL Xtitle$
3590    MOVE 7,40
3600    LORG 2
3610    LDIR 90
3620    LABEL Ytitle$
3630    RAD
3640    SETUU
3650    LDIR 0
3660    RETURN
3670 Angle_test:    !
3680    Sum=0
3690    FOR I=1 TO Nn
3700    IF Ar(I)=0 THEN 3760
3710    DEG
3720    Angle=ATN((Omega+Ai(I))/Ar(I))
3730    RAD
3740    IF Ar(I)<0 THEN 3790
3750    GOTO 3810
3760    Angle=90
3770    IF Omega+Ai(I)<0 THEN Angle=-90
3780    GOTO 3810
3790    IF Omega+Ai(I)<0 THEN Angle=Angle-180
3800    IF Omega+Ai(I)>0 THEN Angle=Angle+180
3810    Sum=Sum+Angle
3820    NEXT I
3830    RETURN
3840 Polarplot:    !
3850    PRINT LIN(E(2)),"The NYQUIST PLOT is a polar plot in which you will"
3860    PRINT "specify the MAGNITUDE RANGE of the transfer function."
3870    PRINT "Due to space & clarity a maximum of 3 DECADES can be covered."
3880    PRINT LIN(1),"Examples: 0.1<=MAGNITUDE<=100 or 100<=MAGNITUDE<=100000"
3890    PRINT LIN(1),"Remember: To get the unit circle ,1.0 must be enclosed by"
3900    PRINT "your limits."
3910    PRINT LIN(1),"The NYQUIST PLOT covers 600 points in a FREQUENCY RANGE 0.1<
=100000,"
3920    PRINT "so it may make several MINUTES to develop."
3930    INPUT "What is the MINIMUM MAGNITUDE of your plot (Must be a power of 10)",Gh
min
3940    IF Ghmin<0 THEN 3930
3950    Value=FRACT(LGT(Ghmin))
3960    IF ABS(Value)>.000001 THEN 3930
3970    INPUT "what is the MAXIMUM MAGNITUDE of your plot (Must be a power of 10.)",G
hmax
3980    IF Ghmax<0 THEN 3970
3990    Value=FRACT(LGT(Ghmax))
4000    IF ABS(Value)>.000001 THEN 3970
4010    PRINT PAGE
4020    IF Ghmin>=Ghmax THEN 4090
4030    IF LGT(Ghmax)-LGT(Ghmin)>3 THEN 4110
4040    PRINT LIN(1),"MAGNITUDE RANGE of plot is ";Ghmin;" <=MAGNITUDE<= ";Ghmax
4050    INPUT "Is this correct (Y/N)?",Gonogo$
4060    IF (Gonogo$="Y") OR (Gonogo$="y") THEN 4130
4070    IF (Gonogo$="N") OR (Gonogo$="n") THEN 3930
4080    GOTO 4050
4090    PRINT LIN(1),"Your MINIMUM value >= Your MAXIMUM...Try again"
4100    GOTO 3930
4110    PRINT LIN(2),"You have chosen more than 3 DECADES...Try again"
4120    GOTO 3930
4130    PRINT LIN(1),"Always hit the CONT key to continue."
```

```
4140  PAUSE
4150  Ndecade=LGT(Ghmax)-LGT(Ghmin)
4160  IF Iplot=1 THEN 4180
4170  PLOTTER IS 13,"GRAPHICS"
4180  GRAPHICS
4190  LOCATE 20,106,8,94
4200  SCALE -Ndecade,Ndecade,-Ndecade,Ndecade
4210  LINE TYPE 3
4220  Logcircles:  !
4230  MOVE 0,0
4240  DEG
4250  FOR Decade=0 TO Ndecade
4260  FOR Interval=1 TO 9
4270  IF (Decade=0) AND (Interval=1) THEN 4360
4280  MOVE Decade+LGT(Interval),0
4290  LINE TYPE 3
4300  FOR Angle=0 TO 360 STEP 15
4310  X=(Decade+LGT(Interval))*COS(Angle)
4320  Y=(Decade+LGT(Interval))*SIN(Angle)
4330  DRAW X,Y
4340  NEXT Angle
4350  IF (Decade=Ndecade) AND (Interval=1) THEN 4370
4360  NEXT Interval
4370  NEXT Decade
4380  LINE TYPE 1
4390  FOR Angle=0 TO 360 STEP 15
4400  MOVE 0,0
4410  PDIR Angle
4420  RPLOT Ndecade,0,-1
4430  NEXT Angle
4440  MOVE 0,0
4450  LORG 6
4460  FOR Decade=0 TO Ndecade
4470  MOVE Decade,-.05
4480  Number=Ghmin*10.0^Decade
4490  LABEL USING "M4D.DDD";Number
4500  NEXT Decade
4510  LORG 2
4520  CSIZE 2.5
4530  FOR Angle=0 TO 345 STEP 15
4540  LDIR Angle
4550  MOVE (Ndecade+.05)*COS(Angle),(Ndecade+.05)*SIN(Angle)
4560  LABEL Angle
4570  NEXT Angle
4580  LDIR 0
4590  RETURN
```

13. Example

This program constructs BODE & NYQUIST plots for CONT. & DIS.-TIME transfer functions of up to 10 poles & 10 zeros. Functions must be of the form   K(s+Z1)(s+Z2).../(s+P1)(s+P2)
   OR   K(z+Z1)(z+Z2).../(z+P1)(z+P2) , where K is the GAIN, Z's are the zeros & -P's are the poles.

All poles & zeros should be in first-order form. Those which are not should be factored until they are. Multiple poles or zeros should be entered as if they were separate.

You will be asked to enter information after which you should hit the CONT key to continue.

You have asked for BODE plots

Your function has   1   zero(s) &   3   pole(s)

The GAIN of the system is   1000

Enter the :-ZEROS in complex form. If either the real or
imaginary part is 0.0, enter as such. Reminder : The -ZERO in a
term such as (s+3-j2) or (z+3-j2) is 3-j2.(Note:-ZERO=0=0+j0)

-Z 1   REAL =
-Z 1   IMAGINARY =
-Z 1   =   3   + j 0

Enter the :-POLES in complex form. If either the real or
imaginary part is 0.0, enter as such. Reminder : The pole in a
term such as (s+3-j2) or (z+3-j2) is 3-j2.(Note:-POLE=0=0+j0)

-P 1   REAL =
-P 1   IMAGINARY =
-P 1   =   0   + j 0
-P 2   REAL =
-P 2   IMAGINARY =
-P 2   =   12 + j 0
-P 3   REAL =
-P 3   IMAGINARY =
-P 3   =   50 + j 0

Transfer function summary for BODE plots

The GAIN is   1000
The :-ZEROS are

-Z 1 =   3   +j 0

The -POLES are

-P 1 =   0   +j 0
-P 2 =   12 +j 0
-P 3 =   50 +j 0

Hit the CONT key to continue

Specify a FREQUENCY RANGE you wish to plot over
Example: 0.1<=W<=1000.

The FREQUENCY RANGE of your plot is   .1   <=W<=   1000

For the MAGNITUDE BODE PLOT, Specify a DECIBEL RANGE you
wish to plot over. Example: -40<=DB<=60..Must be a multiple of 20.

FREQUENCY-RESPONSE METHODS

The DECIBEL RANGE of your PLOT is -60 <=DB<= 40

Always hit the CONT key to continue
For the PHASE BODE PLOT, specify a PHASE ANGLE RANGE you wish
to plot over   Example:.-90<=PHASE<=135..Must be a multiple of 45.

The PHASE ANGLE RANGE of your plot is -180   <=PHASE<= -45

Transfer function summary for BODE plots

The GAIN is   1000
The :-ZEROS are

-Z 1 =   3 +j 0

The -POLES are

-P 1 =   0 +j 0
-P 2 = 12 +j 0
-P 3 = 50 +j 0

Hit the CONT key to continue

Transfer function summary for BODE plots

The GAIN is   1
The -ZEROS are

The -POLES are

-P 1 = 1 +j 0
-P 2 = 2 +j 0
-P 3 = 5 +j 0

Hit the CONT key to continue

Specify a FREQUENCY RANGE you wish to plot over
Example: 0.1<=W<=1000.

The FREQUENCY RANGE of your plot is .1 <=W<= 100

For the MAGNITUDE BODE PLOT, Specify a DECIBEL RANGE you
wish to plot over. Example: -40<=DB<=60..Must be a multiple of 20.

The DECIBEL RANGE of your PLOT is -60 <=DB<= 0

```
Always hit the CONT key to continue
For the PHASE BODE PLOT,specify a PHASE ANGLE RANGE you wish
to plot over   Example:-90<=PHASE<=135..Must be a multiple of 45.

The PHASE ANGLE RANGE of your plot is -270   <=PHASE<=   0
```
The corresponding computer plots are shown in Figs. 7.7.19 to 7.7.22.

7.8 THE CIRCLE CRITERION

The Nyquist stability criterion as stated in the previous section applies only to a special class of systems - l.t.i. SISO systems. It can be generalized to apply to a broader class of systems, namely, to l.t.i. systems with time-varying feedback gains shown in Fig. 1. One extension of the Nyquist stability criterion states that if the Nyquist diagram of $G(s)$ lies inside the unit circle and if $k(t)$ is periodic with magnitude less than one for all t, then the system of Fig. 1 will be stable [7.18]. The so-called *circle criterion* provides a more general extension of the Nyquist stability criterion. Namely, it allows a time-varying gain $k(t)$ in the feedback path where $k(t)$ is bounded by two real numbers α and β, $\alpha < k(t) < \beta$, but replaces the point (-1,j0) in the Nyquist stability criterion by a "circle" or disk $D(\alpha,\beta)$ centered on the real axis and defined in terms of α and β. Before giving the formal criterion, let us define $D(\alpha,\beta)$.

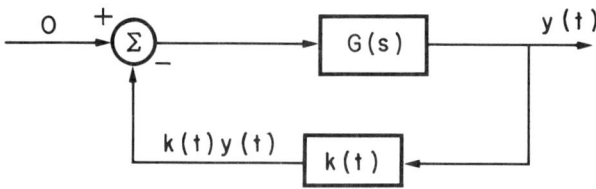

Fig. 7.8.1. A linear feedback system with a time-varying gain

1. Definition [7.6]. Let α and β be real numbers and $\alpha < \beta$. Then $D(\alpha,\beta)$ is a region in the complex plane defined as follows:

(i) If $\alpha\beta > 0$, then

$$D(\alpha,\beta) + \{\sigma + j\omega : [\sigma + \frac{1}{2}(\frac{1}{\alpha} + \frac{1}{\beta})]^2 + \omega^2 < \frac{1}{4}|\frac{1}{\alpha} - \frac{1}{\beta}|^2\} \qquad (1)$$

which is a circle of radius $\frac{1}{2}|\frac{1}{\alpha} - \frac{1}{\beta}|$ (see Fig. 2a).

(ii) If $\alpha\beta < 0$, then

$$D(\alpha,\beta) + \{\sigma + j\omega : [\sigma + \frac{1}{2}(\frac{1}{\alpha} + \frac{1}{\beta})]^2 + \omega^2 > \frac{1}{4}(\frac{1}{\alpha} + \frac{1}{\beta})^2\} \qquad (2)$$

which is the complement of a circle as shown in Fig. 2b.

FREQUENCY-RESPONSE METHODS

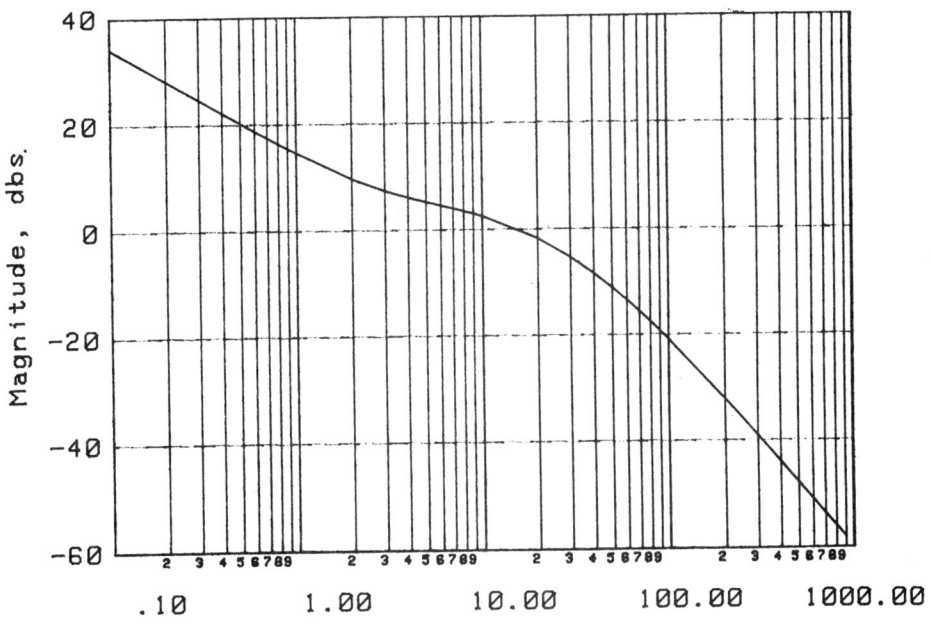

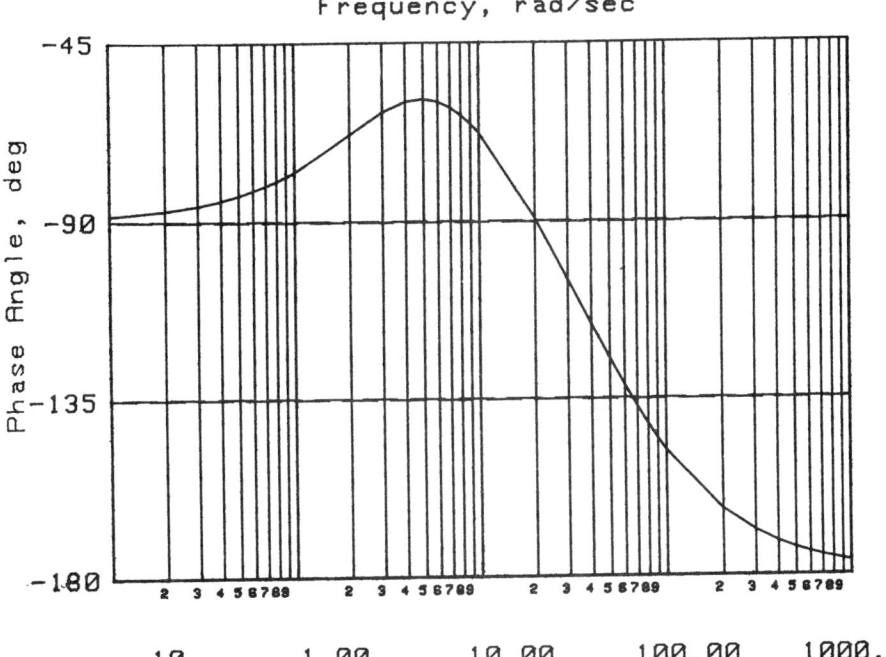

Fig. 7.7.19. Bode plots for Example 7.7.13 — Case a

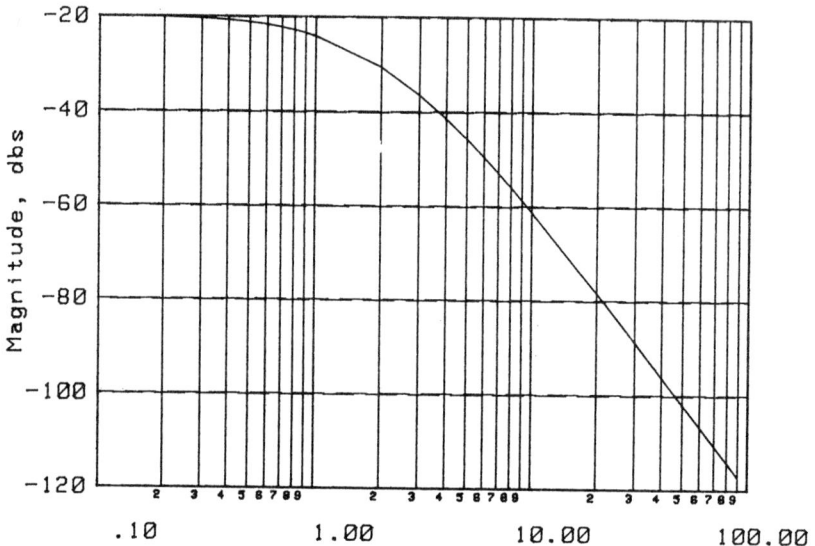

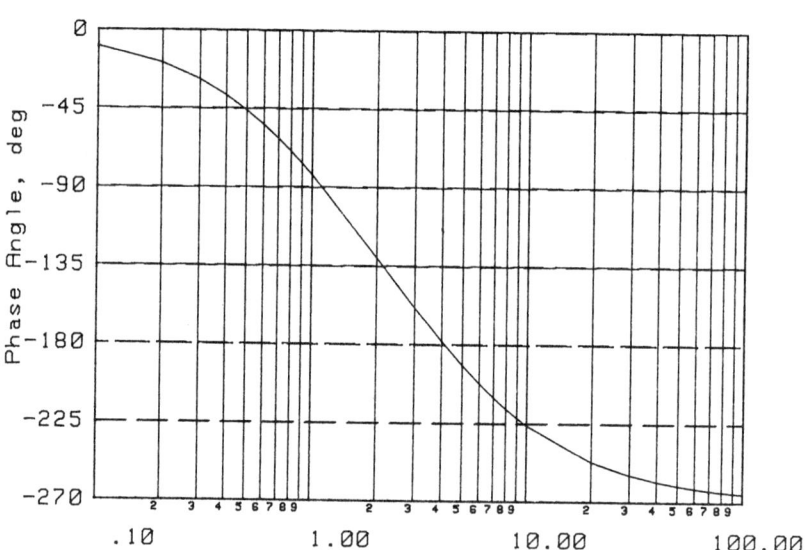

Fig. 7.7.20. Bode plots for Example 7.7.13 — Case b

FREQUENCY-RESPONSE METHODS

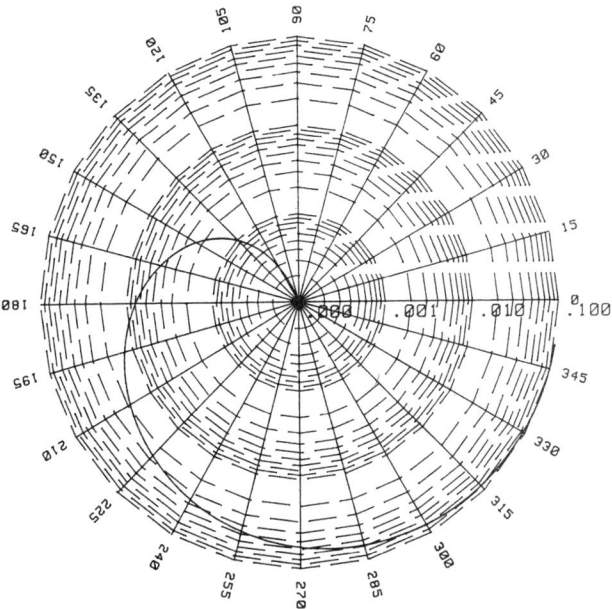

Fig. 7.7.21. Nyquist (Polar) plot for Example 7.7.13 — Case a

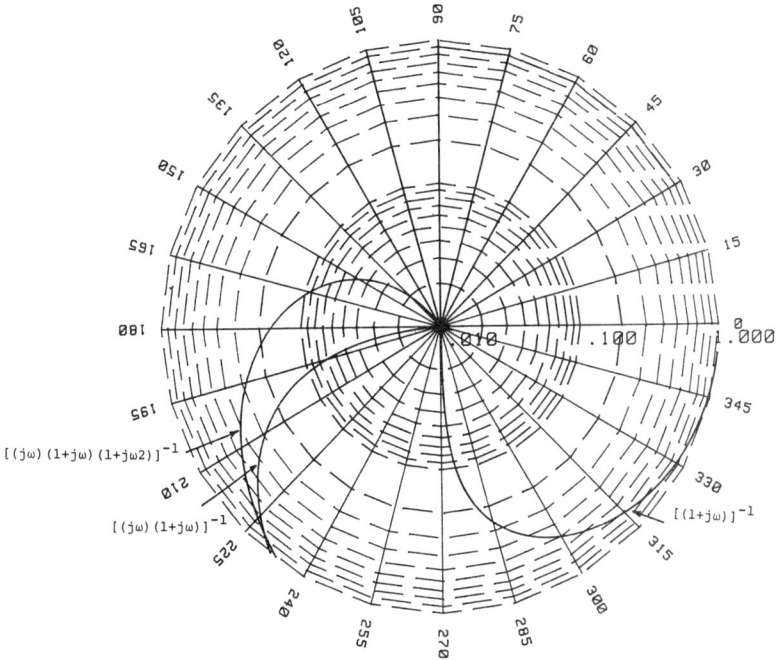

Fig. 7.7.22. Nyquist (Polar) plot for Example 7.7.13 — Case b

(iii) If $\alpha=0$, then

$$D(0,\beta)=\{\sigma+j\omega : \sigma < -\frac{1}{\beta}\} \tag{3}$$

which is a half plane contained in the the *l.h.p.* (see Fig. 2c).

(iv) If $\beta=0$, then

$$D(\alpha,0)=\{\sigma+j\omega : \sigma > -\frac{1}{\alpha}\} \tag{4}$$

which is a half plane contained in the *r.h.p.* (see Fig. 2d).

If in (1), (2), (3) and (4), the signs $>$ and $<$ are replaced by $\geq$ and $\leq$, respectively, the corresponding disks will be denoted $D[\alpha,\beta]$.

2. Theorem. (The circle criterion) Consider the linear time-varying feedback system shown in Fig. 1 where $G(s)$ has p_R poles in the *r.h.p.* and $k(t)$ is a piecewise continuous function such that $\alpha < k(t) < \beta$ for all t. Then the system is asymptotically stable if the polar plot of $G(j\omega)$ encircles the disk $D(\alpha,\beta)$ exactly p_R times in the *CCW* direction (and does not intersect it). Furthermore, the system is unstable if the polar plot of $G(j\omega)$ encircles the disk $D[\alpha,\beta]$ fewer than p_R times in the *CCW* direction (and does not intersect it). △

The proof of this theorem will not be present here. It can be found in Reference [7.6].

3. Example. Consider the system of Fig. 1 where

$$G(s) = \frac{k}{s(1+s\tau_1)(1+s\tau_2)}, \tau_1 > \tau_2 > 0 \tag{5}$$

This transfer function was considered in Example 7.7.5 and its polar plot was shown in Fig. 7.7.9. Since $G(s)$ has no poles in the *r.h.p.*, $p_R=0$. Thus this system is asymptotically stable for $\alpha < k(t) < \beta$ for *all* t if the polar plot of (5) does not encircle the disk $D(\alpha,\beta)$ where

$$-\frac{1}{\beta} < \omega_2 = \frac{1}{\sqrt{\tau_1\tau_2}} \Rightarrow \beta > -\sqrt{\tau_1\tau_2} \tag{6}$$

and $\alpha < \beta$.

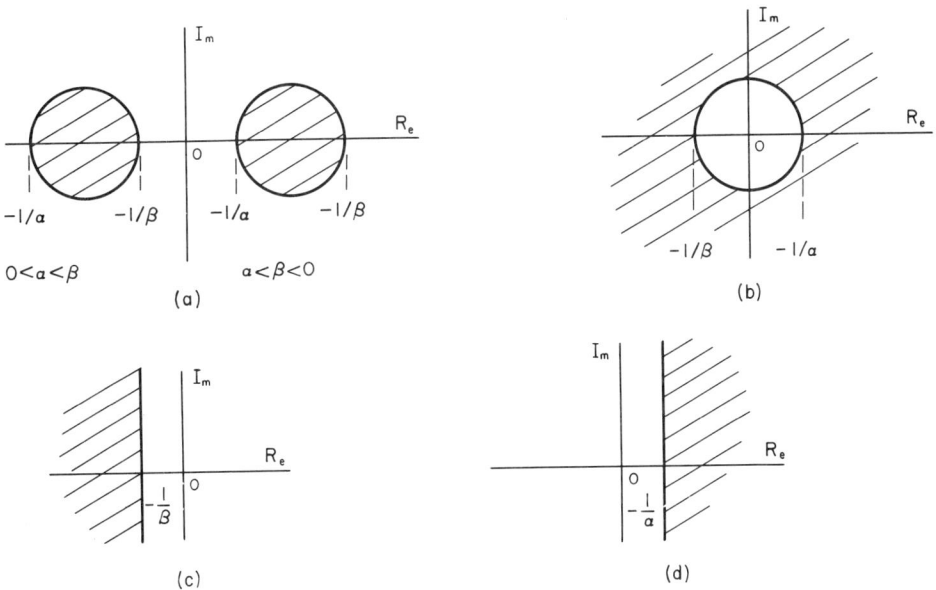

**Fig. 7.8.2. The disk $D(\alpha,\beta)$ (a) $\alpha\beta>0$
(b) $\alpha\beta<0$, (c) $\alpha = 0$, $\beta>0$, (d) $\alpha<0$, $\beta = 0$**

7.9 THE LYAPUNOV'S METHOD

The Lyapunov's method generates sufficient conditions for stability without having to determine system response. It is most useful for nonlinear and/or time-varying systems. This approach, which is based on Lyapunov's work [7.19], has its origin in classical dynamics. It is based on the observation that in a stable system with positive energy E, the rate of change of energy, $\dot{E}$, is negative.

To apply the Lyapunov's method an "energy like" function, referred to as a *Lyapunov function* must be generated. Stability can then be characterized by studying this function and its derivative w.r.t. time. The following example is presented for motivation.

1. Example. Consider the *l.t.i.* network shown in Fig. 1. The initial conditions are the voltage across the capacitor, v_o, and the current through the inductor, i_o. No input is applied to the network. If we chose state variables $x_1(t)=i(t)$ and $x_2(t)=v(t)$, the state equation of this system will be

$$\dot{\mathbf{x}} = \begin{bmatrix} \dot{x}_1 \\ \dot{x}_2 \end{bmatrix} = \begin{bmatrix} -R/L & -1/L \\ 1/C & 0 \end{bmatrix} \begin{bmatrix} x_1 \\ x_2 \end{bmatrix}, t \geq 0, \begin{bmatrix} x_1(0) \\ x_2(0) \end{bmatrix} = \begin{bmatrix} i_o \\ v_o \end{bmatrix} \qquad (1)$$

LINEAR CONTROL SYSTEMS

The total electric and magnetic energy stored in this system at time t is

$$E(t) = 1/2Li^2(t) + 1/2Cv^2(t) = x'(t)Qx(t) \qquad (2)$$

where

$$Q = \begin{bmatrix} L/2 & 0 \\ 0 & C/2 \end{bmatrix} \qquad (3)$$

Note that since $L>0$ and $C>0$, $E(t)$ is a positive-definite quadratic function of $\mathbf{x}$. Also the time derivative of E is given by

$$\frac{d}{dt}E(t) = Li\frac{di}{dt} + Cv\frac{dv}{dt} = Lx_1\dot{x}_1 + Cx_2\dot{x}_2 = -Rx_1^2 \qquad (4)$$

Now if $R \neq 0$ the system is clearly asymptotically stable. Not that in this case $\frac{d}{dt}E(t)$ is negative definite. If $R=0$, the stored energy in the system remains constant and the system is clearly stable isL. Note that in this case $\frac{d}{dt}E(t)=0$.

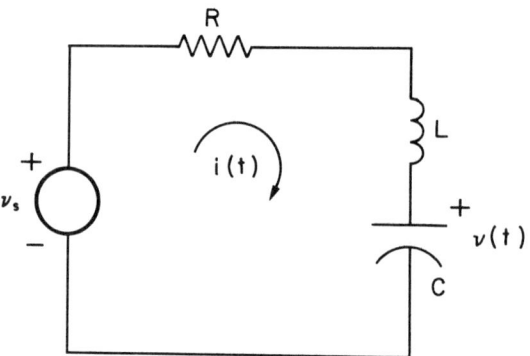

Fig. 7.9.1. Network of Example 7.9.1

2. Definition. A scalar function $V(\mathbf{x},t)$ is called a *continuous-time Lyapunov function* if for all $t \geq t_o$ and all vectors $\mathbf{x}$ in the neighborhood of $\mathbf{0}$, it satisfies the following conditions:

(i) $V(\mathbf{x},t)$ and its first partial derivatives w.r.t. t and components x_i, $i=1,2,...,n$ of $\mathbf{x}$ all exist and are continuous,

(ii) $V(\mathbf{0},t) = 0$,

(iii) A continuous nondecreasing function $\alpha(.)$ exists where $\alpha(0) = 0$, such that $V(\mathbf{x},t) \geq \alpha(||\mathbf{x}||) > 0$ for $\mathbf{x} \neq 0$ and $t \geq t_o$,

(iv) $\dfrac{dV(\mathbf{x},t)}{dt} = \dfrac{dV(\mathbf{x},t)}{d\mathbf{x}} \dot{\mathbf{x}} = \sum_{i=1}^{n} \dfrac{\partial V}{\partial x_i} \dot{x}_i + \dfrac{\partial V}{\partial t} < 0$ for $\mathbf{x} \neq \mathbf{0}$, $t \geq t_o$.

3. Theorem. Consider the system $\dot{\mathbf{x}} = \mathbf{f}(\mathbf{x},t)$ where $\mathbf{f}(\mathbf{0},t) = \mathbf{0}$. If a continuous-time Lyapunov function can be found for this system, the equilibrium state $\mathbf{x} = \mathbf{0}$ will be asymptotically stable.

The proof of this theorem will not be given here. It can be found in reference [7.20] where further restrictions and generalizations are also considered. Note that Lyapunov function is not unique for a particular system. If *any* Lyapunov function can be found for a given system, then by the above theorem stability will be guaranteed. However, the main problem is finding a Lyapunov function. No general method is available to do this.

For uniform asymptotic stability, the conditions of definition 2 must hold for all t_o and the following additional condition must also be satisfied:

(v) A continuous nondecreasing function $\beta(.)$ exists where $\beta(0) = 0$, such that $V(\mathbf{x},t) \leq \beta(||\mathbf{x}||)$.

Also for uniform asymptotic stability in the large, the expression "in the neighborhood of **0**" in Definition 2 must be replaced by "everywhere" and the function $\alpha(.)$ is condition (iii) of that definition must be "increasing" rather than "nondecreasing".

4. Example. Consider the linear time-varying system

$$\begin{bmatrix} \dot{x}_1 \\ \dot{x}_2 \end{bmatrix} = \begin{bmatrix} 0 & 1 \\ -1 & -(t+1)^2 \end{bmatrix} \begin{bmatrix} x_1 \\ x_2 \end{bmatrix}, \quad t \geq 0 \tag{5}$$

Note that $\mathbf{x} = \mathbf{0}$ is an equilibrium state of this system. Let $V(\mathbf{x},t) = (\mathbf{x},\mathbf{x}) = ||\mathbf{x}||^2 = x_1^2 + x_2^2$. Then

$$\dot{V}(\mathbf{x},t) = \dfrac{d}{dt}(x_1^2 + x_2^2) = 2x_1\dot{x}_1 + 2x_2\dot{x}_2 = -2(t+1)^2 x_2^2(t) \tag{6}$$

which is negative for $\mathbf{x} \neq \mathbf{0}$ and $t \geq 0$. Therefore, by Theorem 3, the system is asymptotically stable.

Note that the stability is not uniform (e.g. if $t_o = -1$, then $\dot{v}(\mathbf{x},t)$ would become zero for $\mathbf{x} \neq \mathbf{0}$ and $t = t_o$). However, due to linearity of the system, stability is in the large. △

A parallel development holds for discrete-time systems.

5. Definition. A scalar function $V(\mathbf{x},k)$ is called a *discrete-time Lyapunov function* if for all integer $k > k_o$ and all vectors $\mathbf{x}$ in the neighborhood of **0**, it satisfies the following conditions:

(i) $V(\mathbf{x},k)$ is continuous,

(ii) $V(\mathbf{0},k) = 0$,

(iii) $V(\mathbf{x},k) \geq \alpha(||\mathbf{x}||) > 0$ for $\mathbf{x} \neq \mathbf{0}$ and $k \geq k_o$ where $\alpha(.)$ is a continuous nondecreasing function and $\alpha(0) = 0$,

(iv) $\Delta V(\mathbf{x},k) \triangleq V(\mathbf{x}(k+1), k+1) - V(\mathbf{x}(k),k) < 0$ for $\mathbf{x} \neq \mathbf{0}$.

6. Theorem. Consider the system $x(k+1) = f(x,k)$ where $f(0,k)=0$. If a discrete-time Lyapunov function can be found for this system, the equilibrium state $x = 0$ will be asymptotically stable. △

Discussions similar to those following Theorem 3 apply here too.

7. Example. Consider the system $x(k+1) = Ax(k)$, $k \geqslant 0$ where

$$A = \begin{bmatrix} -1 & -1/4 \\ 2 & 0 \end{bmatrix} \tag{7}$$

The eigenvalues of A are $\lambda_{1,2} = -1/2 \pm j/2$ which are inside the unit circle in the complex plane. Thus the system is asymptotically stable. Let us verify this by the Lyapunov's method. Take

$$V(x,k) = x'(k)Px(k) = [x_1(k), x_2(k)]$$

Then

$$\begin{bmatrix} p_{11} & p_{12} \\ p_{12} & p_{22} \end{bmatrix} \begin{bmatrix} x_1(k) \\ x_2(k) \end{bmatrix} = p_{11}x_1^2 + 2p_{12}x_1x_2 + p_{22}x_2^2 \tag{8}$$

where P is a symmetric positive definite matrix, used in the definition of a candidate Lyapunov function. Note the V has no explicit dependence on k. Therefore, we must have (See Section 2.11.)

$$p_{11} > 0, \quad \det(P) = p_{11}p_{22} - p_{12}^2 > 0 \tag{9}$$

Then we have

$$\Delta V(x) = V(x(k+1)) - V(x(k)) = (4p_{22} - 4p_{12})x_1^2(k)$$

$$+ \left(\frac{p_{11}}{2} - 3p_{12}\right)x_1(k)x_2(k) + \left(\frac{p_{11}}{16} - p_{22}\right)x_2^2(k) \tag{10}$$

Let

$$p_{11} = 6p_{12}, \quad p_{11} > 6p_{22}, \quad p_{11} < 16p_{22} \tag{11}$$

The above conditions, respectively, make the coefficient of $x_1(k)x_2(k)$ in (10) equal to zero and the coefficients of $x_1^2(k)$ and $x_2^2(k)$ in (10) negative so that $\Delta V(\mathbf{x})<0$ $\forall \mathbf{x} \neq \mathbf{0}$. As an example, let $p_{22}=1$ and $p_{11}=12$. Thus

$$\Delta V(\mathbf{x}) = -4x_1^2 - \frac{1}{4}x_2^2 < 0 \quad \forall \mathbf{x} \neq \mathbf{0} \tag{12}$$

Therefore, by Theorem 6, the system is asymptotically stable (in the large, by linearity). △

Note that the existence of a Lyapunov function is a *sufficient* condition for stability. That is, if a Lyapunov function cannot be found for a system, we cannot be certain that the system is unstable. The following theorem provides a sufficient condition for instability.

8. Theorem. If a scalar continuous function $V(\mathbf{x},t)$ with continuous first partial derivatives w.r.t.t and components of $\mathbf{x}$ exist which satisfies the following conditions:

(i) $V(\mathbf{0},t) = 0$.

(ii) $V(\mathbf{x},t) > 0$ for *all* $\mathbf{x} \neq \mathbf{0}$ in the neighborhood of $\mathbf{0}$,

(iii) $\dot{V}(\mathbf{0},t) = 0$ and $\dot{V}(\mathbf{x},t) > 0$ for *all* $\mathbf{x} \neq \mathbf{0}$ in the neighborhood of $\mathbf{0}$.

Then the equilibrium state $\mathbf{x} = \mathbf{0}$ is unstable. △

For linear systems it has been shown [7.6, 7.20] that if the system is asymptotically stable, then a Lyapunov function indeed exists. The following theorem provides necessary and sufficient conditions for uniform asymptotic stability of linear systems.

9. Theorem. The system $\dot{\mathbf{x}}(t) = \mathbf{A}(t)\mathbf{x}(t)$ where $\mathbf{A}(t)$ is piecewise continuous and bounded for $t \geq t_o$ is uniformly asymptotically stable if and only if given any continuous symmetric positive definite real matrix $\mathbf{Q}(t)$, bounded on $t \geq t_o$, there exists a symmetric positive definite real matrix $\mathbf{P}(t)$ which is the solution to

$$\dot{\mathbf{P}}(t) + \mathbf{A}'(t)\mathbf{P}(t) + \mathbf{P}(t)\mathbf{A}(t) + \mathbf{Q}(t) = \mathbf{0}, \; t \geq t_o \tag{13}$$

Proof. Only an outline of the sufficiency proof will be given here. For the complete proof the reader is referred to [7.20]. If $\mathbf{P}(t)$ is positive definite, $V(\mathbf{x}) = \mathbf{x}'\mathbf{P}(t)\mathbf{x}$ is a Lyapunov function. Then

$$\dot{V}(\mathbf{x}) = \dot{\mathbf{x}}'\mathbf{P}\mathbf{x} + \mathbf{x}'\dot{\mathbf{P}}\mathbf{x} + \mathbf{x}'\mathbf{P}\dot{\mathbf{x}}$$

$$= \mathbf{x}'(\mathbf{A}'\mathbf{P} + \mathbf{P}\mathbf{A} + \dot{\mathbf{P}})\mathbf{x} = -\mathbf{x}'\mathbf{Q}(t)\mathbf{x} \tag{14}$$

by (13). Since $\mathbf{Q}(t)$ is positive definite by assumption, then $\dot{V}(\mathbf{x})$ is a negative definite quadratic form. Thus by Theorem 3, the system is asymptotically stable. △

The matrix differential equation (13), which is a special case of *Riccati differential equation*, provides a means of generating Lyapunov functions for linear time-varying systems. An extensive review for the solutions of the Riccati equations can be found in [7.21].

In the *l.t.i.* case, (13) simply becomes

$$A'P + PA = -Q \tag{15}$$

which is an algebraic equation, referred to as the *Lyapunov equation*. In this case it can be shown [7.22] that the system $\dot{x} = Ax$, where A is a constant matrix, is asymptotically stable if and only if for any symmetric positive definite (constant) real matrix Q, a *unique* symmetric positive definite solution P to (15) always exists. Thus, to test the stability of a *l.t.i.* system one can set up equation (15) and check to see if its solution P is positive definite.

This procedure is, in effect, equivalent to Routh-Hurwitz stability criterion.

10. Example. Consider the system $\dot{x} = \begin{bmatrix} -1 & -4 \\ 1 & -1 \end{bmatrix} x$ where the eigenvalues of the system matrix are -1 and -2. Therefore, by Corollary 7.3.5 the system is asymptotically stable. Let us choose $Q = I$ (identity matrix, which is positive definite). Then from (15), $P = \begin{bmatrix} p_{11} & p_{12} \\ p_{12} & p_{22} \end{bmatrix}$ is the solution to

$$\begin{bmatrix} -1 & 1 \\ -4 & -1 \end{bmatrix} \begin{bmatrix} p_{11} & p_{12} \\ p_{12} & p_{22} \end{bmatrix}$$

$$+ \begin{bmatrix} p_{11} & p_{12} \\ p_{12} & p_{22} \end{bmatrix} \begin{bmatrix} -1 & -4 \\ 1 & -1 \end{bmatrix} = - \begin{bmatrix} 1 & 0 \\ 0 & 1 \end{bmatrix} \tag{16}$$

The solution is $P = \dfrac{1}{20} \begin{bmatrix} 7 & -3 \\ -3 & 22 \end{bmatrix}$ which is positive definite since $p_{11} > 0$ and $\det(P) > 0$. (See Theorem 2.11.6.) △

Program "LYAP" solves equation (15) using an iterative method due to Davison and Man [7.23] which is summarized below:

11. Algorithm. Solution of the Lyapunov Equation

Step 1: For the solution of a Lyapunov equation

$$A'L + LA + S = 0 \tag{17}$$

where A is a stable matrix, choose a step size $h = 10^{-4}/(2\|A\|)$ and set $L_o = hS$.

Step 2: Calculate the matrix

$$E=(I-hA/2+h^2A^2/12)^{-1}(I+hA/2+h^2A^2/12) \qquad (18)$$

where **I** is an identity matrix.

Step 3: Find the next value of **L**, i.e.,

$$L_{i+1}=(E\ ')^{2i}L_i E^{2i}+L_i \qquad (19)$$

Step 4: Check if $\|\Delta L\| = \|L_{i+1}-L_i\| < \epsilon$, a prespecified tolerance. If not, set $i=i+1$ and go to Step 3.

Step 5: Stop.

It is emphasized that the above algorithm converges only if all the eigenvalues of A have negative real parts. In computational experiences with Program "LYAP", it was found that in only ten iterations the algorithm converged to within six digits of accuracy. Program "LYAP" and two examples are given below.

12. "LYAP"

```
10    ! PROGRAM NAME : <<LYAP>>  PROG
20    PRINT "This Program solves the matrix LYAPUNOV equation:"
30    PRINT " ***    A'L + LA + C = 0   ***"
40    PRINT " by an Iterative technique. For further reading, see:"
50    PRINT " E.J.Davison & F.T.Man, IEEE Trans. AC.,vol.AC-13,p 448,1968."
60    OPTION BASE 1
70    DIM A(9,9),C(9,9),Asq(9,9),P(9,9),Pt(9,9),Lyap(9,9),Psq(9,9)
80    DIM Ptsq(9,9),Phil(9,9),Phi2(9,9),Ptemp(9,9),Ptempt(9,9)
90    PRINT "Note: step size h is calculated as follows:"
100   PRINT "          h=10^-4/(2||A||)            "
110   INPUT "Have SUBs <<Lyap>> , <<Norm>> & <<Mat>> been LINKED ? (Y/N)",R$
120   IF (R$="Y") OR (R$="y") THEN 180
130   IF (R$="N") OR (R$="n") THEN 150
140   GOTO 110
150   LINK "Lyap",1500,160
160   LINK "Norm",4000,170
170   LINK "Mat",4500,180
180   PRINT "Matrix A(nxn)"
190   INPUT "System order n",N
200   PRINT "n=";N
210   REDIM A(N,N),C(N,N),Asq(N,N),P(N,N),Pt(N,N),Lyap(N,N),Psq(N,N)
220   REDIM Ptsq(N,N),Phil(N,N),Phi2(N,N),Ptemp(N,N),Ptempt(N,N)
230   CALL Mat(A(*),N,N,"A")
240   PRINT "Matrix C(nxn)"
250   CALL Mat(C(*),N,N,"C")
260   CALL Lyap(A(*),C(*),N,25,Lyap(*))
270   END
```

13. Example

```
This Program solves the matrix LYAPUNOV equation:
***    A'L + LA + C = 0    ***
by an Iterative technique. For further reading, see:
E.J.Davison & F.T.Man, IEEE Trans. AC.,vol.AC-18,p 665,1973.
n= 2
Matrix A(nxn)

Matrix A( 2 x 2 ):
```

-1.0000E+00   -4.0000E+00

 1.0000E+00   -1.0000E+00

Matrix C(nxn)

Matrix C(2 x 2):

 1.0000E+00   0.0000E+00

 0.0000E+00   1.0000E+00

Lyapunov matrix
  .350011339317   -.149999992584

 -.149999992584   1.10001081414

14. Example

```
This Program solves the matrix LYAPUNOV equation:
***    A'L + LA + C = 0    ***
by an Iterative technique. For further reading, see:
E.J.Davison & F.T.Man, IEEE Trans. AC.,vol.AC-18,p 665,1973.
n= 4
Matrix A(nxn)

Matrix A( 4 x 4 ):
```

-1.0000E+00   2.5000E-01   1.0000E-01   5.0000E-01

 1.0000E+00 -2.0000E+00   5.0000E-01   1.0000E+00

 2.5000E-01 -1.0000E+00 -2.0000E+00   1.0000E+00

 0.0000E+00 -2.0000E+00   2.5000E-01 -3.0000E+00

Matrix C (nxn)

Matrix C(4 x 4):

```
1.000 0E+00    0.0C00E+00    0.000 0E+00    0.000 0E+00
0.0C00E+00    2.0C00E+00    0.0000E+00    0.0000E+00
0.0C00E+00    0.0C00E+00    3.0C00E+00    0.000 0E+00
0.0C00E+00    0.0C00E+00    0.0C00E+00    4.0C00E+00
```

Lyapunov matrix
```
.709825314742    .191388334C06    7.37096146305E-02    .124853223466
.191388334061    .760062609312    -.159664476836      -.156287 38479
7.37096146305E-02 -.159664476836   .731645978979       .142778407667
.124853223466    -.15628738479    .142778407668       .683011119686
```

The following Theorem provides an extension to the above result.

15. Theorem. The real parts of the eigenvalues of **A** are less than σ if and only if given any symmetric positive definite matrix **Q**, there exists a symmetric positive definite matrix **P** which is the unique solution to.

$$A'P + PA - 2\sigma P = -Q \qquad (20)$$

(See Problem 7.36.) △

For discrete-time systems, the analog of condition (15) is provided by the following theorem.

16. Theorem. System $x(k+1) = Ax(k)$ where **A** is a constant matrix is asymptotically stable if and only if given any symmetric positive definite real matrix **Q**, there exists a symmetric positive definite matrix **P** which is the unique solution to

$$A'PA - P = -Q \qquad (21)$$

17. Example. Let us reconsider Example 7. Let **Q=I**. Then from (17) $P = \begin{bmatrix} p_{11} & p_{12} \\ p_{12} & p_{22} \end{bmatrix}$ is the solution to

$$\begin{bmatrix} -1 & 2 \\ -\frac{1}{4} & 0 \end{bmatrix} \begin{bmatrix} p_{11} & p_{12} \\ p_{12} & p_{22} \end{bmatrix} \begin{bmatrix} -1 & -\frac{1}{4} \\ 2 & 0 \end{bmatrix} - \begin{bmatrix} p_{11} & p_{12} \\ p_{12} & p_{22} \end{bmatrix} = -\begin{bmatrix} 1 & 0 \\ 0 & 1 \end{bmatrix} \qquad (22)$$

The solution is $\mathbf{P} = \begin{bmatrix} 12 & 2 \\ 2 & 7/4 \end{bmatrix}$ which is positive definite. This, by Theorem 15, confirms that the system is asymptotically stable.

PROBLEMS

7.1 Give an example of a l.t.i. continuous-time system which has equilibrium states other than $\mathbf{x}_e = 0$.

7.2 Consider a nonlinear system described by

$$\dot{x}(t) = x^3(t) + 6x^2(t) + 11x(t) + 6$$

a) Find the equilibrium states of this system.

b) Determine new state variables such that each equilibrium state becomes $\mathbf{x}_e = 0$.

7.3 Determine all the equilibrium states of the discrete-time system

$$\mathbf{x}(k+1) = \begin{bmatrix} 1 & 0 \\ 0 & 2 \end{bmatrix} \mathbf{x}(k).$$

7.4 Show that the equilibrium state $\mathbf{x} = 0$ of the n-dimensional system $\dot{\mathbf{x}}(t) = \mathbf{A}(t)\mathbf{x}(t)$ is stable *isL* if and only if for each $t_o \geq 0$ and each $\mathbf{x}(t_o) \in R^n$, there exists a finite constant $M(t_o)$ such that $\| \mathbf{x}(t) \| \leq M(t_o) \| \mathbf{x}(t_o) \|$ for all $t \geq t_o$.

7.5 If the linear system $\dot{\mathbf{x}}(t) = \mathbf{A}(t)\mathbf{x}(t)$ is asymptotically stable, show that for any initial state $\mathbf{x}_o$ and for any given positive constant $\mathbf{M}$, there exists a time $t \geq t_o$ which depends on $\mathbf{x}_o$ such that $\| \mathbf{x}(t,\mathbf{x}_o,t_o) \| < \mathbf{M}$.

7.6 [7.5] Consider the two-dimensional nonlinear system described by the equations $\dot{r} = r^2 - r$ $r \geq 0$ $\dot{\theta} = r$ where polar coordinates are used for the state space. Show that the equilibrium state $(r,\theta)_e = (0,0)$ is asymptotically stable for $r(t_o) < 1$ but unstable for $r(t_o) > 1$.

7.7 Show that the following is equivalent to Definition 7.2.3. System $\dot{\mathbf{x}} = \mathbf{A}(t)\mathbf{x}(t)$ is asymptotically stable if for any sufficiently small $\delta > 0$ and any $\epsilon > 0$ there exists a time T such that $\| \mathbf{x}_o \| < \delta$ implies $\| \mathbf{x}(t) \| < \epsilon$ $\forall t \geq t_o + T$. Further, show that δ and T are independent of t_o for uniform asymptotic stability.

7.8 Consider the l.t.i. discrete-time system (7.3.19) where $\mathbf{A}$ does not have linearly independent eigenvectors. Use the Jordan cononical form of $\mathbf{A}$ to extend the argument leading to Corollaries No. 9 and No. 10 for this case.

PROBLEMS

7.9 Use program "STABCD" to determine zero input stability of the following systems:

a) $\dot{x}(t) = \begin{bmatrix} 0 & 1 & 0 & 0 \\ 0 & 0 & 1 & 0 \\ 0 & 0 & 0 & 1 \\ -1 & -9 & -3 & -4 \end{bmatrix} x(t), \quad t \geq 0$

b) $x(t), \quad t \geq 0 \begin{bmatrix} 0 & 1 & 2 & -4 \\ 2 & 4 & -2 & 0 \\ 1 & -8 & 0 & 3 \\ 4 & 2 & -1 & 1 \end{bmatrix} x(k), \quad k \geq 0.$

7.10 Consider the MIMO l.t.i. system

$$\dot{x} = \begin{bmatrix} 0 & 0 & 0 & 0 \\ 1 & -1 & 0 & a_3 \\ 0 & a_1 & 0 & a_4 \\ 0 & a_2 & 1 & 1 \end{bmatrix} x + \begin{bmatrix} a_5 & a_6 \\ 0 & 0 \\ a_7 & a_8 \\ 0 & 0 \end{bmatrix} u$$

Find ranges for the constants a_i, $i=1,2,...,8$ such that this system is

i) stable isL,

ii) asymptotically stable, and

iii) bibo stable. Verify your results by choosing some values in the corresponding ranges and applying "STABCD".

7.11 [7.5] Show that the linear time-varying system $\dot{x} = \dfrac{-x}{2t} + u(t), \, t \geq 1; \, y(t) = x(t)$ is asymptotically stable but is not bibo stable.

7.12 Give an example of a l.t.i. system which is stable isL but is not bibo stable. Hint: Consider a l.t.i. network with natural frequencies $\pm j\omega$. Apply a sinusoidal input $u(t) = \text{Sin } \omega t$.

7.13 Use the Routh-Hurwitz criterion to determine the stability of each of the following systems:

(a) $G(s) = 2(s-2)/s^2 (s+2), \quad H(s) = 1$

(b) $G(s) = (s+4)/(s^2+2s+2), \quad H(s) = 1/(s+2).$

374   LINEAR CONTROL SYSTEMS

7.14 Determine the stability of the systems whose characteristic equations are given below:

(a) $8s^3 + 6s^2 + 12s + 9 = 0$

(b) $s^5 + 3s^4 + 14s^2 + 200s + 100 = 0$

(c) $s^5 + 2s^3 + 2s = 0$

7.15 Find the range of value(s) of K for the system, $G(s)H(s) = K(s+1)/(s^2(s+2)(s+5))$ to be stable.

7.16 Consider a unity feedback $(H(s)=1)$ system

$$G(s) = K(1+Ts)/(s^2(s+1))$$

Find the region in the $(T-K)$ plane where this system would be stable.

7.17 A feedback control system has open-loop and feedback transfer functions:

$$G(s) = K(s+4)/(s(s+1)), \quad H(s) = 1/(s+2).$$

(a) Determine the value of K which would just make system unstable.

(b) Find the imaginary closed-loop poles of the system corresponding to the gain K which puts the system on the verge of instability.

(c) Reduce this value of gain by a factor of 2 and determine the relative stability of the system by shifting the $j\omega$-axis and applying the Routh-Hurwitz criterion.

7.18 Use program "ROUTH" for problems 7-12 and 7.13.

7.19 Determine the stability conditions of discrete-time systems described by the following characteristic equations:

(a) $z^4 + 9z^3 + 5z^2 + 9z + 1 = 0$

(b) $z^3 - 2.5z^2 - 2z + 2 = 0$

(c) $z^3 + z^2 + z + 1 = 0$

PROBLEMS

7.20 Repeat problem 7.19 using program "JURY".

7.21 Find the values of the gain K which would make a discrete-time system with the following characteristic equation unstable.

$$z^3+Kz^2+1.5Kz-(K+2)=0$$

7.22 A discrete-time system is represented by the following state equation,

$$\mathbf{x}(k+1) = \begin{bmatrix} 0 & 1 \\ -1 & 0 \end{bmatrix} \mathbf{x}(k)$$

with a specified initial state. Determine whether this system is stable by Jury-Blanchard's test. Hint: The system's characteristic equation is $\det(I-z^{-1}A) = 0$.

7.23 Draw the root locus for each of the following transfer functions:

(a) $G(s)H(s) = K/((s+1)(s+3))$

(b) $G(s)H(s) = K(s+2)/(s(s^2+2s+2))$

(c) $G(z)H(z) = K(1+z)/((z(1+2z)(1+5z))$

(d) $G(s)H(s) = K/((s+1)(s^2+4s+7)(s+4))$

7.24 Consider an equation

$$s^3+5s^2+12s+8 = 0$$

Can you use root locus to find the roots of this equation?

Hint: Find a transfer function $G(s)H(s)$ whose characteristic equation is the above equation.

7.25 Use "ROOTLC" to draw the root loci of Problem 7.23.

7.26 Consider that a system has a transfer function, $G(s)H(s) = K/(s(s^2+2s+2))$ and is operating at $K=2$. Find the gain margin of this system using root locus method.

7.27 For each of the transfer functions,

a) $G(s) = 10/(s(1+0.25s)(1+0.1s))$

b) $G(s) = 20(2+s)/((1+s)(1+0.1s)$

c) $G(s) = 40(s+1)/(s(s^2+2s+5))$

draw the Bode diagram. In each case determine the gain and phase margins.

7.28 Consider the log magnitude plots shown in Fig. P7.28.

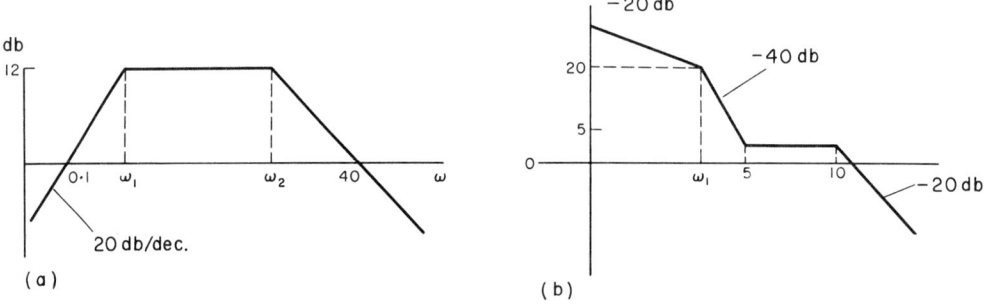

Fig. P7.28

Determine the corresponding transfer functions.

7.29 Draw the polar plot of the systems described in Problem 7-27 and find phase and gain margins. Which one of the systems is stable?

7.30 Polar plots of four system along with their open-loop r.h.p poles are shown. Complete the Nyquist diagrams and apply the stability criterion to determine whether they are stable. Let $H(s)=1$.

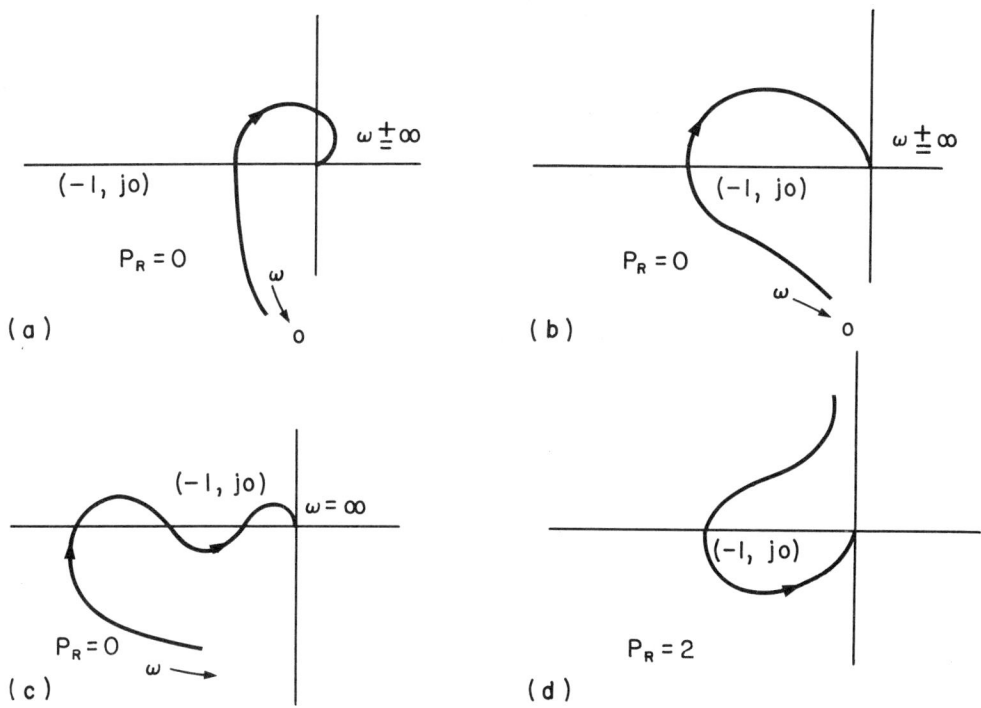

Fig. P7.30

7.31 Use "BNPLOT" to plot the Bode and Nyquist diagram for Problems 7-27.

7.32 Consider the system of Example 7.7.5 with $\tau_1 = 0.5$ and $\tau_1 = 1$. Draw the Bode diagram, Nyquist diagram and the root locus of the system and compare the stability conclusion for the system on each plot in terms of the gain margin and the Routh-Hurwitz criterion.

7.33 A conditionally stable system is given below,

$$G(s)H(s) = K(s+2)(s+3)/(s^2(s+1)(s+16)(s+32))$$

determine the ranges of values of K for which the system is stable and unstable. Sketch an approximate root locus to confirm your result.

7.34 An alternative frequency response is the plot of log-magnitude of $G(s)H(s)$ (in db's) versus its phase angle, called *Nichol's plot*. This plot is essentially obtained by eliminating the frequency ω from the magnitude and phase plots of a Bode diagram. Draw the Nichol's plot for the system of Example 7.7.5.

7.35 Consider the system shown in Fig. 7.8.1 where $G(s) = K \dfrac{s+1}{s(s+2)}$, $K>0$. Use the circle criterion to find conditions on K and $k(t)$ for asymptotic stability of this system.

378   LINEAR CONTROL SYSTEMS

7.36 Consider the nonlinear system

$$\dot{x}_1 = x_2 - 2x_1(x_1^2 + x_2^2)$$

$$\dot{x}_2 = -x_1 - 2x_2(x_1^2 + x_2^2)$$

Find a Lyapunov function for this system to show that the equilibrium state $x_e = 0$ is asymptotically stable in the large.

Hint: Try $V(x) = ax_1^2 + x_2^2$ where a, b>0.

7.37 Show that if Q is symmetric positive definite, $PA + A'P - 2\sigma P = -Q$ has positive definite solution P is and only if the eigenvalues of A have real parts less than σ.

Hint: Let $A = \hat{A} - \sigma I$ and apply Theorem 7.9.9 to $\hat{A}$.

7.38 Consider the system $\dot{x}(t) = A(t)x(t)$ with initial state $x(t_o)=x_o$ where $A(t)$ is continuous. Let $\lambda_{\min}(t)$ and $\lambda_{\max}(t)$ be the smallest and largest eigenvalues of $H = \frac{1}{2}(A + A')$ for each t. Use the Lyapunov function $V(x) = \| x \|_2^2$ to show Wazewski's inequality [7.5]:

$$\| x_o \|_2 \exp(\int_{t_o}^{t} \lambda_{\min}(\tau)d\tau) \leq \| x \|_2 \leq \| x_o \|_2 \exp(\int_{t_o}^{t} \lambda_{\max}(\tau)d\tau)$$

7.39 If in system $\dot{x} = A(t)x$, $A(t)$ is an $n \times n$ normal matrix with eigenvalues $\lambda_j(t)$, $i=1,2,...n$, show that: (i) The system is uniformly asymptotically stable if $\text{Re}[\lambda_j(t)]<0$ for all i and t. (ii) The System is unstable if $\text{Re}[\lambda_j(t)]>0$ for all i and t.

7.40 Use the Lyapunov's method to derive necessary and sufficient conditions for the asymptotic stability of the following l.t.i. systems: (i) $\ddot{x} + a\dot{x} + bx = 0$, (ii) $x(k+2) + ax(k+1) + bx(k) = 0$.

7.41 Study the stability of the system

$$\dot{x} = \begin{bmatrix} 0 & 1 & 0 \\ 0 & 0 & 1 \\ -2 & -5 & -4 \end{bmatrix} x.$$

Use "LYAP" to verify your result by using the Lyapunov's method.

REFERENCES

[7.1] W. Hahn, *Stability of Motion,* Springer-Verlag, New York, 1967

[7.2] C.A. Desoer, "Slowly varying system $\dot{x}=A(t)x$, *IEEE Trans. Auto. Control,* Vol. AC-14, pp. 780-781, Dec., 1969

[7.3] M. Malek-Zavarei, " The stability of linear time-varying systems", *Int. J. Control,* Vol. 27, no. 5, pp. 809-815, 1978

[7.4] H.H. Rosenbrock, "The stability of linear time- dependent control systems", *J. Electronics and Control,* Vol. 15, p. 73-80, July 1963

[7.5] L. A. Zadeh and C. A. Desoer, *Linear System Theory,* McGraw-Hill, 1963

[7.6] R. W. Brocket, *Finite Dimensional Linear Systems,* Wiley, 1970

[7.7] A. Hurwitz, " On the Conditions Under which an Equation Has Only Roots with Negative Real Parts", *Mathematisch Annalen,* Vol. 46, 1895, pp. 273-284. Also in selected papers on *Mathematical Trends in Control Theory,* Dover, New York, 1964, pp. 70-82

[7.8] E. J. Routh, *Dynamics of a System of Rigid Bodies,* Macmillan, New York, 1982

[7.9] E. I. Jury and B. H. Bharucha, "Notes on the stability criterion for linear discrete systems ", *IRE Trans. Auto. Contr.,* Vol. AC-6, Feb, 1961, pp. 88-90

[7.10] R. H. Raible, "A simplification of Jury's tabular form," *IEEE Trans. Auto. Contr.,* Vol. AC-19, June, 1974, pp. 248-250

[7.11] W. R. Evans, "Graphical Analysis of Control Systems", *Trans. of the AIEE,* Vol. 67, 1948, pp. 547-551. Also in *Automatic Control:Classical Linear Theory,* G. L. Thaler, Ed. Dowden, Hutchinson, and Ross, Inc. Stroudsburg, PA , 1974, pp. 417-421

[7.12] W. R. Evans, "Control System Synthesis by Root Locus Method", *Transactions of the AIEE,* Vol. 69, 1950; pp. 1-4, Also in *Automatic Control: Classical Linear Theory, G. J. Thaler, ed. Dowden, Hutchinson, and Ross, Inc., Stroudsburg, PA., 1974, pp. 423-425*

[7.13] W. R. Evans, Control Systems Dynamics, McGraw-Hill, 1954

[7.14] R. C. Dorf, *Modern Control Systems,* 3rd Edition, Addison-Wesley Publ. Co., Reading, MA, 1980

[7.15] J. J. D'Azzo and C. H. Houpis, *Linear Control System Analysis and Design, McGraw-Hill, New York, 1981*

[7.16] H. W. Bode, "Feedback - The History of and Idea", in *Selected papers in On Mathematical Trends in Control Theory,* Dover, New York, 1964, pp. 106-123

[7.17] H. Nyquist, "Regeneration Theory", *Bell Systems Technical Journal,* Vol 11, Jan., 1932, pp. 126-147. Also in *Automatic Control: Dowden, Hutchinson, and Ross, Inc., Stroudsburg, PA, 1974, pp. 105-126*

[7.18] J. J. Bongiorno, Jr., and D. Graham, "An extension of the Nyquist Barkhauser Stability criterion to linear-lumped- parameter systems with time-varying elements", *IEEE Trans Auto. Contr.,* Vol. AC-8, pp. 166-170, April 1963

[7.19] A. M. Lyapunov, "Problems general de la stabilite du movement", *Ann. Fac. Sci. Toulouse,* Vol. 9, pp. 203-474, 1907. Also reprinted as Vol. 17, *Ann. Math. Studies,* Princeton, NJ 1949

[7.20] R. E. Kalman and J. E. Bertram, "Control System Design via the second method of Lyapunov", *ASME J. Basic Engineering,* Vol. 82, pp. 371-400, 1960

[7.21] M. Jamshidi, "An Overview on the Algebraic Matrix Ricatti Equation and Related Problems", *J. Large Scale Systems,* Vol. 1, pp. 167-172, 1980

[7.22] K. Ogata, *State Space Analysis of Control Systems,* Prentice Hall, 1967

[7.23] E. J. Davison and F. T. Man, "The numerical solution of $A^T Q + QA = -C$", *IEEE Trans. Auto. Contr.,* Vol. AC-13, pp. 448-449, 1968.

CHAPTER 8

CONTROLLABILITY AND OBSERVABILITY

8.1 INTRODUCTION

In a linear system representation (A,B,C,D), matrices A and C describe the unforced or the zero-input system behavior, while matrix B characterizes the effect of input (or control) on the system dynamics. Matrix D represents direct transmission from input to output.

The concepts of controllability and observability were introduced by Kalman [8.1-8.3]. They deal, respectively, with the relationships between the input and the state and between the state and the output of a system. More specifically, system controllability addresses the following question: does a control (or input) u always exist which can transfer the initial state x_o of the system to any desired state x_1 in a finite time? System observability addresses the following question: can the initial state x_o of the system be always identified by observing the output y and the input u over a finite time?

Both of the above questions are of extreme importance in control systems. They can be answered most easily by using the state space methods. In a linear system, the concepts of controllability and observability can be characterized by the properties of the matrices A, B, C, and D. From the above discussion, one would expect that matrices A and B characterize controllability and matrices A and C characterize observability of a linear system. For this reason, often we refer to the controllability of the pair (A,B) and the observability of the pair (A,C).

In this chapter we will begin with a discussion on the concept of linear independence of functions. This concept will be needed in later sections. A section on controllability will then present formal definitions of controllability and the related criteria. A similar section on observability will follow. The relationship between controllability and observability and the concept of duality will then be discussed. Finally, the canonical decomposition of the state space will be explained.

8.2 LINEAR INDEPENDENCE OF FUNCTIONS

1. Definition. A set of functions $f_i(t)$, $i=1, 2, ..., n$ are said to be *linearly independent on* $[t_o, t_1]$ if $\sum_{i=1}^{n} c_i f_i(t) = 0$ for all $t \epsilon [t_o, t_1]$ implies that $c_i=0$, $i=1, 2, ..., n$. If even one of the constants c_i is nonzero, the functions are said to be *linearly dependent on* $[t_o, t_1]$.

2. Theorem. A set of continuous complex-valued functions $f_i(t)$, $i=1, 2, ..., n$ are linearly independent on $[t_0, t_1]$ if and only if $det[G(t_0, t_1)] \neq 0$ where

$$G(t_0, t_1) = \int_{t_0}^{t_1} \mathbf{f}(t)\mathbf{f}^*(t)\,dt \tag{1}$$

and $\mathbf{f}(t)$ is the column vector whose components are $f_i(t)$, $i=1, 2, ..., n$. $G(t_0, t_1)$ is referred to as the *Gram matrix* or the *Grammian* of the functions $f_i(t)$, $i=1, 2, ..., n$.

Proof. We will equivalently prove that the functions $f_i(t)$ are linearly dependent on $[t_0, t_1]$ if and only if $det[G(t_0, t_1)] = 0$.

Necessity. By Definition 4 if $f_i(t)$, $i=1, 2, ..., n$ are linearly dependent on $[t_0, t_1]$, a constant vector $\lambda \neq 0$ exists such that $\lambda^* \mathbf{f}(t) = 0$ for all $t \in [t_0, t_1]$. Thus we have

$$\lambda^* G(t_0, t_1) = \int_{t_0}^{t_1} \lambda^* \mathbf{f}(t)\mathbf{f}^*(t)\,dt = \int_{t_0}^{t_1} 0.\mathbf{f}^*(t)\,dt = 0 \tag{2}$$

which implies $det[G(t_0, t_1)] = 0$.

Sufficiency. If $det[G(t_0, t_1)] = 0$, a constant column vector $\lambda \neq 0$ exists such that $\lambda^* G(t_0, t_1) = 0$. Thus we have

$$0 = 0.\lambda = \lambda^* G(t_0, t_1)\lambda = \int_{t_0}^{t_1} \lambda^* \mathbf{f}(t)\mathbf{f}^*(t)\lambda\,dt = \int_{t_0}^{t_1} \|\lambda^* \mathbf{f}(t)\|^2\,dt \tag{3}$$

which implies $\lambda^* \mathbf{f}(t) = 0$ for $t \in [t_0, t_1]$; i.e. $f_i(t)$, $i=1, 2, ..., n$ are linearly dependent. ▽

Theorem 2 can be easily generalized to the case where "functions" are replaced by "vector functions" as described below.

3. Theorem. A set of continuous complex-valued row vector functions $\mathbf{f}_i(t)$, $i=1, 2, ..., n$ are linearly independent on $[t_0, t_1]$ if an only if $det[G(t_0, t_1)] \neq 0$ where

$$G(t_0, t_1) = \int_{t_0}^{t_1} \mathbf{f}(t)\mathbf{f}^*(t)\,dt \tag{4}$$

and $\mathbf{f}(t)$ is the matrix whose rows are $\mathbf{f}_i(t)$, $i=1, 2, ..., n$.

4. Example. Consider the following vector functions on $[0,1]$:

$$\mathbf{f}_1(t) = [t, 1], \quad \mathbf{f}_2(t) = [t^2, t] \tag{5}$$

We have

$$\mathbf{f}(t) = \begin{bmatrix} \mathbf{f}_1(t) \\ \mathbf{f}_2(t) \end{bmatrix} = \begin{bmatrix} t & 1 \\ t^2 & t \end{bmatrix} \quad (6)$$

and

$$\mathbf{G}(0,1) = \int_0^1 \mathbf{f}(t)\mathbf{f}^*(t)\,dt = \begin{bmatrix} \frac{4}{3} & \frac{3}{4} \\ \frac{3}{3} & \frac{4}{8} \\ \frac{3}{4} & \frac{8}{15} \end{bmatrix} \quad (7)$$

which is nonsingular. Indeed, $\mathbf{G}(0,1)$ is positive definite. (See Problem 8.2.) Thus $\mathbf{f}_1$ and $\mathbf{f}_2$ are linearly independent on $[0,1]$. ▽

The following theorem, which will be presented without proof, provides a sufficient condition for linear independence of vector functions.

5. Theorem (the Wroskian test). Consider a set of continuous complex-valued row vector functions $f_i(t)$, $i=1, 2, \ldots, n$ which are continuously differentiable up to order $n-1$ on $[t_o, t_1]$. These functions are linearly independent on $[t_o, t_1]$ if some $t \in [t_o, t_1]$ exists such that the matrix $[\mathbf{f}(t), \dot{\mathbf{f}}(t), \mathbf{f}^{(2)}(t), \ldots, \mathbf{f}^{(n-1)}(t)]$ has rank n.

6. Example. Again consider the vector functions $\mathbf{f}_1(t)$ and $\mathbf{f}_2(t)$ in Example 4. We have

$$[\mathbf{f}(t), \dot{\mathbf{f}}(t)] = \begin{bmatrix} t & 1 & 1 & 0 \\ t^2 & t & 2t & 1 \end{bmatrix} \quad (8)$$

which has rank $n=2$ for any $t \in [0, 1]$. Thus $\mathbf{f}_1$ and $\mathbf{f}_2$ in (8) are linearly independent on $[0,1]$. ▽

The following two theorems are the discrete-time versions of Theorems 3 and 5. They will be stated without proofs.

7. Theorem. A set of discrete complex-valued row vector functions $\mathbf{f}_i(k)$, $i=1, 2, \ldots, n$ are linearly independent on the interval $[k_o, k_1]$ if and only if $\det[\mathbf{G}(k_o, k_1)] \neq 0$ where

$$\mathbf{G}(k_o, k_1) = \sum_{k=k_o}^{k_1-1} \mathbf{f}(k)\mathbf{f}^*(k) \quad (9)$$

and $\mathbf{f}(k)$ is the matrix whose rows are $\mathbf{f}_i(k)$, $i=1, 2, \ldots, n$.

8. Theorem. Consider a set of discrete complex-valued row vector functions $\mathbf{f}_i(k)$, $i=1, 2, \ldots, n$ which have n-1 differences on $[k_o, k_1]$. These functions are linearly independent on $[k_o, k_1]$ if some $k \in [k_o, k_1]$ exists such that the matrix $[\mathbf{f}(k), \mathbf{f}(k+1), \mathbf{f}(k+2), \ldots, \mathbf{f}(k+n-1)]$ has rank n.

9. Example. Consider the discrete vector functions

$$\mathbf{f}_1(k) = [1,k], \quad \mathbf{f}_2(k) = [k+1,0], \quad \mathbf{f}_3(k) = [0,5] \tag{10}$$

over the interval [0,3]. We have

$$\mathbf{f}(k) = \begin{bmatrix} \mathbf{f}_1(k) \\ \mathbf{f}_2(k) \\ \mathbf{f}_3(k) \end{bmatrix} = \begin{bmatrix} 1 & k \\ k+1 & 0 \\ 0 & 5 \end{bmatrix} \tag{11}$$

and

$$[\mathbf{f}(k),\mathbf{f}(k+1),\mathbf{f}(k+2)] = \begin{bmatrix} 1 & k & 1 & k+1 & 1 & K \\ k+1 & 0 & k+2 & 0 & k+3 & 0 \\ 0 & 5 & 0 & 5 & 0 & 5 \end{bmatrix} \tag{12}$$

The above matrix has rank $n=3$ for all k. Thus the vector functions in (10) are linearly independent over any interval. (See Problem 8.4.)

8.3 CONTROLLABILITY

8.3.1 Definitions and Examples

Consider the linear continuous-time system $(\mathbf{A}(t),\mathbf{B}(t),\mathbf{C}(t),\mathbf{D}(t))$, i. e. the system represented by the state and output equations

$$\dot{\mathbf{x}}(t)=\mathbf{A}(t)\mathbf{x}(t)+\mathbf{B}(t)\mathbf{u}(t), \quad t \geqslant t_o; \quad \mathbf{x}(t_o)=\mathbf{x}_o \tag{1a}$$

$$\mathbf{y}(t)=\mathbf{C}(t)\mathbf{x}(t)+\mathbf{D}(t)\mathbf{u}(t) \tag{1b}$$

where **A**, **B**, **C** and **D** are continuous functions of time. Suppose that for some input $\mathbf{u}(t)$, $t \in [t_o, t_1]$ and for the initial state $\mathbf{x}_o$, the state at time t_1 is $\mathbf{x}_1$. We say that input **u** *transfers* $\mathbf{x}_o$ (at t_o) to $\mathbf{x}_1$ (at t_1).

1. Definition. The state $\mathbf{x}_o$ of system (1) is said to be *controllable* on the interval $[t_o, t_1]$ where t_1 is a finite time if some control **u** over $[t_o, t_1]$ exists which transfers $\mathbf{x}_o$ (at t_o) to the origin of the state space at t_1. Otherwise, the state $\mathbf{x}_o$ is said to be *uncontrollable* on $[t_o, t_1]$. ▽

Note that any initial state $\mathbf{x}_o$ can be transferred to the state $\mathbf{x}_1=0$ at finite time t_1 by some control **u** on $[t_o, t_1]$ if and only if any $\mathbf{x}_o$ can be transferred to any *arbitrary* state $\mathbf{x}_1$ at t_1 by some control **u** on $[t_o, t_1]$. (See Problem 8.5.) Also note that the control that transfers $\mathbf{x}_o$ to $\mathbf{x}_1$ is not necessarily unique.

2. Definition. If every state $\mathbf{x}(t_o)$ of a system is controllable on $[t_o, t_1]$, the system is said to be *completely controllable* on $[t_o, t_1]$.

3. Example.
Consider the system (A, B, C, D) where

$$A = \begin{bmatrix} -1 & 0 \\ 0 & -1 \end{bmatrix}, B = \begin{bmatrix} 1 \\ 0 \end{bmatrix}, C = [1, 1], D = 0 \qquad (2)$$

We have

$$\dot{x}_1 = -x_1 + u, \quad x_1(0) = x_{1o} \qquad (3a)$$

$$\dot{x}_2 = -x_2, \quad x_2(0) = x_{2o} \qquad (3b)$$

$$y = x_1 + x_2 \qquad (3c)$$

From (3a) note that an input u on $[0, t_1]$ exists which transfers any x_{1o} to any arbitrary $x_1(t_1)$. This, however, is not the case for x_2 because x_2 is not coupled to the input u. Using Definition 1 we conclude that in this system only the states of the type $[x_1 \ 0]'$ are controllable. Thus, by Definition 2, this system is not completely controllable. Figure 1 shows a representation of this system. ▽

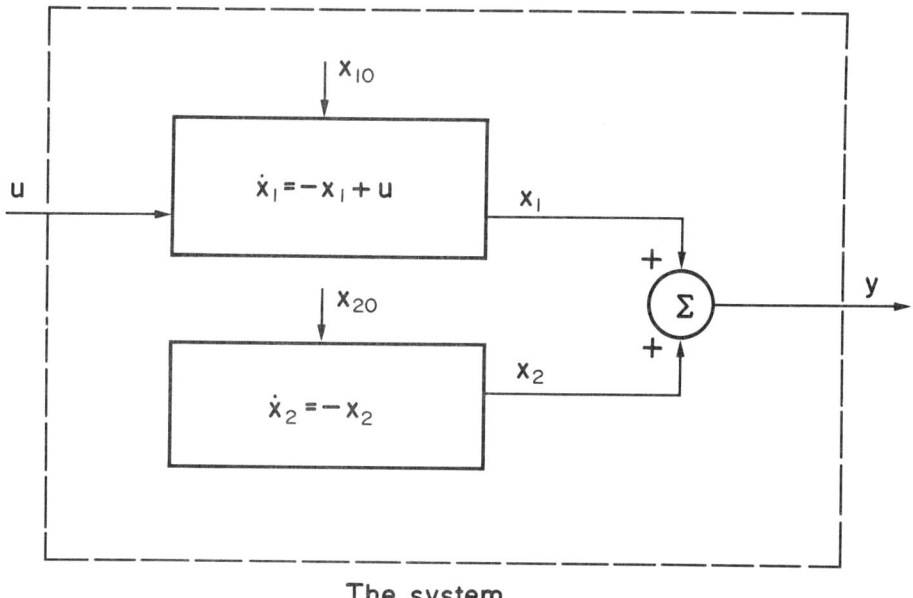

Fig. 8.3.1. Example of a system which is not completely controllable

If instead of t we use k (i.e. replace t_o by k_o, $[t_o, t_1]$ by $[k_o, k_1]$, etc.), Definitions 1 and 2 also apply to linear discrete-time systems.

8.3.2 Characterization of Controllability

In this section we will present controllability criteria for linear systems. First consider the linear time-varying continuous-time system

$$\dot{x}(t) = A(t)x(t) + B(t)u(t), \quad t \geq t_o \tag{4}$$

Consider the following theorem.

4. Theorem. System (4) is completely controllable on $[t_o, t_1]$ if and only if the rows of the matrix $\Phi^{-1}(t)B$ are linearly independent on $[t_o, t_1]$ where $\Phi(t)$ is any fundamental matrix of the unforced system (i.e. when $u=0$); or equivalently, if and only if the matrix

$$M(t_o, t_1) = \int_{t_o}^{t_1} \Phi(t_1,\tau)B(\tau)B^*(\tau)\Phi^*(t_1,\tau)d\tau \tag{5}$$

is nonsingular. Furthermore, the control $u(t)$ which transfers the state of the completely controllable system from $x(t_o)=x_o$ to $x(t_1)=0$ is

$$u(t) = -B^*(t)\Phi^*(t_1, t)M^{-1}(t_o, t_1)\Phi(t_1, t_o)x_o \tag{6}$$

Proof. We will prove that system (4) is completely controllable if and only if matrix $M(t_o, t_1)$ in (5) is nonsingular. The equivalence between the above two conditions follows from Theorem 8.2.3 and the transition property of the transition matrix: $\Phi(t,\tau)=\Phi(t)\Phi^{-1}(\tau)$. If a control $u(t)$, $t\in[t_o, t_1]$ transfers $x(t_o)=x_o$ to $x(t_1)=0$, from (6.6.2) we have

$$-\Phi(t_1, t_o)x_o = \int_{t_o}^{t_1} \Phi(t_1,\tau)B(\tau)u(\tau)d\tau \tag{7}$$

Necessity. We will prove necessity by contradiction. Suppose that the system is completely controllable but $M(t_o, t_1)$ in (5) is singular. Then a vector $\lambda \neq 0$ exists such that

$$\lambda^*\Phi(t_1, \tau)B(\tau)=0 \text{ for all } \tau\in[t_o, t_1]. \tag{8}$$

Let $x_o = \Phi^*(t_1, t_o)\lambda$. By the assumption of complete controllability, an input u characterized by (7) exists which can transfer this x_o to the origin of the state space in a finite time t_1-t_o. Therefore, from (7) we have

$$-\lambda^*\Phi(t_1, t_o)\Phi^*(t_1, t_o)\lambda = \lambda^*\int_{t_o}^{t_1} \Phi(t_1,\tau)B(\tau)u(\tau)d\tau \tag{9}$$

which by (8) is identically zero, implying $\|\lambda^*\Phi(t_1, t_o)\| = 0$ or $\lambda^*\Phi(t_1, t_o)=0$. Since the transition matrix $\Phi(t_1, t_o)$ is nonsingular, this implies $\lambda^*=0$ which contradicts the original assumption $\lambda \neq 0$.

Therefore, $\mathbf{M}(t_o, t_1)$ is nonsingular.

Sufficiency. Assume that $\mathbf{M}(t_o, t_1)$ in (5) is nonsingular. Multiply both sides of (5) on the right by $-\mathbf{M}^{-1}(t_o, t_1)\Phi(t_1, t_o)\mathbf{x}_o$ to obtain

$$-\Phi(t_1, t_o)\mathbf{x}_o = -\int_{t_0}^{t_1} \Phi(t_1, \tau)\mathbf{B}(\tau)\mathbf{B}^*(\tau)\Phi^*(t_1, \tau)\mathbf{M}^{-1}(t_o, t_1)\Phi(t_1, t_o)\mathbf{x}_o \, d\tau \qquad (10)$$

Comparison of (10) with (7) yields the control given in (6) which transfers any initial state $\mathbf{x}_o$ to $\mathbf{x}_1 = \mathbf{0}$. Thus the system is completely controllable. ▽

The following theorem characterizes complete controllability of the linear time-varying discrete-time system

$$\mathbf{x}(k+1) = \mathbf{A}(k)\mathbf{x}(k) + \mathbf{B}(k)\mathbf{u}(k), \quad k \geqslant k_o \qquad (11)$$

It is analogous to the above theorem and will be stated without proof.

5. Theorem. System (11) is completely controllable on $[k_o, k_1]$ if and only if the rows of the matrix $\Phi^{-1}(k+1)\mathbf{B}(k)$ are linearly independent on $[k_o, k_1]$ where $\Phi(k)$ is any fundamental matrix of the unforced system (i.e. when $\mathbf{u}=\mathbf{0}$); or equivalently, if and only if the matrix

$$\mathbf{M}(k_o, k_1) = \sum_{j=k_o}^{k_1} \Phi(k_1, j+1)\mathbf{B}(j)\mathbf{B}^*(j)\Phi^*(k_1, j+1) \qquad (12)$$

is nonsingular. Furthermore, the control $\mathbf{u}(k)$ which transfers the state of the completely controllable system from $\mathbf{x}(k_o) = \mathbf{x}_o$ to $\mathbf{x}(k_1) = \mathbf{0}$ is

$$\mathbf{u}(k) = -\mathbf{B}^*(k)\Phi^*(k_1, k+1)\mathbf{M}^{-1}(k_o, k_1)\Phi(k_1, k_o)\mathbf{x}_o \qquad (13)$$

where $\Phi(k, k_o) = \prod_{j=k_o}^{k-1} \mathbf{A}(j)$ is the transition matrix of the system. ▽

Theorems 4 and 5 are powerful in that they provide necessary and sufficient conditions for complete controllability as well as a method to compute the control. However, their application to time-varying systems is tedious because they require the calculation of the transition matrix of the system and the evaluation of the Gram determinant. (See Problems 8.7 and 8.8.)

For *l.t.i.* systems, Theorems 4 and 5 translate into easily applicable criteria on matrices **A** and **B**. Note that for such systems, due to time invariance, we need not specify controllability on an interval $[t_o, t_1]$ or $[k_o, k_1]$ as in the time-varying case. That is, if a *l.t.i.* system is completely controllable on some interval, then it is completely controllable on any interval.

6. Theorem. The system $\dot{x}(t)=Ax(t)+Bu(t)$ where A and B are $n \times n$ and $n \times r$ constant matrices, respectively, is completely controllable if and only if the $n \times nr$ *controllability matrix*

$$Q_c = [B, AB, A^2 B, \ldots, A^{n-1} B] \tag{14}$$

has rank n. (See Problem 8.12.)

Proof. Necessity. Assume that the system is completely controllable. We will prove, by contradiction, that Q_c has full rank. Suppose that $\rho(Q_c) < n$. Then a nonzero vector λ exists such that $\lambda^* Q_c = 0$ or

$$\lambda^* A^j B = 0, \quad j = 0, 1, 2, \ldots, n-1 \tag{15}$$

This, by Cayley-Hamilton theorem (Theorem 2.9.4)) implies that

$$\lambda^* A^j B = 0 \text{ for } any \; interger \; j \geq 0 \tag{16}$$

Thus we have

$$\lambda^* \sum_{j=0}^{\infty} \frac{A^j (-t)^j}{j!} B = \lambda^* e^{-At} B = 0 \tag{17}$$

Since e^{At} is a fundamental matrix of the system, by Theorem 4, (17) implies that the system is not completely controllable. This is a contradiction; therefore Q_c has rank n.

Sufficiency. We will prove sufficiency also by contradiction. Let $\rho(Q_c) = n$ but assume that the system is not completely controllable. Therefore, form Theorem 4, the rows of $e^{-At} B$ are linearly dependent implying that a vector $\lambda \neq 0$ exists such that $\lambda^* e^{-At} B = 0$ for all t. In particular, $t=0$ implies that $\lambda^* B = 0$. Also

$$\frac{d^k}{dt^k}(\lambda^* e^{-At} B)\Big|_{t=0} = (-1)^k \lambda^* A^k B = 0, \; k=1,2,\ldots,n-1 \tag{18}$$

or $\lambda^* Q_c = 0$. Thus $\rho(Q_c) < n$, which is a contradiction

From Theorem 5 it can be easily proved that the necessary and sufficient condition for complete controllability of the *l.t.i.* discrete-time system $x(k+1) = Ax(k) + Bu(k)$ is that Q_c in (14) has full rank. (See Problem 8.10.)

7. Example. Consider the pair (A, B) where

$$A = \text{diag}(\lambda_1, \lambda_2, \lambda_3), \quad B = \begin{bmatrix} b_1 \\ b_2 \\ b_3 \end{bmatrix} \tag{19}$$

Then the controllability matrix corresponding to this pair is

$$Q_c = [B, AB, A^2B] = \begin{bmatrix} b_1 & \lambda_1 b_1 & \lambda_1^2 b_1 \\ b_2 & \lambda_2 b_2 & \lambda_2^2 b_2 \\ b_3 & \lambda_3 b_3 & \lambda_3^2 b_3 \end{bmatrix}$$

$$= \begin{bmatrix} b_1 & 0 & 0 \\ 0 & b_2 & 0 \\ 0 & 0 & b_3 \end{bmatrix} \begin{bmatrix} 1 & \lambda_1 & \lambda_1^2 \\ 1 & \lambda_2 & \lambda_2^2 \\ 1 & \lambda_3 & \lambda_3^2 \end{bmatrix} \quad (20)$$

Note that the last matrix on the r.h.s. is a Vandermonde matrix. Thus the pair (A,B) in (19) is completely controllable if and only if λ_i, $i=1, 2, 3$ are distinct and b_i, $i=1, 2, 3$ are nonzero.

Program "CONOBS" determines the controllability (and the observability) of MIMO *l.t.i.* continuous-time or discrete-time systems. The source listing and sample test runs of this program appears at the end of Section 8.4.

8.4 OBSERVABILITY

8.4.1 Definitions and Examples

Consider the linear continuous-time system represented by the state and output equations

$$\dot{x}(t) = A(t)x(t) + B(t)u(t), \quad t \geq t_o, \quad x(t_o) = x_o \quad (1a)$$

$$y(t) = C(t)x(t) + D(t)u(t) \quad (1b)$$

where matrices A, B, C and D are continuous functions of time. The complete solution to (1) was given in (6.6.8) and is repeated here for convenience:

$$y(t) = C(t)\Phi(t, t_o)x_o + C(t)\int_{t_o}^{t} \Phi(t,\tau)B(\tau)u(\tau)d\tau + D(t)u(t), \quad t \geq t_o \quad (2)$$

The concept of observability is concerned with the following problem: given system (1) and its input $u(t)$ and output $y(t)$ over a finite interval $[t_o, t_1]$, calculate the initial state x_o. Formally, we have the following definitions.

1. Definition. The state $x_o \neq 0$ of system (1) is said to be *observable* on the interval $[t_o, t_1]$ where t_1 is a finite time if the knowledge of the input $u(t)$ and the output $y(t)$ over $[t_o, t_1]$ suffice to determine x_o. Otherwise, the state x_o is said to be *unobservable* on $[t_o, t_1]$.

2. Definition. If every state of a system is observable on $[t_o, t_1]$, the system is said to be *completely observable* on $[t_o, t_1]$. ▽

Since the zero-state response (the terms containing u in (2)) can be calculated directly, the problem of system observability can be addressed when **u=0**. That is, the problem becomes: given system

(1) and its zero-input response $C(t)\Phi(t, t_o)x_o$ over the finite interval $[t_o, t_1]$, find the initial state x_o. This immediately implies that only matrices **A** and **C** in system representation (1) are involved in characterizing system observability.

3. Example. Consider the system (A,B,C,D) where

$$A=\begin{bmatrix} 1 & 0 \\ 0 & 2 \end{bmatrix}, \quad B=\begin{bmatrix} 1 \\ 1 \end{bmatrix} \quad C=[1, 0], \quad D=0 \tag{3}$$

We have

$$\dot{x}_1 = x_1 + u, \quad x_1(0) = x_{1o} \tag{4a}$$

$$\dot{x}_2 = 2x_2 + u, \quad x_2(0) = x_{2o} \tag{4b}$$

$$y = x_1 \tag{4c}$$

From (4c) and (4a) note that y depends only on x_1 but x_1 is completely independent of x_2. That is, the knowledge of u and y over a finite interval $[0,t_1]$ is sufficient to determine x_{1o} but not x_{2o}. Using Definition 1 we conclude that in this system only the states of the type $[x_{1o}\ 0]'$ are observable. Thus, by Definition 2, this system is not completely observable. Figure 1 shows a representation of this system. Note that we could let $u=0$ with no effect on the above discussion.
▽

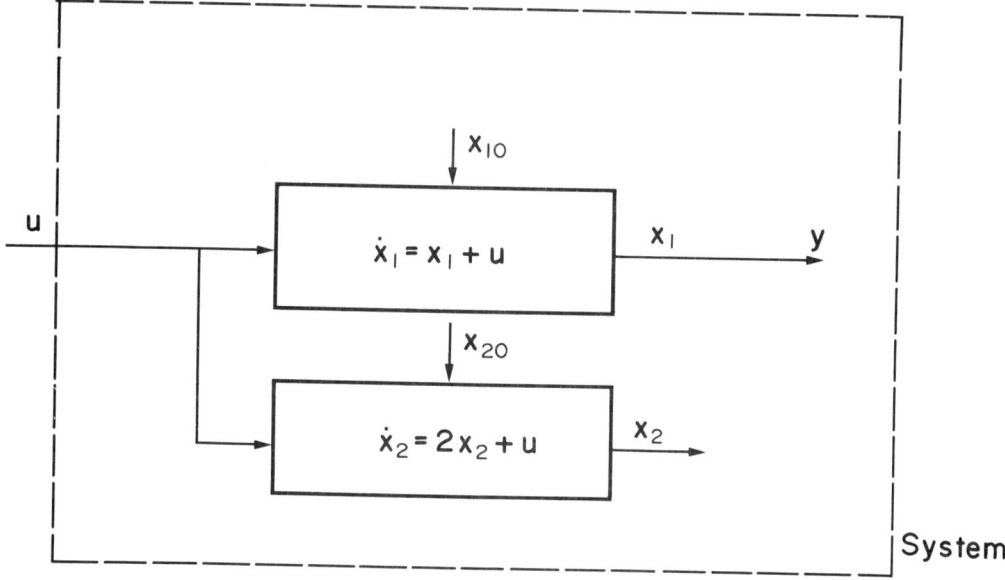

Fig. 8.4.1. Example of a system which is not completely observable

OBSERVABILITY

If instead of t we use k (i.e. replace t_o by k_o, $[t_o, t_1]$ by $[k_o, k_1]$, etc.), Definitions 1 and 2 also apply to linear discrete-time systems.

8.4.2 Characterization of Observability

In this section we will present observability criteria for linear systems. First consider the unforced linear time-varying continuous-time system

$$\dot{x}(t)=A(t)x(t), \quad t \geqslant t_o, \quad x(t_o)=x_o \qquad (5a)$$

$$y(t)=C(t)x(t) \qquad (5b)$$

4. Theorem. System (5) is completely observable on $[t_o, t_1]$ if and only if the columns of the matrix $C(t)\Phi(t)$ are linearly independent on $[t_o, t_1]$ where $\Phi(t)$ is any fundamental matrix of (5a), or equivalently, if and only if the matrix

$$N(t_o, t_1) = \int_{t_o}^{t_1} \Phi^*(t, t_o) C^*(t) C(t) \Phi(t, t_o) dt \qquad (6)$$

is nonsingular. Furthermore, the initial state is

$$x_o = N^{-1}(t_o, t_1) \int_{t_o}^{t_1} \Phi^*(t, t_o) C^*(t) y(t) d \qquad (7)$$

Proof. The equivalence between the above two conditions follows from Theorem 8.2.3 and the transition property of $\Phi(t,\tau)$. Note that for system (5) we have

$$y(t) = C(t)\Phi(t, t_o)x_o \qquad (8)$$

Necessity. We will prove necessity by contradiction. Assume that system (5) is completely observable but $N(t_o, t_1)$ in (6) is singular. From (8) we have

$$\int_{t_o}^{t_1} y^*(t)y(t)dt = \int_{t_o}^{t_1} x_o^* \Phi^*(t, t_o) C^*(t) C(t) \Phi(t, t_o) x_o dt$$

$$= x_o^* N(t_o, t_1) x_o \qquad (9)$$

Since $N(t_o, t_1)$ is singular, a state $x_o \neq 0$ exists such that the quadratic form on the r.h.s. of (9) is 0, implying that $y(t)=0$ for $t \in [t_o, t_1]$. Therefore such an x_o cannot be determined, contradicting our original assumption of complete observability. Thus $N(t_o, t_1)$ is nonsingular.

Sufficiency. Multiplying both sides of (8) by $\Phi^*(t, t_o)C^*(t)$, integrating from t_o to t_1 and using (6) yields

$$\int_{t_o}^{t_1} \Phi^*(t, t_o)C^*(t)y(t)dt = N(t_o, t_1)x_o \tag{10}$$

If $N(t_o, t_1)$ is nonsingular, we can find x_o from (10) as given in (7). ▽

The following theorem characterizes complete observability of the unforced linear time-varying discrete-time systems

$$x(k+1) = A(k)x(k), \quad k \geq k_o \tag{11a}$$

$$y(k) = C(k)x(k) \tag{11b}$$

It is analogous to the above theorem and will be stated without proof.

5. Theorem. System (11) is completely observable on $[k_o, k_1]$ if and only if the columns of the matrix $C(k)\Phi(k)$ are linearly independent on $[k_o, k_1]$ where $\Phi(k)$ is any fundamental matrix of (11a). ▽

For l.t.i. systems, Theorems 4 and 5 translate into easily applicable criteria on matrices A and C. Again note that for such systems, due to time-invariance, we need not specify observability on an interval $[t_o, t_1]$ or $[k_o, k_1]$, as we do in the time-varying case.

6. Theorem. The unforced l.t.i. systems $\dot{x}(t) = Ax(t), y(t) = Cx(t)$ and $x(k+1) = Ax(k), y(k) = Cx(k)$ where A and C are $n \times n$ and $m \times n$ constant matrices, respectively, are completely observable if and only if the $mn \times n$ observability matrix

$$Q_o = \begin{bmatrix} C \\ CA \\ CA^2 \\ \cdot \\ \cdot \\ CA^{n-1} \end{bmatrix} \tag{12}$$

has rank n. (See Problem 8.12.) ▽

The above theorem can be proved by using the results of Theorems 4 and 5.

7. Example. Consider a system described by the following observable companion form:

$$\dot{x} = \begin{bmatrix} 0 & 0 & -6 \\ 1 & 0 & -5 \\ 0 & 1 & -7 \end{bmatrix} x, \quad y = [0 \ 0 \ 1]x \tag{13}$$

We have

$$Q_o = \begin{bmatrix} C \\ CA \\ CA^2 \end{bmatrix} = \begin{bmatrix} 0 & 0 & 1 \\ 0 & 1 & -7 \\ 1 & -7 & 44 \end{bmatrix} \quad (14)$$

which has full rank; thus the system is completely observable. This justifies the adjective "observable" for the above companion form. (See Section 4.4.3.) ▽

Program "CONOBS" determines, the controllability and the observability of MIMO *l.t.i.* systems. The source listing of "CONOBS" followed by two application examples follows.

8. "CONOBS"

```
10    !    PROGRAM NAME : "CONOBS" PROG
20         INPUT "Give your PRINT option: For CRT input 1; for LINE PRINTER input 2",
                 Pr
30         IF (Pr=1) OR (Pr=2) THEN 50
40         GOTO 20
50         IF Pr=1 THEN PRINTER IS 16
60         IF Pr=2 THEN PRINTER IS 7,1
70         PRINT " 'CONOBS' DETERMINES WHETHRE A MULTIVARIABLE LINEAR TIME-INVARIANT"
80         PRINT "CONTINUOUS-TIME SYSTEM dx/dt=Ax+Bu, y=Cx+Du OR DISCRETE-TIME SYSTEM"
90         PRINT "x(k)=Ax(k)+Bu(k), y(k)=Cx(k)+Du(k) IS COMPLETELY CONTROLLABLE OR"
100        PRINT "OBSERVABLE.";LIN(1)
110        PRINT "Matrix dimensions are:   A:NxN,   B:NxR,   C:MxN,   D:MxR"
120        PRINT "Matrix D does not influence system controllability or observability.
                 ";LIN(2)
130        INPUT "Have SUB's <<Mat>>, <<Rank>>, <<Cont>> & <<Obs>> already been LINKED
                 Y/N)?",C$
140        IF C$="Y" THEN 210
150        IF C$="N" THEN 170
160        GOTO 130
170        LINK "Mat",1000,180
180        LINK "Rank",2100,190
190        LINK "Cont",3500,200
200        LINK "Obs",4000,210
210        INPUT "Give matrix dimensions N, R, M (R<=N, M<=N):",N,R,M
220        PRINT "Matrix dimensions :   N=";N;",   R=";R;",   M=";M;LIN(2)
230        Nc=(N-R+1)*R
240        No=(N-M+1)*M
250        CALL C(N,R,M,Nc,No)
260        PRINTER IS 16
270        END
280        SUB C(N,R,M,Nc,No)
290        OPTION BASE 1
300        DIM A(N,N),B(N,R),C(M,N),Qc(N,Nc),Qo(No,N)
310        CALL Mat(A(*),N,N,"A")
320        CALL Mat(B(*),N,R,"B")
330        CALL Mat(C(*),M,N,"C")
340        !    Construct the CONTROLLABILITY matrix:
350        CALL Cont(A(*),B(*),N,R,Qc(*))
360        PRINT "CONTROLLABILITY Matrix Qc=",LIN(1)
370        FLOAT 4
380        MAT PRINT Qc
390        !    Construct the OBSERVABILITY matrix:
400        CALL Obs(A(*),C(*),N,M,Qo(*))
410        PRINT LIN(2),"OBSERVABILITY Matrix Qo=",LIN(1)
420        FLOAT 4
430        MAT PRINT Qo
440        CALL Rank(Qc(*),N,Nc,Rc)
450        CALL Rank(Qo(*),No,N,Ro)
```

```
460    IF Rc<N THEN 510
470    PRINT "The system IS completely controllable."
480    FIXED 0
490    PRINT "The controllability matrix has rank";Rc;"= N";LIN(1)
500    GOTO 540
510    PRINT "The system IS NOT completely controllable."
520    FIXED 0
530    PRINT "The controllability matrix has rank";Pc;LIN(1)
540    IF Ro<N THEN 610
550    FLOAT 4
560    PRINT "The system IS completely observable."
570    FIXED 0
580    PRINT "The observability matrix has rank";Ro;"= N"
590    GOTO 640
600    FLOAT 4
610    PRINT "The system IS NOT completely observable."
620    FIXED 0
630    PRINT "The observability matrix has rank";Fo
640    SUBEND
```

9. Example.

´CONOBS´ DETERMINES WHETHRE A MULTIVARIABLE LINEAR TIME-INVARIANT
CONTINUOUS-TIME SYSTEM dx/dt=Ax+Bu, y=Cx+Du OR DISCRETE-TIME SYSTEM
x(k)=Ax(k)+Bu(k), y(k)=Cx(k)+Du(k) IS COMPLETELY CONTROLLABLE OR
COMPLETETLY OBSERVABLE.

Matrix dimensions are: A:NxN,   B:NxR,   C:MxN,   D:MxR . Matrix D
does not influence system controllability or observability.

Matrix dimensions : N= 4 , R= 3 , M= 2

Matrix A(4 x 4):

  1.0000E+00   0.0000E+00   2.0000E+00   3.0000E+00

  4.0000E+00   0.0000E+00   6.0000E+00   5.0000E+00

 -1.0000E+00 -9.0000E+00   3.0000E+00   0.0000E+00

 -1.0000E+00   2.0000E+00   0.0000E+00   6.0000E+00

Matrix B(4 x 3):

  1.0000E+00   0.0000E+00 -3.0000E+00

  2.0000E+00   1.0000E+00   5.0000E+00

 -1.0000E+00   0.0000E+00   6.0000E+00

 -4.0000E+00   2.0000E+00 -3.0000E+00

OBSERVABILITY

Matrix C(2 x 4):

```
 2.0000E+00   0.0000E+00  -1.0000E+00   5.0000E+00
 6.0000E+00  -3.0000E+00   0.0000E+00   2.0000E+00
```

CONTROLLABILITY Matrix Qc=

```
 1.0000E+00   0.0000E+00  -3.0000E+00  -1.3000E+01   6.0000E+00   0.0000E+00
 2.0000E+00   1.0000E+00   5.0000E+00  -2.2000E+01   1.0000E+01   9.0000E+00
-1.0000E+00   0.0000E+00   6.0000E+00  -2.2000E+01  -9.0000E+00  -2.4000E+01
-4.0000E+00   2.0000E+00  -3.0000E+00  -2.1000E+01   1.4000E+01  -5.0000E+00
```

OBSERVABILITY Matrix Qo=

```
 2.0000E+00   0.0000E+00  -1.0000E+00   5.0000E+00
 6.0000E+00  -3.0000E+00   0.0000E+00   2.0000E+00
-2.0000E+00   1.9000E+01   1.0000E+00   3.6000E+01
-8.0000E+00   4.0000E+00  -6.0000E+00   1.5000E+01
 3.7000E+01   6.3000E+01   1.1300E+02   3.0500E+02
-1.0000E+00   8.4000E+01  -1.0000E+01   8.6000E+01
```

The system IS completely controllable.
The controllability matrix has rank 4 = N

The system IS completely observable.
The observability matrix has rank 4 = N

10. Example.

Matrix dimensions :   N= 2 ,   R= 1 ,   M= 1

Matrix A(2 x 2):

```
 1.0000E+00   0.0000E+00
-5.0000E-01   5.0000E-01
```

Column Vector B(2):

```
 1.0000E+00
-1.0000E+00
```

Row Vector C(2):

5.0000E+00 1.0000E+00

CONTROLLABILITY Matrix Qc=

1.0000E+00 1.0000E+00

-1.0000E+00 -1.0000E+00

OBSERVABILITY Matrix Qo=

5.0000E+00 1.0000E+00

4.5000E+00 5.0000E-01

The system IS NOT completely controllable.
The controllability matrix has rank 1

The system IS completely observable.
The observability matrix has rank 2 = N

8.5 DUALITY

The theorems of controllability and observability and the arguments used for their proofs have been strikingly similar. They are indeed related through the concept of *duality* which was first noted by Kalman [8.6]. We introduced duality in Section 6.7, where we defined the dual of the linear continuous-time system representation $S=[A(.),B(.),C(.),D(.)]$ as $\bar{S}=[-A^*(.),C^*(.),B^*(.),D^*(.)]$ and the dual of the linear discrete-time system representation $S=[A(.),B(.),C(.),D(.)]$ as $\bar{S}=[A^{*-1}(.),C^*(.),B^*(.),D^*(.)]$.

1. Theorem (Duality Theorem of Kalman). The linear continuous-time system S is completely controllable (observable) on the interval $[t_0, t_1]$ if an only if its dual system $\bar{S}$ is completely observable (controllable) on $[t_0, t_1]$.

Proof. We showed in Theorem 6.7.3 that if $\Phi(t)$ is a fundamental matrix of system S, then $\Psi(t)=[\Phi^{-1}(t)]^*$ is a fundamental matrix of its dual system $\bar{S}$. Now by Theorem 8.3.4, system S is completely controllable on $[t_0, t_1]$ if and only if the rows of the matrix $\Phi^{-1}(t)B(t)$ are linearly independent on $[t_0, t_1]$. But $\Phi^{-1}(t)B(t) = \Psi^*(t)B(t)$ and the rows of $\Phi^{-1}(t)B(t)$ are the columns of $[\Phi^{-1}(t)B(t)]^* = B^*\Psi(t)$. Therefore, Theorem 8.4.4 implies that system $\bar{S}$ is completely observable on $[t_0, t_1]$. The reverse can be similarly argued. ▽

Using Theorems 8.3.5 and 8.4.5, a similar duality theorem can be established for linear discrete-time systems. The concept of duality is very important in that for every property related to controllability of linear systems, it implies a dual property related to observability of such systems.

8.6 CANONICAL DECOMPOSITION OF THE STATE SPACE

Consider a *l.t.i.* system representation $S=(A,B,C,D)$ of dimension n (i.e. A is $n \times n$). If S is not completely controllable, it can be shown that a system representation of lower dimension exists which is completely controllable and is zero-state equivalent to S. Similarly, if S is not completely observable, it can be shown that a system representation of lower order exists which is completely observable and is zero-state equivalent to S. Finally, if S is neither CC nor CO, the canonical decomposition theorem of Kalman [8.3] decomposes it into four subsystems with different controllability and observability properties. We will discuss the above topics in this section. Note that the following discussion applies to both continuous-time and discrete-time *l.t.i.* systems. Let us first prove the following useful lemma.

1. Lemma. Consider two system representations $S=(A,B,C,D)$ and $\hat{S}=(\hat{A},\hat{B},\hat{C},\hat{D})$ which are algebraically equivalent. Then S is completely controllable if and only if $\hat{S}$ is completely controllable, and S is completely observable if and only if $\hat{S}$ is completely observable.

Proof. Let us indicate the state vector of system S by x and that of system $\hat{S}$ by $\hat{x}$ and assume that they are both of dimension n. Since S and $\hat{S}$ are algebraically equivalent, a nonsingular constant matrix T exists such that $\hat{x}=Tx$ implies

$$\hat{A}=TAT^{-1}, \quad \hat{B}=TB, \quad \hat{C}=CT^{-1}, \quad \hat{D}=D \tag{1}$$

Now the controllability matrix of $\hat{S}$ is

$$\hat{Q}_c = [\hat{B}, \hat{A}\hat{B}, \hat{A}^2\hat{B}, \ldots, \hat{A}^{n-1}\hat{B}]$$

$$= [TB, (TAT^{-1})TB, (TAT^{-1})^2 TB, \ldots, (TAT^{-1})^{n-1} TB]$$

$$= [TB, TAB, TA^2B, \ldots, TA^{n-1}B] = TQ_c \tag{2}$$

where Q_c is the controllability matrix of S. Since T is nonsingular, $\hat{Q}_c$ has rank n if and only if Q_c has rank n. Thus, by Theorem 8.3.6, $\hat{S}$ is completely controllable if and only if S is completely controllable. Similarly, it can be shown that $\hat{Q}_o = Q_o T^{-1}$ which implies that Q_o and $\hat{Q}_o$ have the same rank. Thus, by Theorem 8.4.6. S is completely observable if and only if $\hat{S}$ is completely observable.

2. Example. In Example 8.3.7 we developed a criterion for complete controllability of *l.t.i.* systems with diagonal system matrices. Given any *l.t.i.* system whose system matrix is diagonalizable, we can apply a similarity transformation to obtain a new system with a diagonal system matrix. By Lemma 1, the original system is completely controllable if and only if the new system with diagonal system matrix is completely controllable. Thus, the criterion of Example 8.3.7 can be used to check controllability. (See Problem 8.16.)

8.6.1 Separation of the Controllable Part

Consider a *l.t.i.* system representation $S = (A,B,C,D)$ where A is $n \times n$. If S is not completely controllable, we want to find a subsystem of S which is completely controllable and is zero-state equivalent to S.

3. Lemma. The range space of the controllability matrix Q_c, $R(Q_c)$, is a subspace which is invariant under A. (See Definition 2.4.11.)

Proof. We have

$$Q_c = [B, AB, A^2B, \ldots, A^{n-1}B]. \tag{3}$$

Thus the range space of Q_c, is a subspace spanned by the columns of B, AB, A^2B, $\ldots$, $A^{n-1}B$. Consider a vector $x \in R(Q_c)$. Thus x is a linear combination of the columns of B, AB, ^2B, $\ldots$, $A^{n-1}B$ and Ax is a linear combination of the columns of AB, A^2B, A^3B, $\ldots$, A^nB. Since by Cayley-Hamilton Theorem $A^n = \sum_{i=1}^{n-1} \alpha_j A^j$ (See Section 2.10.) Ax is also a linear combination of the columns of B, AB, A^2B, $\ldots$, $A^{n-1}B$. Therefore, $Ax \in R(Q_c)$ which proves the lemma.

4. Theorem. Consider a *l.t.i.* system representation $S = (A,B,C,D)$ where A, B, C and D are $n \times n$, $n \times r$, $m \times n$ and $m \times r$ matrices, respectively. If $\rho(Q_c) = q \leq n$, where Q_c is given in (3), then a constant nonsingular matrix T_c exists which transforms S into $\bar{S} = (\bar{A}, \bar{B}, \bar{C}, \bar{D})$ given below:

$$\begin{bmatrix} \dot{\bar{x}}_1 \\ \dot{\bar{x}}_2 \end{bmatrix} = \begin{bmatrix} \bar{A}_{11} & \bar{A}_{12} \\ 0 & \bar{A}_{22} \end{bmatrix} \begin{bmatrix} \bar{x}_1 \\ \bar{x}_2 \end{bmatrix} + \begin{bmatrix} \bar{B}_1 \\ 0 \end{bmatrix} u \tag{4a}$$

$$y = [\bar{C}_1 \ \bar{C}_2] \begin{bmatrix} \bar{x}_1 \\ \bar{x}_2 \end{bmatrix} + Du \tag{4b}$$

where $\bar{x}_1$ and $\bar{x}_2$ are q-dimensional and $(n-q)$-dimensional vectors, respectively. Furthermore, the subsystem $\bar{S}_1 = (\bar{A}_{11}, \bar{B}_1, \bar{C}_1, D)$ is completely controllable and is zero-state equivalent to S.

Proof. Since $\rho(Q_c) < n$, $R(Q_c) \subset C^n$ and we have $C^n = R(Q_c) \oplus R(Q_c^\perp)$ where $R(Q_c^\perp)$ is the orthogonal complement of $R(Q_c)$. Let us pick a basis for C^n such that the first q basis vectors form a basis for $R(Q_c)$ and the remaining $n-q$ basis vectors constitute a basis for $R(Q_c^\perp)$. Since by Lemma 3 $R(Q_c)$ is invariant under A, the system matrix w.r.t. this new basis has the form $\bar{A}$ shown in (4). (See Theorem 2.4.14.) Also, since the columns of B are in $R(Q_c)$, $\bar{B}$ has the form shown in (4) w.r.t. this basis. To show that the subsystem $\bar{S}_1 = (\bar{A}_{11}, \bar{B}_1, \bar{C}_1, D)$ is completely controllable, let $\bar{x} = T_c x$ where $\bar{x} = \bar{x}_1 + \bar{x}_2$, $\bar{x}_1 \in R(Q_c)$ and $\bar{x}_2 \in R(Q_c^\perp)$. Then we have $\bar{A} = T_c A T_c^{-1}$, $\bar{B} = T_c B$, $\bar{C} = C T_c^{-1}$ and $\bar{D} = D$. Thus the controllability matrix of $\bar{S}$ is

$$\bar{Q}_c = [\bar{B}, \bar{A}\bar{B}, \bar{A}^2\bar{B}, \ldots, \bar{A}^{n-1}\bar{B}]$$

$$= \begin{bmatrix} \bar{B}_1 & \bar{A}_{11}\bar{B}_1 & \bar{A}_{11}^2\bar{B}_1 & \bar{A}_{11}^{n-1}\bar{B}_1 \\ 0 & 0 & 0 & 0 \end{bmatrix} \begin{matrix} q \\ n-q \end{matrix} \tag{5}$$

and $\rho(\bar{Q}_c)=q=\rho(Q_c)$. This, due to Cayley-Hamilton Theorem (Theorem 2.9.4), implies that $\rho(\bar{Q}_{c1})=q$ where

$$\bar{Q}_{c1}=[\bar{B}_1,\bar{A}_{11}\bar{B}_1,\bar{A}_{11}^2\bar{B}_1,\ldots,\bar{A}_{11}^{q-1}\bar{B}_1] \tag{6}$$

This, by Theorem 8.3.6 proves that system $\bar{S}_1$ is completely controllable. To show that $\bar{S}_1$ is zero-state equivalent to S, we need to prove that the corresponding transfer function matrices are the same. Since S and $\bar{S}$ are algebraically equivalent, their transfer function matrices are the same: $H(s)=\bar{H}(s)$. But we have

$$\bar{H}(s)=\bar{C}(sI-\bar{A})^{-1}\bar{B}+\bar{D} \tag{7}$$

which by direct calculation, using (4), becomes

$$\bar{H}(s)=\bar{C}_1(sI-\bar{A}_{11})^{-1}\bar{B}_1+\bar{D} \tag{8}$$

That is, the transfer function of $\bar{S}_1$.

System $\bar{S}$ can be represented as shown in Fig. 1. Note that the transfer function matrix of $\bar{S}$ depends only on the matrices $\bar{A}_{11}$, $\bar{B}_1$ and $\bar{C}_1$ because $\bar{x}_2(0)=0$ implies $\bar{x}_2(t)=0$ for all $t>0$ and the transfer function depends only on the zero-state response.

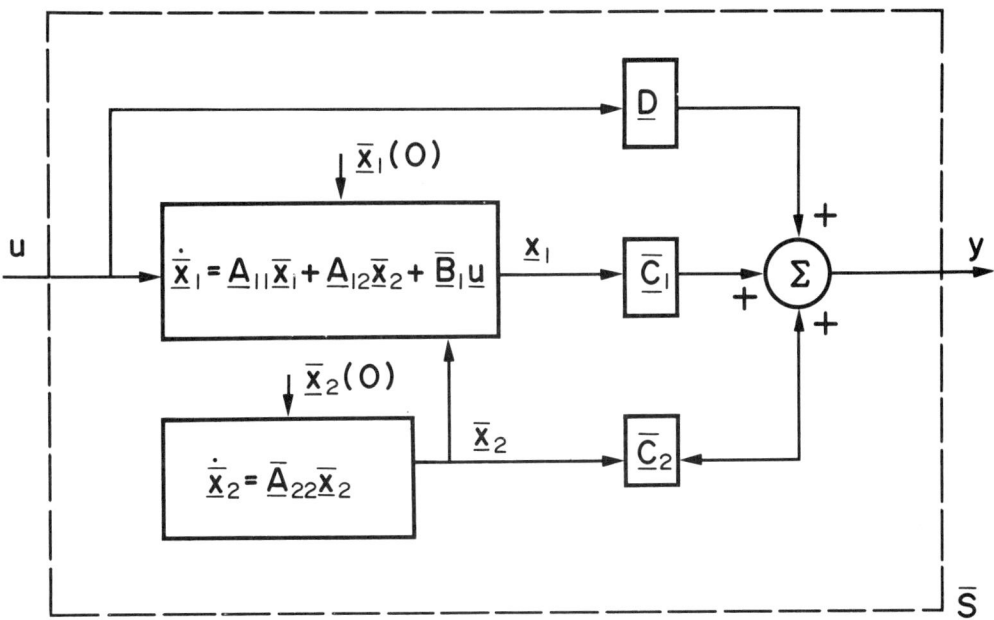

Fig. 8.6.1. Decomposition of the state space of system $\bar{S}$ into controllable state $\bar{x}_1$ and uncontrollable states $\bar{x}_2$

A method to find a suitable transformation matrix T_c is as follows: let $T_c=[u_1,u_2]$ where

$$u_1 = \text{any } q \text{ linearly independent columns of } Q_c \tag{9a}$$

$$u_2 = n-q \text{ arbitrary columns such that } T_c \text{ becomes nonsingular} \tag{9b}$$

5. Example. Consider the system $S=(A,B,C,D)$ where

$$A = \begin{bmatrix} -1 & 1 & 0 \\ 0 & -1 & 0 \\ 4 & -1 & 3 \end{bmatrix}, \quad B = \begin{bmatrix} -1 & 2 \\ 1 & -1 \\ 1 & -2 \end{bmatrix}, \quad C = [1\ 1\ 0], \quad D = [0\ 0] \tag{10}$$

where

$$Q_c = [B, AB, A^2B] = \begin{bmatrix} -1 & 2 & 2 & -3 & -3 & 4 \\ 1 & -1 & -1 & 1 & 1 & -1 \\ 1 & -2 & -2 & 3 & 3 & -4 \end{bmatrix} \tag{11}$$

which has rank 2. Therefore system S is not completely controllable. Let $T_c = [u_1, u_2]$ where $u_1 = B$ and $u_2 = [0\ 0\ 1]'$. Thus

$$T_c = \begin{bmatrix} -1 & 2 & 0 \\ 1 & -1 & 0 \\ 1 & -2 & 1 \end{bmatrix} \quad T_c^{-1} = \begin{bmatrix} 1 & 2 & 0 \\ 1 & 1 & 0 \\ 1 & 0 & 1 \end{bmatrix} \tag{12}$$

and

$$\bar{A} = T_c A T_c^{-1} = \begin{bmatrix} 0 & -1 & 0 \\ 1 & -2 & 0 \\ 0 & 0 & 3 \end{bmatrix}, \quad \bar{B} = T_c B = \begin{bmatrix} 1 & 0 \\ 0 & 1 \\ 0 & 0 \end{bmatrix} \tag{13}$$

$$\bar{C} = C T^{-1} = [0\ 1\ 0], \quad \bar{D} = 0$$

Therefore, the subsystem $\bar{S}_1 = (\bar{A}_{11}, \bar{B}_1, \bar{C}_1, \bar{D})$ where

$$\bar{A}_{11} = \begin{bmatrix} 0 & -1 \\ 1 & -2 \end{bmatrix}, \quad \bar{B}_1 = \begin{bmatrix} 1 & 0 \\ 0 & 1 \end{bmatrix}, \quad \bar{C}_1 = [0,\ 1], \quad \bar{D} = 0 \tag{14}$$

is completely controllable and is zero-state equivalent to S. ▽

Program "SEPCON" uses the method described above to determine the completely controllable subsystem of any MIMO *l.t.i.* system. The source listing of "SEPCON" and an application example are given below

CANONICAL DECOMPOSITION OF THE STATE SPACE

6. "SEPCON"

```
10   !    PROGRAM NAME : "SEPCON" PROG
20        INPUT "Give your PRINT option: For CRT input 1; for LINE PRINTER input 2",
30        IF (Pr=1) OR (Pr=2) THEN 50
40        GOTO 20
50        IF Pr=1 THEN PRINTER IS 16
60        IF Pr=2 THEN PRINTER IS 7,1
70        PRINT "'SEPCON' DETERMINES THE COMPLETELY CONTROLLABLE SUBSYSTEM OF ANY"
80        PRINT "MULTIVARIABLE LINEAR TIME-INVARIANT SYSTEM dx/dt=Ax+Bu.";LIN(1)
90        PRINT "Matrix dimensions are:    A:NxN,    B:NxR";LIN(2)
100       INPUT "Have SUB's <<Mat>>, <<Rank>>, <<Cont>> & <<Sepcob>> already been LIN
     D (Y/N)?",C$
110       IF C$="Y" THEN 180
120       IF C$="N" THEN 140
130       GOTO 100
140       LINK "Mat",1000,150
150       LINK "Rank",2100,160
160       LINK "Cont",3500,170
170       LINK "Sepcob",4000,180
180       INPUT "Give matrix dimensions N, R (R<=N):",N,R
190       FIXED 0
200       PRINT "Matrix dimensions :   N=";N;",    R=";R;LIN(2)
210       Nc=(N-R+1)*R
220       CALL C(N,R,Nc)
230       PRINTER IS 16
240       END
250       SUB C(N,R,Nc)
260       OPTION BASE 1
270       DIM A(N,N),B(N,R),Qc(N,Nc),W2(N,N),W3(N,N),W4(N,R),W5(N,N)
280       CALL Mat(A(*),N,N,"A")
290       CALL Mat(B(*),N,R,"B")
300       CALL Cont(A(*),B(*),N,R,Qc(*))
310       PRINT "CONTROLLABILITY Matrix Qc=",LIN(1)
320       FLOAT 4
330       MAT PRINT Qc
340       Nx=N
350       Ncx=Nc
360       CALL Rank(Qc(*),Nx,Ncx,Rc)
370       IF Rc=N THEN 660
380       FIXED 0
390       PRINT "The controllability matrix has rank";Rc;".";LIN(2)
400       CALL Sepcob(A(*),B(*),N,R,Nc,W2(*),W3(*),W5(*))
410       FLOAT 4
420       PRINT "The transformation matrix is T=";LIN(1)
430       MAT PRINT W5
440       MAT W3=W5*A
450       MAT A=W3*W2
460       MAT W4=W5*B
470       FOR I=1 TO N
480        FOR J=1 TO N
490         IF ABS(A(I,J))<.0005 THEN A(I,J)=0
500        NEXT J
510        FOR K=1 TO R
520         IF ABS(B(I,K))<.0005 THEN B(I,K)=0
530        NEXT K
540       NEXT I
550       PRINT "The new system matrix is A'=";LIN(1)
560       MAT PRINT A
570       PRINT "The new B matrix is B'=";LIN(1)
580       MAT PRINT W4
590       FIXED 0
600       PRINT "The COMPLETELY CONTROLLABLE subsystem:";LIN(2)
610       PRINT "    The system matrix of the completely controllabe subsystem: "
620       PRINT "      Ac=the top left ";Rc;"x";Rc;"block of A'.";LIN(1)
630       PRINT "    The B matrix of the completely controllabe subsystem: "
640       PRINT "      Bc=the top  ";Rc;"x";R;"block of B'."
```

```
650  GOTO 670
660  PRINT "The system IS completely controllable."
670  SUBEND
```

7. Example

´SEPCON´ DETERMINES THE COMPLETELY CONTROLLABLE SUBSYSTEM OF ANY MIMO LINEAR TIV CONTINUOUS-TIME OR DISCRETE-TIME SYSTEM (A,B,C,D)
Matrix dimensions :   N= 3 ,   R= 1

Matrix A(3 x 3):

0.0000E+00   2.0000E+00   -1.0000E+00

3.0000E+00   0.0000E+00   1.0000E+00

0.0000E+00   0.0000E+00   2.0000E+00

Column Vector B(3):

0.0000E+00

1.0000E+00

2.0000E+00

CONTROLLABILITY Matrix Qc=

0.0000E+00   0.0000E+00   0.0000E+00

000E+00   0.0000E+00   0.0000E+00

0.0000E+00   1.0000E+00   -5.0000E-01

The new system matrix is A´=

-2.0000E+00   0.0000E+00   0.0000E+00

0.0000E+00   0.0000E+00   2.0000E+00

0.0000E+00   3.0000E+00   0.0000E+00

The new B matrix is B´=

1.0000E+00

0.0000E+00

0.0000E+00

CANONICAL DECOMPOSITION OF THE STATE SPACE   403

The COMPLETELY CONTROLLABLE subsystem:

```
The system matrix of the completely controllabe subsystem:
   Ac=the top left   1 x 1 block of A´.

The B matrix of the completely controllabe subsystem:
   Bc=the top    1 x 1 block of B´.
```

8.6.2 Separation of the Observable Part

The discussions in this section are essentially the dual of those in Section 8.6.1. Consider a *l.t.i.* system representation $S=(A,B,C,D)$ where A is $n \times n$. If S is not completely observable, we want to find a subsystem of S which is completely observable and is zero-state equivalent to S.

8. Lemma. The null space of the observability matrix, $N(Q_o)$, is a subspace which is invariant under A.

Proof. The proof is similar to that of Lemma 3 and will not be given.

9. Theorem. Consider a *l.t.i.* system $S=(A,B,C,D)$ where A, B, C and D are $n \times n$, $n \times r$, $m \times n$ and $m \times r$, respectively. If $\rho(Q_o) = q < n$, where Q_o is the observability matrix of S defined in (8.4.12), then a constant nonsingular matrix T_o exists which transforms S into $\bar{S} = (\bar{A}, \bar{B}, \bar{C}, D)$ given below:

$$\begin{bmatrix} \dot{\bar{x}}_1 \\ \dot{\bar{x}}_1 \end{bmatrix} = \begin{bmatrix} \bar{A}_{11} & 0 \\ \bar{A}_{21} & \bar{A}_{22} \end{bmatrix} \begin{bmatrix} \bar{x}_1 \\ \bar{x}_2 \end{bmatrix} + \begin{bmatrix} \bar{B}_1 \\ \bar{B}_2 \end{bmatrix} u \quad (15a)$$

$$y = [\bar{C}_1, 0] \begin{bmatrix} \bar{x}_1 \\ \bar{x}_2 \end{bmatrix} + Du \quad (15b)$$

where x_1 and x_2 are q-dimensional and $n-q$-dimensional vectors, respectively. Furthermore, the subsystem $S_1 = (\bar{A}_{11}, \bar{B}_1, \bar{C}_1, D)$ is completely observable and is zero-state equivalent to S. ▽

The proof of the above Theorem is the dual of the proof of Theorem 8.6.4 and will not be presented. The representation of system S is shown in Fig. 2.

A method to find a suitable transformation matrix T_o is as follows:

$$T_o = \begin{bmatrix} T_1 \\ T_2 \end{bmatrix} \quad (16a)$$

where

$T_1 =$ *any q linearly independent rows of* Q_o \quad (16b)

$T_2 = n-q$ *arbitrary rows such that* T_o *becomes nonsingular* \quad (16c)

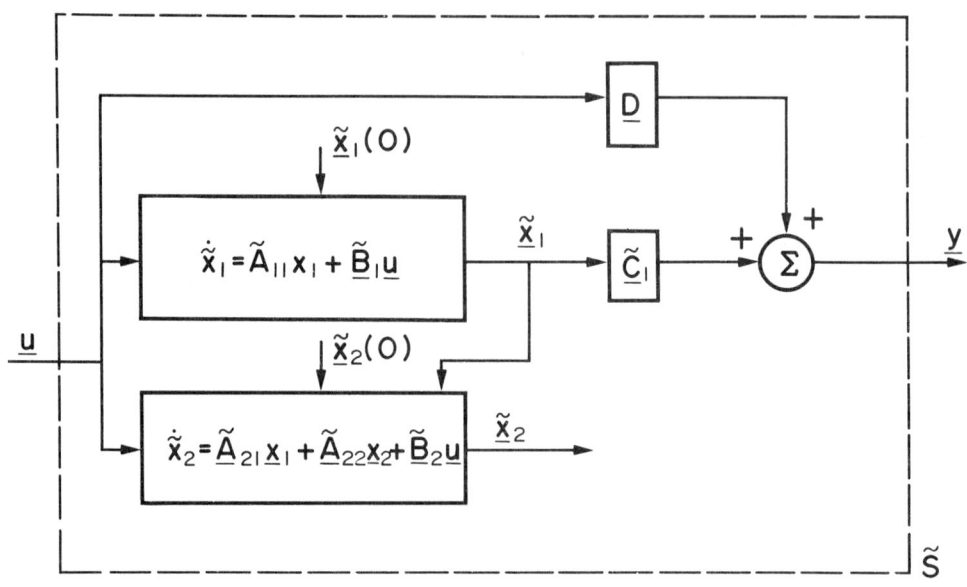

Fig. 8.6.2. Decomposition of the state space of system S into observable states x_1 and unobservable states x_2

10. Example. Consider the system $S=(A,B,C,D)$ where

$$A = \begin{bmatrix} 0 & 2 & -1 \\ 3 & 0 & 1 \\ 0 & 0 & 2 \end{bmatrix}, \ B=0, \ C=[0 \ -2 \ 1], \ D=0 \tag{17}$$

We have

$$Q_o = \begin{bmatrix} C \\ CA \\ CA^2 \end{bmatrix} = \begin{bmatrix} 0 & -2 & 1 \\ -6 & 0 & 0 \\ 0 & -12 & 6 \end{bmatrix} \tag{18}$$

which has rank 2. Therefore system S is not completely observable. Let

$$T_o = \begin{bmatrix} T_1 \\ T_2 \end{bmatrix}, \ T_1 = \begin{bmatrix} 0 & -2 & 1 \\ -6 & 0 & 0 \end{bmatrix}, \ T_2 = [0, 0, 1] \tag{19}$$

which yields

$$\bar{A} = T_o A T_o^{-1} = \begin{bmatrix} 0 & 1 & 0 \\ 6 & 0 & 0 \\ 0 & 0 & 2 \end{bmatrix}, \ \bar{B} = T_o B = 0, \ \bar{C} = CT_o^{-1} = [1,0:0], \ \bar{D} = D = 0 \tag{20}$$

Therefore, the subsystem $\bar{S}_1 = (\bar{A}_{11}, \bar{B}_1, \bar{C}_1, \bar{D})$ where

$$\bar{A}_{11} = \begin{bmatrix} 0 & 1 \\ 6 & 0 \end{bmatrix}, \, \bar{B}_1 = 0, \, \bar{C}_1 = [1,0], \, \bar{D} = 0 \qquad (21)$$

is completely observable and is zero-state equivalent to S. $\triangledown$

Program "SEPOBS" uses the method described above to determine the completely observable subsystem of any MIMO *l.t.i.* system. The source listing of this program followed by an example is given below.

11. "SEPOBS"

```
10    !    PROGRAM NAME : "SEPOBS" PROG
20    INPUT "Give your PRINT option: For CRT input 1; for LINE PRINTER input 2",
30    IF (Pr=1) OR (Pr=2) THEN 50
40    GOTO 20
50    IF Pr=1 THEN PRINTER IS 16
60    IF Pr=2 THEN PRINTER IS 7,1
70    PRINT " 'SEPOBS' DETERMINES THE COMPLETELY OBSERVABLE SUBSYSTEM OF ANY"
80    PRINT "MIMO LINEAR TIV CONTINOUS-TIME OR DISCRETE-TIME SYSTEM (A,B,C,D)"
100   INPUT "Have SUB's <<Mat>>, <<Rank>>, <<Obs>> & <<Sepcob>> already been LINK
ED (Y/N)?",C$
110   IF C$="Y" THEN 180
120   IF C$="N" THEN 140
130   GOTO 100
140   LINK "Mat",1000,150
150   LINK "Rank",2100,160
160   LINK "Obs",3500,170
170   LINK "Sepcob",4000,180
180   INPUT "Give matrix dimensions N, M (M<=N):",N,M
190   FIXED 0
200   PRINT "Matrix dimensions : N=";N;", M=";M;LIN(2)
210   No=(N-M+1)*M
220   CALL C(N,M,No)
230   PRINTER IS 16
240   END
250   SUB C(N,M,No)
260   OPTION BASE 1
270   DIM A(N,N),C(M,N),Qo(No,N),Wo(N,M),W2(N,N),W3(N,N),W4(M,N),W5(N,N)
271   DIM To(N,N),At(N,N)
280   CALL Mat(A(*),N,N,"A")
290   CALL Mat(C(*),M,N,"C")
300   CALL Obs(A(*),C(*),N,M,Qo(*))
310   PRINT "OBSERVABILITY Matrix Qo=",LIN(1)
320   FLOAT 4
330   MAT PRINT Qo;
340   Nx=N
350   Nox=No
360   CALL Rank(Qo(*),Nox,Nx,Ro)
370   IF Ro=N THEN 660
380   FIXED 0
390   PRINT "The observability matrix has rank";Ro;".";LIN(2)
391   MAT Wo=TRN(C)
392   MAT At=TRN(A)
400   CALL Sepcob(At(*),Wo(*),N,M,No,W2(*),W3(*),W5(*))
410   FLOAT 4
411   MAT To=TRN(W5)
420   PRINT "The transformation matrix is To=";LIN(1)
430   MAT PRINT To;
440   MAT W3=To*A
441   MAT At=TRN(W2)
450   MAT A=W3*At
```

```
460   MAT W4=C*At
470    FOR I=1 TO N
480     FOR J=1 TO N
490      IF ABS(A(I,J))<.0005 THEN A(I,J)=0
500     NEXT J
510     FOR K=1 TO R
520      IF ABS(C(K,I))<.0005 THEN C(K,I)=0
530     NEXT K
540    NEXT I
550   PRINT "The new system matrix is A'=";LIN(1)
560   MAT PRINT A;
570   PRINT "The new C matrix is C'=";LIN(1)
580   MAT PRINT W4;
590   FIXED 0
600   PRINT "The COMPLETELY OBSERVABLE subsystem:";LIN(2)
610   PRINT "   The system matrix of the completely observable subsystem: "
620   PRINT "     Ao=the top left ";Ro;"x";Ro;"block of A'.";LIN(1)
630   PRINT "   The C matrix of the completely observable subsystem: "
640   PRINT "     Co=the top    ";M;"x";Ro;"block of C'."
650   GOTO 670
660   PRINT "The system IS completely observable."
670   SUBEND
```

12. Example

``SEPOBS´ DETERMINES THE COMPLETELY OBSERVABLE SUBSYSTEM OF ANY MIMO LINEAR TIV CONTINOUS-TIME OR DISCRETE-TIME SYSTEM (A,B,C,D)
Matrix dimensions :   N= 2 ,   M= 1

Matrix A(2 x 2):

  1.0000E+00   0.0000E+00

  1.0000E+00   3.0000E+00

Row Vector C(2):

  3.0000E+00   4.0000E+00

OBSERVABILITY Matrix Qo=

  3.0000E+00   4.0000E+00

  7.0000E+00   1.2000E+01

The system IS completely observable.

´SEPOBS´ DETERMINES THE COMPLETELY OBSERVABLE SUBSYSTEM OF ANY MIMO LINEAR TIV CONTINOUS-TIME OR DISCRETE-TIME SYSTEM (A,B,C,D)
Matrix dimensions :   N= 2 ,   M= 1

CANONICAL DECOMPOSITION OF THE STATE SPACE

Matrix A(2 x 2):

1.0000E+00 0.0000E+00

1.0000E+00 3.0000E+00

Row Vector C(2):

1.0000E+00 2.0000E+00

OBSERVABILITY Matrix Q_o=

1.0000E+00 2.0000E+00

3.0000E+00 6.0000E+00

The observability matrix has rank 1.

The transformation matrix is T_c=

1.0000E+00 0.0000E+00

0.0000E+00 1.0000E+00

The new system matrix is $\bar{A}'$=

1.0000E+00 0.0000E+00

1.0000E+00 3.0000E+00

The new C matrix is C'=

1.0000E+00 2.0000E+00

The COMPLETELY OBSERVABLE subsystem:

    The system matrix of the completely observable subsystem:
      A_o= the top left 1 x 1 block of A'.

    The C matrix of the completely observable subsystem:
      C_o= the top 1 x 1 block of C'.

8.6.3 Canonical Structure of Systems

By combining the results of Subsections 8.6.1 and 8.6.2 we can now state the following theorem due to Kalman [8.3], known as the *canonical structure theorem* or the *canonical decomposition theorem*.

13. Theorem. Consider a l.t.i. system $S=(A,B,C,D)$ where A, B, C and D are $n \times n$, $n \times r$, $m \times n$ and $m \times r$, respectively. Then S is algebraically equivalent to system $\hat{S}=(\hat{A},\hat{B},\hat{C},\hat{D})$ given below:

$$\begin{bmatrix} \dot{x}_1 \\ \dot{x}_2 \\ \dot{x}_3 \\ \dot{x}_4 \end{bmatrix} = \begin{bmatrix} \hat{A}_{11} & 0 & \hat{A}_{13} & 0 \\ \hat{A}_{21} & \hat{A}_{22} & \hat{A}_{23} & \hat{A}_{24} \\ 0 & 0 & \hat{A}_{33} & 0 \\ 0 & 0 & \hat{A}_{43} & \hat{A}_{44} \end{bmatrix} \begin{bmatrix} x_1 \\ x_2 \\ x_3 \\ x_4 \end{bmatrix} + \begin{bmatrix} \hat{B}_1 \\ \hat{B}_2 \\ 0 \\ 0 \end{bmatrix} u \qquad (22a)$$

$$y = [\hat{C}_1 \; 0 \; \hat{C}_3 \; 0] \begin{bmatrix} x_1 \\ x_2 \\ x_3 \\ x_4 \end{bmatrix} + Du \qquad (22b)$$

where the controllability/observability of the four subsystems is shown in Fig. 3. Subsystem $\hat{S}_1=(\hat{A}_{11},\hat{B}_1,\hat{C}_1,D)$ is completely controllable, completely observable and zero-state equivalent to S.

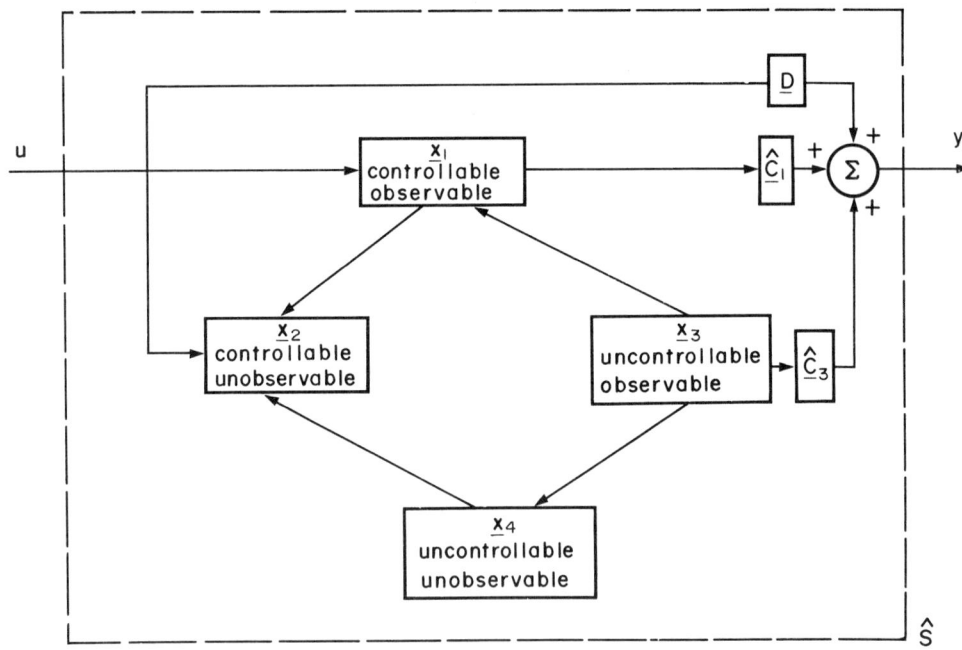

Fig. 8.6.3. Canonical decomposition of the state space of system $\hat{S}$

8.7 MINIMAL REALIZATIONS

In Section 4.5 we defined a *l.t.i.* system representation $S=(A,B,C,D)$ to be a *realization* of a transfer function matrix $H(s)$ if

$$H(s)=C(sI-A)^{-1}B+D \qquad (1)$$

We also observed that a given transfer function has many realizations.

1. Definition. A realization S of dimension n of a transfer function matrix $H(s)$ is said to be *a minimum realization* or a *realization of minimal-order* if no other realization of $H(s)$ of dimension lower than n exists. ▽

Note that minimal realization of a transfer function is not unique. In this section we will investigate the relationship between controllability/observability and minimal realizations. We will see that a *l.t.i.* system $S=(A,B,C,D)$ is uncontrollable, unobservable or both if cancellations occur in the numerator and the denominator of its transfer function matrix (1). That is, the transfer functions of a *l.t.i.* system (after all possible cancellations) represents only the controllable and observable part of the system.

2. Theorem. A realization $S=(A,B,C,D)$ of a transfer function matrix $H(s)$ is minimal if and only if it is both completely controllable and completely observable.

Proof. Necessity. Suppose that S is a minimal representation of $H(s)$ but it is not completely controllable (or completely observable). Then, using Theorem 8.6.4 (or Theorem 8.6.9) we could find a representation of lower dimension which would be completely controllable (or completely observable) and would be zero-state equivalent to S, i.e. would have the same transfer function matrix as S. This would imply that S is not a minimal realization, leading to a contradiction. Therefore, S is completely controllable (and completely observable).

Sufficiency. Let system $S=(A,B,C,D)$ of dimension n be both completely controllable and completely observable. We want to show that no system of lower dimension exists whose transfer function matrix is $H(s)=C(sI-A)^{-1}B+D$. Assume otherwise; then a system $\hat{S}=(\hat{A},\hat{B},\hat{C},D)$ of dimension q, $q<n$, would exist such that

$$\hat{C}(sI-\hat{A})^{-1}\hat{B}+D=C(sI-A)^{-1}B+D \quad \text{for all } s \qquad (2)$$

or

$$\hat{C}e^{\hat{A}t}\hat{B}=Ce^{At}B \quad \text{for all } t,$$

or, equivalently,

$$\hat{C}\hat{A}^k\hat{B}=CA^kB, \quad k=0,1,2,\ldots \qquad (3)$$

We will show that this would lead to a contradiction, thus proving that S is a minimal realization of $H(s)$. Consider the product of the observability and controllability matrices of system S:

$$Q_oQ_c = \begin{bmatrix} C \\ CA \\ \vdots \\ CA^{n-1} \end{bmatrix} [B, AB, \ldots, A^{n-1}B] = \begin{bmatrix} CB & CAB & \cdots & CA^{n-1}B \\ CAB & CA^2B & \cdots & CA^nB \\ \vdots & \vdots & \cdots & \vdots \\ CA^{n-1}B & CA^nB & \cdots & CA^{2(n-1)}B \end{bmatrix} \quad (4)$$

By assumption both Q_o and Q_c have rank n. Therefore, Q_oQ_c has rank n. Now, by using (3) we can write

$$Q_oQ_c = \begin{bmatrix} \hat{C}\hat{B} & \hat{C}\hat{A}\hat{B} & \cdots & \hat{C}\hat{A}^{n-1}\hat{B} \\ \hat{C}\hat{A}\hat{B} & \hat{C}\hat{A}^2\hat{B} & \cdots & \hat{C}\hat{A}^n\hat{B} \\ \vdots & \vdots & \cdots & \vdots \\ \hat{C}\hat{A}^{n-1}\hat{B} & \hat{C}\hat{A}^n\hat{B} & \cdots & \hat{C}\hat{A}^{2(n-1)}\hat{B} \end{bmatrix} =$$

$$\begin{bmatrix} \hat{C} \\ \hat{C}\hat{A} \\ \vdots \\ \hat{C}\hat{A}^{n-1}\hat{B} \end{bmatrix} [\hat{B}, \hat{A}\hat{B}, \ldots, \hat{A}^{n-1}\hat{B}] = Q_oQ_c \quad (5)$$

The matrices Q_o and Q_c can have ranks of at most q. This contradicts the assumption that $q < n$.

3. Example. Consider the system representation $S = (A, B, C, D)$ where

$$A = \begin{bmatrix} 00 & -6 \\ 10 & -11 \\ 01 & -6 \end{bmatrix}, B = \begin{bmatrix} 3 \\ 1 \\ 0 \end{bmatrix}, C = [0, 0, 1], D = 0 \quad (6)$$

For this system we have

$$Q_c = [B, AB, A^2B] = \begin{bmatrix} 3 & 0 & -6 \\ 1 & 3 & -11 \\ 0 & 1 & 3 \end{bmatrix} \quad (7a)$$

$$Q_o = \begin{bmatrix} C \\ CA \\ CA^2 \end{bmatrix} = \begin{bmatrix} 0 & 0 & 1 \\ 0 & 1 & -6 \\ 1 & -6 & 25 \end{bmatrix} \quad (7b)$$

Q_o is nonsingular but Q_c is singular. Therefore, system S is completely observable but uncontrollable. Let us check the transfer function of S:

$$H(s) = C(sI-A)^{-1}B + D = \frac{s+3}{(s+1)(s+2)(s+3)} = \frac{1}{(s+1)(s+2)} \quad (8)$$

Note that a cancellation occurs in the numerator and the denominator of H(s), as expected. A minimal realization of H(s) in (8) (after cancellation of the common term) can be found as follows. The differential equation characterizing the transfer function (8) is

$$\ddot{y} + 3\dot{y} + 2y = u \tag{9}$$

where u and y indicate input and output, respectively. A canonical form for the state and output equations of the system characterized by differential equation (9) is $\bar{S} = (\bar{A}, \bar{B}, \bar{C}, \bar{D})$ (See Section 4.4.3.):

$$\dot{\mathbf{x}} = \begin{bmatrix} 0 & 1 \\ -2 & -3 \end{bmatrix} \mathbf{x} + \begin{bmatrix} 0 \\ 1 \end{bmatrix} u, \ y = [1 \ 0] \mathbf{x} \tag{10}$$

Note that for system (10) we have

$$Q_c = [\bar{B}, \bar{A}\bar{B}] = \begin{bmatrix} 0 & 1 \\ 1 & -3 \end{bmatrix}, \tag{11a}$$

$$Q_o = \begin{bmatrix} \bar{C} \\ \bar{C}\bar{A} \end{bmatrix} = \begin{bmatrix} 1 & 0 \\ 0 & 1 \end{bmatrix} \tag{11b}$$

both of which are nonsingular. Therefore, system $\bar{S}$ is both completely controllable and completely observable, as expected.

PROBLEMS

8.1 Prove Theorem 8.2.3.

8.2 Consider the Gram matrices $G(t_0, t_1)$ in (8.2.1) and (8.2.4). Show that the following statement are equivalent:

i) $G(t_0, t_1)$ is nonsingular.

ii) $G(t_0, t_1)$ is positive definite.

8.3 Consider the vector functions $\mathbf{f}_1(t) = [1, t]$ and $\mathbf{f}_2(t) = [t^2, t]$. Use Theorem 8.2.3 to show that they are linearly independent on [0,1]. Can Theorem 8.2.4 be used to show this?

8.4 Verify the linear independence of functions $\mathbf{f}_i(k)$, i=1, 2, 3 in (8.2.10) by using Theorem 8.2.6.

8.5 Show that for a linear system $(A(t), B(t), C(t), D(t))$ the following two statements are equivalent:

412   LINEAR CONTROL SYSTEMS

    i) Any initial state $x_o = x(t_o)$ can be transferred to any arbitrary state x_1 at t_1 by some control u on $[t_o, t_1]$.

    ii) Any initial state $x_o = x(t_o)$ can be transferred to the origin of the state space $(x_1 = 0)$ at t_1 by some control u or $[t_o, t_1]$.

8.6 Prove Theorems 8.3.5 and 8.4.5.

8.7 Show that the system $\dot{x}(t) = A(t)x(t) + B(t)u(t)$ where $x \epsilon R^n$ and matrices $A(t)$ and $B(t)$ are, respectively, $n-2$ and $n-1$ times differentiable is completely controllable on $[t_o, t_1]$ if and only if the matrix $Q(t)$ given below has rank n for $t \epsilon [t_o, t_1]$:

$$Q(t) = [Q_1, Q_2, \ldots, Q_n]$$

where $Q_1 = B(t)$ and

$$Q_{k+1} = -A(t)Q_k + \dot{Q}_k, \quad k = 1, 2, \ldots, n-1.$$

8.8 Show that the system $x(k+1) = A(k)x(k) + B(k)u(k)$ where $x \epsilon R^n$ is completely controllable on $[k_o, k_1]$ if matrix $Q(k_1)$ has rank n where $Q(k_1) = [Q_1(k_1), Q_2(k_1), \ldots, Q_n(k_1)]$, $Q_1(k) = B(k)$ and $Q_{j+1}(k) = A^{-1}(k+1)Q_j(k+1)$, $j = 1, 2, \ldots, n-1$.

8.9 Verify that for *l.t.i.* systems, the conditions of Problems 8.7 and 8.8 reduce to that of Theorem 8.3.6.

8.10 Prove the following directly and by using Theorem 8.3.5: for every initial state x_o, there exists a control sequence $u(k)$, $k = 0, 1, 2, \ldots, n-1$ which drives the system $x(k+1) = Ax(k) + Bu(k)$ to $x(k_1) = x_1$ where x_1 is any arbitrary state, if and only if the controllability matrix Q_c in (8.3.14) has full rank. Furthermore, a control sequence which makes this transfer possible is given by

$$\begin{bmatrix} u(n-1) \\ u(n-2) \\ \vdots \\ u(n) \end{bmatrix} = (Q_c Q_c)^{-1} Q_c \, (x_1 - A^n x_o)$$

8.11 Study the controllability and the observability of the *l.t.i.* system (A, B, C, D) where

$$A = \begin{bmatrix} -1 & 1 & 0 \\ 0 & -1 & 0 \\ 4 & -1 & 3 \end{bmatrix}, B = \begin{bmatrix} -1 & 2 \\ 1 & -1 \\ 1 & -2 \end{bmatrix}, C = [1 \ 1 \ 0], D = 0$$

Verify your results by using the program "CONOBS".

8.12 If **A** is $n \times n$, **B** is $n \times r$ and **C** is $m \times n$, show that

$$\rho[\mathbf{B},\mathbf{AB},\mathbf{A}^2\mathbf{B}, \ldots, \mathbf{A}^{n-1}\mathbf{B}] = \rho[\mathbf{B},\mathbf{AB},\mathbf{A}^2\mathbf{B}, \ldots, \mathbf{A}^{n-r}\mathbf{B}]$$

$$\rho \begin{bmatrix} \mathbf{C} \\ \mathbf{CA} \\ \mathbf{CA}^2 \\ \vdots \\ \mathbf{CA}^{n-1} \end{bmatrix} = \rho \begin{bmatrix} \mathbf{C} \\ \mathbf{CA} \\ \mathbf{CA}^2 \\ \vdots \\ \mathbf{CA}^{n-m} \end{bmatrix}$$

8.13 The *controllable companion form* and the *observable companion form* of state equations were discussed in Section 4.4.3. Show that the system described by the state equation (4.4.32) is completely controllable. Also show that the observable companion form with output equation (4.4.34) is completely observable. Use "CONOBS" to verify this on several examples.

8.14 Show that system (8.4.11) where **A** and **C** are $n \times n$ and $n \times m$ matrices is completely observable on $[k_o, k_1]$ if and only if the $mn \times n$ matrix

$$\begin{bmatrix} \mathbf{C}(k_o) \\ \mathbf{C}(k_o+1)\Phi(k_o+1,k_o) \\ \vdots \\ \mathbf{C}(k_o+n-1)\Phi(k_o+n-1,k_o) \end{bmatrix}$$

Apply this result to study the observability of a discrete-time system has rank n where $\mathbf{A} = \begin{bmatrix} -1 & 0 \\ k & 1 \end{bmatrix}$ and $\mathbf{C} = [k, 1]$.

8.15 Use duality Theorem 8.5.1 and the result of Theorem 8.3.6 to prove Theorem 8.4.6.

8.16 Use the result of Lemma 8.6.1 to develop a test for complete observability of *l.t.i.* systems whose system matrices are diagonalizable.

8.17 Prove Lemma 8.6.7 and Theorem 8.6.8.

8.18 Consider the SISO *l.t.i.* system $S=(\mathbf{A},\mathbf{B},\mathbf{C},0)$ where **A** is an $n \times n$ matrix. If S is completely controllable, show that it is algebraically equivalent to system $\bar{S}=(\bar{\mathbf{A}},\bar{\mathbf{B}},\bar{\mathbf{C}},0)$ such that

$$\bar{B} = \begin{bmatrix} 1 \\ 0 \\ 0 \\ \vdots \\ 0 \end{bmatrix}, \quad Q_c^{-1}A^nB = \begin{bmatrix} -a_0 \\ -a_1 \\ \vdots \\ -a_{n-1} \end{bmatrix}, \quad \bar{A} = \begin{bmatrix} 0 & 0 & \cdots & -a_0 \\ 1 & 0 & \cdots & -a_1 \\ 0 & 1 & \cdots & \\ 0 & 0 & \cdots & \\ \vdots & \vdots & & \vdots \\ 0 & 0 & \cdots & 1 & -a_{n-1} \end{bmatrix}$$

where a_i, $i=0,1,\ldots,n-1$ are the coefficients of the characteristic polynominal of A and Q_c is the controllability matrix of S.

Hint: Use Q_c^{-1} as the transformation matrix.

8.19 Let $S=(A,B,C,D)$ be any realization of the transfer function matrix $H(s)$. Let Q_c and Q_o be the controllability and observability matrices of S. Show that the dimension of the minimal realization of $H(s)$ is $\rho(Q_o Q_c)$.

8.20 Consider the MIMO l.t.i. system $S=(A,B,C,0)$ where

$$A = \begin{bmatrix} -1 & 1 & 0 \\ 0 & -1 & 0 \\ 4 & -1 & 3 \end{bmatrix}, \quad B = \begin{bmatrix} -1 & 2 \\ 1 & -1 \\ 1 & -2 \end{bmatrix}, \quad C = [1, 1, 0]$$

i) Use "CONOBS" to investigate controllability and observability of S.

ii) If S is not completely controllable, use "SEPCON" to find a completely controllable subsystem of S which is zero-state equivalent to it.

iii) If S is not completely observable, use "SEPOBS" to find a completely observable subsystem of S which is zero-state equivalent to it.

8.21 Consider the system $S=(A,B,C,0)$ where

$$A = \begin{bmatrix} -3 & -3 & 0 & 1 \\ 26 & 36 & -3 & -25 \\ 30 & 39 & -2 & -27 \\ 30 & 43 & -3 & -32 \end{bmatrix}, \quad B = \begin{bmatrix} 3 & 3 \\ -2 & -1 \\ 0 & 0 \\ 0 & 1 \end{bmatrix}, \quad C = [-5, -8, 1, 5]$$

Use "CONOBS" to check controllability and observability of S. If S is not completely controllable, use "SEPCON" to determine a completely controllable subsystem $\tilde{S}$ of S of minimum dimension. If S is not completely observable, use "SEPOBS" to find a completely observable subsystem $\hat{S}$ of S of minimum dimension. Verify that the order of the minimal subsystem of S which is both completely controllable and completely observable is $\rho(Q_o Q_c)$.

REFERENCES

[8.1] R. E. Kalman, "On the general theory of control systems", *Proc. IFAC First Intl. Cong.*, Moscow, 1960, Vol. 1, pp 481-493, Butterworth & Co., Ltd., London, 1961

[8.2] R.E. Kalman, Y. C. Ho and K. S. Narendra, "Controllability of linear dynamical systems", *Contributions to Differential Equations,* Vol. 1, Interscience Publishers, Inc., New York, 1962

[8.3] R. E. Kalman, "Canonical structure of linear dynamical systems", *Proc. Natl. Acad. Sci., U.S.A.,* Vol. 48, No. 4, pp. 596-600, 1962

[8.4] L. M. Silverman and H. E. Meadows, " Controllability and observability in time-variable linear systems," *SIAM J. Control,* Vol. 5, pp. 64-73, 1967

[8.5] C.A. Desoer, *Notes For a Second Course in Linear Systems,* Van Nostrand Reinhold, 1970

[8.6] R. E. Kalman, "Mathematical description of linear dynamical systems", *SIAM J. Control,* Vol. 1, pp. 152-192, 1963

[8.7] L. M. Silverman and B. D. O. Anderson, "Controllability, observability, and stability of linear systems," *SIAM J. Control,* Vol. 6, pp. 121-130, 1968

PART III

DESIGN

CHAPTER 9

DESIGN OF LINEAR CONTROL SYSTEMS BY COMPENSATION

9.1 INTRODUCTION

This is the first chapter of the text's third part - design. The general idea behind the design of any system is to add a controller (control element) to the system so that certain system's performance specifications (time or frequency responses) would be achieved. It must be noted that, although this simplified definition of "design" is general, it would be modified in accordance with the design's criterion and system's model. The emphasis in the present chapter is the design of conventional (classical) *SISO* control systems via compensation which will be defined shortly. In Chapter 11, on the other hand, optimal adaptive control systems design from the points of view classical and modern control will be discussed.

Suppose that a *SISO* system has undesirable characteristics or performance specifications and a controller is desired to alter the system to achieve specific characteristics. The addition of equipment in the system to achieve the above change in performance is commonly known as *compensation*. If the original system is unstable then the process of system compensation is also called *stabilization*. In this chapter, the frequency-domain techniques of root locus, Bode and Nyquist diagrams discussed in Chapter 7 will be used to compensate *SISO* linear control systems. The compensating element or simply compensator can be in either cascade or feedback form as shown in Fig. 1. The choice between cascade and feedback compensators depend on many factors such as system's form (electrical, hydraulic, etc.), signals, environment, economics, noise, time response and so on. For further detail on these distinctions, the reader can refer to the book by D'Azzo and Houpis [9.1].

The design of MIMO systems continuous-time and discrete-time within the context of pole placement via state and output feedback is discussed in Chapter 10. Optimal design of *SISO* continuous- time and discrete-time systems with respect to plant parameters is given in Chapter 11.

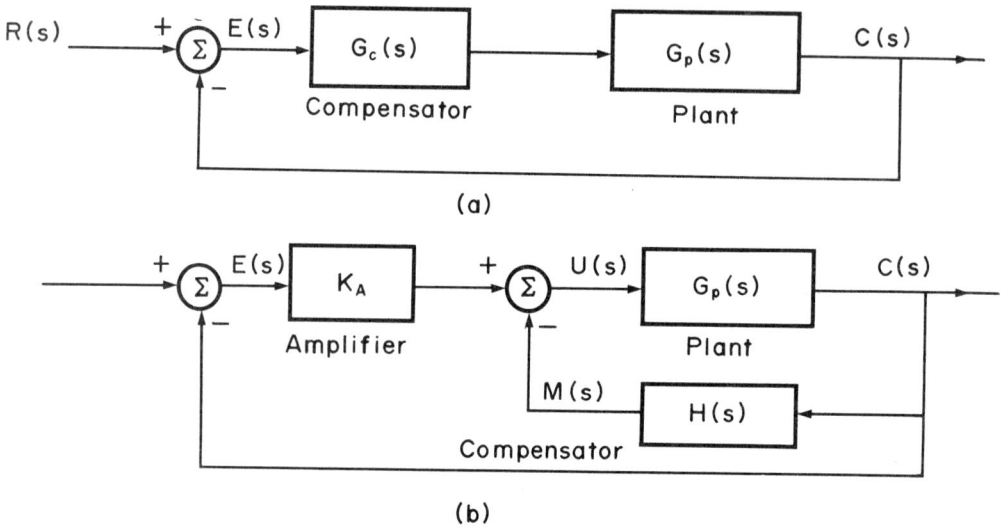

Fig. 9.1.1. Block-diagrams for compensated *SISO* systems
(a) Cascade Compensation,
(b) Feedback Compensation

9.2 DOMINANT POLE ASSUMPTION

The distinction between open- and closed-loop poles was made in Section 4.5. It was noted that the location of closed-loop poles (roots of the characteristic equation) has a direct impact on the system's stability. On the other hand, as noted from the Nyquist Stability Criterion, it is possible to have r.h.p. open-loop poles and the closed-loop system be still stable.

In several cases an nth order system may have two of its closed-loop poles closer to the $j\omega$-axis as compared with the remaining poles and zeros, as seen in Fig. 1a. In such a situation it is said that the system has a pair of *dominant poles*. Note that as the system's third pole (which is real) moves closer to the $j\omega$-axis (Fig. 1b,c), the system's time response deviates further from that of a second order system. In view of the discussions made on the reduction of large-scale linear systems in Chapter 5, a system with a pair of dominant poles can be thought of a second-order modally aggregated system (see Section 5.3.2). The dominant pole assumption in nth order linear *SISO* systems will be used rather extensively in the design techniques which follow in this Chapter. In other words if an nth order system possesses a pair of dominant poles then the design process can be performed as though the system is a *2nd* order.

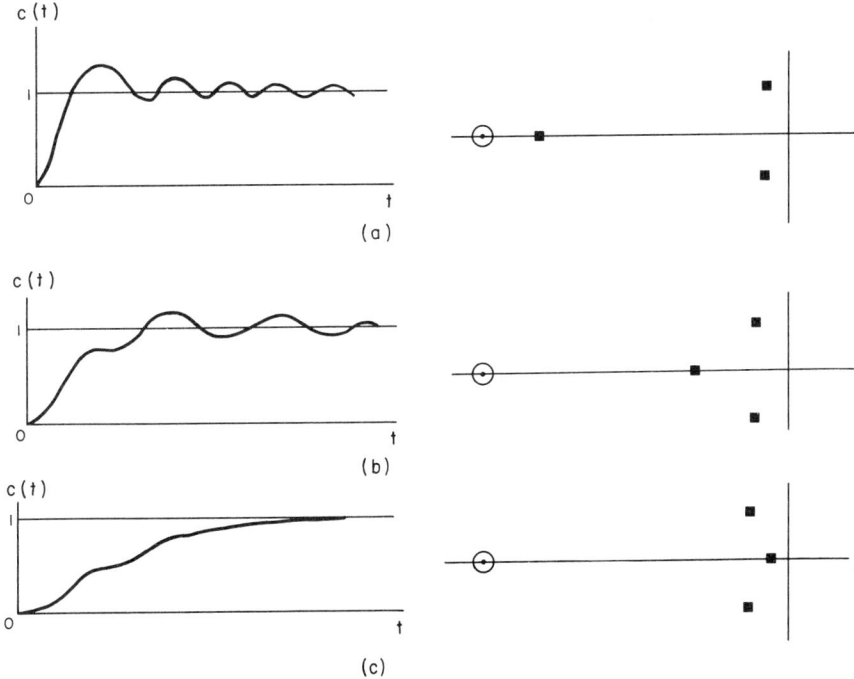

Fig. 9.2.1. Typical responses for a third order system with
(a) a pair of dominant poles,
(b) a closer location for the third (real) pole, and
(c) a third pole closest to $j\omega$-axis

9.3 STEADY-STATE ERRORS AND STATIC ERROR COEFFICIENTS

9.3.1 Steady-State Error

One of the more important parameters in the behavior of *SISO* systems is the steady-state error denoted e_{ss}, which is a measure of input - output deviation. In order to evaluate it, consider the block diagram of a *SISO* system shown in Fig. 1. The closed-loop transfer function of this system is

$$C(s)/R(s) = G(s)/(1+G(s)H(s)). \tag{1}$$

Replacing $C(s)$ in (1) by $G(s)E(s)$, the input-to-error transfer function is given by

$$E(s)/R(s) = 1/(1+G(s)H(s)). \tag{2}$$

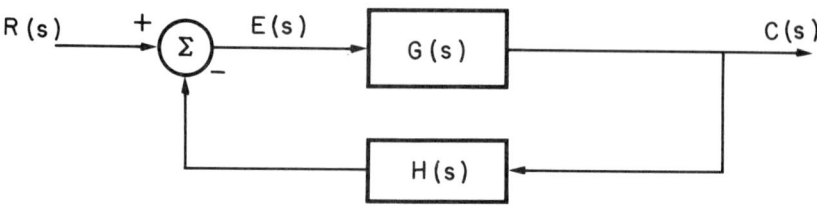

Fig. 9.3.1. A control system

Hence, in view of the final-value theorem of the Laplace transform (See Section 3.3.), the steady-state error is

$$e_{ss} = \lim_{t \to \infty} e(t) = \lim_{s \to 0} sE(s) = \lim_{s \to 0} sR(s)/(1+G(s)H(s)) \qquad (3)$$

Figures of merit for a measure of the steady-state error are the static error coefficients which are considered next.

9.3.2 Static Error Coefficients

Let the open-loop transfer function of a *SISO* system be represented by

$$G(s)H(s) = K \prod_{i=1}^{m}(1+T_i s)/(s^N \prod_{j=1}^{q}(1+\tau_j s)) \qquad (4)$$

where $N+q \triangleq n$ is the system order and m represents the number of system's zeros. It is a common practice to classify the system depending on the value of N. A system is said to be of *type 0, type 1* or *type 2* ,... if $N=0,1,2$, etc..., respectively. Now we consider the steady-state error value defined by (3) with a unit step input, i.e. $R(s) = 1/s$,

$$e_{ss} = \lim_{s \to 0}(s(1/s)/(1+G(s)H(s))) = 1/(1+G(0)H(0)) \qquad (5)$$

The *static position error coefficient* K_p is defined by

$$K_p \triangleq \lim_{s \to 0} G(s)H(s) = G(0)H(0). \qquad (6)$$

A comparison of (5) and (6) indicates that the steady-state error for a unit step input is given by

$$e_{ss} = 1/(1+K_p). \qquad (7)$$

STEADY-STATE ERRORS AND STATIC ERROR COEFFICIENTS

In view of (4), for a type 0 system, K_p is given by

$$K_p = \lim_{s \to 0} (K \prod_{i=1}^{m}(1+T_i s)/\prod_{j=1}^{q}(1+\tau_j s)) = K \tag{8}$$

For systems of types 1 and higher, the value of K_p is given by,

$$K_p = \lim_{s \to 0} K \prod_{i=1}^{m}(1+T_i s)/(s^N \prod_{j=1}^{q}(1+\tau_j s)) = \infty. \tag{9}$$

Therefore, the static position error coefficient is finite for systems of type 0 and infinite for systems of type 1 and higher. It must be noted that the attribute "position" is given to the system's output traditionally [9.2] and has no important significance on the present analysis. In fact, since the system's output can be position, velocity, pressure, temperature, etc. other coefficients can be defined as will be shown. To summarize the relations between the steady-state error and static position error coefficient for unit step input,

$$e_{ss} = 1/(1+K) \quad - \text{ type 0 system}$$

$$= 0 \text{ - type 1 or higher system}$$

The next error coefficient called *static velocity error coefficient* K_v is defined. Consider the steady-state error of a unit-ramp input system,

$$e_{ss} = \lim_{s \to 0} (s (1/s^2)/(1+G(s)H(s)))$$

$$= \lim_{s \to 0} 1/(sG(s)H(s)). \tag{10}$$

Now let K_v be defined by

$$K_v \triangleq \lim_{s \to 0} sG(s)H(s) \tag{11}$$

which indicates that the steady-steady error for unit-ramp input is given by

$$e_{ss} = 1/K_v \tag{12}$$

i.e. the reciprocal of the static velocity error coefficient. Now turning attention to system types, K_v

is given by

Type 0:

$$K_v = \lim_{s \to 0}(sK\prod_{i=1}^{m}(1+T_i s)/\prod_{j=1}^{q}(1+\tau_j s)) = 0 \tag{13}$$

Type 1:

$$K_v = \lim_{s \to 0}(sK\prod_{i=1}^{m}(1+T_i s)/(s\prod_{j=1}^{q}(1+\tau_j s))) = K \tag{14}$$

Type 2 and higher ($N \geq 2$):

$$K_v = \lim_{s \to 0}(sK\prod_{i=1}^{m}(1+T_i s)/(s^N \prod_{j=1}^{q}(1+\tau_j s))) = \infty. \tag{15}$$

Hence, the steady-state error would become $\infty, 1/K$ and 0, respectively for types 0, 1, 2 and higher systems.

The final error coefficient, called *static acceleration error coefficient*, K_a is defined as

$$K_a \triangleq \lim_{s \to 0} s^2 G(s)H(s) \tag{16}$$

which would provide a unit acceleration input (i.e. $r(t) = t^2/2$) steady-state error of

$$e_{ss} = \lim_{s \to 0}(s(1/s^3)/(1+G(s)H(s))) = 1/K_a. \tag{17}$$

It is therefore easy to see (Problem 9.2) that this error would be ∞, $1/K$ and 0 depending on the system's type being 0, 1, 2 or 3 and higher, respectively. Table 1 summarizes the above development on the steady-state error and the three static error coefficients and the three unit inputs. A pictorial representation of the steady-state error is shown in Fig. 2.

Table 9.3.1 Steady-state error as a function of error coefficients

System Type	Unit Step Input $r(t) = 1$	Unit Ramp Input $r(t) = t$	Unit acceleration Input $r(t) = t^2/2$
N = 0	$1/(1+K_p)$	∞	∞
N = 1	0	$1/K_v$	∞
N = 2	0	0	$1/K_a$

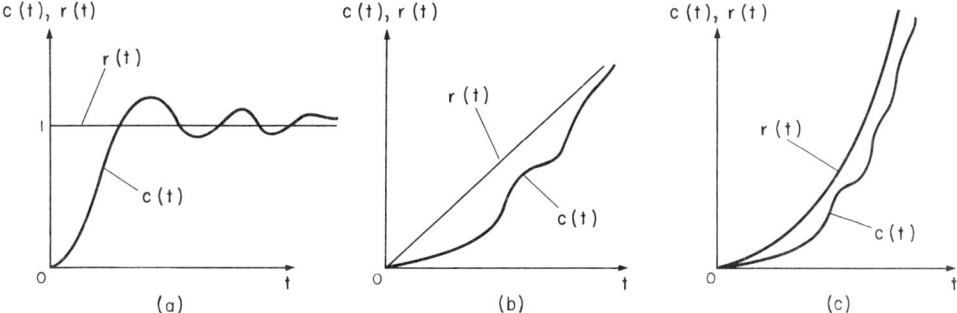

Fig. 9.3.2. Response of a unity feedback system to
(a) Unit-step and type 0,
(b) Unit ramp and type 1 and
(c) Unit acceleration and type 2

The use of error function within the context of an optimizing criterion will be discussed in Chapter 11. The following example illustrates how the error coefficients can be obtained.

1. Example. Consider an automatic speed control system in the cruise mode given by

$$G(s) = K_a/(1+\tau_a s) \tag{18}$$

which is being controlled by a throttle controller given by

$$G_c(s) = K_1 + K_2/s \tag{19}$$

It is desired to evaluate its static error coefficients for a unit step input and a unity feedback.

Let $K_2 = 0$ for now, i.e. consider a proportional controller first, then the steady-state error is given by

$$e_{ss} = \lim_{s \to 0} sE(s) = 1/(1+K_p) = 1/(1+K_1 K_a) \tag{20}$$

when $K_2 > 0$, the system would become type 1 with

$$GG_c H = \frac{K_a(K_1 s + K_2)}{s(1+\tau_a s)} \tag{21}$$

and $e_{ss} = 0$ for a unit step input. If the input was a ramp, then $e_{ss} = 1/K_v$, where

$$K_v = \lim_{s \to 0} sG_c(s)G(s)H(s) = K_a K_2. \tag{22}$$

9.4 COMPENSATING NETWORKS

Consider a compensated control system shown in Fig. 9.1.1a. The compensator's transfer function is indicated by $G_c(s)$. In this section, three popular compensating networks will be described.

9.4.1 Lead Network

Figure 1 show an RC lead network. The transfer function of the compensator can be easily seen to be,

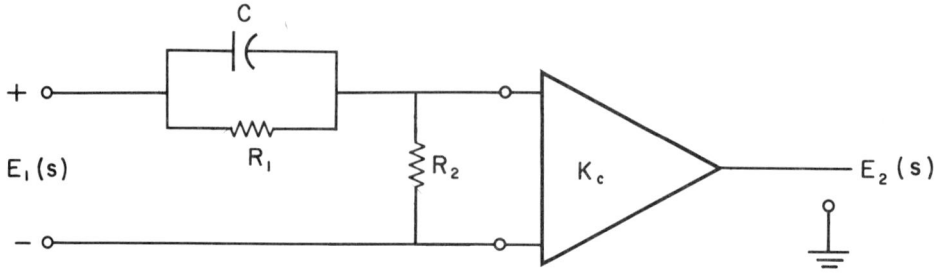

Fig. 9.4.1. A Lead Compensator

$$G_c(s) = E_2(s)/E_1(s) = K_c\alpha(1+Ts)/(1+\alpha Ts)$$

$$= K_c(s+1/T)/(s+1/\alpha T) = K_c(s-z_c)/(s-p_c) \quad (1)$$

where $\alpha = R_2/(R_1+R_2) < 1$ and $T = R_1 C$. Thus, the addition of a lead compensator would add a zero and a pole at $s = -1/T$ and $s = -1/\alpha T$, respectively. The Bode and Nyquist plots of and pole-zero locations of a lead compensator is shown in Fig. 2.

It is noted that the transfer functions of lead compensator can be rewritten as

$$G_c(j\omega) = K_c\alpha(1+j\omega T)/(1+j\omega\alpha T) \quad (2)$$

$$= K_c\alpha((1+\omega T)^2)/(1+(\omega\alpha T)^2))^{\frac{1}{2}} \angle\phi$$

where

$$\phi = \tan^{-1}(\omega T) - \tan^{-1}(\omega\alpha T) \quad (3)$$

which would directly result in the responses shown in Fig. 2.

COMPENSATING NETWORKS

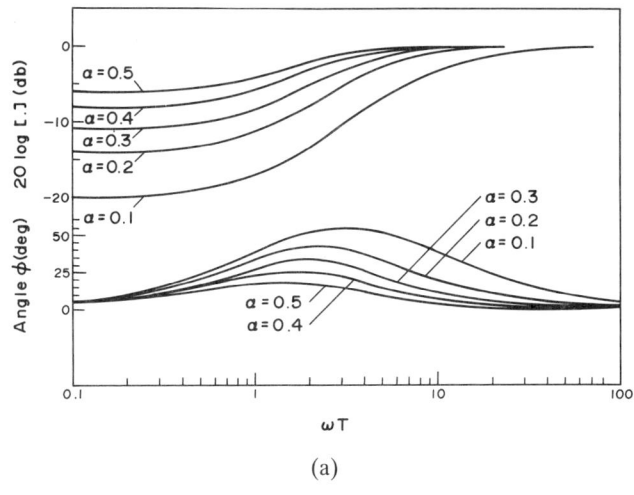

(a)

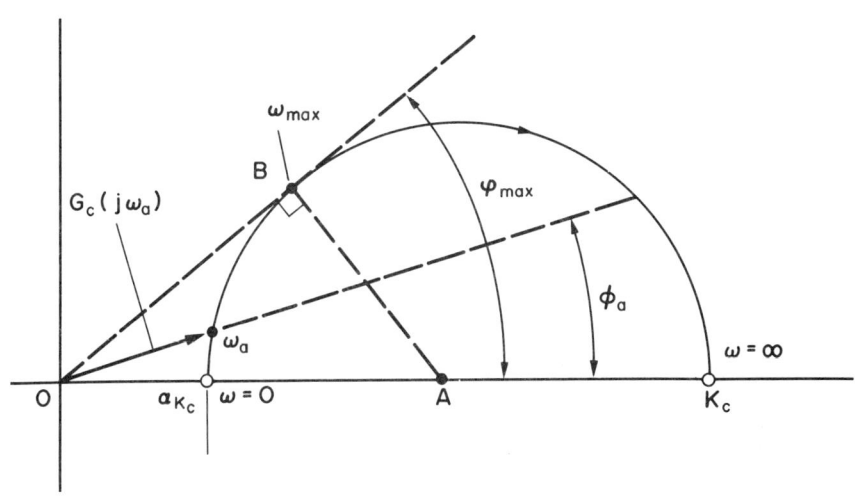

(b)

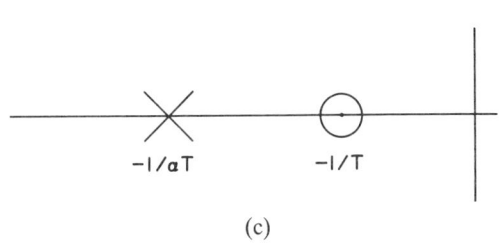

(c)

Fig. 9.4.2. Frequency responses for a lead compensator
(a) Bode diagram,
(b) Nyquist diagram
and (c) Pole-zero locations

9.4.2 Lag Network

An alternative RC network called lag, and sometimes integrator compensator is shown in Fig. 3.

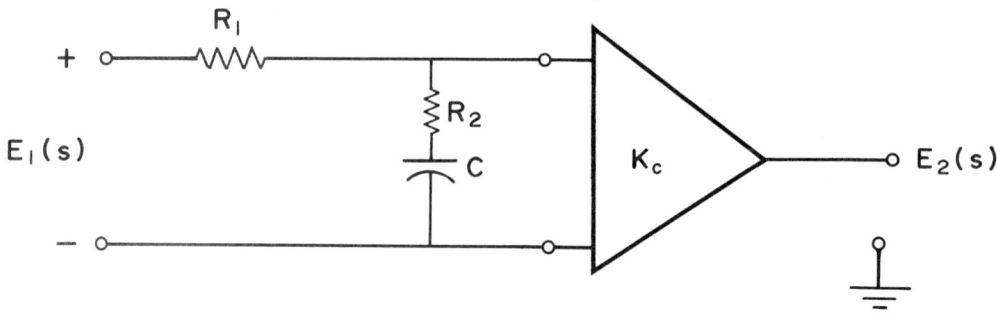

Fig. 9.4.3. A lag compensator

The transfer function of this compensator is given by

$$G_c(s) = E_2(s)/E_1(s) = K_c(1+Ts)/(1+\alpha Ts) =$$

$$(K_c/\alpha)(s+1/T)/(s+1/\alpha T) = (K_c/\alpha)(s+z_c)/(s+p_c) \qquad (4)$$

where $\alpha = (R_1+R_2)/R_2 > 1$ and $T = R_2C$. In this case, the network would also add a zero and a pole to the system. However, a lag compensator's pole is closer to the $j\omega$-axis than its zero. The Bode and Nyquist plots and pole-zero locations of a lag compensator are shown in Fig. 4.

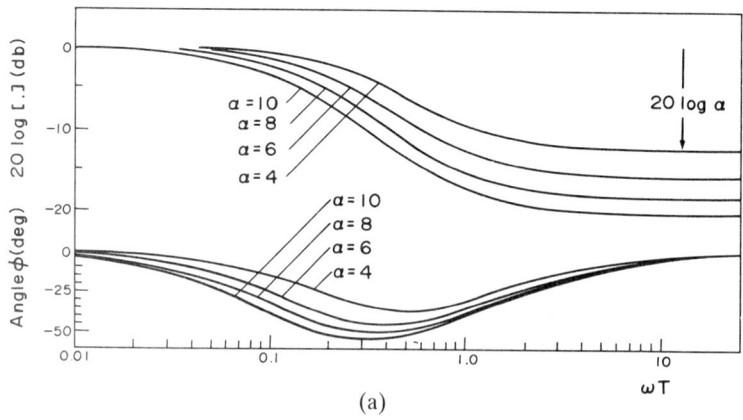

Fig. 9.4.4. Frequency responses for a lag compensator
(a) Bode diagram

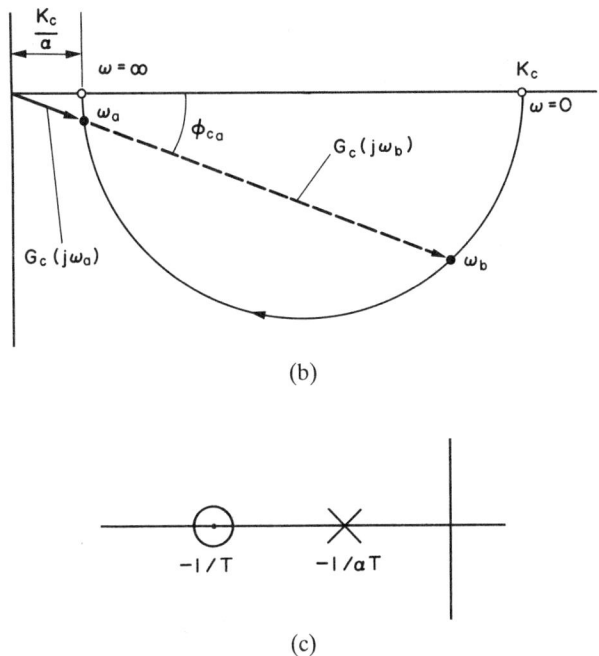

Fig. 9.4.4. Frequency responses for a lag compensator
(b) Nyquist diagram and
(c) Pole-zero locations

9.4.3 Lag-Lead Network

A third RC network used for system compensation is a lag-lead compensator shown in Fig. 5.

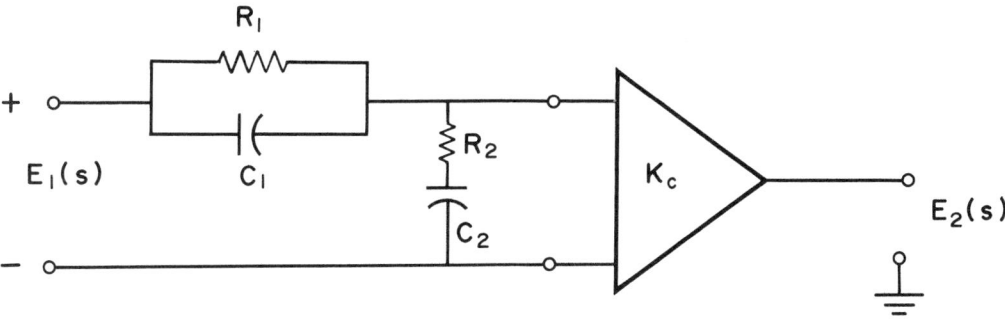

Fig. 9.4.5. A lag-lead compensator

430   LINEAR CONTROL SYSTEMS

The transfer function of this compensator can be easily seen to be,

$$G_c(s) = E_2(s)/E_1(s) = K_c(1+T_1s)(1+T_2s)/((1+\alpha T_1s)(1+(T_2/\alpha)s))$$

$$= K_c(s+1/T_1)(s+1/T_2)/((s+1/\alpha T_1)(s+\alpha/T_2)) \qquad (5)$$

where $T_1 = R_1C_1$, $T_2 = R_2C_2$ and $\alpha T_1+T_2/\alpha = R_1C_1+R_2C_2+R_1C_2$, $\alpha>1$ and $T_1>T_2$. The fractions $(1+T_1s)/(1+\alpha T_1s)$ and $(1+T_2s)/(1+(T_2/\alpha)s)$ represent the lag and lead compensator, respectively. Thus, there are two poles and two zeros added to the system. For this particular set of relations between T_1 and T_2, the zeros are at $-1/T_1$ and $-1/T_2$ while the poles are at $-1/\alpha T_1$ and $-\alpha/T_2$. The Bode and Nyquist diagrams for a lag-lead compensator are shown in Fig. 6.

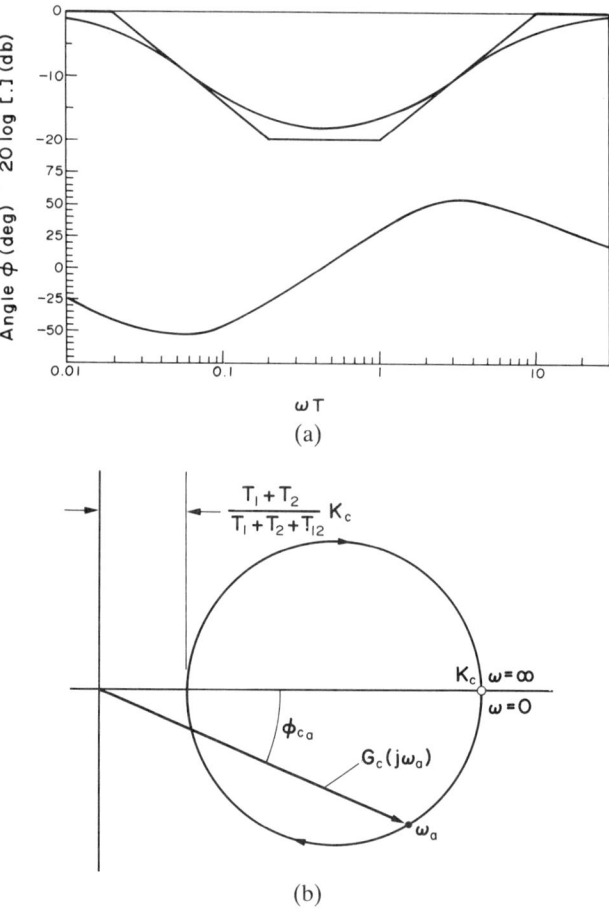

Fig. 9.4.6. Frequency responses for a lag-lead compensator
(a) Bode plots
(b) Nyquist plot

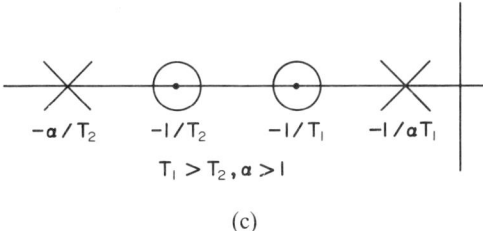

$-a/T_2 \quad -1/T_2 \quad -1/T_1 \quad -1/aT_1$

$T_1 > T_2, a > 1$

(c)

Fig. 9.4.6. Frequency responses for a lag-lead compensator
(c) Poles-zero locations

9.5 PERFORMANCE SPECIFICATIONS

It has been mentioned already that when a system's behavior is not desirable, it is often required to add cascade and/or feedback compensators such that the system would have certain desirable characteristics. In this section, two sets of "performance specifications" for time- and frequency-domain purposes are discussed.

9.5.1 Time-Domain Specifications

If an *nth* order system can be assumed to have a pair of complex conjugate dominant poles as discussed in Section 9.1, then the system can be designed as though it is a second-order system. Let the closed-loop transfer function of a second-order system be,

$$C(s)/R(s) = \omega_n^2/(s^2+2\xi\omega_n s+\omega_n^2) \tag{1}$$

where ω_n and ξ are known as the *undamped natural frequency* and *damping ratio* of the system, respectively. If the input $r(t) = u(t)$, a unit step, then it is easy to see that (See Table 3.4.1), through inversion of the Laplace transform, the time response of $c(t)$ becomes

$$c(t) = 1 - (1/\sqrt{1-\xi^2})\, e^{-\xi\omega_n t} Sin(\omega_d t + \phi) \tag{2}$$

where $\omega_d = \omega_n\sqrt{1-\xi^2}$ is known as the *damped natural frequency* and $\phi = \cos^{-1}\xi$. This response corresponds to the case of a pair of complex poles on the root locus of a unity - feedback system with an open-loop transfer function $G(s) = \omega_n^2/(s(s+2\xi\omega_n))$ (see Example 7.4.1). Figure 1 shows the coordinates of the roots of (1) (closed-loop poles) of a second-order system.

It is easily seen that the cosine of the angle ϕ of the line segment OA with the negative real axis is the damping ratio ξ. Thus a radial line passing through the origin making an angle ϕ with the negative real axis is a locus of constant $\xi = \cos\theta$. In a similar fashion a circle whose center is at the origin and whose radius is ω_n represents a constant undamped natural frequency locus. Figure 2 shows a family of loci for constant ξ, ω_n, $\sigma = \xi\omega_n$ and ω_d.

LINEAR CONTROL SYSTEMS

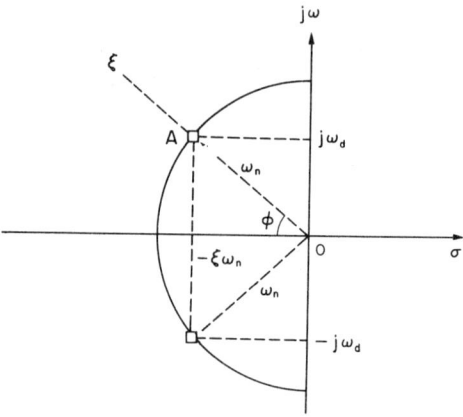

Fig. 9.5.1. Coordinates of a pair of complex-conjugate poles

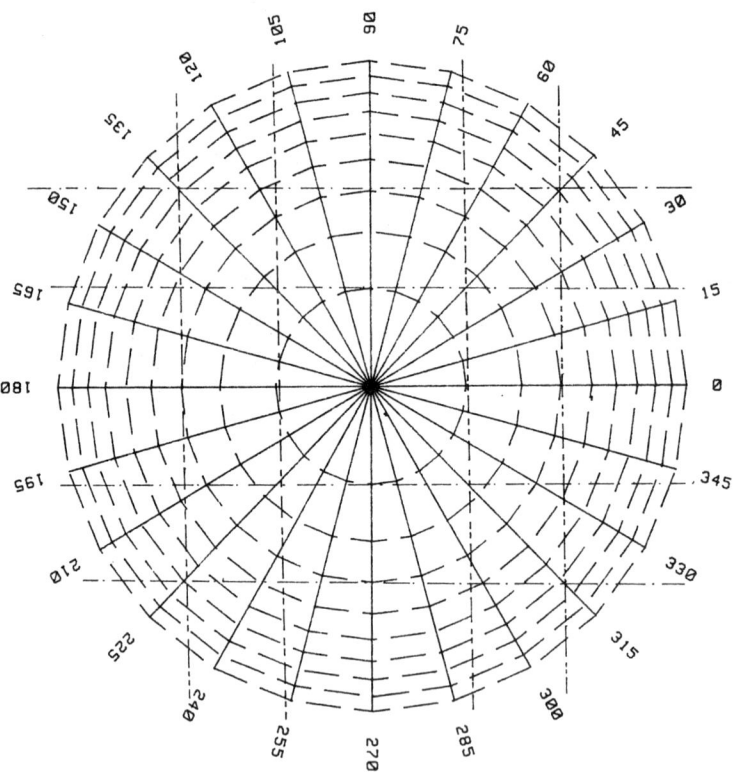

Fig. 9.5.2. Loci of constant ξ, ω_n, ω_d and σ for a 2nd-order system

PERFORMANCE SPECIFICATIONS   433

The time response of this second-order system is shown in Fig. 3. As seen the response has an overshoot and settles down sinusoidally with a certain decay rate towards its steady-state value. It is easy to note (See Problem 9.6.) that the peak time and peak value (maximum value of c(t)) of the output are,

$$c_p = 1 + e^{-\xi\pi/\sqrt{1-\xi^2}}, \quad T_p = \pi/\omega_n\sqrt{1-\xi^2}. \tag{3}$$

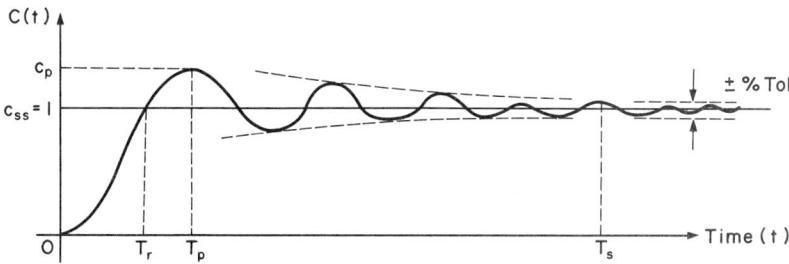

Fig. 9.5.3. Time response of an underdamped second-order system

Peak time is the time when c(t) reaches its maximum value. The time where the output response c(t) reaches to within plus or minus of some tolerance value, say 5%, is called the *settling time*, T_s. The degree of approachability of c(t) depends on the time constant $T = 1/\xi\omega_n$ of the response. It is clear from Fig. 3 that the response is to within 36.8% ($100e^{-1}$) of its final steady-state value after four time constants Similarly, when three time constants have elapsed, the response is to within 5% ($100e^{-3}$) or 2% ($100e^{-4}$) of the steady-state value after four time constants. Hence, the settling time for the assumed tolerances would be

$$T_s = \begin{cases} 3T = 3/\xi\omega_n & \cdots \pm 5\% \\ 4T = 4/\xi\omega_n & \cdots \pm 2\% \end{cases} \tag{4}$$

A fourth parameter, shown in Fig. 3, is the *rise time*, T_r which is the time it takes for the response to reach its steady-state value for the first time. Some texts use other definitions such as time elapsed where c(t) is between 10% to 90% of its steady-state value. However, we feel that those definitions would make system design unnecessarily cumbersome. Another important time-domain performance specification is the *percent overshoot*, M_o defined by

$$M_o\% = 100(c_p - c_{ss})/c_{ss} = 100e^{-\xi\pi/\sqrt{1-\xi^2}} \tag{5}$$

Figure 4 shows a response of $M_o\%$ versus damping ratio. It is clear from this figure that for no damping ($\xi = 0$) one has 100% overshoot corresponding to pure oscillation and a pair of purely imaginary closed-loop poles on the $j\omega$-axis. As ξ approaches one, the percent overshoot reduces until $\xi = 1$ where there is no overshoot corresponding to a critically damped situation.

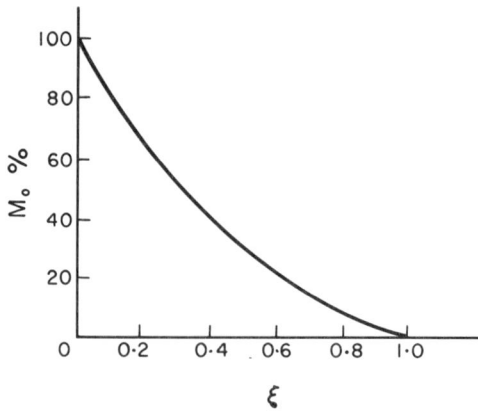

Fig. 9.5.4. A plot of percent overshoot versus damping ratio

To summarize our discussion, the time-response performance specifications are given by Table 1. The only entry in this table which was not accounted for in this section is the steady-state error which was discussed in Section 9.3. The following example illustrates how these specifications may be used in a design situation.

Table 9.5.1. Time-Domain Performance Specifications

Symbol	Meaning
C_p	Maximum overshoot
$M_o\%$	percent overshoot
T_s	settling time
T_p	peak time
T_r	rise time
ξ	damping ratio
ω_n	undamped natural frequency
ω_d	damped natural frequency
e_{ss}	steady-state error

1. Example. Let us assume that a system is to be designed (compensated) such that its dominant closed-loop poles provide a response with $M_o \leqslant 10\%$ and a settling time of $T_s = 3$ seconds. Following the plot of $M_o\%$ versus ξ (Fig. 4), it is clear that for a $M_o \leqslant 10\%$ it requires that $0.6 \leqslant \xi < 1$. Moreover, for a $T_s = 3$ seconds, by (4), it follows that $\xi\omega_n = 3/T_s = 1$. Thus for the above two conditions to hold simultaneously, it is necessary to have $1/\bar{\xi} \leqslant \omega_n \leqslant 1/\xi$ or $1 \leqslant \omega_n \leqslant 1/0.6 = 1.67$. The ranges of ξ and ω_n can be superimposed on the complex plane as shown in Fig. 5.

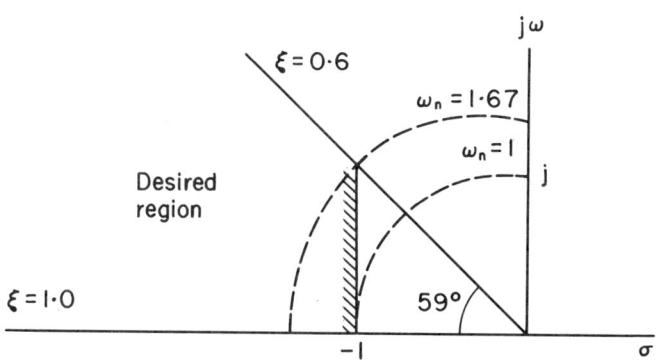

Fig. 9.5.5. Desired region for closed-loop poles for $M_0 \leqslant 10\%$ and $T_s = 3$ seconds

9.5.2 Frequency-Domain Specifications

Once again, let us consider transfer function of a second-order system given by (9.4.1) and rewrite it in terms of magnitude-phase quantity,

$$C(j\omega)/R(j\omega) = |C(j\omega)/R(j\omega)| \angle C(j\omega)/R(j\omega). \tag{6}$$

The frequency response of this system depends on the value of damping ratio ξ, as shown in Fig. 7.7.3. If the peak value of the $|C(j\omega)/R(j\omega)|$ is denoted by M_m, it is easy to see that its value is given by (Problem 9.8),

$$M_m = (1/(2\xi\sqrt{1-\xi^2})) \tag{7}$$

at

$$\omega_m = \omega_n\sqrt{1-2\xi^2} \tag{8}$$

as long as $\xi < 0.707$. Figure 6 shows plots of M_m and ω_m versus ξ.

Now, consider a plot of $M(\omega) = |C(j\omega)/R(j\omega)|$ versus ω shown in Fig. 7. From this diagram, another important specification can be defined for frequency-response of a second-order system. The *bandwidth* or *passband* is the range of frequencies from 0 up to the frequency ω_b where $M(\omega_b) = 0.707M(0)$. In practice it is more convenient to use ω_m as a figure of merit instead of ω_b.

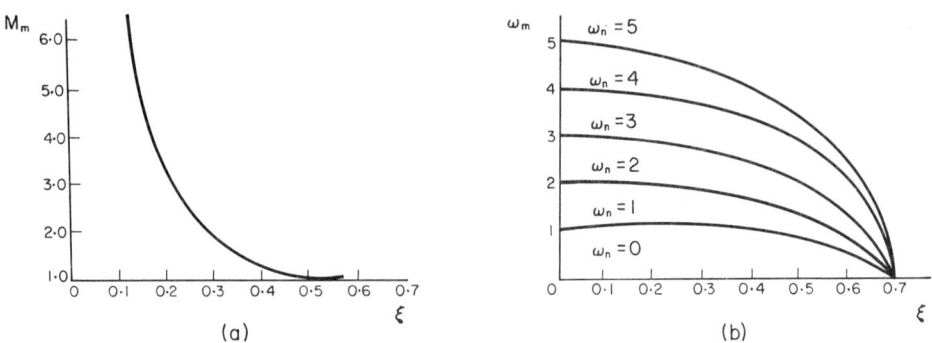

Fig. 9.5.6. Plots of peak magnitude and peak frequency versus damping ratio for a second-order system (a) M_m and (b) ω_m

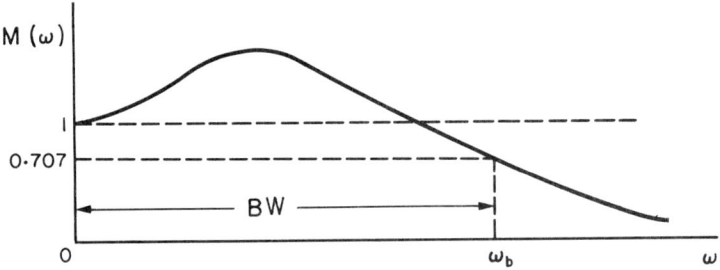

Fig. 9.5.7. Definition of the bandwidth for a second-order system

Thus far, three frequency-domain performance specifications, i.e. M_m, ω_m and ω_b are given. Two other important figures of merit, phase and gain margins were defined and discussed in Section 7.7.2. In sequel, two other often used figures of merit for frequency-domain design are presented.

An important aspect is the desire to design a system for a specific gain and phase. This would lead to two family of circles on the complex plane for loci of constant magnitude and phase of the system. They are known as constant M and constant α contours, respectively. In order to derive these contours, let the open-loop transfer function of a unity feedback system be,

$$G(j\omega) = x + jy \tag{9}$$

Then,

$$M(\omega) = |C(j\omega)/R(j\omega)| = |G(j\omega)/(1+G(j\omega))|$$

$$= |x+jy|/|1+x+jy| = [(x^2+y^2)/((1+x)^2+y^2)]^{1/2}. \tag{10}$$

PERFORMANCE SPECIFICATIONS

Squaring both sides of (10), collecting terms and completing a quadratic term involving x, it is easy to see that (Problem 9.11)

$$(x+M^2/(M^2-1))^2+y^2 = M^2/(M^2-1)^2 \tag{11}$$

which represents a circle with center at $(-M^2/(M^2-1),0)$ and a radius $r = |M/(M^2-1)|$. As M varies, both the center and radius of the circles would change. Figure 8 shows a family of these $M(\omega)$ circles for $0.4 \leqslant M \leqslant 2.0$. The behavior of the closed-loop frequency response can be readily predicted by superimposing the polar plot of $G(j\omega)$ on the $M(\omega)$ circles. (See Problem 9.9.)

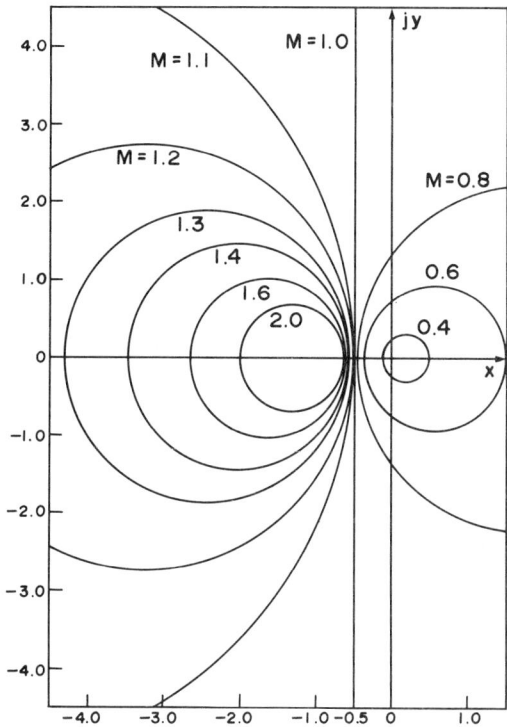

Fig. 9.5.8. Constant $M(\omega)$ circles

When gain of the system is to be set at a certain value, the tangent line from the origin to an M circle corresponding to the value of the particular gain plays an important role. Consider an M circle with the polar plot of unity feedback system.

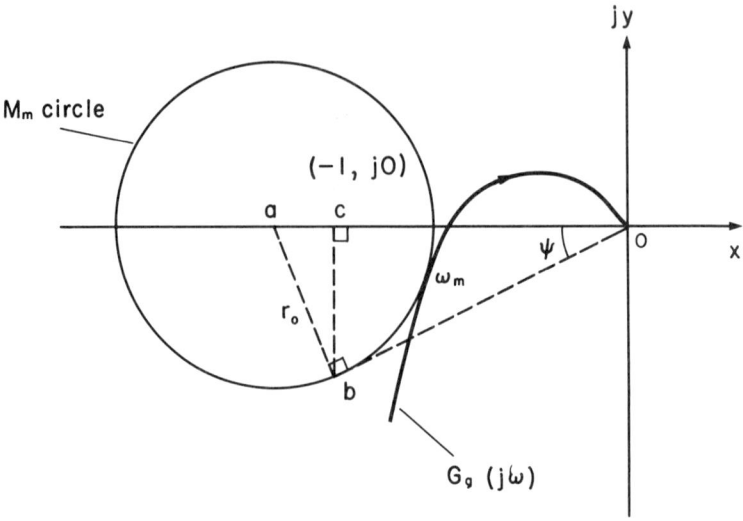

Fig. 9.5.9. Gain adjustment of a unity feedback system using M circle

The M circle shown is tangent to $G(j\omega)$, hence it corresponds to the maximum magnitude of the gain, i.e. M_m. The tangent line ob makes an angle ψ with the negative real axis whose line is given by $\sin \psi = ab/oa = r_o/x_o = 1/M$. Furthermore the perpendicular line segment drawn from b to the negative real axis, i.e. bc passes through the point $(-1, j0)$. This can be easily seen from the similarity of the two right triangles oab and obc and the fact that line ob $= M/\sqrt{M^2-1}$. Now, suppose that the d.c. gain of the system is K_g then the polar plot of another transfer function,

$$G'_g(j\omega) = (1/K_g)G_g(j\omega) = (x+jy)/K_g = x'+jy' \tag{12}$$

would be identical to that of $G_g(j\omega)$ except for the scales on the two axes. However, the M circle (old M_m circle for $G_g(j\omega)$) is no longer an M_m for $G'_g(j\omega)$ since its corresponding line segment oc' on the negative real axis (See Fig. 9) is $(-1/K_g, 0)$ instead of $(-1, j0)$. In order to adjust the gain of the original system to achieve the old value of M as the maximum of gain's magnitude M_m, the gain K_g must be adjusted by a $= K_m/K_g$, where $K_m = 1/oc'$. This application of M circles for gain adjustment will be used when a SISO system is designed using polar plots in Section 9.7.

The derivation of the constant phase can be determined by noting the transfer function (9) and evaluating,

$$\alpha = \tan^{-1}(C(j\omega)/R(j\omega))$$

$$= \tan^{-1}((x+jy)/(1+x+jy)) = \tan^{-1}(y/x) - \tan^{-1}(y/(1+x)) = \tan^{-1}(y/(x^2+x+y^2)) \tag{13}$$

Now let $N = \tan\alpha$. Then by rearranging the expression

$$N = y/(x^2+x+y^2) \tag{14}$$

one obtains

$$(x+\frac{1}{2})^2+(y-\frac{1}{2N})^2 = (N^2+1)/4N^2 \tag{15}$$

which represent a circle with center at $(x_o, y_o) = (-1/2, 1/2N)$ with radius $R = (N^2+1)^{1/2}/2N$. Figure 10 shows a family of N circles.

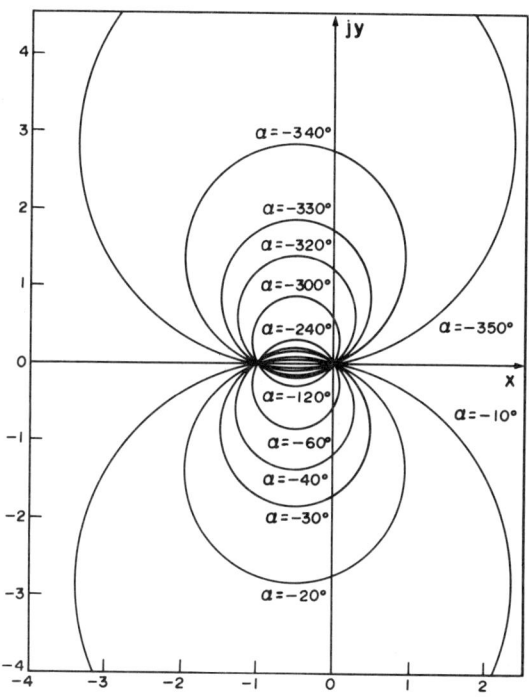

Fig. 9.5.10. Constant $\alpha(\omega)$ circles

The phase angle in Table 2 summarizes frequency-domain performance specifications.

Program "MACRCL" provides a computer code to plot M and α circles along with the constant ψ angle line.

Table 9.5.2. Frequency-domain Performance Specifications

Frequency-Domain Specifications	
Symbol	Meaning
ω_m	Frequency of Peak magnitude
M_m	peak value of magnitude
γ	phase margin
G	gain margin
$M(\omega)$	constant M magnitude
$\alpha(\omega)$	constant phase angle

2. Program MACRCL

```
10    OPTION BASE 0
20    DIM A(20),B(20)
30    COM X0,Y0,X1,Y1,Xmin,Xmax,Ymin,Ymax
40    COM Paxes,P$,Xa,Ya,X$,Y$,Xti,Yti,Xni,Yni,Grid,Print,Pen,Lt,Pt
50    PRINT "THIS PROGRAM PLOTS A POLAR PLOT FOR A GIVEN TRANSFER "
60    PRINT "FUNCTION.IT THEN SUPERIMPOSES M AND ALPHA CIRCLES OF "
70    PRINT "SPECIFIED VALUES ON THE POLAR   PLOT. IN ADDITION IT PLOTS"
80    PRINT "THE LINE THAT CORRESPONDS TO Psi=ARCSIN(1/M).THE ANGLE Psi"
90    PRINT "IS THE ANGLE BETWEEN THE PLOTTED LINE AND THE X AXIS."
100   PRINT "_____"
110   INPUT "HAVE ALL NECESSARY SUBROUTINES BEEN LINKED? Y/N",A$
120   IF A$<>"N" THEN GOTO 220
130   LINK "Transf",1300,140
140   LINK "Polpro",1790,150
150   LINK "Bdnync",2010,160
160   LINK "Spoly",2920,170
170   LINK "P_sul",3390,180
180   LINK "P_sb",3960,190
190   LINK "P_xslg",4870,200
200   LINK "P_xlin",6170,210
210   LINK "Vec",7250,220
220   INPUT "ENTER ORDER OF NUMERATOR,N, AND DENOMINATOR,M,AS N,M",N,M
230   INPUT "ANY CHANGES? Y/N",A$
240   IF A$<>"N" THEN GOTO 220
250   REDIM A(N),B(M)
260   PRINT "YOU MUST NOW PREPARE TO ENTER COEFFICIENTS OF A(S)"
270   PRINT "AND B(S) IN ASCENDING ORDER. DO NOT FORGET TO "
280   PRINT "INCORPORATE THE VALUE OF GAIN K."
290   PRINT "----------------------------------"
300        FOR I=0 TO N STEP 1
310          DISP "ENTER A(";I;")";
320          INPUT A(I)
330        NEXT I
340   INPUT "ANY CHANGES? Y/N",A$
350   IF A$<>"N" THEN GOTO 300
360          FOR I=0 TO M
370            DISP "ENTER B(";I;")";
380            INPUT B(I)
390          NEXT I
```

PERFORMANCE SPECIFICATIONS

```
400     INPUT "ANY CHANGES? Y/N ",A$
410     IF A$<>"N" THEN GOTO 360
420     DIM X(121),Y(121)
430     CALL Bdnync(A(*),N,B(*),M,2)
440     INPUT "DO YOU WANT TO PLOT AN M CIRCLE? Y/N",A$
450     IF A$<>"Y" THEN GOTO 990
460     PRINT "^^^^^^^^^^^^^^^^^^^^^^^^^^^^^^^^^^^^^^^^^^^^^"
470     PRINT "ENTER PRACTICAL   VALUE OF M FOR M CIRCLE. ABS(M)=1 YIELDS"
480     PRINT "A VERTICAL LINE AT X=-.5. _____     "
490     PRINT "IF YOU ARE NOT GETTING A PLOT OF YOUR M CIRCLE IT COULD BE"
500     PRINT "BECAUSE OF THE SCALE OF YOUR POLAR PLOT!! PLEASE CHECK."
510     INPUT "WHAT VALUE OF M DO YOU WISH TO USE? ",M
520     IF M=0 THEN GOTO 470
530     IF ABS(M)<>1 THEN GOTO 680
540     Ycor=0
550             FOR I=0 TO 60 STEP 1
560             X(I)=-.5
570             Y(I)=.2+Ycor
580             Ycor=Y(I)+Ycor
590             NEXT I
600     Ycor=0
610             FOR I=61 TO 120.0 STEP 1
620             X(I)=-.5
630             Y(I)=-.2+Ycor
640             Ycor=Y(I)+Ycor
650             NEXT I
660     CALL P_sb(X(*),Y(*),121)
670     GOTO 970
680     C=-M^2/(M^2-1)
690     R=ABS(M/(M^2-1))
700     PRINT "CENTER OF CIRCLE IS AT      Y=0,    X=",C
710     PRINT "RADIUS OF M CIRCLE IS                R=",R
720     PRINT "*************************************************"
730     DEG
740     I=0
750             FOR A=0 TO 360 STEP 3
760             X(I)=C+R*COS(A)
770             Y(I)=R*SIN(A)
780             I=I+1
790             NEXT A
800     CALL P_sb(X(*),Y(*),121)
810     IF ABS(1/M)>1 THEN GOTO 830
820     Psi=ASN(1/M)
830     PRINT "     NOTE: PROGRAM CANNOT PLOT "
840     PRINT "THE LINE THAT INDICATES Psi IF ABS(1/M)>1."
850     INPUT "DO YOU WISH TO PLOT THE LINE FOR Psi=ASN(1/M)? Y/N",A$
860     IF A$<>"Y" THEN GOTO 970
870     IF (1/M>1) OR (1/M<-1) THEN GOTO 960
880     Delx=(Xmax-Xmin)/120.0
890     X(0)=Xmin
900     Y(0)=Xmin*TAN(Psi)
910             FOR I=1 TO 120.0
920             X(I)=X(I-1)+Delx
930             Y(I)=X(I)*TAN(Psi)
940             NEXT I
950     CALL P_sb(X(*),Y(*),121)
960     PRINT "PROGRAM CANNOT PLOT LINE FOR THIS VALUE OF M."
970     INPUT "DO YOU WANT TO PLOT ANOTHER M CIRCLE? (Y/N)",A$
980     IF A$<>"N" THEN GOTO 470
990     INPUT "DO YOU WANT TO PLOT AN ALPHA CIRCLE? Y/N ",A$
1000    IF A$<>"Y" THEN GOTO 1270
1010    PRINT "TO PLOT AN ALPHA CIRCLE YOU MUST GIVE A VALID VALUE"
1020    PRINT "FOR ALPHA IN DEGREES.DO NOT USE VALUES OF ALPHA    "
1030    PRINT "EQUAL TO -90,90, (OR MULTIPLES OF) OR 0. ############"
1040    INPUT "INDICATE VALUE OF ALPHA",Alpha
1050    IF (Alpha=-90) OR (Alpha=90) THEN GOTO 1010
1060    IF (Alpha=180) OR (Alpha=-180) THEN GOTO 1010
1070    IF (Alpha=-270) OR (Alpha=270) THEN GOTO 1010
1080    IF (Alpha=-360) OR (Alpha=450) THEN GOTO 1010
1090    IF (Alpha=0) OR (Alpha=360) THEN GOTO 1010
```

```
1100 IF ABS(Alpha)<.25 THEN GOTO 1010
1110 Na=TAN(Alpha)
1120 Rad=.5*SQR((Na^2+1)/Na^2)
1130 Ycor=1/(2*Na)
1140 PRINT "......................................"
1150 PRINT "All alpha circles are centered at X = -.5   "
1160 PRINT "Radius of Alpha circle is ",Rad,"Ycor is ",Ycor
1170 DEG
1180 I=0
1190      FOR A=0 TO 360 STEP 3
1200      X(I)=-.5+Rad*COS(A)
1210      Y(I)=Rad*SIN(A)+Ycor
1220      I=I+1
1230      NEXT A
1240 CALL P_sb(X(*),Y(*),121)
1250 INPUT "DO YOU WANT TO PLOT ANOTHER ALPHA CIRCLE? Y/N",A$
1260 IF A$="Y" THEN GOTO 1010
1270 PRINT "-------------- END OF PROGRAM -----------------"
1280 STOP
1290 END
1300 ! Transf - TRANSFER FUNCTION
1310 SUB Transf(A(*),N,B(*),M,C(*),K,D(*),L,Ac(*),Bc(*),E(*),F(*))
1320 PRINT "DETERMINES THE CLOSED-LOOP TRANSFER FUNCTION OF LINEAR,"
1330 PRINT "SINGLE-LOOP, FEEDBACK SYSTEMS.",LIN(1)
1340 PRINT "FORWARD transfer function : G(s)=A(s)/B(s)"
1350 PRINT "FEEDBACK transfer function : H(s)=C(s)/D(s)"
1360 PRINT "OPEN-LOOP transfer function: GH=E(s)/F(s)"
1370 PRINT "CLOSED-LOOP transfer function : C_r(s)=G(s)/[1+G(s)H(s)]=Ac(s)/Bc(s)

ZLIN(1)
1380 PRINT "                                         2          N"
1390 PRINT " where   A(s)=A(0) + A(1)s + A(2)s +...+A(N)s"
1400 N1=MAX(M+L,N+K)
1410 CALL R(A(*),N,B(*),M,C(*),K,D(*),L,Ac(*),Bc(*),E(*),F(*))
1420 PRINTER IS 16
1430 SUBEXIT
1440 SUBEND
1450 SUB R(A(*),N,B(*),M,C(*),K,D(*),L,Ac(*),Bc(*),E(*),F(*))
1460 N2=MIN(M+L,N+K)
1461 N1=MAX(M+L,N+K)
1480 PRINT "Give coefficients of the NUMERATOR polynomial of G(s) in ASCENDING o

Zer:";
1490 CALL Vec(A(*),N,"A")
1500 PRINT "Give coefficients of the DENOMINATOR polynomial of G(s) in ASCENDING

Zrder:"
1510 CALL Vec(B(*),M,"B")
1520 PRINT "Give coefficients of the NUMERATOR polynomial of H(s) in ASCENDING o

Zer:"
1530 CALL Vec(C(*),K,"C")
1540 PRINT "Give coefficients of the DENOMINATOR polynomial of H(s) in ASCENDING

Zrder:"
1550 CALL Vec(D(*),L,"D")
1560 PRINT "Coefficients of the NUMERATOR polynomial of C_r(s) in ASCENDING orde

Z",LIN(1)
1570 CALL Polpro(A(*),D(*),N,L,Ac(*))
1580 FOR I=0 TO L+N
1590 PRINT "Ac(";I;")=";Ac(I)
1600 NEXT I
1610 PRINT LIN(1),"Coefficients of the DENOMINATOR polynomial of C_r(s) in ASCEN

ZNG order:",LIN(1)
1620 CALL Polpro(B(*),D(*),M,L,F(*))
1630 CALL Polpro(A(*),C(*),N,K,E(*))
1640 FOR I=0 TO N2
1650 Bc(I)=E(I)+F(I)
```

PERFORMANCE SPECIFICATIONS

```
1660 NEXT I
1670 IF N1=M+L THEN 1710
1680 FOR I=N2+1 TO N+K
1690 Bc(I)=E(I)
1700 NEXT I
1710 FOR I=N2+1 TO M+L
1720 Bc(I)=F(I)
1730 NEXT I
1740 FOR I=0 TO N1
1750 PRINT "Bc(";I;")=";Bc(I)
1760 NEXT I
1770 SUBEXIT
1780 SUBEND
1790 SUB Polpro(A(*),B(*),N,M,R(*))
1800 !    "Polpro" CALCULATES THE PRODUCT OF TWO POLYNOMIALS.
1810 DIM A1(M+N),B1(M+N)
1820 FOR I=0 TO N
1830 A1(I)=A(I)
1840 NEXT I
1850 FOR I=N+1 TO M+N
1860 A1(I)=0
1870 NEXT I
1880 FOR I=0 TO M
1890 B1(I)=B(I)
1900 NEXT I
1910 FOR I=M+1 TO M+N
1920 B1(I)=0
1930 NEXT I
1940 FOR I=0 TO M+N
1950 R(I)=0
1960 FOR J=0 TO I
1970 R(I)=R(I)+A1(J)*B1(I-J)
1980 NEXT J
1990 NEXT I
2000 SUBEND
2010 ! Bodnyq - Bode, Nyquist, Nichols plotter
2020 SUB Bdnync(A(*),N,B(*),M,Type)
2030 OPTION BASE 0
2040 !
2050 COM X0,Y0,X1,Y1,Xmin,Xmax,Ymin,Ymax
2060 COM Paxes,P$,Xa,Ya,X$,Y$,Xti,Yti,Xni,Yni,Grid,Print,Pen,Lt,Pt
2070 !
2080 ! Plots Bode, Nyquist, or Nichols diagram of G(s)=A(s)/B(s)
2090 !                              2          N
2100 !  where A(s)=A(0) + A(1)s + A(2)s + ... + A(N)s
2110 !  B(s) is similar
2120 ! Type = 1-Bode , 2-Nyquist , 3-Nichols
2130 !
2140 INPUT "Do you want to clear the old graph? (Y/N)",A$
2150 IF A$<>"Y" THEN 2180
2160 PLOTTER IS 13,"GRAPHICS"
2170 IF P$="9872A" THEN PLOTTER IS "9872A"
2180 PRINT "The frequency F will vary from"
2190 PRINT " F(start) to F(end) (Hz.) according to the equation:"
2200 PRINT "      F(new)=A*F(old)"
2210 PRINT
2220 DIM X(200),Y(200),Y1(200)
2230 I2:INPUT "Enter F(start),F(end),A",Fs,Fe,A
2240 Np=INT((LGT(Fe)-LGT(Fs))/LGT(A)+1)    ! number of points
2250 REDIM X(Np),Y(Np),Y1(Np)
2260 PRINT "There will be ";Np;" points."
2270 INPUT "Is this O.K.?(Y/N)",A$
2280 IF A$<>"Y" THEN I2
2290 !
2300 DEG
2310 F=Fs
2320 I=0
2330 Loop:W=F*3.14159/180         ! Convert to radians
2340    CALL Spoly(A(*),N,0,W,Ar,Ai,Amag,Aang)
2350    CALL Spoly(B(*),M,0,W,Br,Bi,Bmag,Bang)
```

```
2360      Mag=Amag/Bmag
2370      Ang=Aang-Bang          ! value of G(s)
2380      Gr=Mag*COS(Ang)
2390      Gi=Mag*SIN(Ang)        ! value of G = Gr + j(Gi)
2400      ON Type GOTO Bode,Nyq,Nich
2410 Bode:X(I)=F
2420      Y(I)=20*LGT(Mag)       ! in db
2430      Y1(I)=Ang
2440      GOTO C1
2450 Nyq:X(I)=Gr
2460      Y(I)=Gi
2470      GOTO C1
2480 Nich:X(I)=Ang
2490      Y(I)=20*LGT(Mag)
2500      GOTO C1
2510 C1:F=A*F                    ! next frequency value
2520      I=I+1                  ! next
2530      IF F<=Fe THEN Loop     ! next point
2540      PRINT
2550      P$="CRT"
2560      INPUT "Do you want the plot on the 9872A plotter? (Y/N)",A$
2570      IF A$="Y" THEN P$="9872A"
2580      ON Type GOTO Bode2,Nyq2,Nich2
2590 Bode2:PRINT "BODE plot"
2600      CALL P_sul(X(*),Y(*),Np)
2610      X$="Frequency,Hz"
2620      Y$="Mag,db"
2630      ! Magnitude plot
2640      Paxes=2                ! semi-log plot
2641      Y$="Phase, deg."
2650      CALL P_xslg
2660      CALL P_sb(X(*),Y(*),Np)
2670      ! Phase plot
2680      Paxes=2
2690      CALL P_sul(X(*),Y1(*),Np)
2700      CALL P_xslg
2710      CALL P_sb(X(*),Y1(*),Np)
2720      GOTO End
2730 Nyq2: PRINT "NYQUIST PLOT"
2740      CALL P_sul(X(*),Y(*),Np)
2750      X$="REAL"
2760      Y$="IMAGINARY"
2770      Paxes=1
2780      CALL P_xlin
2790      CALL P_sb(X(*),Y(*),Np)
2800      GOTO End
2810 Nich2: PRINT "NICHOL'S CHART"
2820      CALL P_sul(X(*),Y(*),Np)
2830      Paxes=1                ! semilog
2840      X$="Phase,deg."
2850      Y$="Mag.,db"
2860      CALL P_xlin
2870      CALL P_sb(X(*),Y(*),Np)
2880      GOTO End
2890 End: SUBEXIT
2900 SUBEND
2920 ! Spoly - polynomial of s=X+jY
2930 SUB Spoly(P(*),N,Sig,W,Real,Imag,Mag,Phase)
2940 !                          2            N
2950 ! P(s)= p(0) + p(1)s + p(2)s +...+ p(N)s
2960 !       where s=Sig+jW, p(i) are real numbers
2970 ! P(s)=(Real)+j(Imag)=Mag/Phase (Phase in degrees)
2980 DEF FNMod(X,Y)=X-Y*INT(X/Y)
2990 OPTION BASE 0
3000 ! ---------SUM-----------
3001 DEG
3010 Ms=SQR(Sig^2+W^2)
3020 As=FNAtan(Sig,W)
3030 Real=0
3040 Imag=0
```

```
3050 FOR I=0 TO N STEP 1
3060    M=Ms^I*P(I)          ! s^i=Ms^i @ As*i
3070    A=As*I
3080    Real=Real+M*COS(A)
3090    Imag=Imag+M*SIN(A)
3100 NEXT I
3110 ! -----END OF SUM---------
3120 Mag=SQR(Real^2+Imag^2)
3130 Phase=FNAtan(Real,Imag)           ! arctan considering quadrant
3140 SUBEXIT
3150 SUBEND
3160 !
3170 ! Atan - arctan considering quadrant
3180 DEF FNAtan(X,Y)
3190 DEG          ! angle in degrees
3200 IF X<>0 THEN A2
3210    IF Y>0 THEN Ang=90
3220    IF Y<0 THEN Ang=270
3230    IF Y=0 THEN Ang=0
3240    GOTO End
3250 A2:Ang=ATN(Y/X)
3260 IF Y/X>=0 THEN Pos
3270 Neg:IF Y<0 THEN Q4
3280 Q2:Ang=Ang+180            !. quadrant 2
3290    GOTO End
3300 Q4:Ang=Ang+360
3310    GOTO End
3320 Pos:IF X>=0 THEN Q1
3330 Q3:Ang=Ang+180
3340    GOTO End
3350 Q1:Ang=Ang
3360 End:RETURN Ang
3370 FNEND
3390 !. P_sul - plotter setup
3400 !        sets ranges, labels, ticks
3410 SUB P_sul(X(*),Y(*),N)
3420 OPTION BASE 0
3430 !
3440 COM X0,Y0,X1,Y1,Xmin,Xmax,Ymin,Ymax
3450 COM Paxes,P$,Xa,Ya,X$,Y$,Xti,Yti,Xni,Yni,Grid,Print,Pen,Lt,Pt
3460 !
3470 !    LOCATE PLOT ON PAPER
3480 EXIT GRAPHICS
3490 PRINT
3500 PRINT "Please digitize X0,Y0 - then X1,Y1"
3510 WAIT 1000
3520 SETGU
3530 DIGITIZE X0,Y0
3540 MOVE X0,Y0
3550 DRAW X0,Y0          ! Put a point there
3560 WAIT 1000
3570 DIGITIZE X1,Y1
3580 MOVE X1,Y1
3590 DRAW X1,Y1
3600 PENUP
3610 SETUU
3620 !
3630 Maxx=Minx=X(0)
3640 Maxy=Miny=Y(0)
3650 FOR I=1 TO N-1
3660    IF X(I)>Maxx THEN Maxx=X(I)
3670    IF X(I)<Minx THEN Minx=X(I)
3680    IF Y(I)>Maxy THEN Maxy=Y(I)
3690    IF Y(I)<Miny THEN Miny=Y(I)
3700 NEXT I
3710 PRINT
3720 PRINT "Minx,Maxx=";Minx;Maxx
3730 PRINT "Miny,Maxy=";Miny;Maxy
3740 PRINT
3750 !
```

```
3760 M1: INPUT "Enter Xmin,Xmax,Ymin,Ymax.",Xmin,Xmax,Ymin,Ymax
3770 PRINT "Xmin=";Xmin;"Xmax=";Xmax;"Ymin=";Ymin;"Ymax=";Ymax
3780 INPUT "Changes ?(Y/N).",A$
3790 IF A$="Y" THEN M1
3800 M2:INPUT "Enter Xti,Xni,Yti,Yni,Xa,Ya.",Xti,Xni,Yti,Yni,Xa,Ya ! ti=tick int
val
3810 PRINT "Xti=";Xti;"Xni=";Xni;"Yti=";Yti;"Yni=";Yni;"Xa=";Xa;"Ya=";Ya
3820 INPUT "Changes ?(Y/N).",A$                    ! ni=number interval
3830 IF A$="Y" THEN M2
3840 !. DEFAULT VALUES
3850 Paxes=1
3860 Pen=1
3870 Lt=1
3880 Pt=1
3890 Grid=0
3900 Print=1
3901 INPUT "Would you like a grid?(Y/N)",A$
3902 IF A$="Y" THEN Grid=1
3910 !.
3920 SUBEXIT
3930 SUBEND
3960 ! P_sb    - LOW LEVEL PLOTTER PROGRAM
3970 !            will now do semi-log plots
3980 SUB P_sb(X(*),Y(*),N)
3990 !
4000 ! X,Y - INDEPENDENT AND DEPENDENT VARIABLES (sorted in ascending X)
4010 ! N - the number of points in X,Y
4020 !. X0,Y0 - lower left corner of plot in GDU's
4030 !. X1,Y1 - upper right corner of plot in GDU's
4040 !. Xmin,Xmax,Ymin,Ymax - desired ranges of plot in UDU's
4050 ! P$ - plotter string
4060 ! Paxes - axes type (0=no axes, 1=lin-lin,2=lin-log)
4070 !
4080 OPTION BASE 0
4090 !
4100 COM X0,Y0,X1,Y1,Xmin,Xmax,Ymin,Ymax
4110 COM Paxes,P$,Xa,Ya,X$,Y$,Xti,Yti,Xni,Yni,Grid,Print,Pen,Lt,Pt
4120 !
4130 ON ERROR GOTO Error1
4140 Plotter$="9872A"
4150 GRAPHICS
4160 SETGU                       ! GDU's
4170 LOCATE X0,X1,Y0,Y1
4180 Xl=Xmin
4190 Xh=Xmax
4200 Yl=Ymin
4210 Yh=Ymax
4220 IF Paxes<>2 THEN P6
4230    Xl=LGT(Xmin)      ! SEMILOG, CHANGE SCALING
4240    Xh=LGT(Xmax)
4250 P6: SCALE Xl,Xh,Yl,Yh              ! UDU's(assumes log values not given for semi
g)
4260 Plotit: IF Lt>0 THEN C2
4270    IF (Pt<>1) AND (Pt<>2) AND (Pt<>3) THEN 4380
4280 C2: IF P$=Plotter$ THEN PEN Pen
4290 IF Lt<=0 THEN Point          ! NEGATIVE TYPE->POINTS and LINES
4300 Line:LINE TYPE Lt
4310    H=X(0)             !. FIRST POINT
4320    V=Y(0)
4330    IF Paxes<>2 THEN L2
4340    H=LGT(H)
4350 L2:MOVE H,V            ! LINE PLOT
4360    FOR I=1 TO N-1      ! REST OF POINTS
4370    H=X(I)
4380    V=Y(I)
4390    IF Paxes<>2 THEN L3
4400       H=LGT(H)
4410 L3: DRAW H,V
4420    NEXT I
4430    GOTO End
```

```
4440 Point:Ssize=.65             ! POINT PLOT
4450       DEG
4460       LINE TYPE 1
4470       FOR I=0 TO N-1
4480          H=X(I)
4490          V=Y(I)
4500          IF Paxes<>2 THEN P2
4510             H=LGT(H)
4520 P2:     MOVE H,V           ! PUT A POINT
4530         DRAW H,V
4540         MOVE H,V
4550         SETGU
4560         ON Pt GOSUB Circle,Triangl,Box      ! DRAW FIGURE
4570         SETUU
4580       NEXT I
4590       IF Lt=0 THEN End     ! just points
4600       Lt=ABS(Lt)           ! now connect the lines
4610       GOTO Line
4620       SUBEXIT
4630 End:  EXIT GRAPHICS
4640       SUBEND
4650 !.
4660 Circle:  Size=Ssize
4670       FOR Arc=0 TO 360 STEP 30
4680          PDIR Arc
4690          RPLOT Size,0        . ! RELATIVE PLOT
4700       NEXT Arc
4710       PENUP
4720       RETURN
4730 Triangl: Size=Ssize
4740       FOR Arc=90 TO 450 STEP 120
4750          PDIR Arc
4760          RPLOT Size,0
4770       NEXT Arc
4780       PENUP
4790       RETURN
4800 Box:  Size=Ssize
4810       FOR Arc=45 TO 405 STEP 90
4820          PDIR Arc
4830          RPLOT Size,0
4840       NEXT Arc
4850       PENUP
4860       RETURN
4870 ! P_xslg - SEMILOG AXES
4880 SUB P_xslg
4890 !
4900 !. X0,Y0 - lower left corner of plot in GDU's
4910 ! X1,Y1 - upper right corner of plot in GDU's
4920 ! Xmin,Xmax,Ymin,Ymax - desired ranges of plot in UDU's
4930 !. P$ - name of plotter
4940 ! Paxes - PRINT AXES CODE (0=NO,1=LINEAR,2=SEMILOG)
4950 !       SEMILOG - Xni doesn't matter, Xmin,Xmax=actual values
4960 ! Xa,Ya - location of x-axis and yaxis in UDU's
4970 ! X$,Y$ - titles for X,Y axes
4980 ! Yti - y tick interval in UDU's
4990 ! Yni - Y number interval
5000 ! Grid - grid code (0= no grid, 1=grid)
5010 ! Print - print code (0=don't print values , 1=print values)
5020 !        prints only in evenly spaced increments
5030 !
5040 OPTION BASE 0
5050 !
5060 COM X0,Y0,X1,Y1,Xmin,Xmax,Ymin,Ymax
5070 COM Paxes,P$,Xa,Ya,X$,Y$,Xti,Yti,Xni,Yni,Grid,Print,Pen,Lt,Pt
5080 !.
5090 ON ERROR GOTO Errorl
5100 GRAPHICS
5110 SETGU                      ! GDU's
5120 LOCATE X0,X1,Y0,Y1
5130 X1=LGT(Xmin)
```

```
5140 Xh=LGT(Xmax)
5150 Yl=Ymin
5160 Yh=Ymax
5170 P6:Xrg=Xl-X0          !  X range in GDU's
5180 Yrg=Yl-Y0             !  Y
5190 Xru=Xh-Xl
5200 Yru=Yh-Yl
5210 SCALE Xl,Xh,Yl,Yh          ! UDU's(assumes log values not given for semilog)
5220 Mtcy=Yni/Yti
5230 IF P$<>Plotter$ THEN C4
5240 PEN Pen
5250 C4:LINE TYPE 1              ! AXES FOR SEMILOG
5260   SETUU
5270   Yas=LGT(Ya)              ! Y AXIS SEMILOG CONVERSION
5280   IF Print=1 THEN Printit
5290   SUBEXIT
5300 !
5310 Error1:PRINT ERRM$
5320 BEEP
5330 PAUSE
5340 GOTO Plotit
5350 !
5360 ! Prints axes and axes labels
5370 !
5380 Printit: CSIZE 2.5
5390         LINE TYPE 1
5400 ! ************AXES NUMBERS********************
5410    IF P$<>Plotter$ THEN Nopen
5420    PEN Pen
5430 Nopen:    LINE TYPE 1
5440          SETUU                    ! UDU's
5450 ! ----------------X----------------
5460 Labelx: LORG 6                .
5470         DEG
5480         LDIR 0
5490 Semilog: FOR D=Xl TO Xh STEP 1
5500   FOR I=1 TO 9      ! DO THESE LINES BETWEEN DECADES
5510   H=LGT(10.0^D*I)      ! Horizontal position
5520   IF Grid<>1 THEN S2
5530   LINE TYPE 3           ! DASHED
5540   IF I=1 THEN LINE TYPE 4 ! DASH-DOT FOR DECADE
5550   MOVE H,Yl      ! GRID LINE
5560   DRAW H,Yh
5570   LINE TYPE 1
5580   GOTO S3
5590 S2:Tl=.5
5600 IF I=1 THEN Tl=1   ! longer tick for decade
5610   MOVE H,Xa+Tl*Yru/Yrg   ! TICK
5620   DRAW H,Xa-Tl*Yru/Yrg
5630 S3:  Yp=1        !. Y POSITION
5640   CSIZE 2.5
5650   V=Xa-Yp*Yru/Yrg
5670   MOVE H,V
5680   X=10.0^H       ! ACTUAL VALUE
5690   IF I=1 THEN LABEL USING "K";X       ! LABEL DECADE ONLY
5700 S4:IF D=Xh THEN Labely         ! STOP AT LAST DECADE
5710  NEXT I
5720 NEXT D
5730 ! ----------------Y----------------
5740 Labely: LORG 8
5750     LDIR 0
5760     FOR Yax=Yl TO Yh STEP Yti       ! SEMILOG TICKS
5770     IF Grid<>1 THEN Ly4
5780     LINE TYPE 3
5790     MOVE Xl,Yax       ! PUT A GRID LINE IF SEMILOG WANTS
5800     DRAW Xh,Yax
5810     LINE TYPE 1
5820     GOTO Ly3
5830 Ly4: MOVE Yas-1*Xru/Xrg,Yax  ! TICK MARK
5840     DRAW Yas+1*Xru/Xrg,Yax
```

```
5850 Ly3: NEXT Yax
5860     MOVE X1,Xa                  ! AXES
5870     DRAW Xh,Xa
5880     MOVE Yas,Yl
5890     DRAW Yas,Yh
5900 Ly2: FOR Yax=Yl TO Yh STEP Yni  ! NUMBERS
5910     MOVE Yas-1.5*Xru/Xrg,Yax    !1.5 GDU's LEFT
5920     IF Yax<>0 THEN Ly1
5930     LABEL USING "K";Yax
5940     GOTO Nexty
5950 Ly1:    IF (ABS(Yax)>=1E5) OR (ABS(Yax)<=1E-5) THEN Toobigy
5960    LABEL USING "K";Yax
5970    GOTO Nexty
5980 Toobigy:   LABEL USING "MZ.DDE";Yax
5990 Nexty: NEXT Yax
6000 ! ------------------------------------------------------------
6010 Xtitle:Csize=2
6020       LINE TYPE 1
6030       Tsize=2.5
6040        SETGU
6050        LORG 6
6060        LDIR 0
6070        Xx=X0+Xrg/2
6080        Yx=Y0-5         ! X LABEL POSITION
6090        MOVE Xx,Yx
6100        SETUU
6110        CSIZE Tsize
6120        LABEL USING "K";X$
6130 Ytitle:LORG 4
6140       LDIR 90
6150       SETGU
6160       Xy=X0-9           ! Y LABEL POSITION
6170 ! P_xlin - LOW LEVEL LINEAR  AXES
6180 SUB P_xlin
6190 !
6200 ! X0,Y0 - lower left corner of plot in GDU's
6210 ! X1,Y1 - upper right corner of plot in GDU's
6220 ! Xmin,Xmax,Ymin,Ymax - desired ranges of plot in UDU's
6230 ! P$ - name of plotter
6240 ! Xa,Ya - location of x-axis and yaxis in UDU's
6250 ! X$,Y$ - titles for X,Y axes
6260 ! Xti,Yti - x,y tick interval in UDU's
6270 ! Xni,Yni - X,Y number interval
6280 ! Grid - grid code (0= no grid, 1=grid)
6290 ! Print - print code (0=don't print values , 1=print values)
6300 !         prints only in evenly spaced increments
6310 !
6320 OPTION BASE 0
6330 !
6340 COM X0,Y0,X1,Y1,Xmin,Xmax,Ymin,Ymax
6350 COM Paxes,P$,Xa,Ya,X$,Y$,Xti,Yti,Xni,Yni,Grid,Print,Pen,Lt,Pt
6360 !
6370 ON ERROR GOTO Error1
6380 Plotter$="9872A"
6390 GRAPHICS
6400 SETGU                           ! GDU's
6410 LOCATE X0,X1,Y0,Y1
6420 P6:Xrg=X1-X0        !  X range in GDU's
6430 Yrg=Y1-Y0           !  Y
6440 Xru=Xmax-Xmin
6450 Yru=Ymax-Ymin
6460 Xl=Xmin
6470 Xh=Xmax
6480 Yl=Ymin
6490 Yh=Ymax
6500 SCALE Xmin,Xmax,Ymin,Ymax   ! UDU's
6510 Mtcx=Xni/Xti
6520 Mtcy=Yni/Yti
6530 AXES Xti,Yti,Ya,Xa,Mtcx,Mtcy
6540 IF Grid=1 THEN GRID Xni,Yni,Ya,Xa,Mtcx,Mtcy
```

```
6550 GOTO Printit
6560 !
6570 Error1: PRINT ERRM$
6580 BEEP
6590 PAUSE
6600 !
6610 ! Prints axes and axes labels
6620 !
6630 Printit: CSIZE 2.5    ! FOR GRAPH PAPER
6640           LINE TYPE 1
6650 !***********AXES NUMBERS********************
6660    IF P$<>Plotter$ THEN Nopen
6670    PEN Pen
6680 Nopen:    LINE TYPE 1
6690           SETUU                  ! UDU's
6700 Labelx:LORG 6
6710         DEG
6720         LDIR 0
6730    FOR Xax=Xmin TO Xmax STEP Xni
6740    H=Xax
6750    V=Xa-2*Yru/Yrg              ! 2 GDU's BELOW BOTTOM
6760      MOVE H,V
6770      IF Xax<>0 THEN Lx1
6780      LABEL USING "K";Xax  ! SHORTENED ZERO
6790      GOTO Nextx
6800 Lx1:IF (ABS(Xax)>=1E5) OR (ABS(Xax)<=1E-5) THEN Toobigx
6810 LABEL USING "K";Xax    ! SHORT VERSION
6820 GOTO Nextx
6830 Toobigx: LABEL USING "MZ.DDE";Xax
6840 Nextx: NEXT Xax
6850 ! -----------------Y------------------
6860 Labely: LORG 8
6870     LDIR 0
6880   FOR Yax=Yl TO Yh STEP Yni  ! NUMBERS
6890     MOVE Ya-1*Xru/Xrg,Yax      !2 GDU's LEFT
6900     IF Yax<>0 THEN Ly1
6910     LABEL USING "K";Yax
6920     GOTO Nexty
6930 Ly1:   IF (ABS(Yax)>=1E5) OR (ABS(Yax)<=1E-5) THEN Toobigy
6940    LABEL USING "K";Yax
6950    GOTO Nexty
6960 Toobigy:   LABEL USING "MZ.DDE";Yax
6970 Nexty: NEXT Yax
6980 IF Print=1 THEN Xtitle
6990 GOTO End
7000 ! ----------------------------------------------------------
7010 Xtitle:Csize=2
7020       LINE TYPE 1
7030       Tsize=2.5
7040         SETGU
7050         LORG 6
7060         LDIR 0
7070         Xx=X0+Xrg/2
7080         Yx=Y0-5           ! X LABEL POSITION
7090         MOVE Xx,Yx
7100         SETUU
7110         CSIZE Tsize
7120         LABEL USING "K";X$
7130 Ytitle:LORG 4
7140        LDIR 90
7150        SETGU
7160        Xy=X0-9            ! Y LABEL POSITION
7170        Yy=Y0+Yrg/2
7180        MOVE Xy,Yy
7190        SETUU
7200        LABEL USING "K";Y$
7210        PENUP
7220 End:  EXIT GRAPHICS
7230        SUBEXIT
7240         SUBEND
```

```
7250 SUB Vec(A(*),N,A$)
7260 !    "Vec"   INPUTS A VECTOR IN OPTION BASE 0
7270 FIXED 0
7280 FOR I=0 TO N
7290 DISP A$;"(";I;")=";
7300 INPUT A(I)
7310 NEXT I
7320 FOR I=0 TO N
7330 PRINT A$;"(";I;")=";
7340 FLOAT 4
7350 PRINT A(I)
7360 FIXED 0
7370 NEXT I
7380 PRINT LIN(1)
7390 INPUT "Any Changes(Y/N)",X$
7400 IF (X$="Y") OR (X$="y") THEN 7430
7410 IF (X$="N") OR (X$="n") THEN 7510
7420 GOTO 7390
7430 INPUT "Element Number to be Changed ",In
7440 FIXED 0
7450 DISP A$;"(";In;")=";
7460 INPUT A(In)
7470 PRINT A$;"(";In;")=";
7480 FLOAT 4
7490 PRINT A(In)
7500 GOTO 7390
7510 SUBEND
```

9.6 DESIGN VIA ROOT LOCUS

This section is concerned with the design of a linear *SISO* system with the aid of the root locus plot (Section 7.6) and time-domain performance specifications discussed in Section 9.5.1. In sequel, a lead compensator is used to stabilize an oscillatory system with the aid of the root locus.

1. Example. Consider a unity-feedback linear *SISO* system with an open-loop transfer function,

$$G(s) = K/s^2, \quad K=2 \tag{1}$$

It is desired to compensate this system such that it has the following time-domain specifications:

$$\text{Percent overshout}, M_o\% \leqslant 10\% \quad \text{Settling Time}, T_s = 5 \text{ seconds}. \tag{2}$$

The characteristic equation of the uncompensated system is

$$1+G(s) = 1+K/s^2 = 0 \tag{3}$$

or $s^2+K=0$. The root locus of this system is the $j\omega$−axis and all its closed-loop poles are pure imaginary. In order to compensate this system, it would be desirable to shift the branches of the root locus to the left-half plane. This, as evident from Equation (9.4.1) and Fig. 9.4.c, can be achieved by a lead compensator which has a zero closer to the $j\omega$−axis. Now that is is decided to use a lead compensator, one needs to determine the desirable location of the compensated system's closed-loop poles. Following the development of Example 9.5.2, the desired region can be determined by noting that a $M_o \leqslant 10\%$ requires a damping ratio (see Fig. 9.5.4) to be $0.60 \leqslant \xi \leqslant 1$. Moreover, for a settling time of ±2% tolerance, by virtue of (9.5.4), $4/\xi\omega_n = 4/\sigma = 5$ or $\sigma = \xi\omega_n = 0.8$. The desired region would be similar to that of Fig. 9.5.5 except for the fact that the real part of the dominant poles must be at least $\sigma = 4/5 = 0.8$. Now choosing an arbitrary damping ration of $\xi = 0.65$ which predicts $M_o = 8\%$, a desired pairs of closed-loop dominant poles can be arbitrarily chosen at $-0.8 \pm j\omega_n\sqrt{1-\xi} = -0.8 \pm j(1.23)\sqrt{1-(0.65)^2} = -0.8 \pm j0.935$. In order to find the values $s_{1,2} = -\sigma \pm j\omega_d =$ of T and α of the lead compensator, let the zero be located arbitrarily at $z_c = -1/T = -0.8$ which corresponds to $T = 1.25$ seconds. Using the poles-zero location plot of Fig. 1, we have

$$2\theta_1 + \theta_3 - \theta_2 = 2(130.7) + \theta_3 - 90° = 180° \tag{4}$$

or $\theta_3 = 8.6°$ or $0.935/(1/\alpha T - 0.8) = \tan \theta_3 = 0.151$ or $1/\alpha T = 6.86$. Thus, one possible lead compensator is

$$G_c(s) = K_c(s+0.8)/(s+6.86). \tag{5}$$

In order to find K_c, note that the gain at $s_1 = -0.8 + j0.935$ is given by

$$KK_c = |s^2| \cdot |s+6.96| / |s+0.8| = 10.14 \tag{6}$$

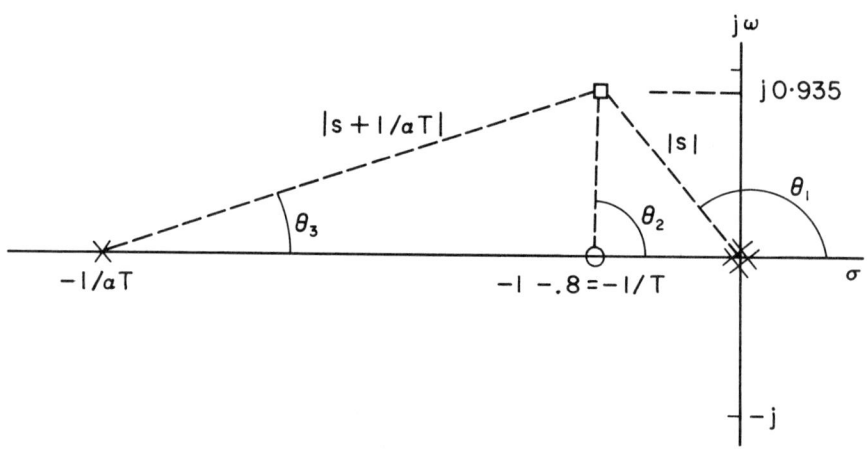

Fig. 9.6.1. Poles-zero locations of system (9.6.1) with a lead compensator

The value of α for this design is approximately 0.715. In an alternative fashion, one can use arbitrary values for $\alpha < 1$ and find new locations for p_c. For example for values of $\alpha = 0.21$ and $\alpha = 0.1$, one obtains $p_c = 3.8$ and $p_c = 8.0$, respectively. The total gains for these two choices are easily seen to be $KK_c = 5.084$ and $KK_c = 11.75$, respectively. The transfer functions of these two lead compensators are thus

$$G_c(s) = 2.542(s+0.8)/(s+3.8) \tag{7a}$$

$$G_c(s) = 5.875(s+0.8)/(s+8) \tag{7b}$$

The root loci of the uncompensated and compensated systems G in (1) and GG_c of (7a), using Program "ROOTLC" (see Section 7.6), are shown in Fig. 2. The time responses of the uncompensated and the three lead compensators suggested by (5) and (7) are shown in Fig. 3. As seen, the system has been stabilized and the original goals of Equation (2) are met. The settling time, T_s turns out to be approximately 5 seconds in all cases. The percent overshoot turn out to be 10.8%, 6.5% and 6.5% respectively for compensators (7a), (5) and (7b). The next example illustrates the use of a lag compensator to design a linear *SISO* system.

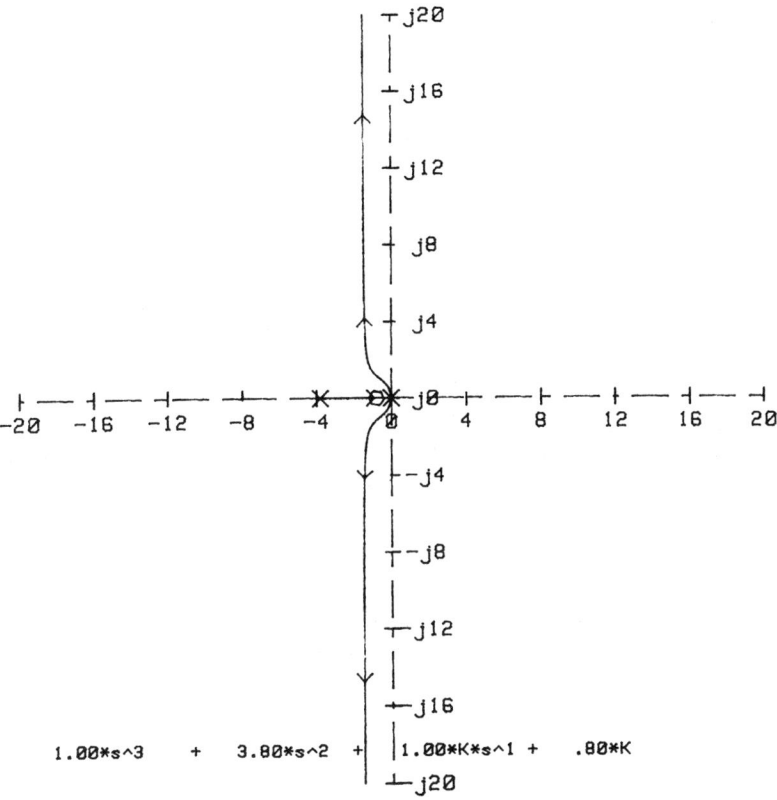

Fig. 9.6.2. Root locus of the uncompensated system of Example 9.6.1.

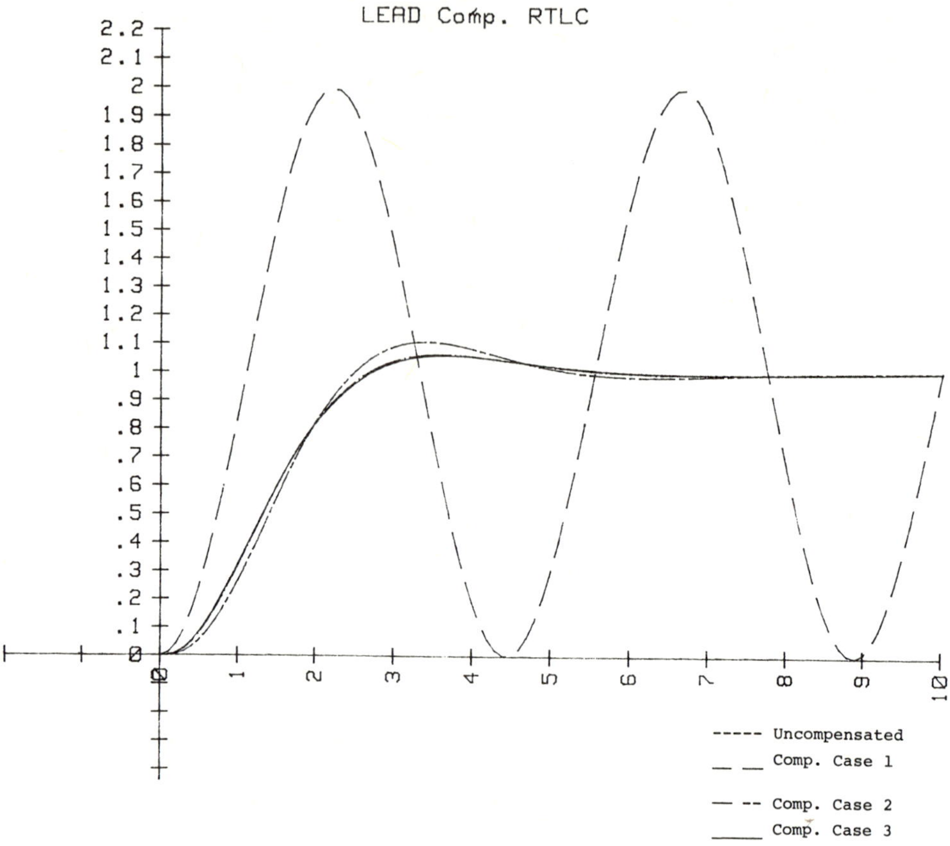

454   LINEAR CONTROL SYSTEMS

Fig. 9.6.3. Time responses of uncompensated and three-lead compensated systems of Example 9.6.1.

2. Example. Consider a unity-feedback control system shown in Fig. 4. The closed-loop transfer function for the system is

$$C(s)/R(s) = 2.05/(s^3+6s^2+5s+2.05) \tag{8}$$

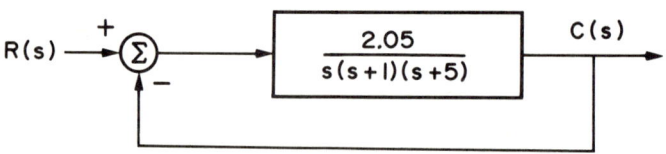

Fig. 9.6.4. A unity feedback control system

which indicates that the closed-loop poles are located at $s_{1,2} = -0.451 \pm j0.446$ and $s_3 = -5.1$. These roots are shown in the root locus of the uncompensated system shown in Fig. 5.

DESIGN VIA ROOT LOCUS

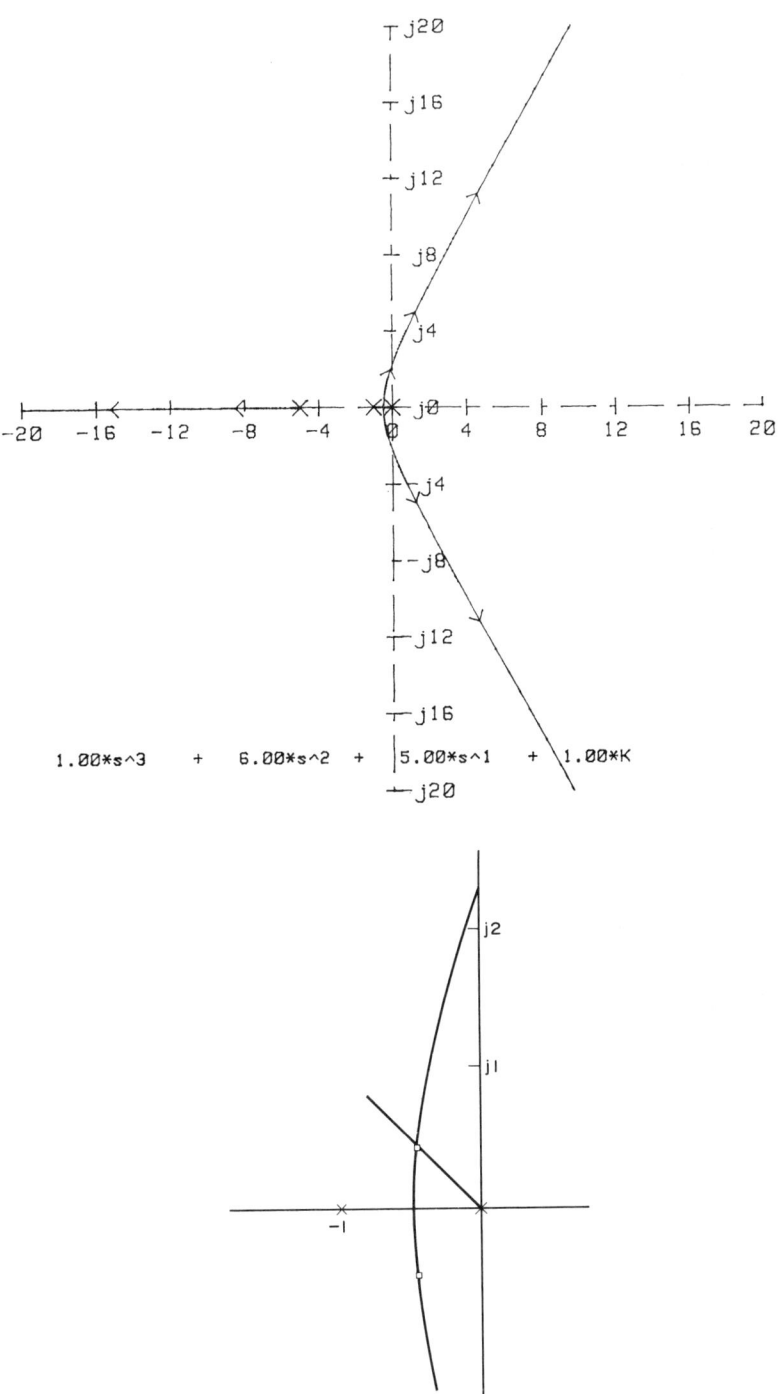

Fig. 9.6.5. Root locus of the uncompensated system of Example 9.6.2

456   LINEAR CONTROL SYSTEMS

The damping ratio corresponding to the dominant poles of this system is $\xi = 0.71$. The undamped natural frequency is $\omega_n = 0.634$. The static velocity, using Equation (9.3.11), is given by $K_v = \lim_{s \to 0} sG(s) = K/5 = 2.05/5 = 0.41$. The remaining important time-domain specifications are $M_o = 4.2\%$, $T_s = 4/\xi\omega_n = 8.87$ and $T_p = \pi/\omega_n\sqrt{1-\xi^2} = 7.03$ seconds. If the system is simulated on the computer, its responses to step and ramp inputs for three different values of $K_v = 0.41, 0.82, 2.0$ can be seen in Fig. 6. Since this system is Type 1, hence its steady-state error to unit step input is zero, while for unit ramp input is $1/K_v$ (see Table 9.3.1). Thus, the steady-state error is $e_{ss} = 1/K_v = 1/0.41 = 2.44$.

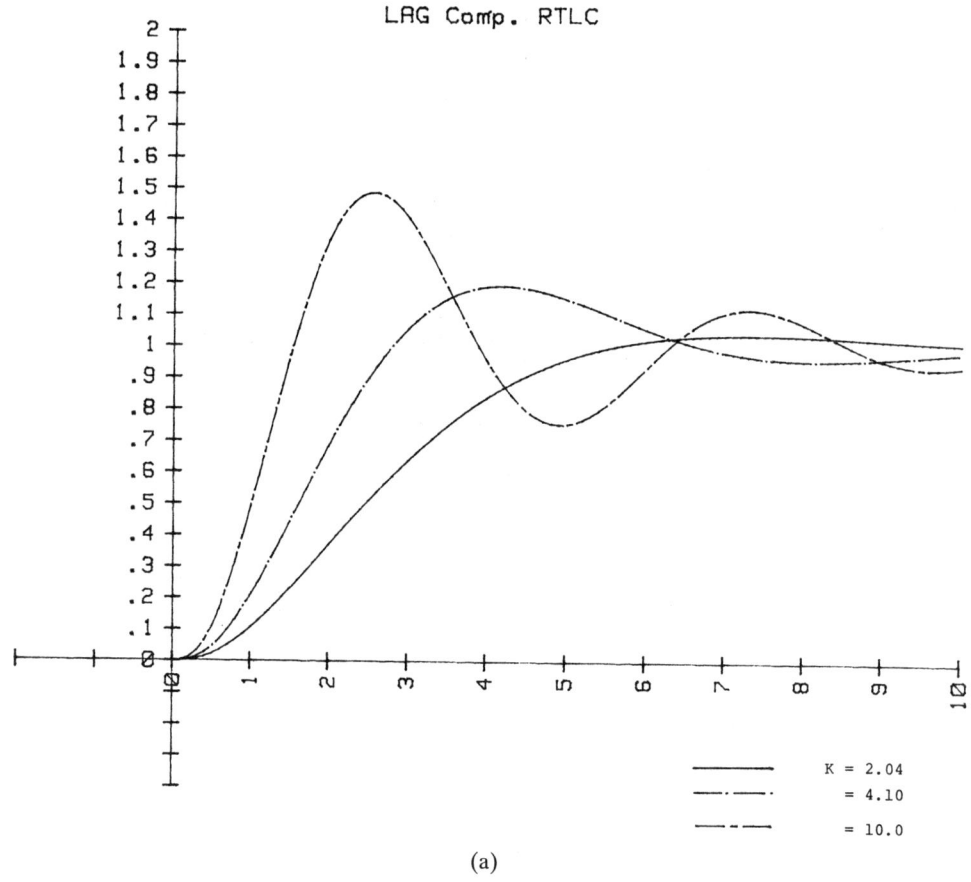

(a)

Fig. 9.6.6. Time responses of uncompensated system of Example 9.6.2 to (a) unit step

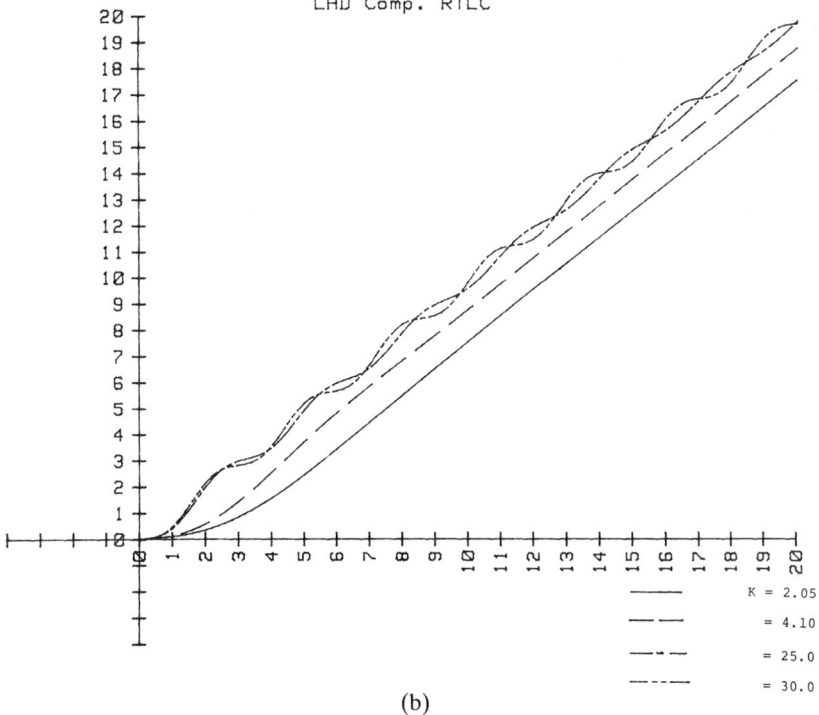

(b)

Fig. 9.6.6. Time responses of uncompensated system of Example 9.6.2 to (b) unit ramp for 3 values of K_v

In an attempt to increase K_v so that e_{ss} would be decreased, it would cause the system's precent overshoot to be increased undesirably. In order to keep the system's transients reasonable, i.e. not changing the location of the dominant poles appreciably, a lag compensator can be employed. Let the desired K_v for the compensated system be $K_v = 4.5$. A lag compensator's transfer function in (9.4.4) can be utilized for an $\alpha = 10$ and a $T = 10$, i.e.

$$G_c(s) = (K_c/10)(s+0.1)/(s+0.01) \tag{9}$$

or a new pole and a new zero are located at -0.01 and -0.1, respectively. The compensated system's open-loop transfer function is thus given by,

$$GG_c = (KK_c/10)(s+0.10)/(s(s+0.01)(s+1)(s+5)). \tag{10}$$

whose root locus is given in Fig. 7. The new locations of the closed-loop poles for $\xi = 0.71$ are at $s_{1,2} = -0.415 \pm j0.4$. The gain of the system is given by,

$$\hat{K} = (|s|.|s+0.01|.|s+1|.|s+5|)/(|s+0.10|)|s_1 = 2.102 \tag{11}$$

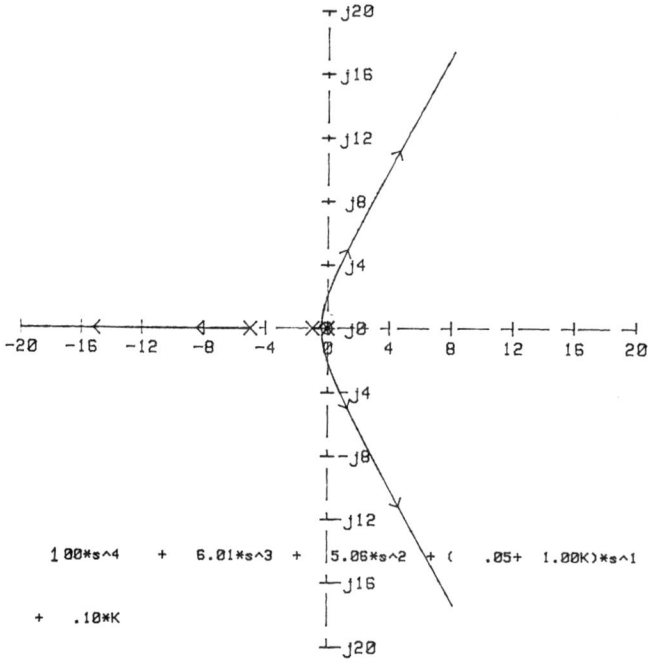
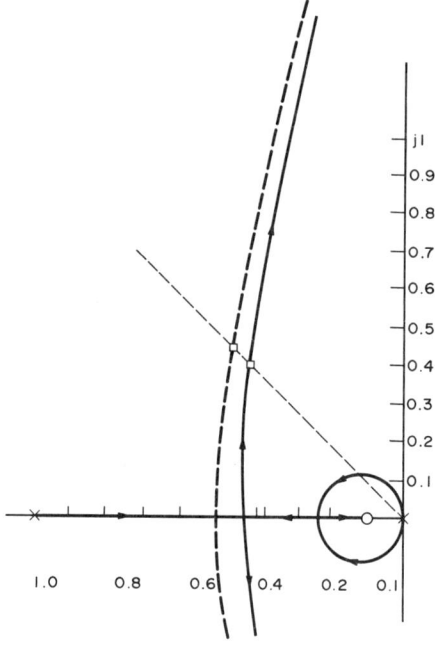

Fig. 9.6.7. Root loci of uncompensated and compensated system of Example 9.6.2

where $\hat{K} \triangleq KK_c/10$. The new value of the static error coefficient is

$$\hat{K}_v = \lim_{s \to 0} sG(s)G_c(s) = 2\hat{K} = 4.204 \tag{12}$$

The change in gain, i.e. isolating amplifier's gain is given by

$$K_c = \hat{K}_v/K_v = 4.204/0.41 = 10.25 \tag{13}$$

The third and fourth poles are at $s_3 \cong -0.126$ and $s_4 = -5.09$ with $\hat{T}_s = 4/\xi\hat{\omega}_n = 9.64 > T_s$. However, the steady-state error has decreased from $e_{ss} = 1/0.41 = 2.44$ to $\hat{e}_{ss} = 1/\hat{K}_v = 1/4.204 = 0.238$ or by a factor of about 10.25 (value of K_c) while sacrificing on the transients a little. The transient responses of the uncompensated and compensated systems to a unit ramp input are shown in Fig. 8 which verifies the above behavior. The effect of the lag compensator is that the net change in the phase of the compensated system is about $5°$ which indicates that the root locus (hence the transients) do not appreciably change.

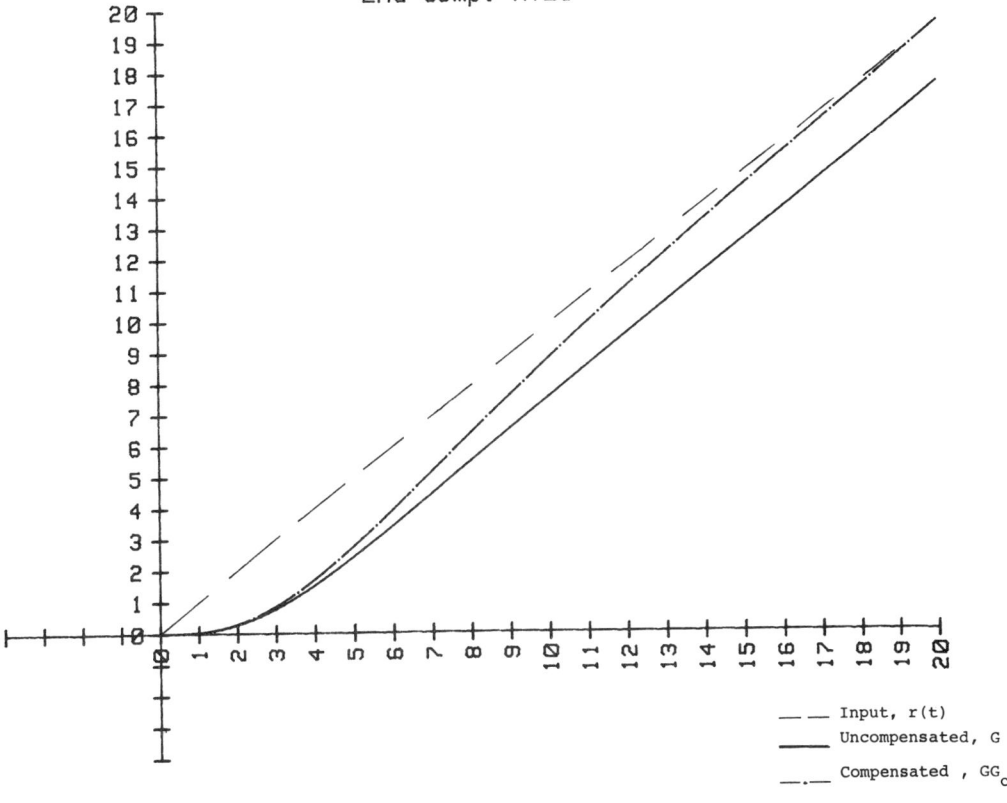

Fig. 9.6.8. Transient responses of uncompensated and compensated system of Example 9.6.2

The next section deals with the use of Bode plots (Section 7.7.1) to design a *SISO* system via cascade compensation.

9.7 DESIGN VIA BODE AND NYQUIST DIAGRAMS

In this section, the log-magnitude and phase angle plots (Bode diagram) and polar plots (Nyquist diagram) are used as a graphical tool to design linear *SISO* systems using compensating networks. Since Bode diagram represents a frequency response for the system, one can make the system's design based on frequency-domain performance specification discussed in Section 9.5.3. It is not always necessary to associate frequency-domain (or time-domain) specifications to design a system with frequency responses (root locus) as a graphical tool. In fact, one can use any one or both of sets of design specifications with any graphical design tools such as root locus, Bode diagram, Nyquist diagram or Nicholes chart (see Problems 9.14 through 9.21). In sequel, a typical example involving Bode diagram, time- and frequency-domain specifications is presented.

1. Example. Consider a unity-feedback, type-1 system whose open-loop transfer function is given by,

$$G(s) = K/(s(s+1)(0.5s+1)). \tag{1}$$

It is desired to design a compensator so that the static velocity error coefficient K_v is $8\ sec^{-1}$, the phase margin is at least $30°$, and the gain margin is at least 10 dbs.

The value of the velocity error coefficient is obtained from (9.3.11),

$$K_v = \lim_{s \to 0} sG(s) = \lim_{s \to 0}(sK/(s(s+1)(0.5s+1))) = K = 8 \tag{2}$$

Thus, if one chooses a value of $K = 8$, the first requirement is satisfied. The range of values of K for stability by, for example, Routh-Hurwitz criterion, can be easily seen to be $0 < K < 3$. This means that, the system is unstable with a pair of its dominant complex-conjugate poles in the right-half plane. Clearly, with a $K_v = 8$, the steady-state performance requirement is of significance and the transient response characteristics, as in Example 9.6.2, is of less concern. In order to secure a reasonable margin for stability (phase and gain margins), one can plot the Bode diagram for the system as shown in Fig. 1. As seen, the phase margin is $-21°$ which reaffirms that the system is unstable. ∆

One may want to reduce the system's dc gain to increase the margins of stability. However, this would increase the steady-state error (recall that $e_{ss} = 1/K_v$) and would violate the first specification. A lead compensator would be inappropriate, since such a network is essentially a high-pass filter (see Fig. 9.4.2a) and can not attenuate the systems' Bode diagram for high frequencies with $\omega_\phi = 2.2$ and $\omega_g = 1.6$ radians per second (see Fig. 1). A lag compensator would be more appropriate to achieve the design's specifications on gain and phase margins. The addition of a lag compensator would modify the phase plot of the Bode diagram, thus one can add on additional $5°$ to $15°$ to the specified gain margin to offset the modification of phase margin. The frequency corresponding to $30°$ phase margin is about 0.85 rad/sec, which indicates that the new gain cross-over frequency of the compensated system should be chosen around this value. A corner frequency $\omega = 1/T$ (zero of the lag compensator) is chosen as $0.1 = 1/10$ radian/sec in order to avoid large time constants for the lag network. Since this corner frequency is not too far below the new gain cross-over frequency, the modification of phase angle due to the compensator may not be insignificant. Therefore one can add $8°$ to the phase margin to make it now $38°$. The phase angle

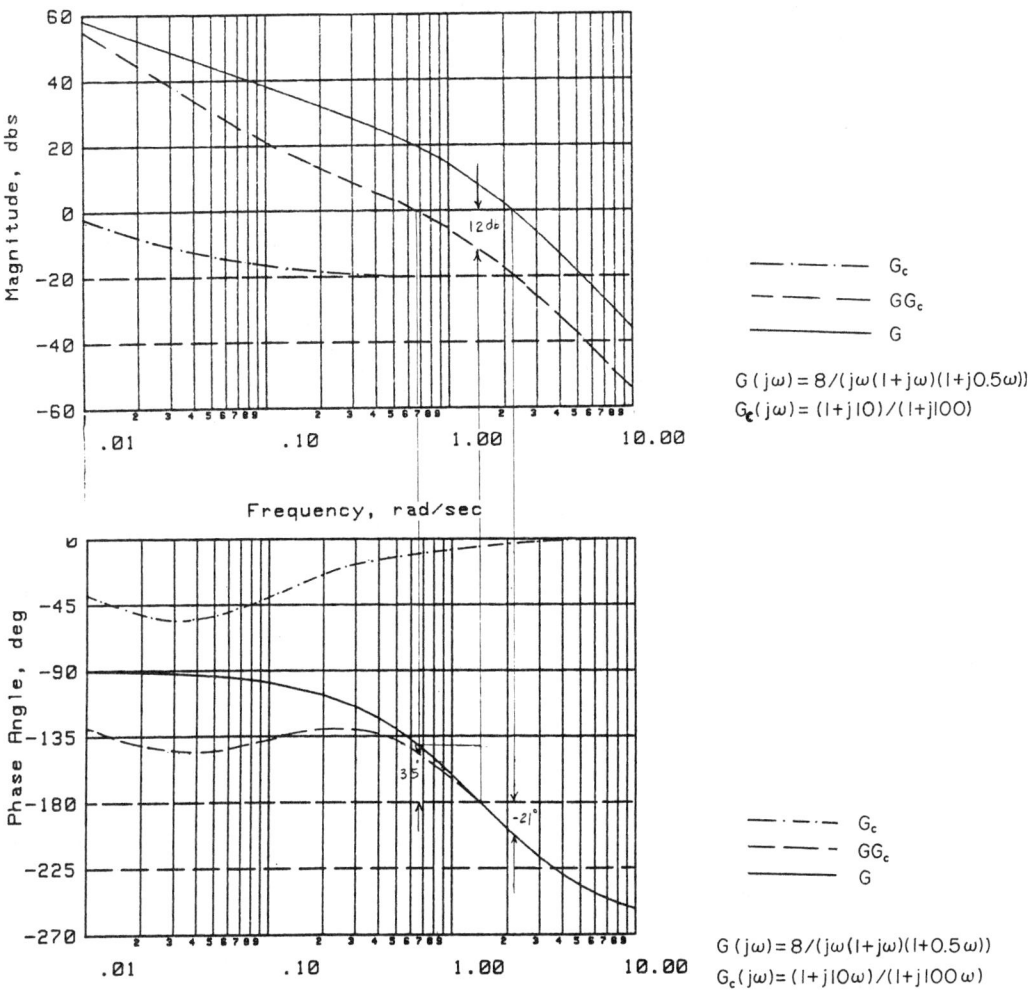

Fig. 9.7.1. Bode diagrams for the uncompensated (*G*), the compensated (*GG_c*) and the compensator (*G_c*) for Example 9.7.1

of the uncompensated system at $\omega=0.57$ rad/sec is $-142°$. Thus, one can choose the new gain cross-over frequency to be 0.57 rad/sec. The gain of the uncompensated system at this frequency is 20 dbs (see Fig. 1) and in order to attenuate this gain to zero db, compensator's gain must be chosen for this purpose. Thus,

$$20 \log (1/\alpha) = -20 \qquad (3)$$

or $\alpha = 10$. The other corner frequency (see Fig. 9.4.4c) is $1/\alpha T = 1/100 = 0.01$ rad/sec. Therefore, the compensator's transfer function is

$$G_c(s) = (1/10)(s+0.1)/(s+0.01). \tag{4a}$$

The open-loop transfer function of the compensated system is thus,

$$GG_c(s) = 8(1+10s)/(s(1+s)(1+0.5s)(1+100s)). \tag{4b}$$

The log-magnitude and phase angle plots of the compensated system and the lag compensator are also shown in Fig. 1. The phase margin of the compensated system is about $35°$, which is $5°$ higher than the required value. The gain margin is about 12 dbs which is also 2 dbs higher than the required 10 dbs value. The static velocity error is 8 sec^{-1} which is exactly the desired value. The new gain crossover frequency has decreased from 2.2 to 0.67 radians/sec. This would indicate that the system's bandwidth (see Fig. 9.5.7) has been decreased. Thus, the compensated system's transient response speed is expected to be slower, as seen in Fig. 2. $\triangle$

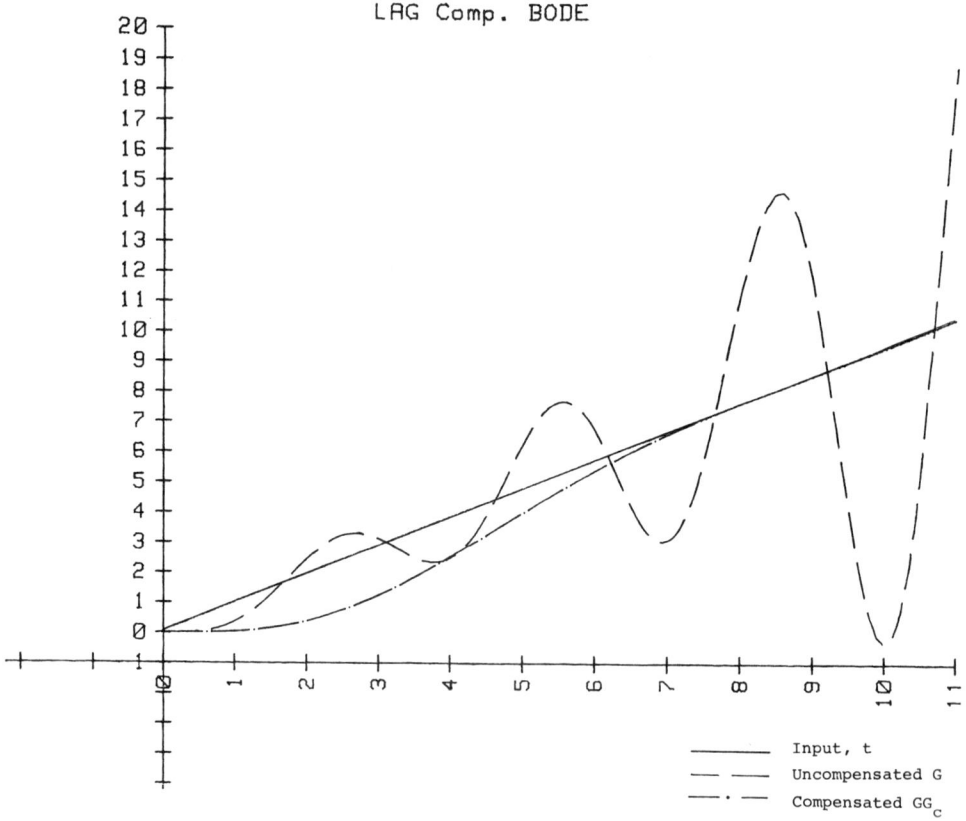

Fig. 9.7.2. Transient responses for Example 9.7.1

The next example illustrates the use of a lead-lag network to compensated a linear *SISO* system.

2. Example. Consider a unity feedback system describe by

$$G(s) = K/(s(1+0.2s)(1+0.1s)). \qquad (5)$$

It is desired to compensate this system to improve the steady-state error for an $M_m = 2.0$ and a phase margin, $\gamma \geqslant 90°$. In order to initiate the design procedure, we need to find a value of K such that $M_m = 2$. This can be done by the procedure discussed in Section 9.5.3 (see Fig. 9.5.9). To repeat the same procedure, let us draw the frequency-sensitive portion of the Nyquist diagram of

$$G_g(j\omega) = 1/(j\omega(1+j0.2\omega)(1+j0.1\omega)) \qquad (6)$$

which is shown in Fig. 3. For an $M_m = 2$ an angle $\psi = \sin^{-1}(1/M_m) = \sin^{-1}(0.5) = 30°$ describes the desired line of tangent to the $M_m = 2$ circle. Following the procedure of Section 9.5.3, by trial and error, the circle tangent to line ob and polar plot of $G_g(j\omega)$ is drawn. The line segment $oc = 0.3$ must be $1/K$ for this circle to be $M_m = 2$. Therefore, the desired gain $K = 3.34$. The value of ω_m can be read off to be $\omega_m = 5.34 \ rad/sec$. Here, we need to choose a compensator which would improve both the transient as well as the steady-state error. Through the previous examples, it was

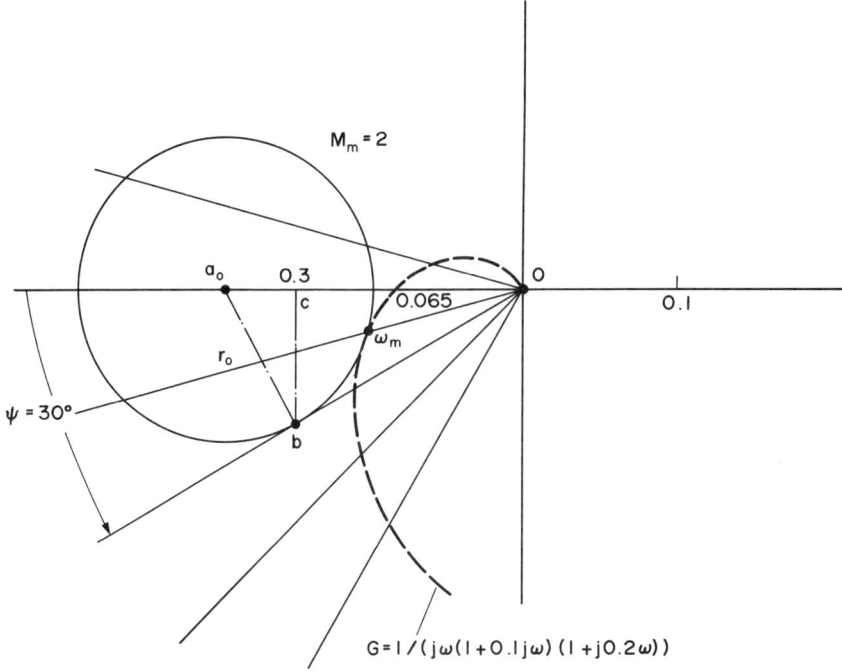

Fig. 9.7.3. Gain adjustment for system of Example 9.7.2

deduced that a lead network would be effective for transients, while the steady-state error would be affected by a lag network. Thus one can use a lead-lag network:

$$G_c(j\omega) = K_c(1+j\omega T_1)(1+j\omega T_2)/((1+j\omega\alpha T_1)(1+j\omega T_2/\alpha)) \qquad (7)$$

where $\alpha > 1$ and $T_1 > T_2$. A good choice for $\alpha = 10$. In order to determine the other two parameters, one can set $T_2 = 1$ and experiment with T_1 by trial and error until satisfactory results can be obtained. For this example a value of $T_1 = 15$ was chosen, leading to

$$G_c(j\omega) = K_c(1+j15\omega)(1+j\omega)/((1+j150\omega)(1+j0.1\omega)). \qquad (8)$$

Now, the compensated system's open-loop transfer function becomes,

$$G(j\omega)G_c(j\omega) = KK_c(1+j\omega)(1+j15\omega)/(j\omega(1+j0.1\omega)^2(1+j0.2\omega)(1+j150\omega)) \qquad (9)$$

The Bode diagrams of the uncompensated, compensated systems along with the compensator are shown in Fig. 4, for $K_c = 2$.

At this point by adjusting the value of compensator's isolating amplifier gain K_c the desired phase margin can be obtained. The final design's parameters are then $K_c = 2$, $\alpha = 10$, $T_1 = 15$ and $T_2 = 1$ which have resulted in an $M_m = 2$ and a phase margin $\gamma = 106°$. These parameters would lead to the following resistors and capacitors (see Section 9.4.3). $R_1 = 10M\Omega$, $R_2 = 1M\Omega$, $C_1 = 0.1\mu f$ and $C_2 = 15\mu f$. Next another example is considered in which Nyquist diagram is used.

3. Example. Consider a system shown in Fig. 5. The open-loop transfer function for this system is given by

$$G(s) = K/(s(1+s)) \qquad (10)$$

It is desired to compensate this system such that its steady-state error for a unity ramp input $r(t) = t$ is 0.05 and its phase margin be at least $50°$.

Since gain and phase margins are specified, one can normally use Bode diagram as a graphical design tool. However, in this example, we demonstrate that Nyquist diagram can be similarly used with ease. To initiate the design process, let us first adjust the value of K such that $e_{ss}(t) = 0.05$. Since the system is type 1 (see Table 9.3.1), $e_{ss}(t) = 1/K_v$, where K_v is the static velocity error coefficient desired to be $1/0.05 = 20\ sec^{-1}$. K_v is given by

$$K_v = \lim_{s \to 0} sG(s) = \lim_{s \to 0}(Ks/(s(s+1))) = K = 20 \qquad (11)$$

or $K = 20$. With this value of K, the steady-state error condition is satisfied. Next consider the polar plot (frequency-sensitivie part of the Nyquist diagram) of the system

$$G(j\omega) = 20/(j\omega(1+j\omega)) \qquad (12)$$

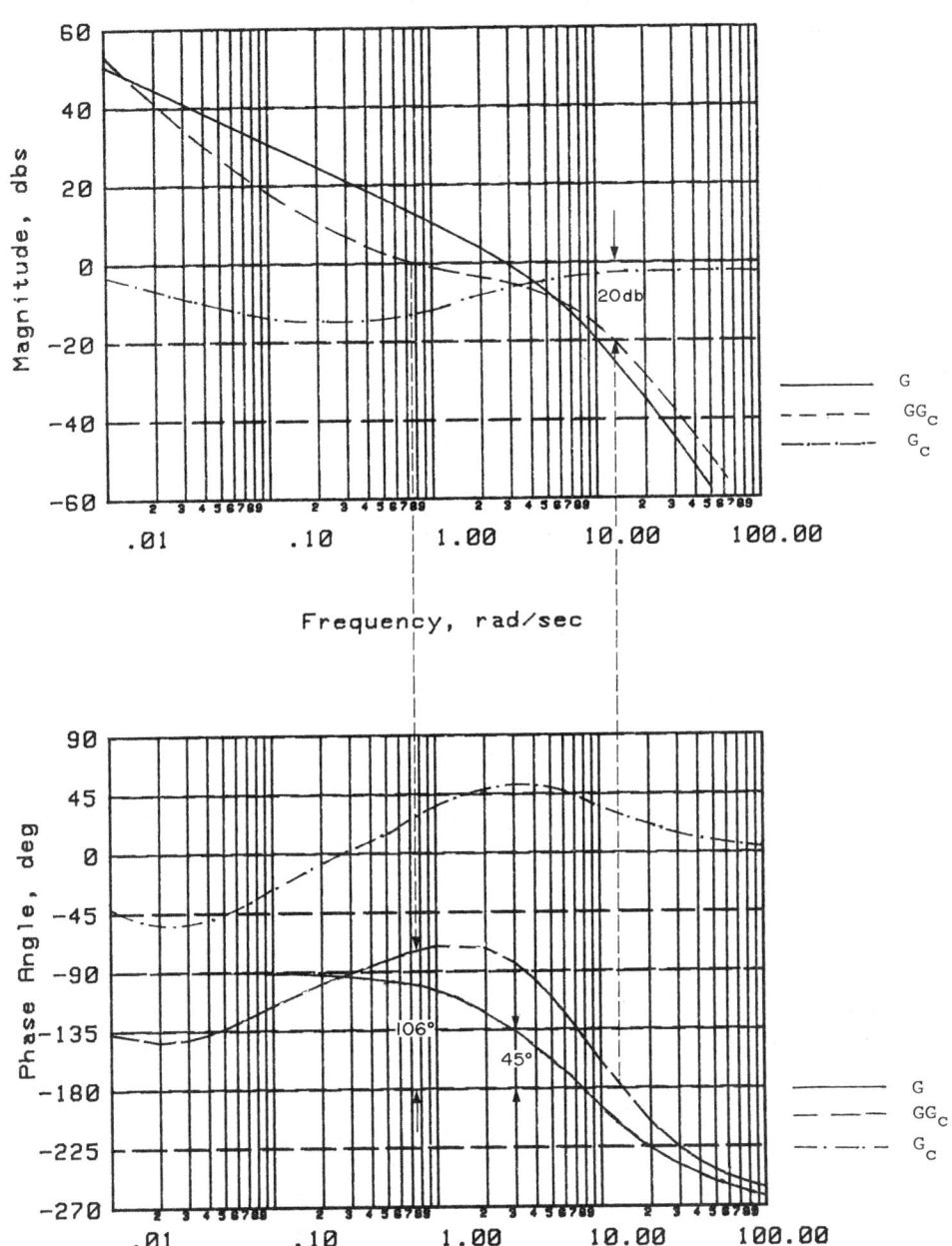

Fig. 9.7.4. Bode diagrams for uncompensated (*G*) compensated (*GG_c*) and compensator (*G_c*) for Example 9.7.2

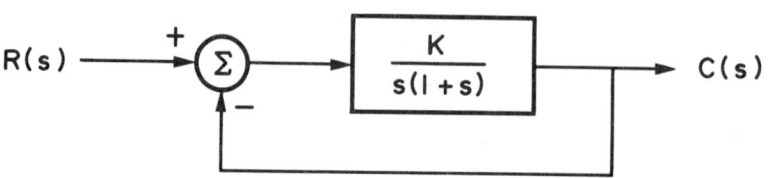

Fig. 9.7.5. A unity feedback control system

shown in Fig. 6. It is clear that, the phase margin of the system is about 12.5°, while the gain margin is ∞. A phase margin of 12.5° indicates that the system is oscillatory. This undesirable transient characteristic is due to the large value of gain K. Since a phase margin of 50° is desired, an additional 37.5° is required to insure the stability margin. In order to add this extra phase without decreasing the value of K, a lead compensator is appropriate (see Fig. 9.4.2).

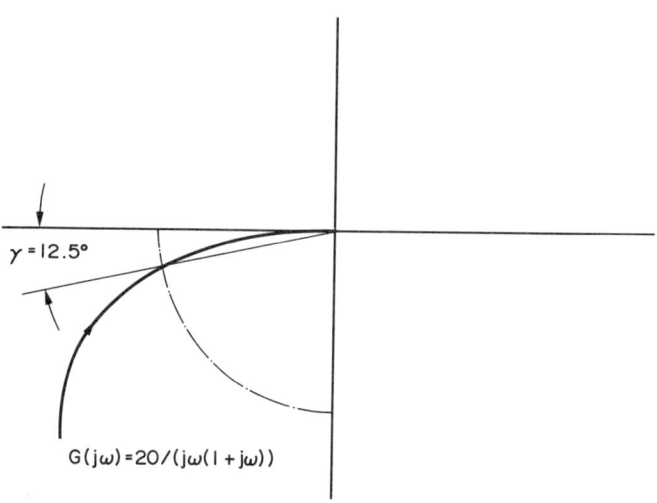

Fig. 9.7.6. Polar plot of transfer function of Example 9.7.3

It is noted that by adding a lead compensator the frequency-sensitive part of the Nyquist diagram would shift closer to the negative imaginary axis (see Fig. 6). This would mean that the new phase margin frequency (gain crossover point) would be higher. Moreover, $G(j\omega)$ would have an additional phase lag which must also be offset by the lead compensator's phase increase. Thus, a phase angle of approximately 37.5° + 5° = 42.5° will be sought for the lead network. Therefore, we can choose the lead compensator's attenuation factor α such that it would provide its maximum phase angle ϕ_{max} (see Fig. 9.4.2b). As seen from this figure the angle ϕ_{max} can be obtained from

$$\phi_{max} = \sin^{-1}(AB/OA) = \sin^{-1}((K_c(1-\alpha)/2)/(K_c(1+\alpha)/2)$$

$$= \sin^{-1}((1-\alpha)/(1+\alpha)) \tag{13}$$

which indicates that

$$\sin \phi_{max} = (1-\alpha)/(1+\alpha). \tag{14}$$

If we take $\phi_{max} = 42.5°$, it would correspond to an attenuation factor $\alpha \cong 0.194$. The remaining parameter for the lead network is the time constant T which determines the corner frequencies $\omega_{c1} = 1/T$ and $\omega_{c2} = 1/\alpha T$. Noting the log-magnitude phase-angle plots of a lead compensator as shown in Fig. 9.4.2a, it is clear that the maximum angle's frequency ω_{max} is the geometric mean of the two corner frequencies [9.2], i.e.

$$\log(\omega_{max}T) = (\log(1) + \log(1/\alpha))/2 = \log(1/(1/\alpha))^{1/2} = \log(1/\sqrt{\alpha}) \tag{15}$$

or $\omega_{max} = 1/(T\sqrt{\alpha})$. The magnitude modification due to the compensator is

$$|(1+j\omega T)/(1+j\omega \alpha T)|_{\omega=1/(T\sqrt{\alpha})} = 1/\sqrt{\alpha} \tag{16}$$

which is equal to $1/\sqrt{0.194} = 2.27 = 7.12$ db. From Fig. 6, it is easily verified that $|G(j\omega)| = 2.27$ corresponds to a frequency $\omega = 2.652$ rad/sec. We can choose this value as the new phase margin (gain crossover point) frequency. Thus,

$$\omega_\phi = 1/(T\sqrt{\alpha}) = 2.27/T \tag{17}$$

or

$$1/T = \omega_\phi \sqrt{\alpha} \cong 1.17 \text{ and } 1/\alpha T = \omega_{Phi}/\sqrt{\alpha} \cong 6.0 \tag{18}$$

The transfer function of the lead compensator is thus,

$$G_c(s) = (s+1.17)/(s+6) \tag{19}$$

The open-loop transfer function of the compensated system becomes,

$$G(s)G_c(s) = 20(s+1.17)/(s(s+1)(s+6)) \tag{20}$$

Figure 7 show the polar plots of the uncompensated and compensated systems' open-loop transfer

468   LINEAR CONTROL SYSTEMS

functions. It is noted that the resonant peaks M_m for the two systems are approximately 3.10 and 1.05, respectively. This indicates that the lead compensator has clearly improved stability.

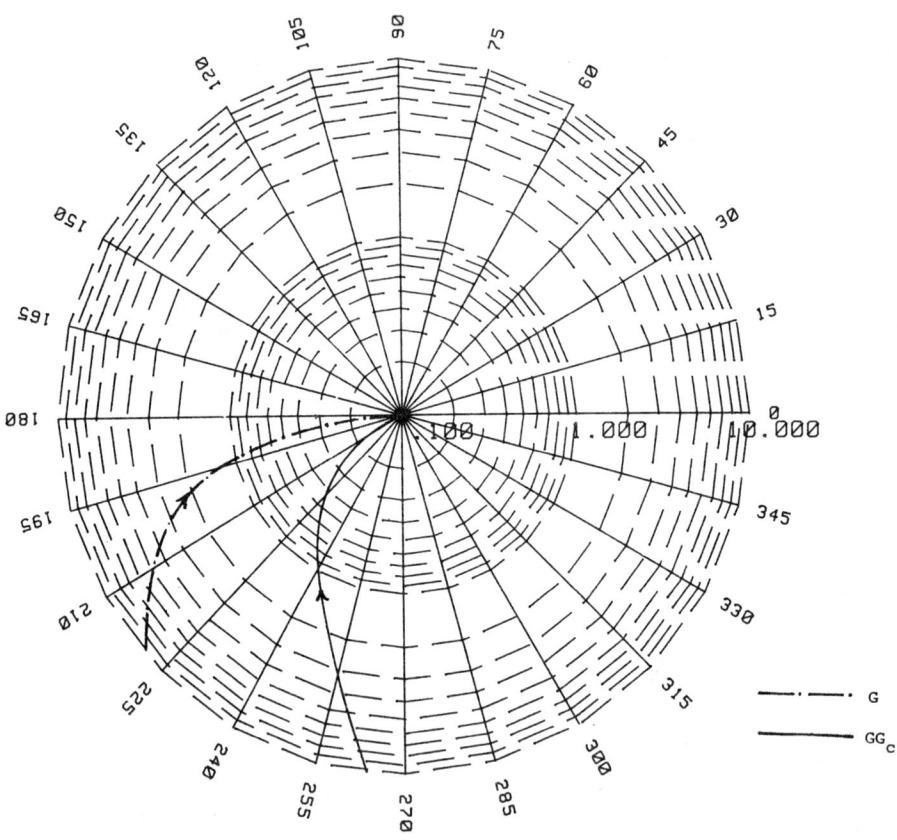

Fig. 9.7.7. Polar plots of uncompensated system (G) and compensated system (GG_c)

9.8 RULES FOR SELECTION OF A COMPENSATOR

Thus far in this chapter system performance specifications, types of compensating networks and graphical schemes such as root locus, Bode and Nyquist diagrams have been presented and a number of design examples have been considered. However, we have not yet presented a set of rules such that parameters of a given type of compensator can be identified. In a general sense a lead network is useful for improving system transients, while a lag network can be used to improve the steady-state response and in particular the steady-state error. A lead-lag compensator can be employed to improve both responses. With three types of compensators (lead, lag and lead-lag), two general graphical tools (root locus and frequency response) and two types of performance specifications (time-domain and frequency-domain) there are numerous combinations of possible

RULES FOR SELECTION OF A COMPENSATOR

design situations. In order to present a reasonable set of rules for designing *SISO* systems by compensation, one can assume that frequency-domain specifications are associated with frequency responses such as Bode or Nyquist diagrams, while time-domain specifications are considered in connection with design using root locus. These assumptions are reasonable when considering the conveniences of each set of specifications for a system with a pair of dominant poles. In sequel, six design situations which are more popular are considered.

9.8.1 Lead Compensation and Root-Locus Scheme

As mentioned earlier, the root locus is a very powerful design tool when used with time-domain specifications such as percent overshoot, settling time and natural frequency. Consider a case when the system is either unstable or its transient response is undesirable, i.e. its dominant closed-loop poles are too close to the $j\omega$-axis. Under such a condition a lead compensator would be very appropriate to be put in cascade. This design procedure will be summarized by the following steps:

(i) Using the time-domain specifications, determine a region in the s-plane for dominant poles locations (see Example 9.4.2).

(ii) Draw the root locus of the uncompensated system and determine whether the desired closed-loop poles can be obtained by gain adjustments alone. If not, determine the angle ϕ for which the uncompensated system has deficiency. This angle can be obtained by checking the root locus' angle condition at a desired dominant pole and subtract $180°$ from it.

(iii) In order to find the value of T, draw a horizontal line through the dominant pole, as shown in Fig. 1, i.e. line SA. Then connect point S to the origin and draw the bisecting line SB of the angle OSA. Then draw two lines SD and SC which make angles $\pm\phi/2$ with line SB. The intersections of lines SC and SD with the negative real axis give the locations of zero and pole respectively. Note that this scheme would provide α and T simultaneously.

(iv) Find the open-loop gain of the compensated system using the root locus' magnitude condition.

When all unknown parameters of the lead compensator are determined, digital computer programs such as "ROOTLC" (Section 7.6), and "COMSCS" (Section 6.6) can be used to check whether the specifications are satisfied. If some are not satisfied a new compensator can be utilized to achieve the desired specifications.

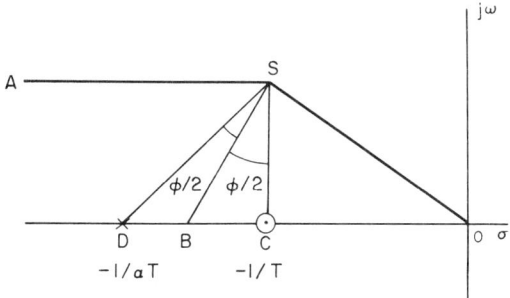

Fig. 9.8.1. Determination of a lead compensator's pole and zero locations

9.8.2 Lag Compensation and Root Locus Scheme

The most suitable situation where a lag compensator can be used with the aid of the root locus is to decrease the steady-state error by increasing static error coefficient (see Table 9.3.1) without changing the transient response appreciably. This can be achieved by locating the zero and pole of the compensator (see Eq. 9.4.4) very close together, i.e. if s_1 is a dominant pole, then $|s_1+1/T|$ and $|s_1+1/\alpha T|$ are almost the same, thus the net angle change $\angle G_c \approx 5°$. Under such conditions, the new dominant poles are rather close to the old ones (see Example 9.6.2 and Fig. 9.6.7). In view of the transfer function of the lag compensator (Eq. 9.4.4) and above discussion on static velocity error coefficient K_v, we can see that the new coefficient is

$$K_v = \lim_{s \to 0} sG(s)G_c(s) = (\lim_{s \to 0} G_c(s))(\lim_{s \to 0} sG(s))$$

$$= \lim_{s \to 0}(K_c(1+Ts)/(1+\alpha Ts))K_v = K_c K_v . \qquad (1)$$

The following steps summarize the lag compensator design using root locus:

(i) Draw the root locus of the uncompensated system and locate the dominant closed-loop poles on the s-plane based on system's transients.

(ii) Obtain the open-loop (dc) gain at dominant poles using magnitude condition

(iii) Evaluate the specified static error coefficient and determine the gain increase necessary to achieve it.

(iv) Find lag compensator's pole and zero locations without changing the dominant poles' locations, while achieving the necessary increase in gain.

(v) Draw the compensated system's root locus and locate the position of the desired closed-loop poles. If the angle contribution of the compensator is a few degrees the new dominant poles are close to the old ones. Otherwise a small discrepancy would occur.

(vi) Adjust the isolating amplifier's gain to obtain the new static error coefficient

(vii) Use program "ROOTLC" and "COMSCS" to check the results.

9.8.3 Lead-Lag Compensation and Root Lucus Scheme

The lead-lag compensator is the most logical type of compensator when improvements on both transient and steady-state responses are desired. Thus, the rules discussed in Sections 9.8.1 and 9.8.2 can be combined for this purpose, leading to the following steps:

(i) Find the location of the new closed-loop poles based on the desired performance specifications.

(ii) Obtain the additional angle ϕ of the lead portion of the compensator to shift the uncompensated system's root locus to go through the desired location (or region) of the new closed-loop poles.

RULES FOR SELECTION OF A COMPENSATOR

(iii) For the lead-lag compensator,

$$G_c(s) = K_c[(s+1/T_1)/(s+\alpha/T_1)][(s+1/T_2)/(s+1/\alpha T_2)] \quad (2)$$

determine the value of K_c needed to obtain new static error coefficient.

(iv) Choose time constant T_2 large enough such that $|(s+1/T_2)/(s+1/\alpha T_2)|$ is approximately unity at dominant poles.

(v) Using the following magnitude - angle conditions,

$$|(s_1+1/T_1)/(s_1+\alpha/T_1)| \, |K_c G(s_1)| = 1 \quad (3)$$

$$< (s+1/T_1)/(s+\alpha/T_1) = 180° \quad (4)$$

find the unknowns α and T_1. The point $s=s_1$ corresponds to the dominant pole coordinate.

(vi) Using the value of α just obtained, determine T_2 such that

$$|(s_1+1/T_2)/(s_1+1/\alpha T_2)| \approx 1 \quad (5)$$

and

$$0< <(s_1+1/T_2)/(s_1+1/\alpha T_2) <3° \quad (6)$$

The above rules can become simpler to apply when programs such as "ROOTLC" (Section 7.6), "ROOTFD" (Appendix 1) and "COMSCS" (Section 6.6) are also utilized. The next three sections provide rules for the selection compensating networks based on frequency-domain specifications.

9.8.4 Lead Compensation and Frequency-Response Schemes

The most important effect of a lead compensator is to reshape the frequency response diagrams (Bode, Nyquist or Nichols) such that sufficient phase lead angle is added to the system to offset its components' phase lags. The following steps summarize this situation (assume that the system is unity feedback):

(i) Find the dc gain K to satisfy the desired static error coefficients.

(ii) Determine the phase margin (see Section 7.7.2) of the system using the value of K thus determined.

(iii) Obtain the necessary phase lead angle ϕ to achieve the desired phase margin.

(iv) Use Equation (9.7.14) to find the attenuation factor α. Obtain the frequency at which $20 \log |G| = -20 \log(1/\sqrt{\alpha})$. Choose this frequency as the new phase margin frequency (gain crossover point). This value of frequency is thus ω_{max} of the maximum phase shift ϕ_{max}.

(v) Using relations like (9.7.15)-(9.7.17), $\omega_1 = 1/T$ and $\omega_2 = 1/\alpha T$ to obtain crossover frequencies.

(vi) Set a gain 1/2 for the isolating amplifier to complete the design

(vii) Use programs "BNPLOT" (Section 7.7), "STATEQ" (Section 4.4) and "COMSCS" (Section 6.6) to check the design.

9.8.5 Lag Compensation and Frequency-Response Schemes

The basic effect of a lag network is to provide sufficient attenuation in the high-frequency range to improve the phase margin of the system. The following steps summarize the design of a lag compensator:

(i) Based on a prespecified value of the static error coefficient, determine the open-loop dc gain.

(ii) Using this value of dc gain, draw the Bode diagram (or polar plot) and determine the phase and gain margins of the uncompensated system.

(iii) If the phase and gain margins are not satisfied, determine the frequency where the phase angle is $-180°$ plus the required phase margin. Add $5°$ to $12°$ to the required phase margin to take into account the compensator's phase angle. Choose this frequency as the new phase-margin (gain crossover) frequency.

(iv) Choose the first corner frequency $\omega = 1/T$ (zero of the compensator) one octave or decade (if the time constant does not become too large) below the new phase-margin frequency.

(v) Corresponding to the necessary attenuation to bring the magnitude curve down to 0db at the new phase-margin frequency. Noting that this attenuation is $-20 \log \alpha$, find the value of α.

It is noted that the above steps can be repeated using a polar plot with gain margin.

9.8.6 Lead-Lag Compensation and Frequency-Response Schemes

This compensator's design procedure is basically a combination of those of lead and lag compensators. The lag portion of the compensator provides attenuation near and above gain-crossover point (phase-margin frequency) hence allowing a low-frequency increase in gain to improve steady-state response. The lead portion of the network increases the phase margin at the gain crossover point.

9.9 FEEDBACK COMPENSATION

Up to now, all the compensator design discussed were concerned with unity feedback systems. This choice was primarily for convenience and posed no serious limitation since one can easily transform a non-unity feedback to a unity feedback system as shown in Fig. 1. In this section effects of a feedback compensation on the system's overall performance is investigated. The general class of *SISO* systems considered for this purpose is shown in Fig. 9.1.1b. The feedback compensator term is assumed to be of rate feedback form, $H(s) = K_t s$, where K_t is the adjustable parameter of the compensator. This term, together with the outer unity feedback loop (see Fig. 9.1b) represent a conventional proportional plus derivative controller, $1+K_t s$. In order to illustrate the use of a feedback compensation the following example is presented.

FEEDBACK COMPENSATION

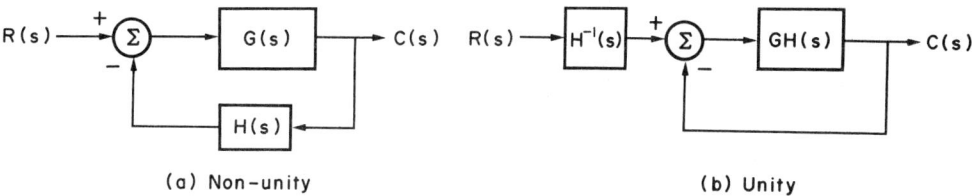

(a) Non-unity   (b) Unity

Fig. 9.9.1. Transformation of a non-unity feedback *SISO* system to a unity feedback system

1. Example. Consider a type 1 third-order system with a transfer function

$$G_p(s) = K/(s(s+0.5)(s+5)). \tag{1}$$

The forward transfer function of the system with the inner loop taken into account becomes

$$G(s) = \frac{K_A G_p(s)}{(1+G_p(s)H(s))} = K_A K/(s(s+0.5)(s+5)+KK_t s). \tag{2}$$

In order to compensate the system the rate feedback constant K_t must be determined. The characteristic equation of the system is given by

$$1+G(s) = 1+K_A K/(s(s+0.5)(s+5)+KK_t s) = 0 \tag{3}$$

which after completing the addition and equating to zero leads to

$$s(s+0.5)(s+5)+KK_t(s+K_A/K_t) = 0. \tag{4}$$

This equation can be partitioned to represent a system with open-loop transfer function suitable for drawing a root locus:

$$KK_t(s+K_A/K_t)/(s(s+0.5)(s+5)) = -1 \tag{5}$$

This form can be used directly to draw the system's root locus. However, before that is done the reader should note that equation (5) *does not* represent the forward transfer function of the system. Rather, it is merely a *reformulation* of the feedback compensation system's characteristic equation whose roots lead to the closed-loop poles of the system. In other words, the effect of the rate feedback compensation is *as though* a simple zero has been added to the uncompensated system's transfer function. This characterizes an effective proportional plus derivative controller put in cascade to the open-loop plant. The root loci of the uncompensated system (1) and the compensated system (5) for three difference values of K_t are shown in Fig. 2.

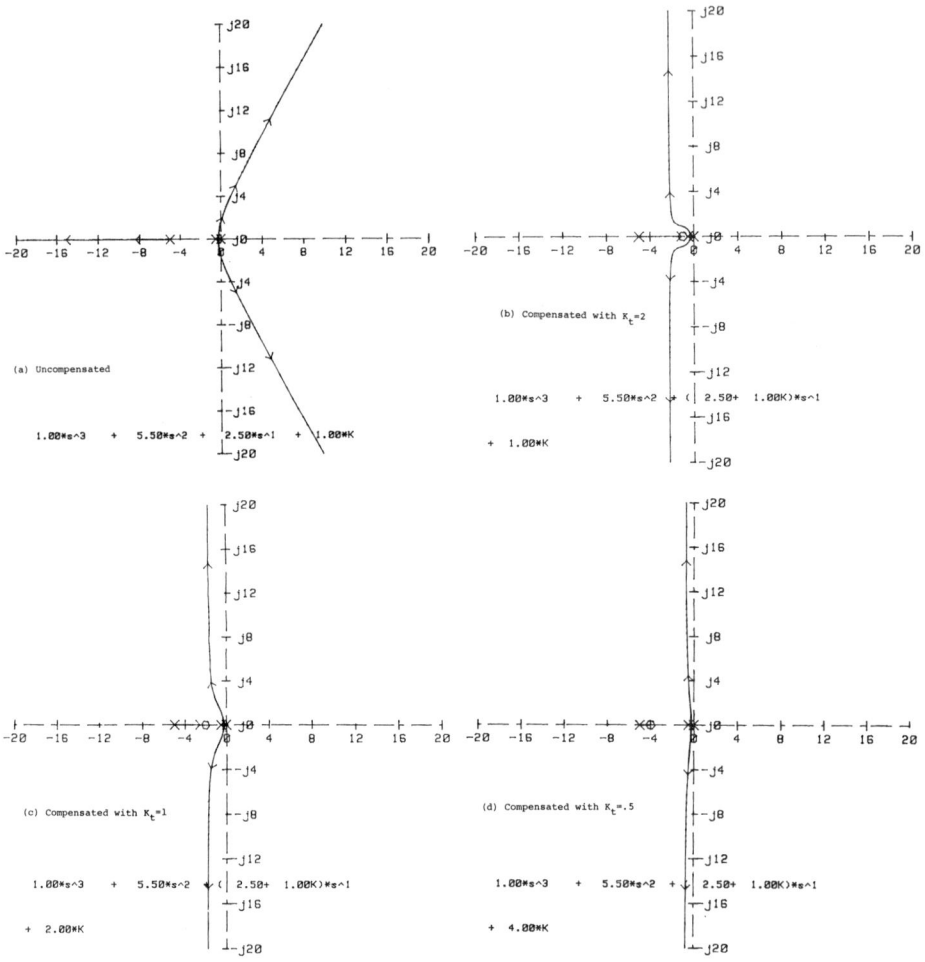

Fig. 9.9.2. Root loci for uncompensated and compensated system of Example 9.9.1

The effect of moving the zero to the left as K_t is decreased from 2 to 0.5 with $K_A = 2$. From these figures it is clear that as K_t is decreased the third pole (real pole) will lose its dominant location as compared to the pair of complex conjugate poles. The time responses of the system's output $c(t)$ for various values of K_t are shown in Fig. 3 for $K = K_A = 2$. The deteriorating behavior of the response for very low values of K_t is clearly shown. In fact if $K_t > 2/5.5 = 0.364$, the asymptotes of the corresponding root locus will be located on the right-half plane indicating that if $K >> 2$, the system would become unstable. The characteristics of the system responses are summarized in Table 1.

FEEDBACK COMPENSATION

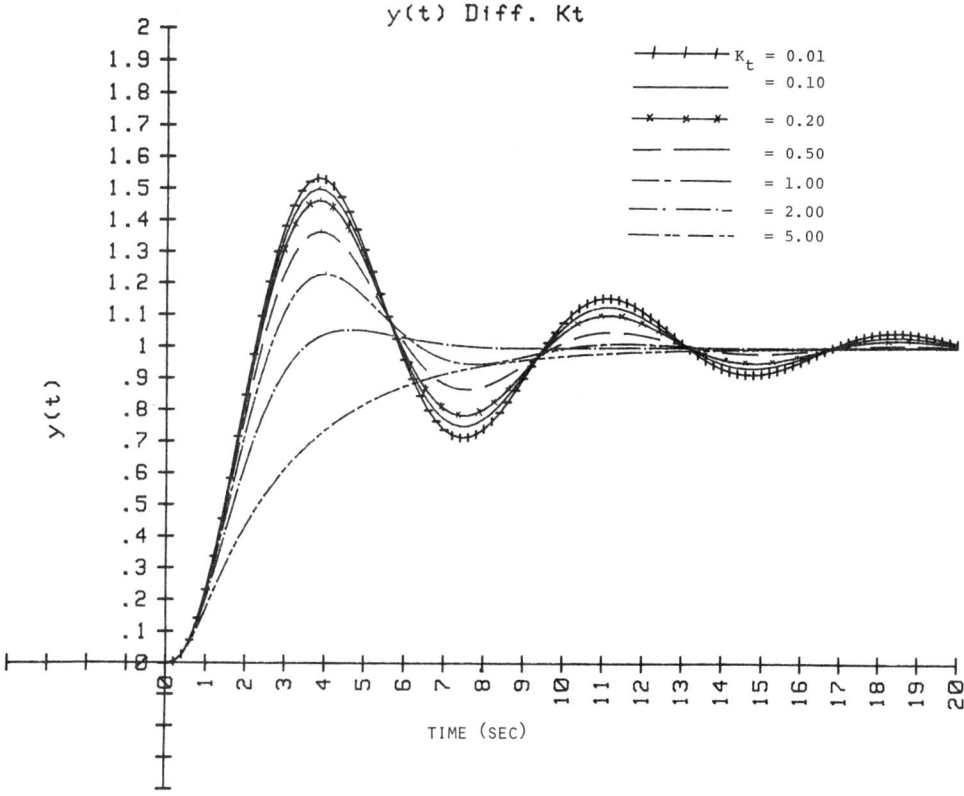

Fig. 9.9.3. Time responses of the system of Example 9.9.1 for various rate feedback constants

Table 9.9.1 Characteristics of the responses of Example 9.9.1

Trial	K_t	KK_t	ξ approx	Roots	$M_o\%$ approx	T_s ($\pm 2\%$)	T_p	K_v
1	.01	.02	.19	$-0.17 \pm j0.864, -5.16$	52.67	19.75	3.65	.80
2	.10	.2	.21	$-0.187 \pm j0.863, -5.12$	51.34	17.35	3.70	.77
3	.2	.4	.25	$-0.207 \pm j0.862, -5.08$	44.67	16.25	3.75	.74
4	.5	1.0	.3	$-0.27 \pm j0.85, -4.95$	37.34	12.82	3.78	.67
5	1.0	2.0	.42	$-0.386 \pm j0.835, -4.73$	22.67	8.80	3.95	.57
6	2.0	4.0	.69	$-0.664 \pm j0.72, -4.17$	5.34	6.32	4.68	.44
7	5.0	10.0	-	$-2.56 \pm j2, -0.38$	0	9.50	*	.267

* overdamped response

The static velocity error coefficient can be obtained by

$$K_v = \lim_{s \to 0} sG(s) = K_A K/(2.5+KK_t) \tag{6}$$

$$= 4/(5+2K_t).$$

The most reasonable case of the feedback compensation corresponds to trial number 6 ($K_t = 2$) where both the percent overshoot and settling time are the least. This corresponds to the root locus of Fig. 2b. The third pole is at $s_3 = -4.17$ which indicates that it is not too dominant as compared with the pair of complex conjugate poles at $s_{1,2} = -0.664 \pm j0.72$. △

In this final section of Chapter 9 it has been shown that feedback compensation through a rate feedback is basically equivalent to an ideal derivative control in cascade. The appearance of a zero in the compensated system root locus had an effect of shifting two of the branches of root locus away from the $j\omega$-axis for certain values of rate feedback constant K_t. This type of compensation like all the previous cascade-type compensators are, to a great extent, based on trial and error and the designer's past experience. A distinct feature of the cascade compensation as compared with the feedback compensation is that in the former the compensated system's order is increased by one or two while in the latter the system's order remains the same. It must be noted that with the aid of a digital computer a designer can overcome the laborious trial and error methods. In cases where a single cascade compensator is not sufficient to achieve the desired characteristics, one may employ additional compensator through feedback. Finally, a closely related topic to compensation in general and feedback compensation in particular in modern control theory is *pole placement* or *eigenvalue assignment* through state or output feedback. This will be discussed in Chapter 10.

PROBLEMS

9.1 The existence of a pair of dominant poles for an *n*th order system corresponds to a second-order aggregated model of a large-scale linear time-invariant system. For a system with

$$G(s) = K/(s(s+1)(s+5)), \quad K = 5$$

are exact aggregation conditions (5.3.4) satisfied?

9.2 Let the open-loop transfer function of a system be

$$G(s) = K \prod_{i=1}^{2}(1+\tau_i s)/s^N \prod_{j=1}^{3}(1+p_i s)$$

determine the static error coefficients for $N = 0, 1$ and 2.

9.3 For an input $r(t) = 1+2t+3t^2$ and a system with an open-loop transfer function

$$G(s) = 50/(s(0.5s+1)).$$

Determine the steady-state error of the system.

9.4 For the following unity feedback systems, determine the position, velocity and acceleration static error coefficients:

(a) $G(s) = 10/(s(s+1)(s+5))$

(b) $G(s) = 40/(s+2)^2$

(c) $G(s) = 8(s+4)/(s^2(s+5)(s+10))$

(d) $G(s) = 35/(s^3(s+4)(1+0.5s))$

9.5 A unity feedback system has a transfer function,

$$G(s) = 5/(s(s+a)(s+2)), \; r(t) = 5t.$$

(a) Determine the value of the steady-state error when $a=1$. (b) Find the value of a such that $e_{ss} \leq 0.1$. (c) Is the system stable for this value of a?

9.6 The underdamped time response of a second order system is given by (9.5.2). Show that its peak value and peak time are given by (9.5.3).

9.7 Determine the region on the s-plane which is feasible for a system's dominant poles corresponding to a percent overshoot $5\% \leq M_o \leq 10\%$ and a settling time $3 \leq T_s \leq 5$ seconds.

9.8 Show that the maximum value M_m of the frequency response for a second-order system of (9.5.6) is given by (9.5.7) at frequency ω_m is given by (9.5.8).

9.9 Determine the closed-loop magnitude, phase angle plots of a closed-loop system whose open-loop transfer function is

$$G(s)H(s) = K/(s(s+1))$$

where $K = 2$ using $M(\omega)$ and $\alpha(\omega)$ circles.

9.10 Consider a second-order system with open-loop transfer function,

$$G(s) = K/((s+1)(s-1)), \quad H(s) = 1.$$

a) Is this system stable? b) Design a compensator for it such that $M_o < 15\%$ and a damped natural frequency of $\omega_d = 1.5$ rad/sec. Use root locus as a design tool.

9.11 Show that the loci of the constant M circles are given by (9.5.11).

9.12 Repeat Problem 9.10 using $1.5 < M_m < 1.7, \omega_m = 2$ rad/sec. and root locus.

Hint: convert the frequency-domain specifications to time-domain first.

9.13 For the oscillatory unity-feedback system

$$G(s) = K/(s^2+1)$$

design a compensator which provides a phase margin $\gamma > 45°$ and a gain margin frequency of $\omega_q > 2$ rad/sec. using Bode diagrams.

9.14 Repeat Problem 9.13 using polar plots and a gain margin frequency of 2 instead of $\omega_q > 2$.

9.15 A system has a transfer function

$$G(s) = K/(s(s+2)(s+6)), \quad H(s) = 1.$$

It is desired to design a compensator which provides a steady-state error of $e_{ss} \leq 0.1$. Would the choice of compensator change if the system had an open-loop pole at -0.5 instead of the one at -2?

9.16 For the system of Problem 9.13, design a compensator which provides a $5 < M_o < 15\%$ and a natural frequency $\omega_n = 2$ rad/sec. using the Nyquist diagram.

Hint: Change the time-domain specifications to frequency domain before the design process.

9.17 For a system with

$$G(s) = K(1+s)/(s(1+2s)(1+4s)), \quad H(s) = 1$$

find (a) type, (b) steady-state error for $r(t) = (2+3t)u(t), u(t)$ is step input and (c) is the corresponding value of the error coefficient.

9.18 A system has an open-loop transfer function, $G = K/(s - 0.5)^2$ with $K = 2$. Let $H(s) = 1$ and $r(t) = u(t)$. It is desired to compensate this system to have an $M_m \leqslant 1.35$ and a $T_s \leqslant 2$ seconds. Use a dominant pole assumption and root locus methods to do that. What kind of compensator do you suggest and why? Give the designed RC-network's actual values and any feasible new closed-loop poles.

9.19 A unity-feedback system has a transfer function

$$G(s) = 1.5/(s^2(1+0.2s)).$$

which is inherently unstable. Design a compensator which would give the system an $M_m = 1.3$ at $\omega_m = 0.8$ rad/sec.

9.20 Repeat Problem 9.19 using Bode diagram and specifications $\gamma \geqslant 50°$ and $G_m \geqslant 1.5$.

9.21 Repeat Problem 9.19 using Nyquist and Nichols plots with an $M_m \leqslant 1.4$ at $\omega_n = 2$ rad/sec.

9.22 Consider a non-unity feedback system

$$G(s) = K/(s(s+1)(s+4))$$

with $H(s) = 1 + K_{t1}s + K_{t2}s^2$. a) Find a region in K_{t2}–K_{t1} plane such that the system is stable. b) Find an equivalent expression of $1+G=0$ and observe the effective changes of the system. c) Determine values of K_{t2} for $K_{t1} = 1$ such that a $M_o \leqslant 10\%$ for the compensated system.

Hint: Follow the development of Example 9.9.1.

REFERENCES

[9.1] J. J. D'Azzo and C. H. Houpis, *Linear Control System Analysis and Design,* Second Edition, McGraw-Hill, New York, 1981

[9.2] K. Ogata, *Modern Control Engineering,* Prentice-Hall, Inc., Englewood Cliffs, NJ, 1970

CHAPTER 10

FEEDBACK CONTROL AND STATE ESTIMATION

10.1 INTRODUCTION

In Chapter 9 we introduced the concept of feedback control and explained how it could be used to achieve desirable system characteristics. In that chapter frequency-domain techniques were used for compensator design and attention was primarily focused on SISO systems. The present chapter is concerned with the application of time-domain methods to the design of linear MIMO feedback control systems.

The design of linear feedback control systems essentially consist of suitably positioning the closed-loop poles in the complex plane. It was explained in Chapters 7 and 9 how system behavior such as stability, response time, sensitivity to disturbances, and other system dynamics relate to closed-loop pole locations. In this chapter we will also examine the effects of controllability and observability on the design of feedback systems. In Section 10.2 we will discuss the concepts of state feedback and output feedback and examine the effect of feedback on system properties. In Section 10.3 we will show how feedback can be used to position the closed-loop poles of a system. The topic of state estimation will be taken up in Section 10.4.

10.2 STATE AND OUTPUT FEEDBACK

Consider a *l.t.i.* system described by

$$\dot{\mathbf{x}}(t) = \mathbf{A}\mathbf{x}(t) + \mathbf{B}\mathbf{u}(t) \tag{1}$$

$$\mathbf{y}(t) = \mathbf{C}\mathbf{x}(t) + \mathbf{D}\mathbf{u}(t) \tag{2}$$

where $\mathbf{x}$, $\mathbf{u}$ and $\mathbf{y}$ are the state vector, the input vector and the output vector, respectively of dimensions n, r and m. Equations (1) and (2) are the state and output equations of the system to be controlled. It was explained in Chapter 9 that it would often be useful to apply a linear feedback from the state $\mathbf{x}(t)$ to the input $\mathbf{u}(t)$ or from the output $\mathbf{y}(t)$ to the input. In the first type of feedback, known as *state feedback* we have

$$\mathbf{u}(t) = \mathbf{r}(t) - \mathbf{Kx}(t) \tag{3}$$

and in the second type of feedback, referred to as *output feedback*, we have

$$\mathbf{u}(t) = \mathbf{r}(t) - \mathbf{K'y}(t) \tag{4}$$

The resulting closed-loop systems in these two cases are shown in Fig. 1. The function $\mathbf{r}(t)$ in (3) and (4) represents an external input which may be a *reference input* or a *disturbance input*. The $r \times n$ matrix $\mathbf{K}$ in (3) and the $r \times m$ matrix $\mathbf{K'}$ in (4) are called *feedback gain matrices*.

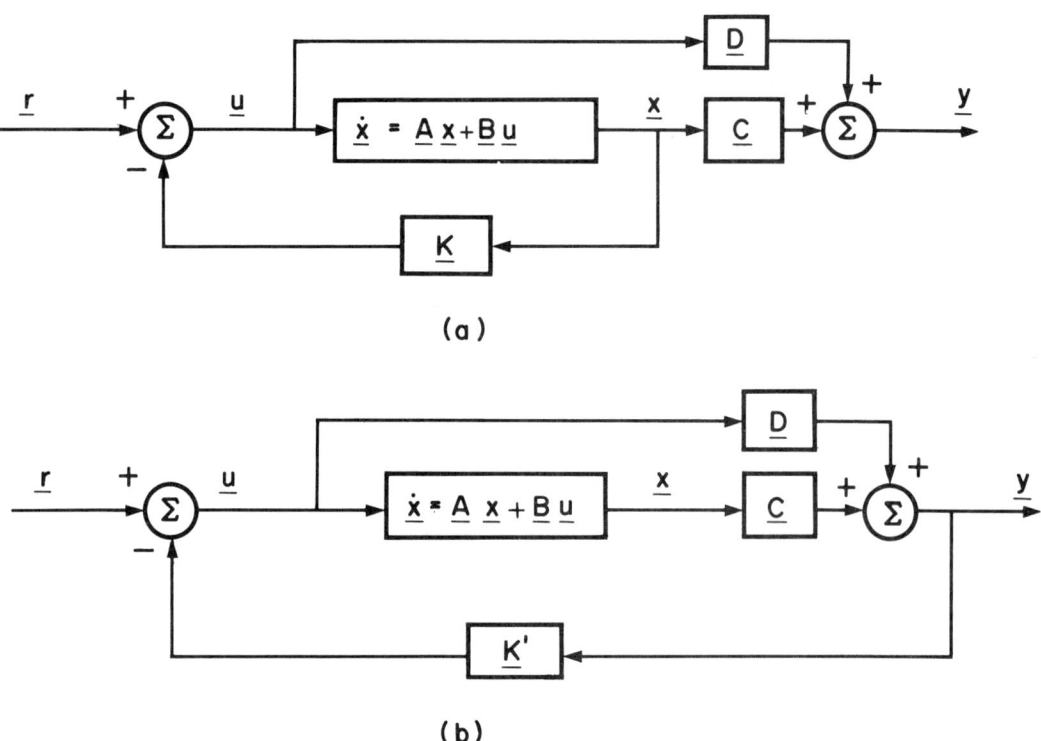

Fig. 10.2.1. Types of feedback a) state feedback, b) output feedback

The state and output equations in the case of state feedback can be obtained by substituting $\mathbf{u}(t)$ from (3) into (1) and (2):

$$\dot{\mathbf{x}}(t) = (\mathbf{A} - \mathbf{BK})\mathbf{x}(t) + \mathbf{Br}(t) \tag{5}$$

$$\mathbf{y}(t) = (\mathbf{C} - \mathbf{DK})\mathbf{x}(t) + \mathbf{Dr}(t) \tag{6}$$

The behavior of the closed-loop system of Fig.1a largely depends on the eigenvalues of matrix **A**−**BK** in (5). For example, the original (open-loop) system may be unstable (i.e. matrix **A** may have eigenvalues in the *r.h.p.*) but the closed-loop system with state feedback can be stable. We will shortly see that it is possible to position the eigenvalues of matrix **A**−**BK** in the complex plane arbitrarily by choosing the feedback gain matrix **K** properly.

Similarly, the state and output equations in the case of output feedback can be obtained by substituting **u**(*t*) from (4) into (1) and (2):

$$\dot{x}(t) = [A - BK'(I+DK')^{-1}C]x(t) + B(I+K'D)^{-1}r(t) \tag{7}$$

$$y(t) = (I+DK')^{-1}Cx(t) + (I+DK')^{-1}Dr(t) \tag{8}$$

Note that state feedback could be considered the same as output feedback when matrix **C** in (2) is an identity matrix. In fact, (7) with **C**=**I** and **D**=**0** becomes identical to (5). Output feedback corresponds to the case where only *some* state variables (or their linear combinations) but not all of them are available to be fed back.

10.3 EFFECT OF FEEDBACK ON SYSTEM PROPERTIES

We are now in a position to examine the effect of state and output feedback on system properties such as stability, controllability and observability. The stability of the closed-loop system is determined by the eigenvalues of the system matrix. Thus, in the case of state feedback stability depends on the eigenvalues of **A**−**BK** and in the case of output feedback it depends on the eigenvalues of **A**−**BK**′(**I**+**DK**′)$^{-1}$**C**. (See (10.2.5) and (10.2.7).) In Section 10.4 we will show that the locations of these eigenvalues in the complex plane can be specified by the choice of feedback gain matrices **K** and **K**′, respectively.

Controllability of the closed-loop system depends on the properties of the matrices in the state equation of the closed-loop system. We will shortly see that controllability is always preserved under state feedback. Observability of the closed-loop system in the case of state feedback depends on the matrix pair **A**−**BK** and **C**−**DK** (See (10.2.5) and (10.2.6).), and in the case of output feedback depends on the matrix pair **A**−**BK**′(**I**+**DK**′)$^{-1}$**C** and (**I**+**DK**′)$^{-1}$**C** (See (10.2.7) and (10.2.8).). Observability is always preserved under output feedback.

1. Theorem. State feedback does not affect controllability; that is, any controllable state of (1) is also a controllable state of (5) and vice versa.

Proof. Suppose that state x_o of (1) is controllable. That is, control **u**(*t*), *t*∈[t_o, t_1] exists which transfers x_o(at t_o) to the origin of the state space at t_l over a trajectory which depends on **u**(*t*). (See Definition 8.3.1.) Now if state feedback (10.2.3) is applied, the state equation will be given by (5). It is clear from (10.2.1) and (10.2.5) that the control

$$r(t) = u(t) + Kx(t) \tag{1}$$

also transfers the state x_o (*at* t_o) to the origin of the state space at t_1 over the *same* trajectory. Thus x_o must also be a controllable state of (10.2.5). Similar argument can be used to show the converse, i.e. any controllable state of (10.2.5) is also a controllable state of (10.2.1). △

For an alternate proof of the above theorem, see Problem 10.4. Theorem 1 also holds for time-varying systems (with constant feedback gain matrices). In the case of output feedback, using (10.2.4) and (10.2.2) we have

$$r(t) = (I+K'D)u(t) + K'Cx(t) \tag{2}$$

which is similar to (1). Thus, output feedback does not affect controllability if matrix **D** is identically zero. To show this simply replace the feedback gain matrix **K** in the proof of Theorem 1 by **K'C**. In general, using the same argument as in the proof of Theorem 1, we conclude that controllability is preserved under output feedback provided that the matrix **I+K'D** is nonsingular.

It can be shown that output feedback does not affect observability. To see this let **D=0** in (10.2.2) with no loss of generality. Then the observability matrix of the output feedback system becomes

$$\hat{Q}_o = \begin{bmatrix} C \\ C(A-BK'C) \\ \vdots \\ C(A-BK'C)^{n-1} \end{bmatrix} \tag{3}$$

which can be shown to have the same rank as the open-loop observability matrix

$$Q_o = \begin{bmatrix} C \\ CA \\ \vdots \\ CA^{n-1} \end{bmatrix} \tag{4}$$

(See Problem 10.6.) Observability, however, is not preserved under state feedback. For example, when **C=DK**, the state feedback system will not be observable even if the open loop system is. (See (10.2.6).)

2. Example. Consider the system

$$\dot{x}(t) = \begin{bmatrix} 0 & 1 \\ -3 & -4 \end{bmatrix} x(t) + \begin{bmatrix} 0 \\ 1 \end{bmatrix} u(t), \quad y(t) = [1 \; -1]x(t) \tag{5}$$

The eigenvalues of the system matrix are $\lambda_1 = -1$ and $\lambda_2 = -3$. Also, we have

$$Q_c = [B, AB] = \begin{bmatrix} 0 & 1 \\ 1 & -4 \end{bmatrix} \tag{6}$$

$$Q_o = \begin{bmatrix} C \\ CA \end{bmatrix} = \begin{bmatrix} 1 & -1 \\ 3 & 5 \end{bmatrix}$$

EFFECT OF FEEDBACK ON SYSTEM PROPERTIES

both of which have full rank. Thus, the system given in (5) without feedback is asymptotically stable, completely controllable and completely observable. Let us apply the following state feedback:

$$u(t)=r(t)-\mathbf{K}\mathbf{x}(t)=r(t)-[k_1\ k_2]\mathbf{x}(t) \tag{7}$$

This results in the following system

$$\dot{\mathbf{x}}(t)=\begin{bmatrix}0 & 1\\ -3-k_1 & -4-k_2\end{bmatrix}\mathbf{x}(t)+\begin{bmatrix}0\\ 1\end{bmatrix}r(t)$$

$$y(t)=[1-1]\mathbf{x}(t) \tag{8}$$

The characteristic polynomial of the new system matrix is

$$\lambda^2+(4+k_2)\lambda+3+k_1=0 \tag{9}$$

which yields both roots in the $l.h.p.$ for $k_1>-3$ and $k_2>-4$. The controllability matrix of the new system is

$$\hat{Q}_c=[\hat{\mathbf{B}},\hat{\mathbf{A}}\hat{\mathbf{B}}]=\begin{bmatrix}0 & 1\\ 1 & -4-k_2\end{bmatrix} \tag{10}$$

which has rank 2 independent of the values of k_1 and k_2. Thus the state-feedback system is completely controllable, as expected. The observability matrix of the new system is

$$\hat{Q}_o=\begin{bmatrix}\hat{\mathbf{C}}\\ \hat{\mathbf{C}}\hat{\mathbf{A}}\end{bmatrix}=\begin{bmatrix}1 & -1\\ 3+k_1 & 5+k_2\end{bmatrix} \tag{11}$$

which has rank 2 only when $k_1+k_2\neq-8$. Thus, the observability of the state-feedback system depends on the feedback gain parameters k_1 and k_2.

3. Example. Consider the system

$$\dot{\mathbf{x}}(t)=\begin{bmatrix}0 & 1\\ -2 & -3\end{bmatrix}\mathbf{x}(t)+\begin{bmatrix}1\\ -1\end{bmatrix}u(t) \tag{12}$$

$$y(t)=[0\ 1]\mathbf{x}(t)+u(t)$$

The eigenvalues of the system matrix are $\lambda_1=-1$ and $\lambda_2=-2$. Also we have

$$Q_c = [B, AB] = \begin{bmatrix} 1 & -1 \\ -1 & 1 \end{bmatrix}, \quad Q_o = \begin{bmatrix} C \\ CA \end{bmatrix} = \begin{bmatrix} 0 & 1 \\ -2 & -3 \end{bmatrix} \quad (13)$$

Thus the system without feedback is asymptotically stable and completely observable, but not completely controllable. Let us apply the following output feedback:

$$u(t) = r(t) - ky(t) \quad (14)$$

This results in the following system

$$\dot{x}(t) = \begin{bmatrix} 0 & \frac{1}{1+k} \\ -2 & \frac{-3-2k}{1+k} \end{bmatrix} x(t) + \frac{1}{1+k} \begin{bmatrix} 1 \\ -1 \end{bmatrix} r(t)$$

$$y(t) = \frac{1}{1+k}[0 \ 1]x(t) + \frac{1}{1+k} r(t)$$

The characteristic polynomial of the new system matrix is

$$(1+k)\lambda^2 + (3+2k)\lambda + 2 = 0 \quad (15)$$

which has both roots in the *l.h.p.* for $k > -1$. The controllability matrix of the new system is

$$\hat{Q}_c = [\hat{B}, \hat{A}\hat{B}] = \frac{1}{1+k} \begin{bmatrix} 1 & -1 \\ \frac{k}{-1} & \frac{1+k}{1} \\ \overline{k} & \frac{1}{1+k} \end{bmatrix} \quad (16)$$

which in singular. Thus the new system is not completely controllable either. But the observability matrix of the new system is

$$\hat{Q}_o = \begin{bmatrix} \hat{C} \\ \hat{C}\hat{A} \end{bmatrix} = \frac{1}{1+k} \begin{bmatrix} 0 & 1 \\ -2 & \frac{-3-2k}{1+k} \end{bmatrix} \quad (17)$$

which has rank 2 independent of the value of k. Thus the output-feedback system is completely observable.

10.4 POLE PLACEMENT

From (10.2.5) and (10.2.7) it is clear that the poles of the closed-loop system depend on the feedback gain matrix. In Examples 10.3.2 and 10.3.3 we showed that it is possible to destabilize a stable system by using state or output feedback. Conversely, it is possible to make an unstable system stable by applying state or output feedback. This process is referred to as *stabilization via* feedback. In addition to stabilization, there are other properties of the system which depend on pole locations. In fact, the dynamic behavior of a system largely depends on the position of its closed-loop poles. In this section we will show that it is possible to position the poles of a closed-loop system arbitrarily in the complex plane by using state or output feedback. This process is known as *pole placement* or *pole assignment*. We will see that controllability and observability of the system play important roles in this process.

The following discussions apply to both continuous-time and discrete-time systems. For example, to stabilize a *l.t.i.* continuous-time system, the poles of the closed-loop system must be located in the left half of the complex plane; but for *l.t.i.* discrete-time systems they must be located inside the unit circle in the complex plane. The same technique, however, will be used for pole placement in continuous-time and discrete-time systems.

10.4.1 Pole Placement Via State Feedback

From (10.2.5), the poles of a closed-loop system with state feedback are the roots of the characteristic polynomial

$$\Delta(\lambda) \triangleq \det(\mathbf{A}-\mathbf{BK}-\lambda\mathbf{I}) \tag{1}$$

Pole placement in this case is the problem of choosing a feedback gain matrix **K** which results in specified roots for (1). This problem has been the subject of considerable research [10.2-10.7]. Brocket [10.2] first demonstrated that for a single-input completely controllable system always a unique **K** exists for each set of specific roots of (1). For a multi-input system, arbitrary pole assignment is possible provided that the open-loop system is completely controllable; however, the corresponding feedback gain matrix is not unique. This is formalized in the following theorem.

1. Theorem. Consider the n-dimensional *l.t.i.* system described by the state equation

$$\dot{\mathbf{x}}(t) = \mathbf{A}\mathbf{x}(t) + \mathbf{B}\mathbf{u}(t) \tag{2}$$

where **A** and **B** are real matrices. Also consider any preassigned n complex numbers. A state feedback control law

$$\mathbf{u}(t) = \mathbf{r}(t) - \mathbf{K}\mathbf{x}(t) \tag{3}$$

exists such that the poles of the closed-loop system

$$\dot{\mathbf{x}}(t) = (\mathbf{A}-\mathbf{BK})\mathbf{x}(t) + \mathbf{B}\mathbf{r}(t) \tag{4}$$

are the above complex numbers if and only if the open-loop system (2) is completely controllable.

Proof. Necessity. If system (2) is not completely controllable, by Theorem 8.6.4 it can be decomposed into controllable and uncontrollable subsystems where input $\mathbf{u}(t)$ has no effect on the dynamics of the uncontrollable subsystem. Thus feedback control law (3) will have no effect on the poles of the uncontrollable subsystem and pole placement will not be possible.

488 LINEAR CONTROL SYSTEMS

Sufficiency. We will prove sufficiency for only the single-input case. For the multi-input case the proof can be found in references [10.3, 10.7, 10.9]. If system (2) is completely controllable, a similarity transformation $\hat{\mathbf{x}}(t) = \mathbf{Q}_c \mathbf{x}(t)$, where $\mathbf{Q}_c$ is the controllability matrix of (2), transforms (2) into the controllable companion form (See Section 4.5.)

$$\dot{\hat{\mathbf{x}}}(t) = \hat{\mathbf{A}}\hat{\mathbf{x}}(t) + \hat{\mathbf{B}}\mathbf{u}(t) \tag{5}$$

where

$$\hat{\mathbf{A}} = \mathbf{Q}_c \mathbf{A} \mathbf{Q}_c^{-1} = \begin{bmatrix} 0 & 1 & 0 & \cdots & 0 \\ 0 & 0 & 1 & \cdots & 0 \\ \vdots & & 0 & & \vdots \\ 0 & 0 & \cdots & & 1 \\ -p_0 & -p_1 & \cdots & & -p_{n-1} \end{bmatrix} \tag{6}$$

$$\hat{\mathbf{B}} = \mathbf{Q}_c \mathbf{B} = \begin{bmatrix} 0 \\ 0 \\ \vdots \\ 0 \\ 1 \end{bmatrix} \tag{7}$$

and the characteristic polynomial of $\hat{\mathbf{A}}$ (*or* $\mathbf{A}$) is

$$\lambda^n + p_{n-1}\lambda^{n-1} + \ldots + p_1\lambda + p_0 \tag{8}$$

We would like to show that row vector $\mathbf{k}$ can be chosen such that the system matrix $\mathbf{A} - \mathbf{Bk}$ will have any prespecified set of eigenvalues. Note that the eigenvalues of the matrices $\mathbf{A} - \mathbf{Bk}$ and

$$\mathbf{Q}_c(\mathbf{A} - \mathbf{Bk})\mathbf{Q}_c^{-1} = \hat{\mathbf{A}} - \hat{\mathbf{B}}\hat{\mathbf{k}} = \hat{\mathbf{A}} - \hat{\mathbf{B}}\mathbf{k}\mathbf{Q}_c^{-1} \tag{9}$$

are the same. Thus we can equivalently show that row vector

$$\hat{\mathbf{k}} = [k_o, k_1, \ldots, k_{n-1}] \tag{10}$$

exists such that matrix $\hat{\mathbf{A}} - \hat{\mathbf{B}}\hat{\mathbf{k}}$ in (9) will have any prespecified set of eigenvalues. From (6), (7) and (10) we have

$$\hat{\mathbf{A}} - \hat{\mathbf{B}}\hat{\mathbf{k}} = \begin{bmatrix} 0 & 1 & \cdots & 0 \\ 0 & 0 & 1 & \\ & 0 & & \vdots \\ \vdots & & & 0 \\ 0 & 0 & \cdots & 1 \\ -k_o - p_0 & -k_1 - p_1 & \cdots & -k_{n-1} - p_{n-1} \end{bmatrix} \tag{11}$$

Thus the characteristic polynomial of $\hat{\mathbf{A}} - \hat{\mathbf{B}}\hat{\mathbf{k}}$ is

$$\lambda^n + (p_{n-1} - k_{n-1})\lambda^{n-1} + \ldots + (p_1 - k_1)\lambda + (p_0 - k_0) \tag{12}$$

It is clear that k_i, $i = 0, 1, 2, \ldots, n-1$ may always be chosen such that (12) will have any set of prespecified roots. Δ

In order to synthesize the closed-loop system with real hardware, all the elements of the feedback gain matrix $\mathbf{k}$ must be real. Since matrices $\mathbf{A}$ and $\mathbf{B}$ are also assumed to be real, then any assigned complex poles must occur in conjugate pairs. Theorem 1 holds for any prespecified set of closed-loop poles with no restriction provided that vector $\mathbf{k}$ is allowed to have complex elements.

2. Example. Consider the system

$$\dot{\mathbf{x}}(t) = \begin{bmatrix} 0 & 1 \\ 2 & -1 \end{bmatrix} \mathbf{x}(t) + \begin{bmatrix} 0 \\ 1 \end{bmatrix} u(t) \tag{13}$$

Suppose that we want to use state feedback to place the closed-loop poles at $\lambda_1 = -1$ and $\lambda_2 = -3$. We have

$$Q_c = [\mathbf{B}, \mathbf{AB}] = \begin{bmatrix} 0 & 1 \\ 1 & -1 \end{bmatrix} \tag{14}$$

which is nonsingular. Thus the system is completely controllable and by Theorem 1 we are guaranteed that a feedback gain vector $\mathbf{k} = [k_1, k_2]$ exists such that the closed-loop poles, i.e. the eigenvalues of

$$\mathbf{A} - \mathbf{Bk} = \begin{bmatrix} 0 & 1 \\ 2 - k_1 & -1 - k_2 \end{bmatrix} \tag{15}$$

will be at -1 and -3. The (unique) values of k_1 and k_2 can be determined by equating the two characteristic polynomials:

$$(\lambda + 1)(\lambda + 3) = \lambda^2 + 4\lambda + 3 \equiv \det(\mathbf{A} - \mathbf{Bk} - \lambda \mathbf{I}) = \lambda^2 + (1 + k_2)\lambda + k_1 - 2 \tag{16}$$

which yields $k_1 = 5$ and $k_2 = 3$. Δ

Note that in a single-input system, the state feedback control law which yields specified closed-loop poles is unique. In a multi-input system, however, many control laws exist which result in the *same* specified closed-loop poles. The designer must choose the *optimal* control law based on other specified objectives.

State feedback is desirable in that it allows the placement of all closed-loop poles and preserves controllability. However, often not all the states will be available to be fed back. In such cases either output feedback (10.2.4) is used or an estimate of the state vector is constructed from the output. The latter method is the subject of Section 10.5. We will discuss pole placement using output feedback next.

Program "POLPLA" is a state-feedback pole placement routine in BASIC. The source listing of "POLPLA" with two examples are given below.

3. "POLPLA"

```
10    !    PROGRAM NAME : "POLPLA" PROG
20       Tste=0
30       PRINT "'POLPLA' is a POLE PLACEMENT Program via STATE feedback for SISO"
40       PRINT " Systems. Given the System: dx/dt=Ax+Bu, it determines a Vector"
50       PRINT " 'K' such that when : u=Kx+w, the System  dx/dt=(A+B*K)x+Bw will"
60       PRINT " have a prescribed Set of Real or Complex Conjugate EIGENVALUES "
70       PRINT " Lambda(i), i=1,2...,n, in the Left Half S-plane."
80       IF Tste=12 THEN 100
90       Tste=12
100      INPUT "Have SUB's <Lamb>,<Mat>,<Resmat>,& <Cont> been LINKED (Y/N)?",C$
110      IF C$<>"N" THEN 160
120      LINK "Lamb",710,130
130      LINK "Mat",2500,140
140      LINK "Cont",4000,150
150      LINK "Resmat",5000,160
160      PRINT "Matrix dimensions are:   A:NxN,      B:Nx1"
170      INPUT "Input Matrix dimension   N:",N
180      CALL R(N,Pr)
190      END
200      SUB R(N,Pr)
210   OPTION BASE 1
220      DIM B(N,1),A(N,N),W(1,N),La(2*N,2)
230      DIM P(N),Q(N,N,N),Qc(N,N),K(1,N),W1(N,N),W2(N,N),M(N,N),Wa(N,1)
240      CALL Mat(A(*),N,N,"A")
250      CALL Mat(B(*),N,1,"B")
260      CALL Cont(A(*),B(*),N,1,Qc(*))
270      Detq=DET(Qc)
280      IF ABS(Detq)<.0005 THEN Notcont
290      CALL Lamb(N,W(*),La(*))
300      PRINT "Matrix `A´:"
310      MAT PRINT A;
320      PRINT "Matrix `B´:"
330      MAT PRINT B;
340      PRINT "Designated Eiganvalues:"
350      FOR I=1 TO N
360      PRINT "Lambda";I;", Re: -";La(I,1);" Im:";La(I,2)
370      NEXT I
380      CALL Resmat(N,A(*),P(*),Q(*))
390       FOR In=1 TO N
400        Pi=(-1)^In
410        K(1,N-In+1)=-Pi*W(1,In)+P(N-In+1)
420       NEXT In
430       FOR I=1 TO N
440        FOR J=1 TO N
450         FOR K=1 TO N
460          M(J,K)=Q(I,J,K)
470         NEXT K
480        NEXT J
490        IF I=N THEN MAT M=IDN
500        MAT Wa=M*B
510        FOR J=1 TO N
520         W2(J,I)=Wa(J,1)
530        NEXT J
540       NEXT I
550      PRINT "The Elements of Row Vector 'K' are:";LIN(1)
560      MAT W1=INV(W2)
570      MAT W=K*W1
```

```
580    FOR I=1 TO N
590      FIXED 0
600      PRINT "K(";I;")=";
610      FLOAT 4
620      PRINT W(1,I)
630    NEXT I
640    PRINT LIN(1),"The NEW System Matrix is: A+B*K =";LIN(1)
650    MAT W1=B*W
660    MAT A=A+W1
670    MAT PRINT A;
680    GOTO 700
690 Notcont: PRINT "System is NOT CONTROLLABLE."
700    SUBEND
710    SUB Lamb(N,W(*),La(*))
720    REM Lamb(da) is a replacement subroutine for `Coeffs´.
730    REM It inputs complex Eiganpoles (forces them to the LHP),
740    REM and computes the coeffs of the polynomial expression (Z-La(i))^N
750    OPTION BASE 1
760    DIM Pn(4*N),Pq(N,3),T(N),Pt(4*N)
770 MAT La=(0)
780    FOR I=1 TO N
790    T(I)=0
800    PRINT "Input Lambda";I
810    INPUT "Re:",Real
820    IF Real<=0 THEN 860
830    BEEP
840    PRINT "POLES must be placed in the L.H.P."
850    PRINT "Lambda";I;", Re = -";Real
860    La(I,1)=ABS(Real)
870    IF I=N THEN Ont
880    INPUT "Im:",Img
890    La(I,2)=ABS(Img)
900    IF La(I,2)=0 THEN Ont
910    T(I)=1
920    I=I+1
930    La(I,1)=La(I-1,1)
940    La(I,2)=-La(I-1,2)
950    T(I)=1
960    PRINT "Lambda";I;" = Re: -";La(I,1),"Im:";La(I,2)
970 Ont: NEXT I
980 PRINT
990 PRINT
1000 FOR I=1 TO N
1010 PRINT "Lambda";I;" = Re: -";La(I,1),"Im:";La(I,2)
1020. NEXT I
1030 INPUT "Any CHANGES    (Y/N)?",C$
1040 IF C$="Y" THEN 770
1050 REM    PUT REAL LAMBDAS LAST
1060 Rel=0
1070 J=1
1080 FOR I=1 TO N
1090 IF T(I)=0 THEN Twn
1100 La(N+J,1)=La(I,1)
1110 La(N+J,2)=La(I,2)
1120 J=J+1
1130 GOTO Trt
1140 Twn:La(2*N-Rel,1)=La(I,1)
1150 Rel=Rel+1
1160 Trt:NEXT I
1170 FOR I=1 TO N
1180 FOR J=1 TO 2
1190 La(I,J)=La(N+I,J)
1200 NEXT J
1210 NEXT I
1220 REM    DO QUADRATICS
1230 J=1
```

```
1240 FOR I=1 TO (N-Rel)/2
1250 Pq(I,1)=La(J,1)^2+La(J,2)^2
1260 Pq(I,2)=La(J,1)*2
1270 Pq(I,3)=1
1280 J=J+2
1290 NEXT I
1300 J=0
1310 FOR I=(N-Rel)/2+1 TO (N-Rel)/2+Rel
1320 J=J+1
1330 Pq(I,1)=La(N+1-J,1)
1340 Pq(I,2)=1
1350 NEXT I
1360 FOR I=3 TO 5
1370 Pt(I)=Pq(1,I-2)
1380 NEXT I
1390 REM    QUADRATIC MULTS
1400 FOR L=2 TO (N-Rel)/2+Rel
1410 FOR J=3 TO 2*L+3
1420 Pn(J)=0
1430 FOR I=1 TO 3
1440 Sum=Pt(J+1-I)*Pq(L,I)
1450 Pn(J)=Pn(J)+Sum
1460 NEXT I
1470 NEXT J
1480 FOR J=3 TO 2*L+3
1490 Pt(J)=Pn(J)
1500 NEXT J
1510 NEXT L
1520 FOR I=1 TO N
1530 W(1,I)=Pt(I+2)*(-1)^I
1540 NEXT I
1550 SUBEND
```

4. Example

'POLPLA' is a POLE PLACEMENT Program via STATE feedback for SISO Systems. Given the System: dx/dt=Ax+Bu, it determines a Vector 'K' such that when : u=Kx+w, the System dx/dt=(A+B*K)x+Bw will have a prescribed Set of Real or Complex Conjugate EIGENVALUES Lambda(i), i=1,2...,n, in the Left Half S-plane.
Matrix dimensions are: A:NxN,   B:Nx1

Matrix A(3 x 3):

0.0000E+00   1.0000E+00   0.0000E+00

0.0000E+00   0.0000E+00   1.0000E+00

1.0000E+00   2.0000E+00   3.0000E+00

Column Vector B(3):

0.0000E+00

0.0000E+00

1.0000E+00

```
Input Lambda 1
Lambda 2  = Re: - 2  Im:-2
Input Lambda 3


Lambda 1  = Re: - 2  Im: 2
Lambda 2  = Re: - 2  Im:-2
Lambda 3  = Re: - 1  Im: 0
Matrix `A´:
 0  1  0

 0  0  1

 1  2  3


Matrix `B´:
 0

 0

 1


Designated Eiganvalues:
Lambda 1 , Re: - 2   Im: 2
Lambda 2 , Re: - 2   Im:-2
Lambda 3 , Re: - 1   Im: 0
The Elements of Row Vector `K´ are:

K( 1 )=-6.0000E+00
K( 2 )=-1.4000E+01
K( 3 )=-1.1000E+01

The NEW System Matrix is: A+B*K =

 0.0000E+00   1.0000E+00   0.0000E+00

 0.0000E+00   0.0000E+00   1.0000E+00

-5.0000E+00  -1.2000E+01  -8.000E+00
```

10.4.2 Pole Placement Via Output Feedback

From (10.2.7), the poles of the closed-loop system with output feedback are the roots of the characteristic polynomial

$$\Delta(\lambda) \triangleq \det[\mathbf{A}-\mathbf{B}\mathbf{K}'(\mathbf{I}+\mathbf{D}\mathbf{K}')^{-1}\mathbf{C}-\lambda\mathbf{I}] \qquad (17)$$

For **D=0**, (17) becomes similar to (1) with **K** replaced by **K'C**. It has been shown [10.10] that if the pair **(A,B)** is completely controllable and if **C** has full rank m ($m \leq n$), then m of the n poles of the closed-loop system can be placed almost exactly as specified. Another result related to pole placement via output feedback is that if the open-loop system is both completely controllable and completely observable and if $\rho(\mathbf{B})+\rho(\mathbf{C})>n$, then feedback control law (10.2.4) with a real feedback gain matrix **K'** exists which places the closed-loop poles arbitrarily close to any prespecified locations in the complex plane [10.11]. (Complex poles occur in conjugate pairs.)

5. Example Consider the completely controllable but unstable system **(A,B,C,D)** where

$$\mathbf{A} = \begin{bmatrix} -5 & 2 \\ 3 & 0 \end{bmatrix}, \mathbf{B} = \begin{bmatrix} 0 \\ 1 \end{bmatrix}, \mathbf{C} = [1, 0], \mathbf{D} = 0 \qquad (18)$$

Suppose that we want to use output feedback to place one of the close-loop poles at $\lambda_1 = -4$. We have

$$\mathbf{A} - \mathbf{B}k'\mathbf{C} = \begin{bmatrix} -5 & 2 \\ 3-k' & 0 \end{bmatrix} \qquad (19)$$

where scalar k' is the output feedback gain. The characteristic polynomials can be equated:

$$\Delta(\lambda) = \det(\mathbf{A}-\mathbf{B}k'\mathbf{C}-\lambda\mathbf{I}) = \lambda^2 + 5\lambda - 6 + 2k'$$

$$\equiv (\lambda+\lambda_1)(\lambda+\lambda_2) = (\lambda+4)(\lambda+\lambda_2) \qquad (20)$$

which yields closed-loop poles at $\lambda_1 = -4$ and $\lambda_2 = -1$ for $k' = 5$. △

If output feedback does not result in a satisfactory dynamic behavior of the system, a *state estimator* can be designed to estimate the state vector so that state feedback may be used. This will be discussed in the next section.

10.5 STATE ESTIMATION - OBSERVERS

In the previous section we showed how the entire state of an open-loop system can be fed back for the exact placement of closed-loop poles. We will also see in Chapter 12 that the implementation of optimal control laws requires measurement of the state of the controlled plant. In many practical situations, however, the entire state vector of the system is not available for measurement. In such cases an *estimate* of the state may be used. The mechanism for estimating the state of a system is called a *state estimator* or an *observer*. We will use these terms interchangeably.

An observer utilizes measurements of the outputs and inputs of the system and provides an estimate of the state of the system. This can then be used to realize a state feedback control. In fact, an observer acts as a *compensator* which can control system dynamics similar to the compensators discussed in Chapter 9 for SISO systems. The remainder of this chapter is concerned with the design of MIMO control systems via state estimation. We will discuss two types of observers: one with the same dimension as the system, referred to as *full-dimension observer*, and the other with a lower dimension then the system, known as *reduced-dimension observer*.

Figure 1 shows the block diagram of a state estimator. Its inputs are the inputs and the outputs of the system. The output of the state estimator is an estimate $x_e(t)$ of the state of the system $x(t)$. One of the main attributes of a "good" state estimator is that it should be able to provide an acceptable estimate of the state in the presence of noise. To emphasize this point, let us first look at a plausible but unacceptable state estimator.

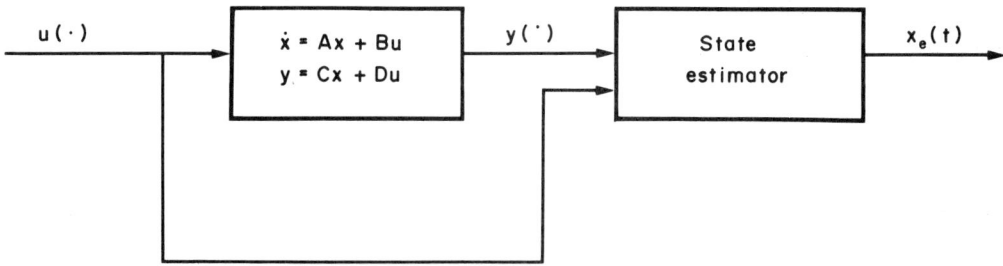

Fig. 10.5.1. Block diagram of a state estimator

Consider a single-output n-dimensional system

$$\dot{x}(t) = Ax(t) + Bu(t) \tag{1a}$$

$$y(t) = Cx(t) \tag{1b}$$

and assume that it is completely observable. Differentiating (1b) w.r.t. time and substituting in (1a) yields

$$\dot{y} - CBu = CAx \tag{2}$$

Differentiating (2) w.r.t. time and substituting in (1a) yields

$$\ddot{y} - CB\dot{u} - CABu = CA^2 x \tag{3}$$

Continuing in this manner, we obtain $z = Q_o x$ where

$$Q_o = \begin{bmatrix} C \\ CA \\ \vdots \\ CA^{n-1} \end{bmatrix} \tag{4}$$

is the observability matrix of (1), which by assumption is nonsingular, and z is a vector whose elements consist of linear combinations of **u**, **y** and their derivatives w.r.t. time. Thus we have $\mathbf{x} = \mathbf{Q}_o^{-1}\mathbf{z}$; that is, in a single-output completely observable system the exact state vector can be expressed in terms of the system's input, output and their derivatives w.r.t. time. However, this solution is not acceptable from the practical point of view. The reason is that noise is always present in **u** and **y** and their differentiation w.r.t time aggravates the situation, leading to considerable errors in the state vector. We will discuss practical state estimators in Sections 10.6 and 10.7. We will assume, with no loss of generality, that $t_o = 0$ and that matrix **D** is identically zero in the system whose state is to be estimated. (If $\mathbf{D} \neq 0$, define output as **y**−**Du**.)

10.6 FULL-DIMENSION OBSERVERS

The state estimator shown in Fig. 10.5.1 is itself a dynamic system. We expect the state estimator of an n-dimensional *l.t.i.* system to be also an n-dimensional *l.t.i.* system whose state equation is of the form

$$\dot{\mathbf{x}}_e(t) = \hat{\mathbf{A}}\mathbf{x}_e(t) + \hat{\mathbf{B}}\mathbf{u}(t) + \mathbf{K}\mathbf{y}(t) \tag{1}$$

Thus we need matrices $\hat{\mathbf{A}}$, $\hat{\mathbf{B}}$ and **K** to build the state estimator shown in Fig. 10.5.1. In this section we will see how these matrices can be calculated.

Let us start with a discussion of a so-called *open-loop state estimator* shown in Fig. 1. The state estimate $\mathbf{x}_e$ is obtained by constructing a *model* of the original system

$$\dot{\mathbf{x}} = \mathbf{A}\mathbf{x} + \mathbf{B}\mathbf{u}, \quad \mathbf{y} = \mathbf{C}\mathbf{x} \tag{2}$$

as

$$\dot{\mathbf{x}}_e = \mathbf{A}\mathbf{x}_e + \mathbf{B}\mathbf{u} \tag{3}$$

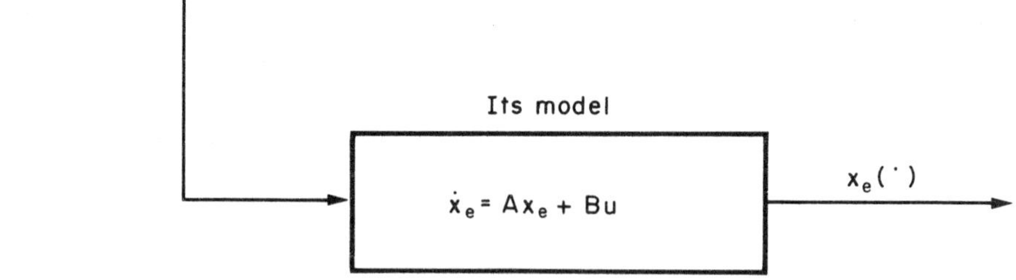

Fig. 10.6.1. An open-loop state estimator

If the initial states $x(0)$ and $x_e(0)$ were the same, (3) would provide an exact estimate of $x(t)$, i.e. $x(0) = x_e(0)$ implies $x(t) = x_e(t)$ for $t > 0$. However, this method has two drawbacks. First, we do not have the initial state $x(0)$ and second, we can measure output y but we have not used this available information in the method. We can make use of y to improve the estimate as follows. Consider the *output error vector*

$$\tilde{y}(t) = y(t) - y_e(t) = Cx(t) - Cx_e(t) = C[x(t) - x_e(t)] \triangleq C\tilde{x}(t) \tag{4}$$

where $\tilde{x}(t)$ is the *state error vector*. The dynamic behavior of $\tilde{x}(t)$ can be determined using (2) and (3):

$$\dot{\tilde{x}}(t) = \dot{x}(t) - \dot{x}_e(t) = Ax + Bu - (\hat{A}x_e + Bu) = A(x - x_e) = A\tilde{x}(t) \tag{5}$$

which yields

$$\tilde{x}(t) = e^{At}\tilde{x}(0), \quad t \geq 0 \tag{6}$$

Thus the state error vector in the open-loop state estimator is completely characterized by the system matrix A. Note that if $\tilde{x}(0) = 0$, then $\tilde{x}(t) = 0$ for $t > 0$ as discussed above. But in general $\tilde{x}(0) \neq 0$. If all the eigenvalues of A are in the *l.h.p.*, i.e. if system (2) is asymptotically stable, then the state error vector $\tilde{x}(t)$ will decay exponentially and the state estimate will improve as time progresses. On the other hand, if A has any eigenvalue in the *r.h.p.*, the initial error $\tilde{x}(0)$ will grow exponentially with time and the state estimate will be unacceptable.

To alleviate this open-loop dependence of the state estimate on A, it is reasonable to close the loop by including a correction term which depends on the output error vector (4):

$$\dot{x}_e(t) = Ax_e(t) + Bu(t) + M\tilde{y}(t), \quad t \geq 0 \tag{7}$$

where matrix M must be determined. Using (4), (7) becomes

$$\dot{x}_e(t) = (A - MC)x_e(t) + Bu(t) + My(t), \quad t \geq 0 \tag{8}$$

In fact, it has been shown [10.15] that every full-dimension observer has this structure. This new arrangement is shown in Fig. 2. Note that now the dynamic behavior of $\tilde{x}(t)$ depends on both A and M:

$$\dot{\tilde{x}}(t) = \dot{x} - \dot{x}_e = Ax + Bu - (Ax_e + Bu + M\tilde{y}) = (A - MC)\tilde{x}(t), \quad t \geq 0 \tag{9}$$

and

$$\tilde{x}(t) = e^{(A-MC)t}\tilde{x}(0), \quad t \geq 0 \tag{10}$$

If matrix **M** is chosen such that the eigenvalues of **A−MC** are in the l.h.p., the state error vector $\tilde{x}(t)$ will decay exponentially with time. Note that the eigenvalues of **A−MC** can be arbitrarily placed by the choice of **M** if and only if system (2) is completely observable. (See Section 10.4.2.) Also note that comparison of (8) and (1) yields

$$\hat{\mathbf{A}} = \mathbf{A} - \mathbf{MC}, \quad \hat{\mathbf{B}} = \mathbf{B}, \quad \mathbf{K} = \mathbf{M} \tag{11}$$

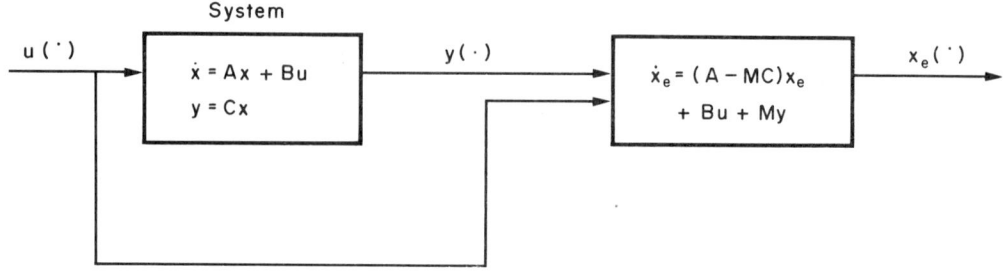

Fig. 10.6.2. A full-dimension observer

Thus we have the following theorem.

1. Theorem. The state of a completely observable *l.t.i.* *n*-dimensional system (**A,B,C,D**) can be estimated with an *n*-dimensional observer described by (8) where matrix **M** can be chosen to place the observer poles arbitrarily. Further, the error in the state estimate decays *exponentially* with time at a rate determined by the observer poles. △

It is important to note that in the observer of Fig. 2 the problem of differentiating any noise associated with u and y will not be encountered. In fact, such noise will be smoothed by the observer. (See Problem 10.15.) Another point is that one might think that matrix **M** should be chosen such that the poles of the observer (the eigenvalues of **A−MC**) have real parts as negative as possible. There are, however, limitations such as the bandwidths of components, on how negative the real parts of observer poles can be. Also, the effective bandwidth of the observer increases as the real parts of its poles become more negative. Thus the noise in the state estimate x_e (due to the noise associated with u and y) will increase as the observer becomes faster.

2. Example. Design a full-dimension observer for the system of Example 10.4.6 such that error in the state estimate will decay at least as fast as e^{-4t}. This means that $\text{Re}(\lambda_1) = -4$ and $\text{Re}(\lambda_2) \leqslant -4$. Let us take $\lambda_1 = -4, \lambda_2 = -5$. Note that the system is completely observable. Therefore, by Theorem 1, it is possible to design an observer with the given dynamics for this system. We have to find a 2-dimensional vector $\mathbf{M} = [m_1, m_2]'$ such that matrix **A−MC** has the prescribed eigenvalues. We have

$$\mathbf{A} - \mathbf{MC} = \begin{bmatrix} -5 & 2 \\ 3 & 0 \end{bmatrix} - \begin{bmatrix} m_1 \\ m_2 \end{bmatrix} [1 \ 0] = \begin{bmatrix} -5-m_1 & 2 \\ 3-m_2 & 0 \end{bmatrix} \tag{12}$$

Equate the characteristic polynomial of $\mathbf{A}-\mathbf{MC}$ with $(\lambda+4)(\lambda+5)$:

$$\Delta(\lambda) = \lambda^2 + (5+m_1)\lambda - 2(3-m_2) \equiv (\lambda+4)(\lambda+5) = \lambda^2 + 9\lambda + 20 \tag{13}$$

This yields $m_1=4$ and $m_2=13$. Thus the desired observer is described by (see (8)):

$$\dot{\mathbf{x}}_e(t) = \begin{bmatrix} -9 & 2 \\ -10 & 0 \end{bmatrix} \mathbf{x}_e(t) + \begin{bmatrix} 0 \\ 1 \end{bmatrix} u(t) + \begin{bmatrix} 4 \\ 13 \end{bmatrix} y(t) \tag{14}$$

10.7 REDUCED-ORDER OBSERVER

The observer described in the previous section is called a *full-dimension observer* or an *identity observer* [10.16] since it estimates *all* the state variables. But some linear combinations of state variables, specified by the rows of matrix **C**, are already available as outputs and need not be estimated. Therefore, it should be possible to reduce the dimension of the observer by the number of outputs of the system. That is, the state vector of an n-dimensional l.t.i. system with m outputs may be estimated by an $n-m$ dimensional observer. In this section we will see how such an observer can be designed.

To illustrate this point consider the observable companion form of the n-dimensional SISO system discussed in Section 4.4.3. In this form the output of the system, as given in (4.4.34), is always equal to the last state variable. Thus, this state variable is directly accessible and needs not be estimated. The observer for such a system should estimate only the remaining $n-1$ state variables. In fact, any n-dimensional system $(\mathbf{A},\mathbf{B},\mathbf{C},\mathbf{0})$ with m outputs where **C** has full rank can be transformed to an algebraically equivalent system where the output vector consists of the last m state variables in the state vector. It is easy to show that such a state transformation is

$$\bar{\mathbf{x}} = \mathbf{T}\mathbf{x} = \begin{bmatrix} \mathbf{R} \\ \mathbf{C} \end{bmatrix} \mathbf{x} \tag{1}$$

where $\bar{\mathbf{x}}$ and $\mathbf{x}$ are the new and the old state vectors, respectively, and $\mathbf{R}$ is any $(n-m) \times n$ matrix such that **T** is nonsingular. (See Problem 10.18.) Thus we can assume, with no loss of generality, that the system whose state is to be estimated has the form

$$\begin{bmatrix} \dot{\mathbf{x}}_1 \\ \dot{\mathbf{x}}_2 \end{bmatrix} = \begin{bmatrix} \mathbf{A}_{11} & \mathbf{A}_{12} \\ \mathbf{A}_{21} & \mathbf{A}_{22} \end{bmatrix} \begin{bmatrix} \mathbf{x}_1 \\ \mathbf{x}_2 \end{bmatrix} + \begin{bmatrix} \mathbf{B}_1 \\ \mathbf{B}_2 \end{bmatrix} \mathbf{u} \tag{2a}$$

$$\mathbf{y} = [\mathbf{0}, \mathbf{I}] \begin{bmatrix} \mathbf{x}_1 \\ \mathbf{x}_2 \end{bmatrix} = \mathbf{x}_2 \tag{2b}$$

Since part of the state vector, $\mathbf{x}_2$, is available as the output, we need to design an observer to estimate only $\mathbf{x}_1$ consisting of $n-m$ state variables. Such an observer will have the form

$$\dot{\mathbf{w}}(t) = \hat{\mathbf{A}}\mathbf{w}(t) + \hat{\mathbf{B}}\mathbf{u}(t) + \mathbf{K}\mathbf{y}(t) \tag{3}$$

where $\hat{A}$ is $(n-m) \times (n-m)$, so that from $w(t)$ and $y(t)$ an estimate $x_e(t)$ of the system state $x(t)$ can be constructed. Figure 1 shows the structure of such an observer, known as a *Luenberger observer* [10.17, 10.18] and also referred to as a *reduced-dimension (or reduced-order) observer*.

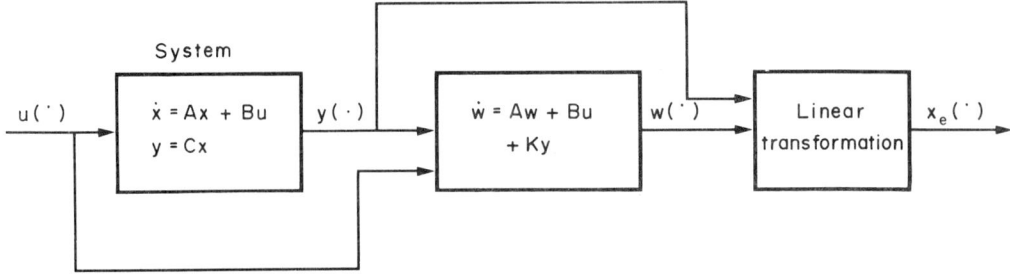

Fig. 10.7.1. A reduced-dimension observer

The above general idea has been translated into a computational scheme. Here we briefly describe the method used in references [10.16, 10.20]. From (2a), the $(n-m)$-dimensional subsystem with state x_1 is obtained as

$$\dot{x}_1(t) = A_{11}x_1(t) + r(t), \quad z(t) = A_{21}x_1(t) \tag{4}$$

where, from (2b),

$$r(t) = A_{12}x_2(t) + B_1u(t) = A_{12}y(t) + B_1u(t) \tag{5}$$

$$z(t) = \dot{x}_2(t) - A_{22}x_2(t) - B_2u(t) = \dot{y}(t) - A_{22}y(t) - B_2u(t) \tag{6}$$

which are measurable and known. Note, however, that (6) involves differentiation of output y which is not desirable due to noise considerations.

In order to design an observer for system (4), first we have to make sure that complete observability of the original system (2) implies complete observability of this subsystem. This can be easily verified. (See Problem 10.19.) Then we have to somehow get around the differentiation of y w.r.t. in (6). From Theorem 10.6.1, an estimate x_{1e} of x_1 can be obtained by the following $(n-m)$-dimensional observer:

$$\dot{x}_{1e}(t) = (A_{11} - MA_{21})x_{1e}(t) + r(t) + Mz(t) \tag{7}$$

where the $(n-m) \times m$ matrix M can be chosen to place the observer poles arbitrarily. Using (5) and (6), (7) becomes

$$\dot{x}_{1e}(t) = (A_{11} - MA_{21})x_{1e}(t) + A_{12}y(t) + B_1u(t) + M[\dot{y}(t) - A_{22}y(t) - B_2u(t)] \tag{8}$$

To avoid differentiation of **y**, define

$$w(t) = x_{1e}(t) - My(t) \tag{9}$$

Thus, from (8), we have

$$\dot{w} = \dot{x}_{1e} - M\dot{y} = (A_{11} - MA_{21})w + (B_1 - MB_2)u + (A_{12} - MA_{22} + A_{11}M - MA_{21}M)y \tag{10}$$

This is an $(n-m)$-dimensional observer which estimates x_1 and does not involve $\dot{y}$. Comparison of (10) with (3) yields

$$\hat{A} = A_{11} - MA_{21}, \quad \hat{B} = B_1 - MB_2, \quad K = A_{12} - MA_{22} + A_{11}M - MA_{21}M \tag{11}$$

Therefore, the complete state vector **x** of (2) is estimated as follows:

$$x_e = \begin{bmatrix} x_{1e} \\ x_{2e} \end{bmatrix} = \begin{bmatrix} w + My \\ y \end{bmatrix} = \begin{bmatrix} I \\ 0 \end{bmatrix} w + \begin{bmatrix} M \\ I \end{bmatrix} y \tag{12}$$

Equation (12) defines the block indicated "linear transformation" in Fig. 1. Note that the last m components of x_e are directly measurable and there is no error in estimating them. Let us define $\tilde{x}_1(t) = x_1(t) - x_{1e}(t)$ as the error vector for the first $n-m$ components of x_e. The dynamic behavior of $\tilde{x}_1(t)$ can be determined using (4), (5) and (8):

$$\dot{\tilde{x}}_1 = \dot{x}_1 - \dot{x}_{1e} = A_{11}x_1 + A_{12}y + B_1u - (A_{11} - MA_{21})x_{1e} - A_{12}y - B_1u - M(\dot{y} - A_{22}y - B_2u)$$

$$= (A_{11} - MA_{21})\tilde{x}_1 \tag{13}$$

which yields

$$\tilde{x}_1(t) = e^{(A_{11} - MA_{21})t} \tilde{x}_1(0), \quad t \geq 0 \tag{14}$$

Due to complete observability of system (4), it is always possible to choose matrix M to place the eigenvalues of $A_{11} - MA_{21}$ arbitrarily. (See Section 10.4.2.) Thus, error (14) can be made to decay at an arbitrary exponential rate (subject to limitations discussed in Section 10.6). An estimate of the state vector of the original system (before transformation (1)) can be found as

$$x_e = T^{-1}\bar{x}_e. \tag{15}$$

(See Problem 10.20.) We can summarize the discussions in this section in the following theorem.

1. Theorem. The state of a completely observable l.t.i. n-dimensional system $(A,B,C,0)$ with m outputs can be estimated with an $(n-m)$-dimensional observer described in (10) where matrix M can be chosen to place the observer poles arbitrarily. Further, the estimate of m of the state variables is exact and the error in the estimate of the remaining $n-m$ state variables decays exponentially with time at a rate determined by the observer poles.

2. Example. Let us reconsider the system $(A,B,C,0)$, where

$$A = \begin{bmatrix} -5 & 2 \\ 3 & 0 \end{bmatrix}, \quad B = \begin{bmatrix} 0 \\ 1 \end{bmatrix}, \quad C = [1, 0], \tag{16}$$

of Example 10.6.2 and design a reduced-dimension observer for it such that the error in the state estimate decays with time at least as fast as e^{-4t}. Note that the system has one output and $\rho(C)=1$. We will first put the system in the form specified by (2), using similarity transformation

$$\bar{x} = Tx = \begin{bmatrix} R \\ C \end{bmatrix} x = \begin{bmatrix} 0 & 1 \\ 1 & 0 \end{bmatrix} x \tag{17}$$

Thus, the matrices of the new system are

$$\bar{A} = TAT^{-1} = \begin{bmatrix} 0 & 1 \\ 1 & 0 \end{bmatrix} \begin{bmatrix} -5 & 2 \\ 3 & 0 \end{bmatrix} \begin{bmatrix} 0 & 1 \\ 1 & 0 \end{bmatrix} = \begin{bmatrix} 0 & 3 \\ 2 & -5 \end{bmatrix} = \begin{bmatrix} A_{11} & A_{12} \\ A_{21} & A_{22} \end{bmatrix} \tag{18a}$$

$$\bar{B} = TB = \begin{bmatrix} 0 & 1 \\ 1 & 0 \end{bmatrix} \begin{bmatrix} 0 \\ 1 \end{bmatrix} = \begin{bmatrix} 1 \\ 0 \end{bmatrix} = \begin{bmatrix} B_1 \\ B_2 \end{bmatrix} \tag{18b}$$

$$\bar{C} = CT^{-1} = [1\ 0] \begin{bmatrix} 0 & 1 \\ 1 & 0 \end{bmatrix} = [0\ 1] \tag{18c}$$

The reduced-dimension observer (10) becomes

$$\dot{w} = -2Mw + u + (3 + 5M - 2M^2)y \tag{19}$$

Note that both w and M are scalars. To ensure that the error in the estimate of the state variable decays at least as fast as e^{-4t}, we must have $-2M \leq -4$ or $M \leq 2$. Let us take $M=2$ which would result in

$$\dot{w} = -4w + u + 5y \tag{20}$$

Now by (12) an estimate of the state vector of the transformed system is

$$\bar{x}_e = \begin{bmatrix} 1 \\ 0 \end{bmatrix} w + \begin{bmatrix} 2 \\ 1 \end{bmatrix} y = \begin{bmatrix} w + 2y \\ y \end{bmatrix} \tag{21}$$

where w is given by (20). Therefore, an estimate of the state vector of the original system is

$$\mathbf{x}_e = \mathbf{T}^{-1}\bar{\mathbf{x}}_e = \begin{bmatrix} 0 & 1 \\ 1 & 0 \end{bmatrix} \begin{bmatrix} w+2y \\ y \end{bmatrix} = \begin{bmatrix} y \\ w+2y \end{bmatrix} \quad (22)$$

PROBLEMS

10.1 Verify that (10.2.4) with (10.2.1) and (10.2.2) yield (10.2.7) and (10.2.8).

10.2 Determine the transfer function matrices of the closed-loop ;systems in Fig. 1 in terms of their open-loop transfer functions.

10.3 Verify (10.2.7) and (10.2.8). *Hint:* Show that $\mathbf{I}-\mathbf{K}'(\mathbf{I}+\mathbf{DK}')^{-1}\mathbf{D} = (\mathbf{I}+\mathbf{K}'\mathbf{D})^{-1}$.

10.4 Show that the state feedback system described by state equation (10.2.5) is completely controllable for any feedback gain matrix $\mathbf{K}$ if and only if the open-loop system described by state equation (10.2.1) is completely controllable. *Hint:* Use elementary column operations to show that the two controllability matrices have the same rank.

10.5 Prove Theorem 10.2.1 for the linear time-varying case.

10.6 Show that the output feedback system described by the state and output equations (10.2.7) and (10.2.8) is completely observable for any feedback gain matrix $\mathbf{K}'$ if and only if the open-loop system described by the state and output equations (10.2.1) and (10.2.2) is completely observable. *Hint:* Let $\mathbf{D}=\mathbf{0}$ with no loss of generality. Use elementary row operations to show that the two observability matrices have the same rank.

10.7 Consider a completely observable *l.t.i.* system $(\mathbf{A},\mathbf{B},\mathbf{C},\mathbf{D})$. Determine a state feedback gain matrix which makes the closed-loop system unobservable. Provide an example.

10.8 Consider a completely controllable *l.t.i.* system $(\mathbf{A},\mathbf{B},\mathbf{C},\mathbf{D})$. Determine an output feedback gain matrix which makes the closed-loop system uncontrollable. Provide an example.

10.9 Show that any uncontrollable system is stabilizable with state feedback if an only if all its uncontrollable modes (poles) are stable.

10.10 Consider the system

$$\dot{\mathbf{x}}(t) = \begin{bmatrix} 0 & 1 \\ 3 & 2 \end{bmatrix} \mathbf{x}(t) + \begin{bmatrix} 0 & 1 \\ -1 & 3 \end{bmatrix} \mathbf{u}(t)$$

Find a state feedback gain matrix which places the closed-loop system poles at $\lambda_1 = -1$ and $\lambda_2 = -3$.

LINEAR CONTROL SYSTEMS

10.11 Repeat Problem 10.10 for $\lambda_{1,2} = -1 \pm j2$.

10.12 Find the state feedback gain matrix **K** which places the closed-loop poles of the discrete-time system

$$x(k+1) = \begin{bmatrix} 0 & 1 & 0 \\ 0 & 0 & 1 \\ 1 & 7 & 1 \\ 2 & 4 & \end{bmatrix} x(k) + \begin{bmatrix} 0 \\ 0 \\ 1 \end{bmatrix} u(k)$$

at $\lambda_1 = \frac{1}{2}$, $\lambda_{2,3} = -\frac{1}{2} \pm j\frac{1}{2}$. Verify your result using "POLPLA".

10.13 Consider a completely controllable, completely observable but unstable continuous-time system (**A,B,C,D**) where

$$A = \begin{bmatrix} 2 & 0 & 0 \\ 0 & -3 & 1 \\ 0 & 0 & -3 \end{bmatrix}, \quad B = \begin{bmatrix} 1 & 0 \\ 0 & 0 \\ 0 & 1 \end{bmatrix}, \quad C = \begin{bmatrix} 1 & 0 & 0 \\ 0 & 1 & 1 \end{bmatrix}, \quad D = \begin{bmatrix} 0 & 0 \\ 0 & 1 \end{bmatrix}$$

Find an output feedback gain matrix which yields two closed-loop poles at -2 and -4. Can all three closed-loop poles be specified?

10.14 Consider a discrete-time system (**A,B,C,D**) where matrices **A, B, C** and **D** are given in Problem 10.13. Is it possible to place the closed-loop poles at $\lambda_1 = -\frac{1}{2}$, $\lambda_{2,3} = \frac{1}{2} \pm j\frac{1}{3}$ using output feedback?

10.15 Use techniques of Section 4.4.2 to draw the detailed simulation diagram of the observer of Fig. 10.6.2. Discuss how this arrangement smooths out noise due to measurement of input and output.

10.16 Design a three-dimensional observer for the system of Problem 10.13 (assuming **D=0**) which places the observer poles at $\lambda_1 = -4$, $\lambda_2 = -5 \pm j2$.

10.17 Consider an unobservable l.t.i. system. Show that it is impossible to estimate all its state variables. Decompose it into observable and unobservable subsystems and show how the observable state variables could be estimated.

10.18 Consider a l.t.i. n-dimensional system (**A,B,C,0**) with m outputs where **C** has full rank. Use state transformation $\bar{x} = Tx + \begin{bmatrix} R \\ C \end{bmatrix} x$ where **R** is any $(n-m) \times n$ matrix such that **T** is nonsingular. Show that in the algebraically equivalent system $(\bar{A},\bar{B},\bar{C},0)$ we have $\bar{C} = [0, I]$.

10.19 Use contradiction to show that complete observability of system (10.7.2) implies complete observability of its subsystem (10.7.4).

10.20 Suppose that an observer of the form (10.7.12) has been designed for the system $(\overline{A},\overline{B},\overline{C},0)$ of Problem 10.18. Find an equivalent observer for the original system $(A,B,C,0)$.

10.21 Design a reduced-dimension observer for the system of Problem 10.13 (assuming $D=0$) which places the observer pole at $\lambda = -4$.

REFERENCES

[10.1] W. L. Brogan, *Modern Control Theory*, Quantum Publishers, 1974

[10.2] R. W. Brocket, "Poles, zeroes and feedback: state space interpretation", *IEEE Trans. Auto. Contr.*, Vol. AC-10, No. 2, pp. 129-135, 1965

[10.3] B.D.O. Anderson and D. G. Luenberger, "Design of multivariable feedback systems", *Proc. IEE*, Vol. 114, pp. 395-399, 1967

[10.4] E. J. Davison, "On pole assignment in multivariable linear systems", *IEEE Trans. Auto. Contr.*, Vol AC-13, No. 6, pp. 747-748, 1968

[10.5] C. T. Chen, *Introduction to Linear Systems Theory*, Holt, Rinehart, and Winston, 1970

[10.6] D. G. Luenberger, "Canonical forms for linear multivariable systems", *IEEE Trans. Auto. Contr.*, Vol. AC-12, pp. 290-293, 1967

[10.7] W. M. Wonham, "On pole assignment in multi-input controllable linear systems," *IEEE Trans. Auto. Contr.*, Vol. AC-12, pp. 660-665, 1967

[10.8] J. J. D'Azzo and C. H. Houpis, *Linear Control Systems: Conventional and Modern*, McGraw-Hill, 1975

[10.9] V. M. Popov, "Hyperstability and optimality of automatic systems with several control functions", *Rev. Roum. Sci.-Electrotechn. et Energ.*, Vol. 9, pp. 629-690, 1964

[10.10] E. J. Davison, "On pole assignment in linear systems with incomplete state feedback", *IEEE Trans. Auto.Contr.*, Vol. AC-15, No. 3, pp. 348-351, 1970.

[10.11] E. J. Davison and S. H. Wang, "On pole assignment in linear multivariable systems using output feedback", IEEE Trans. Auto. Contr., Vol. AC-20, pp. 516-518, 1975

[10.12] B.D.O. Anderson, N. K. Bose and E. I. Jury, "Output feedback stabilization and related problems--solution via decision methods," *IEEE Trans. Aut. Contr.*, Vol. AC-20, pp. 53-66, 1975

[10.13] B.D.O. Anderson and J. B. Moore *Linear Optimal Control*, Prentice-Hall, 1971

[10.14] T. E. Fortmann and K. L. Hitz, *An Introduction to Linear Control Systems*, Marcel Dekker, 1977

[10.15] H. Kwakernaak and R. Sivan, *Linear Optimal Control Theory*, Wiley 1972

[10.16] D. G. Luenberger, "An introduction to observers", *IEEE Trans. Auto. Contr.*, Vol. AC-16, pp. 596-602, 1971

[10.17] D. G. Luenberger, "Observing the state of a linear system," *IEEE Trans. Military Electronics*, Vol. MIL-8, pp. 74-80, 1964

[10.18] D. G. Luenberger, "Observers for multivariable systems", *IEEE Trans. Auto. Contr.*, Vol AC-11, pp. 190-197, 1966

[10.19] J. B. Pearson and C. Y. Ding, "Compensator design for multivariable linear systems", *IEEE Trans. Auto. Contr.*, Vol. AC-14, No. 2, pp. 130-134, 1969

[10.20] B. Gopinath, "On the control of linear multiple input-output systems", *Bell Systems Tech. J.*, Vol 50, pp. 1063-1081, 1971

CHAPTER 11

DESIGN VIA PARAMETER OPTIMIZATION

11.1 INTRODUCTION

Every system, whether it is to be controlled or not, can be represented by a mathematical model which depends, in one way or other, on one or a set of parameters. For example in a simple servomechanism control system, the amplifier gain, the tachogenerator (in general transducers) constant, inertia, friction, time constants, etc. constitute the system's parameters. In an economic system, some of the model's parameters are interest rate and consumer price index. In almost all practical systems the parameters are subject to variations which may adversely affect the system's behavior and properties.

One important consequence of the dependency of a system's model on parameters is that the control system engineer would be faced with a great number of alternatives for the system design. The engineer often has to select the "*best*" of the alternatives while considering parameters variations. The design which is best *w.r.t.* certain value of parameters constitutes a characteristic which is sometimes termed the "*optimum*" [11.1].

The desire for an optimum design is an old idea. It was discussed in Chapter 9 that one can resort to a set of "rules of thumb" to design a compensating network (control law) which would meet a set of so-called "performance specifications". However, what is new in an optimum control system design is that certain mathematical tools are at the disposal of the designer which would help calculate the optimum parameters based on a given "*performance index*" or a "*cost function*". In this chapter we consider the design of control systems through parameter optimization.

11.2 PERFORMANCE INDICES

A *performance index* of a system is a quantitative measure of its performance when it is subjected to parameter variations. The exact choice of the performance index depends on the objectives of the design. One common choice has been to relate the index to system's time response. In many SISO linear control systems it is often of interest to regulate the output, i.e. so that the output response follows a constant reference input. Let the output response and the reference input be denoted by $c(t)$ and $r(t)$, respectively. A measure of regulation is the error signal $e(t) = r(t) - c(t)$.

508   LINEAR CONTROL SYSTEMS

A suitable performance index is the cumulative effect of the error function $e(t)$, often represented by the integral of the square of the error signal (ISE), defined by

$$I_1 = \int_0^\infty e^2(t)\,dt \qquad (1)$$

For a second-order SISO linear system with a closed-loop transfer function

$$C(s)/R(s) = \omega_n^2/(s^2 + 2\xi\omega_n s + \omega_n^2) \qquad (2)$$

with underdamped $(0<\xi<1)$ characteristic, ISE is illustrated in Fig. 1.

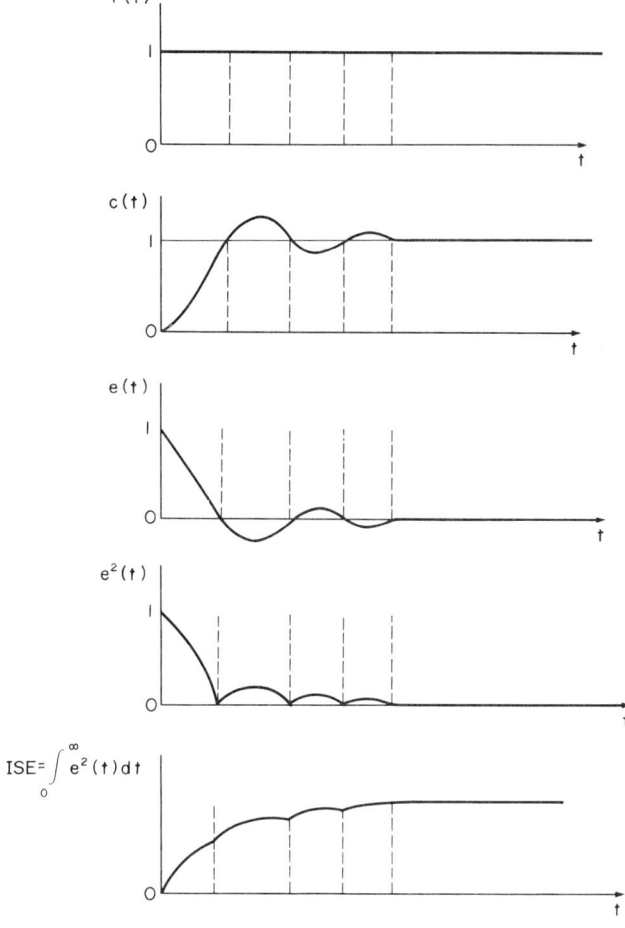

Fig. 11.2.1. A pictorial representation of the ISE for a second order system

As will be seen shortly an important property of the ISE is that it can be represented analytically as a function of plant parameters. There are a number of performance indices other than the ISE. One is the integral of absolute value of error (IAE),

$$I_2 = \int_0^\infty |e(t)| dt \tag{3}$$

Utilizing the magnitude of the error, both negative and positive variations would cause the index to increase. Another performance index which is sometimes used to reduce large initial errors or to put an emphasis on error occurring later in the response is the integral of time multiplied by the absolute value of the error signal (ITAE), i.e.

$$I_3 = \int_0^\infty t|e(t)| dt \tag{4}$$

Other performance indices of general interest are:

(i) integral of time multiplied by squared error (ITSE),

$$I_4 = \int_0^\infty te^2(t) dt \tag{5}$$

(ii) integral of squared time multiplied by squared error (ISTSE),

$$I_5 = \int_0^\infty t^2 e^2(t) dt , \tag{6}$$

and

(iii) integral of squared time multiplied absolute value of error (ISTAE),

$$I_6 = \int_0^\infty t^2 |e(t)| dt \tag{7}$$

In a more general sense, a SISO system can have a performance index of the form,

$$I = \int_0^\infty \sum_{i=1}^N f_i^2(t) dt \tag{8}$$

where $f_i(t)$, $i = 1,2,...,N$ represent various system signals of importance such as error, output, etc. This definition will be used in Section 11.3 to investigate the use of Parseval's Theorem for parameter optimization of both continuous-time and discrete-time systems. For discrete-time systems the general performance index is expressed by

510   LINEAR CONTROL SYSTEMS

$$I = \sum_{k=0}^{\infty} \sum_{i=1}^{N} f_i^2(kT) \tag{9}$$

where $f_i(kT)$, $i = 1,2,...N$ are functions of importance in a discrete-time system. The next section treats the optimization of a SISO system by Parseval's Theorem.

11.3 OPTIMIZATION BY PARSEVAL'S THEOREM

In this section a mathematical result, called Parseval's Theorem [11.1, 11.2] will be used to find the system's performance index in terms of its design (plant) parameters.

11.3.1 Continuous-Time Systems

Let us reconsider the general performance index (11.2.8),

$$I = \int_0^{\infty} \sum_{i=1}^{N} f_i^2(t) \, dt \tag{1}$$

and let the contribution of *jth* function in I be represented by

$$I_j = \int_0^{\infty} f_j^2(t) \, dt \tag{2}$$

If $f_j(t)$ is Laplace-transformable, one can rewrite (2) as

$$I_j = \int_0^{\infty} f_j(t) . f_j(t) \, dt = \int_0^{\infty} f_j(t) [\frac{1}{2\pi j} \int_{c-j\infty}^{c+j\infty} F_j(s) e^{st} \, ds] \, dt \tag{3}$$

Let us further assume that it is possible to interchange the order of the two integrations in (3), i.e.

$$I_j = \frac{1}{2\pi j} \int_{c-j\infty}^{c+j\infty} F_j(s) [\int_0^{\infty} f_j(t) e^{st} \, dt] \, ds = \frac{1}{2\pi j} \int_{c-j\infty}^{c+j\infty} F_j(s) F_j(-s) \, ds \tag{4}$$

Thus,

$$I_j \triangleq \int_0^{\infty} f_j^2(t) \, dt = \frac{1}{2\pi j} \int_{c-j\infty}^{c+j\infty} F_j(s) F_j(-s) \, ds \tag{5}$$

The above relation is generally known as the *Parseval's Theorem*. Calculations of the exact value of I_j in (5) is not an easy matter. Some authors [11.1, 11.2] have provided a table for the determination of the right-hand side of (5). Let the Laplace transform $F_j(s)$ of a linear lumped time-invariant system having inputs with rational transform be expressed by

$$F_j(s) = \frac{A(s)}{B(s)} = \frac{a_{n-1}s^{n-1}+\cdots+a_1 s+a_0}{b_n s^n+\ldots+b_1 s+b_0} \tag{6}$$

where n is the order of the system. The expressions I_j as functions of coefficients a_i, $i = 0,\ldots, n-1$ and $b_j, j = 0,1,\ldots,n$ for up to $n=5$ are given in Table 1. In Reference [11.2], expressions for up to $n=10$ are available.

Table 11.3.1 Evaluation of Parseval's Theorem for continuous-time systems up to 5th order

$$I_n = \frac{1}{2\pi j} \int_{-j\infty}^{j\infty} \{(A(s)A(-s))/(B(s)B(-s))\} ds$$
$$A(s) = a_{n-1}s^{n-1} +\ldots+ a_0$$
$$B(s) = b_n s^n + \cdots + b_0$$

$$I_1 = a_0^2/2b_0 b_1$$
$$I_2 = (a_1^2 b_0 + a_0 2 b_2)/2b_0 b_1 b_2$$
$$I_3 = \{a_2^2 b_0 b_1 + (a_1^2 - 2a_0 a_2)b_0 b_3 + a_0^2 b_2 b_3\}/\{2b_0 b_3(-b_0 b_3 + b_1 b_2)\}$$
$$I_4 = \{[a_3^2(-b_0^2 b_3 + b_0 b_1 b_2) + (a_2^2 - 2a_1 a_3)b_0 b_1 b_4] +$$
$$[(a_1^2 - 2a_0 a_2)b_0 b_3 b_4 + a_0^2(-b_1 b_4^2 + b_2 b_3 b_4)]\}/\Delta_4$$

where

$$\Delta_4 = 2b_0 b_4(-b_0 b_3^2 - b_1^2 b_4 + b_1 b_2 b_3)$$
$$I_5 = [a_4^2 m_0 + (a_3^2 - 2a_2 a_4)m_1 + (a_2^2 - 2a_1 a_3 + 2a_0 a_4)m_2$$
$$+ (a_1^2 - 2a_0 a_2)m_3 + a_0^2 m_4]/2\Delta_5$$

where

$$m_0 = (b_3 m_1 - b_1 m_2)/b_5, \quad m_1 = -b_0 b_3 + b_1 b_2$$
$$m_2 = -b_0 b_5 + b_1 b_4, \quad m_3 = (b_2 m_2 - b_4 m_1)/b_0$$
$$m_4 = (b_2 m_3 - b_4 m_2)/b_0 \text{ and } \Delta_5 = b_0(b_1 m_4 - b_3 m_3 + b_5 m_2)$$

1. Example. Consider a single-axis attitude-control system [11.1] shown in Fig. 1. It is desired to evaluate an expression for its ISE performance index, Equation (11.2.1), in terms of design parameters K_1 and K_2 for a unit step reference input.

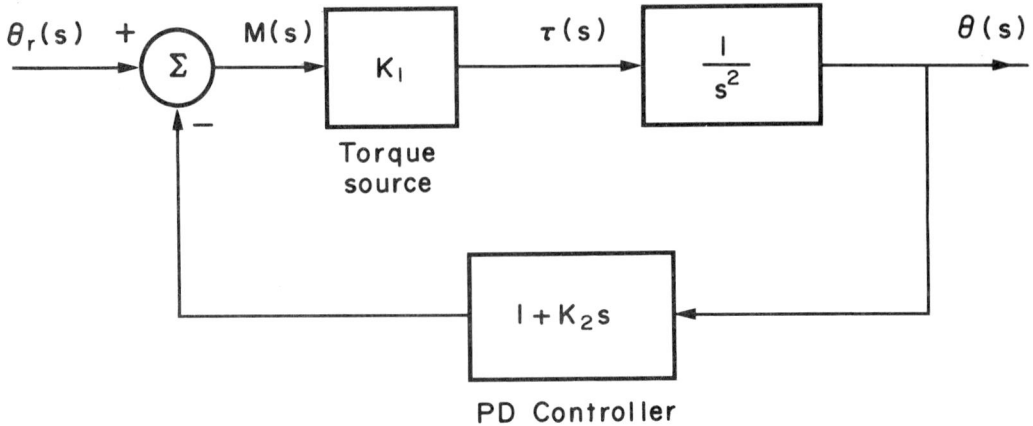

Fig. 11.3.1. A single-axis attitude-control system of Example 11.3.1

The first step is to find and expression for $E(s) = \Theta(s) - \Theta_r(s)$. From Fig. 1 it is clear that

$$E(s) = \Theta_r(s) - \Theta(s) = \Theta_r(s) - \frac{K_1}{s^2} M(s) = \Theta_r(s) - \frac{K_1}{s^2}[\Theta_r(s) - (1+K_2 s)\Theta(s)] \tag{7}$$

Using $\Theta_r(s) = 1/s$, replacing $\Theta(s)$ by $\Theta_r(s) - E(s)$ and simplifying (7) leads to

$$E(s) = (s+K_1 K_2)/(s^2 + K_1 K_2 s + K_1). \tag{8}$$

Now by noting the development of Parseval's Theorem, it is clear that

$$ISE = \int_0^\infty e^2(t)dt = \frac{1}{2\pi j}\int_{c-j\infty}^{c+j\infty} E(s)E(-s)ds \tag{9}$$

Table 1 can be used for $n=2$. Noting (8) and Table 1 it is clear that

$$a_o = K_1 K_2, \; a_1 = 1, \; b_o = K_1, \; b_1 = K_1 K_2, \; b_2 = 1. \tag{10}$$

Hence, $ISE = I_2$ can be easily seen to be

$$I_2(K_1, K_2) = \int_0^\infty e^2(t)dt = (K_1 + K_1^2 K_2^2)/2K_1^2 K_2 = 1/2K_1 K_2 + K_2/2 \tag{11}$$

In general one can then take partial derivatives of $I_2(.)$ w.r.t. K_1 and K_2 and set them equal to zero as the necessary conditions for minimizing ISE of the system. Moreover, one may use Table 1 for as many quadratic terms as there are in the performance index. These points will be demonstrated shortly.

11.3.2 Discrete-Time Systems

Let us rewrite the general performance index (11.2.9) of discrete-time systems

$$I = \sum_{k=0}^{\infty} \sum_{i=1}^{N} f_i^2(kT) \tag{12}$$

and once again consider its *j*th quadratic function. This component can be similarly expressed in terms of the z-transform, i.e.

$$I_j = \sum_{k=0}^{\infty} f_j(kT) [\frac{1}{2\pi j} \int_{\Gamma} F(z) z^{k-1} dz] \tag{13}$$

$$= \frac{1}{2\pi j} \int_{\Gamma} F(z) z^{-1} [\sum_{k=0}^{\infty} f_j(kT) z^k] dz \tag{14}$$

which can be rewritten as

$$I_j = \frac{1}{2\pi j} \int_{\Gamma} F(z) F(z^{-1}) z^{-1} dz \tag{15}$$

The above expression is the discrete-time counterpart of the Parseval's Theorem. Here again, the values of I_j in (15) in terms of $F(z)$ has been obtained and documented in tabular form. Jury [11.3] has presented the value of I_j in (15) for many orders of a linear time-invariant discrete-time systems. Table 2 shows two entries of such performance indices.

2. Example. Consider a first-order SISO discrete-time system with the following state equation:

$$x(k+1) = (1-K_1)x(k) + K_2 r(k) \tag{16}$$

where K_1 and K_2 are plant parameters, $r(k)$ is a unit step reference input and sampling period $T=1$. It is desired to find an expression for

$$I = \sum_{k=0}^{\infty} e^2(kT) = \sum_{k=0}^{\infty} [x(k)-r(k)]^2 \tag{17}$$

LINEAR CONTROL SYSTEMS

Table 11.3.2. Evaluation of Parseval's Theorem for discrete-time systems up to 2nd order

$$I_n = \frac{1}{2\pi j} \oint_\Gamma F(z)F(z^{-1})z^{-1}dz$$

1. $F(z) = (a_o z + a_1)/(b_o z + b_1)$
   $I_1 = [(a_o^2 + a_1^2)b_o - 2a_o a_1 b_1]/[b_o(b_o^2 - b_1^2)]$

2. $F(z) = (a_o z^2 + a_1 z + a_2)/(b_o z^2 + b_1 z + b_2)$
   $I_2 = [A_o b_o e_1 - A_1 b_o b_1 + A_2(b_1^2 - b_2 e_1)]/$
   $\{b_o[(b_o^2 - b_2^2)e_1 - (b_o b_1 - b_1 b_2)b_1]\}$

   where
   $A_o = a_o^2 + a_1^2 + a_2^2, \quad A_1 = 2(a_o a_1 + a_1 a_2)$
   $A_2 = 2a_o a_2 \text{ and } e_1 = b_o + b_2$

Following the development of the discrete-time version of the Parseval's Theorem and Table 2, it is clear that we need to evaluate $E(z)$. Using (16) to find the z-transfer of $x(k)$ we have

$$X(z) = \frac{K_2 z}{(z-1)(z-1+K_1)} \qquad (18)$$

and noting that $R(z) = z/(z-1)$ one obtains

$$E(z) = \frac{-z^2 + (K_2 - K_1 + 1)z}{z^2 + (K_1 - 2)z - K_1 + 1} \qquad (19)$$

or

$$a_o = -1, \; a_1 = 1 + K_2 - K_1, \; a_2 = 0, \qquad (20)$$

$$b_0 = 1, \; b_1 = K_1 - 2, \; b_2 = -K_1 + 1. \qquad (21)$$

Using the second entry of Table 2, one can find

$$A_o = 1+(1+K_2-K_1)^2, \quad A_1 = -2(1+K_2-K_1)$$

$$A_2 = 0, \quad e_1 = 1-K_1 \qquad (22)$$

Then after using I_2 in Table 2 and some simplifications, I in (17) will become

$$I(K_1, K_2) = \frac{K_1^2+K_2^2+2K_2^2-2K_1K_2}{K_1(K_2-K_1)} \qquad (23)$$

11.4 FUNCTIONAL MINIMIZATION TECHNIQUES

Thus far we have defined the various performance indices for SISO systems and one possible way of expressing the index as a function of plant parameters. In this section three functional minimization schemes of multivariable functions are presented. These techniques can be used to optimally design a controller for a SISO system.

11.4.1 Direct Search Method

Here the direct search method of Rosenbrock [11.4] will be briefly discussed and a BASIC program will be given for it. The procedure assumes a unimodal function i.e. one with only one extreme point in the domain of search, thus if the function is known to have more than one minimum, several initial values of the independent variables must be tried. Let the objective function, to be minimized (or maximized), be

$$I(\mathbf{x}) = f(x_1, x_2, \ldots, x_n) \qquad (1)$$

where $\mathbf{x} = (x_1 x_2 \cdots x_n)'$ is the vector of unknown independent variables. The essence of the direct search method is simply to find a minimum of $I(\mathbf{x})$ by changing $x_1, x_2, \ldots, x_n$ in turn, reducing $I(\mathbf{x})$ as far as possible with each variable and then passing on to the next. This point is illustrated in Fig. 1 for a two-parameter function. This method is particularly effective when the contours of $I(\mathbf{x})$ are nearly circular [11.4]. However, when there is interactions among terms, i.e. $x_i x_j$ terms for $i, j=1,2,\ldots,N$ or for the cases when the contours would display a narrow banana-type shape it may not be an effective method. However, this is a typical problem in many gradient-based schemes such as conjugate gradient.

The development of the direct search method depends on two basic steps: (i) Step Size Determination and (ii) Direction of Search.

1. Step Size Determination. The basic functional scheme to choose the value of the step size is to try an arbitrary value Δx. If this step size "succeeds", i.e. if the new and old values of $I(x)$ have the relation $I(x_{old}) \leq I(x_{new})$, Δx is multiplied by a factor $a \geq 1$. If it "fails", then it is decreased by a factor $0 \leq b \leq 1$. Each such change of step size is called a "trial".

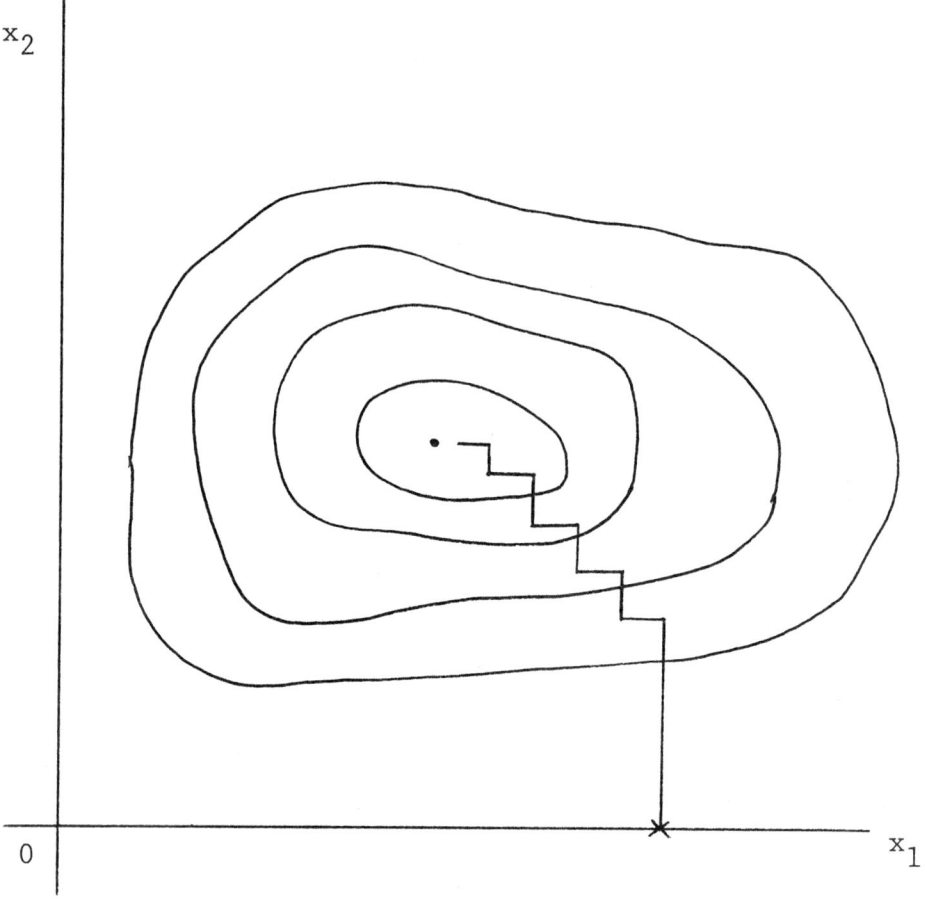

Fig. 11.4.1. An illustration of direct search method for a two-parameter functional minimization problem

2. Direction of Search. The direction of search is chosen by the following scheme. Let d_1 be the algebraic sum of all successful steps Δx_1 in the direction D_1, etc. Then let

$$A_1 = d_1 D_1^0 + d_2 D_2^0 + ... + d_n D_n^0$$

$$A_2 = d_2 D_2^0 + ... + d_n D_n^0 \quad (2)$$

$$\cdot \; \cdot \; \cdot$$

$$A_n = d_n D_n^0$$

Therefore A_1 is the vector joining the initial and final points obtained by use of the vectors $D_1^0, \ldots, D_n^0$; A_2 is the sum of all the advances made in the directions other than the first one, etc. A set of orthogonal unit vectors $D_1^1, D_2^1, \ldots, D_n^1$ are determined in the following way:

$$F_1 = A_1$$

$$D_1^1 = F_1' \mid F_1 \mid$$

$$F_2 = A_2 - A_1 D_1^1 D_1^1$$

$$D_2^1 = F_2 / \mid F_2 \mid$$

$$\vdots$$

$$F_n = A_n - \sum_{j=1}^{n-1} A_n D_j^1 D_j^1$$

$$D_n^1 = F_n' \mid F_n \mid$$

(3)

The above formulation of the direction of search can be briefly described as follows. The process of search for a "success" is followed until at least one trial is successful in each direction, and one has failed. It is noted that a given trial should finally succeed because Δx becomes so small after repeated failures that it would cause no change in I.

The insertion of constraints in the process of minimization, i.e. $\underline{x} \leqslant x \leqslant \bar{x}$ can also be made in this scheme. In Section 11.4.3 one such scheme known as "hill climbing" will be briefly described.

Finally, it must also be noted that the direct search methods is not limited to the work of Rosenbrock [11.4]. In a survey, Fletcher [11.5] notes that Hooke and Jeeves [11.6] have also noticed that the line $x_{n+1} - x_1$ joining the first and last points of a cycle of alternating directions proves to be a favorable search direction. Swann [11.7] has developed Rosenbrock's method to include provisions for linear search. Nedler and Mead [11.8] have proposed a method in which the function is first evaluated at a basis set or "simplex" of $n+1$ points and the set is systematically changed - adding some points and dropping others - until the region of the minimum is reached. Other efforts are in the explicit construction of conjugate gradient directions by Smith [11.9] and Powell [11.10]. The following algorithm describes the direct search method of Rosenbrock [11.11].

3. Algorithm

Step 1. Start with an initial point x_{oi}, $i=1,\ldots,n$, initial step sizes Δx_i, $i=1,\ldots,n$ and evaluate the objective function (1).

Step 2. Increment the first variable x_1 by a value Δx_1 parallel to its axis and evaluate the objective function. If the new value of $I(x)$ is decreased, the step is termed a "success" and the step Δx_1 is increased by a factor a, $a \geqslant 1.0$. If $I(x)$ is increased the step is called a "failure" and Δx_1 is decreased by a factor b, $0 \leqslant b \leqslant 1.0$ and the direction of movement is reversed.

Step 3. The ith variable, $i=2,...n_i$ is stepped a distance Δx_1 parallel to the ith axis. The search for a decrease in $I(x)$ (success) or an increase in $I(x)$ (failure) has been achieved in all n directions.

Step 4. The axes arenow rotated by the following relations:

$$D_{i,j}^{(r+1)} = \frac{F_{i,j}^{(r)}}{[\sum_{k=1}^{n}(F_{k,j}^{(r)})^2]^{1/2}} \tag{4}$$

where

$$F_{i,1}^{(r)} = A_{i,1}^{(r)}$$

$$F_{i,j}^{(r)} = F_{i,j}^{(r)} - \sum_{k=1}^{j-1}[\sum_{m=1}^{j} D_{m,k}^{(r+1)} A_{m,j}^{(r)} D_{i,k}^{(r+1)}],$$

$$j = 2,3,...,n \tag{5}$$

$$A_{i,j}^{(r)} = \sum_{k=j}^{n} d_k^{(r)} D_{i,k}^{(r)}$$

Each rotation of the axes is called a *stage*. In above development, i = variable index = 1,2,...,n; j = direction index = 1,2,...,n, r = stage index, d_i = sum of distances moved in direction i since last axes rotation and $D_{i,j}$ is the normalized direction vector component.

Step 5. In each of the *n* directions search is made using the new coordinate axes:

$$\text{new } x_i^{(r)} = \text{old } x_i^{(r)} + \Delta x_1 \, x_j^{(r)} D_{i,j}^{(r)} \tag{6}$$

Step 6. The procedure terminates when the difference between two consecutive values of $I(x)$ is within a specified tolerance value ϵ.

In every optimization problem the algorithm moves the solution from one value to another on a series of equal function $I(x)$ values known as *contours*. When there are two variables in vector x, these contours constitute closed curves very conveniently. For the purpose of studying such contours a BASIC program was written to plot them on the screen and/or X-Y plotter. The program "CONTUR" and an example follows.

4. "CONTUR"

```
10   ! PROGRAM NAME : <<CONTUR>> PROG
20   PRINT " 'CONTUR' PLOTS UP TO 5 CLOSED CONTOURS FOR A FUNCTION F(x,y)"
30   PRINT " ON THE X-Y PLANE FOR UP TO 5 VALUES OF F(x,y)"
40   INPUT "Would you like INSTRUCTIONS (Y/N)?",C$
50   IF C$="Y" THEN 80
60   IF C$="N" THEN 310
70   GOTO 40
80   PRINT "THE VALUE OF THE FUNCTION F CAN BE ENTERED INTO THE PROGRAM, i.e"
90   PRINT " CONTOURS FOR F(X,Y) =1,2,66,900, etc."
100  PRINT "A MAXIMUM OF 5 CONTOURS MAY BE PLOTTED AT ONCE TO PREVENT"
110  PRINT "CUTTERING OF THE DISPLAY. THE FUNCTION F(x,y) IS DEFINED"
120  PRINT "SUBROUTINE << SUB Func >> WHICH BEGINS AT LINE 3000. A TYPICAL"
130  PRINT " EXAMPLE IS ALREADY INCLUDED THERE. THE ONLY PARAMETERS"
140  PRINT "NECESSARY ARE THE RANGE OF X AND Y, THE NUMBER OF ITERATIONS"
150  PRINT "TO BE USED OVER THIS RANGE, THE NUMBER OF CONTOURS DESIRED, AND"
160  PRINT "THE VALUE OF EACH CONTOUR TO BE PLOTTED. THE NUMBER OF ITERATIONS"
170  PRINT "DETERMINES THE LENGTH OF TIME FOR EACH PLOT. THE LENGTH OF TIME "
180  PRINT "REQUIRED INCREASES AS THE SQUARE OF THE NUMBER OF ITERATIONS FOR"
190  PRINT "i.e. 100 ITERATIONS TAKES 4 TIMES LONGER THAN 50 ITERATIONS TO "
200  PRINT "DRAW THE SAME CONTOUR."
210  DISP "PRESS `CONT' TO CONTINUE THE INSTRUCTIONS"
220  PAUSE
230  PRINT "FOR THE ATTACHED EXAMPLE USE THE FOLLOWING PARAMETERS-",LIN(1)
240  PRINT "Xmin = 0"
250  PRINT "Xmax = 5"
260  PRINT "Ymin = 0"
270  PRINT "Ymax = 10"
280  PRINT "NUMBER OF ITERATIONS = 50"
290  PRINT "CONTOUR VALUES = 3, 4"
300  PRINT LIN(1),"TO FAMILIARIZE YOURSELF WITH THE PROGRAM."
310  DIM C(5),M(2),P(150,2)
320  LINPUT "HAS `Func' ALREADY BEEN LINKED (Y/N)?",C$
330  IF C$="Y" THEN 370
340  IF C$="N" THEN 360
350  GOTO 320
360  LINK "Func",1580,430
370  INPUT "CRT OR XY-PLOTTER? GIVE a C or an X)",P$
380  IF P$="C" THEN 430
390  IF P$="X" THEN 410
400  GOTO 370
410  PLOTTER IS "9872A"
420  GOTO 440
430  PLOTTER IS "GRAPHICS"
440  INPUT "WHAT IS THE DESIRED RANGE OF X (min,max)? ",Xmin,Xmax
450  IF Xmin>=Xmax THEN 440
460  INPUT "WHAT IS THE DESIRED RANGE OF Y (min,max)?",Ymin,Ymax
470  IF Ymin>=Ymax THEN 460
480  INPUT "NUMBER OF X AND Y ITERATIONS (<=150)",Iter
490  INPUT "HOW MANY CONTOURS WOULD YOU LIKE (<=5)?",C
500  IF (C>5) OR (C<=0) THEN 490
510  FOR I=1 TO C
520  DISP "WHAT IS THE VALUE OF CONTOUR #";I;
530  INPUT C(I)
540  NEXT I
550  PRINT "THE VALUE OF Xmin AND Xmax YOU HAVE CHOSEN =";Xmin;",";Xmax
560  PRINT LIN(1),"THE VALUE OF Ymin AND Ymax YOU HAVE CHOSEN =";Ymin;",";Ymax
570  PRINT LIN(1),"THE NUMBER OF ITERATIONS IN X AND Y =";Iter
580  FOR I=1 TO C
590  PRINT LIN(1),"THE VALUE OF CONTOUR #";I;"=";C(I)
600  NEXT I
610  LINPUT "DO YOU WISH TO MAKE ANY CHANGES (Y/N)?",C$
620  IF C$="Y" THEN 440
630  IF C$="N" THEN 650
640  GOTO 610
650  REDIM P(Iter,2)
```

```
660    FOR I=1 TO C
670    MAT P=(0)
680    FOR M=1 TO Iter
690    Count=1
700    Dz=0
710    FOR N=1 TO Iter
720    X=(Xmax-Xmin)/Iter*M+Xmin
730    Y=(Ymax-Ymin)/Iter*N+Ymin
740    CALL Func(X,Y,Val)
750    Z=SGN(Val-C(I))
760    IF N=1 THEN 820
770    IF Z=Z1 THEN 820
780    IF (Z1=0) AND (Count=2) THEN 820
790    P(M,Count)=N
800    Count=Count+1
810    IF Count=3 THEN 840
820    Z1=Z
830    NEXT N
840    NEXT M
850    GOSUB Draw
860    NEXT I
870    LORG 5
880    CSIZE 2.5
890    LINE TYPE 1
900    AXES Iter/10,Iter/10,0,0
910    LINE TYPE 1
920    FOR I=0 TO Iter STEP Iter/10
930    MOVE I,-.06*Iter
940    LABEL USING "K";I*10/Iter*(Xmax-Xmin)/10+Xmin
950    MOVE -.06*Iter,I
960    LABEL USING "K";I*10/Iter*(Ymax-Ymin)/10+Ymin
970    NEXT I
980    EXIT GRAPHICS
990    LINPUT "WOULD YOU LIKE A COPY OF THE CONTOUR (Y/N)?",C$
1000   IF C$="Y" THEN 1030
1010   IF C$="N" THEN 1040
1020   GOTO 990
1030   DUMP GRAPHICS
1040   LINPUT "WOULD YOU LIKE ANOTHER SET OF CONTOURS (Y/N)?",C$
1050   IF C$="Y" THEN 430
1060   IF C$="N" THEN 1080
1070   GOTO 1040
1080   END
1090 Draw: GRAPHICS
1100   MAT M=(0)
1110   FOR Count=1 TO 2
1120   GRAPHICS
1130   SCALE -.15*Iter,1.3*Iter,-.1*Iter,1.1*Iter
1140   J=0
1150   FOR M=1 TO Iter
1160   IF J=1 THEN 1220
1170   IF P(M,Count)=0 THEN 1270
1180   M(Count)=M
1190   MOVE M,P(M,Count)
1200   PEN 1
1210   J=1
1220   IF P(M,Count)=0 THEN 1250
1230   DRAW M,P(M,Count)
1240   GOTO 1270
1250   PENUP
1260   MOVE M,(Count-1)*Iter
1270   NEXT M
1280   PENUP
1290   NEXT Count
1300   IF (M(1)=0) OR (M(2)=0) THEN 1340
1310   IF (M(1)+Iter/10<M(2)) OR (M(1)-Iter/10>M(2)) THEN 1340
1320   MOVE M(1),P(M(1),1)
```

```
1330    DRAW M(2),P(M(2),2)
1340    PENUP
1350    MAT M=(0)
1360    FOR Count=1 TO 2
1370    FOR M=Iter TO 0 STEP -1
1380    IF P(M,Count)=0 THEN 1410
1390    M(Count)=M
1400    GOTO 1420
1410    NEXT M
1420    NEXT Count
1430    IF (M(1)=0) OR (M(2)=0) THEN 1480
1440    IF (M(1)+Iter/10<M(2)) OR (M(1)-Iter/10>M(2)) THEN 1480
1450    MOVE M(1),P(M(1),1)
1460    DRAW M(2),P(M(2),2)
1470    PENUP
1480    LINE TYPE 3
1490    MOVE Iter+.03*Iter,.5*Iter-.1*Iter*(I-1)
1500    DRAW M(1),P(M(1),1)
1510    PENUP
1520    LORG 1
1530    MOVE Iter+.05*Iter,.5*Iter-.1*Iter*(I-1)
1540    LINE TYPE 1
1550    CSIZE 2.5
1560    LABEL "F(X,Y)= ";C(I)
1570    RETURN
3000 SUB Func(X,Y,Val)
3010    Val=-3803.84-138.08*X-232.92*Y+123.08*X^2+203.64*Y^2+182.25*X*Y
3020 SUBEND
```

5. Example. Consider a two-variable function,

$$I(x) = F(x_1,x_2) = \frac{1}{2x_1x_2} + \frac{x_2}{2} + \frac{2x_1}{x_2}. \tag{7}$$

Program "CONTUR" was used to plot the contours of (7) corresponding to $I(x) = 4.25, 3.75, 3.0$ and 2.25. The X-Y plotter plot of these four contours are shown in Fig. 1. The necessary conditions of minimization of (7) are

$$\partial F/\partial x_1 = 0 \quad 4x_1^2 - 1 = 0 \tag{8a}$$

$$\partial F/\partial x_2 = 0 \quad x_1 x_2^2 - 4x_1^2 - 1 = 0 \tag{8b}$$

resulting in $x_1 = 1/2$ and $x_2 = 2$. From Fig. 2 it is clear that this point where $I(x) = 2.0$ does correspond to the minimum point. The following example illustrates the use of programs "FMROS" and "CONTUR".

6. Example. Let us reconsider the system of Example 11.3.1 shown in Fig. 11.3.1. The Integral Square Error of this system is given by Equation (11.3.11). In an attempt to find the minimum value of $I_2(K_1,K_2)$ we can take its first partial derivatives with respect to K_1 and K_2 and set them equate to zero, i.e.

$$\partial I_2(K_1,K_2)/\partial K_1 = -1/2K_1^2K_2 = 0 \tag{9a}$$

$$\partial I_2(K_1,K_2)/\partial K_2 = -1/2K_1K_2^2 + 1/2 = 0 \tag{9b}$$

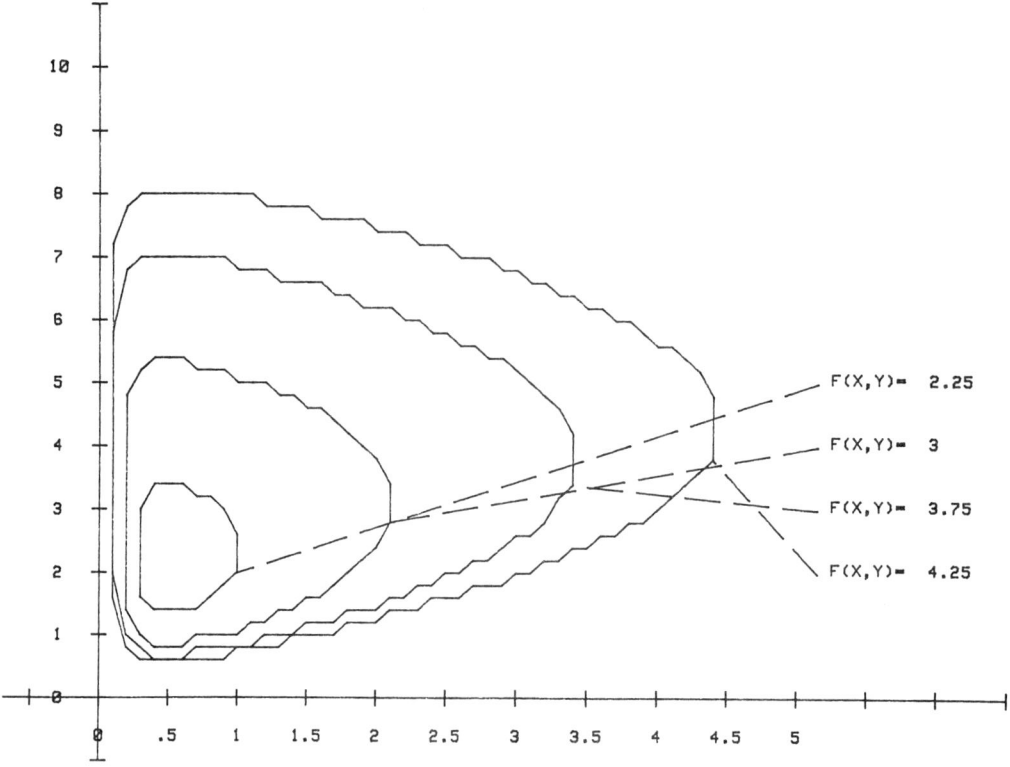

Fig. 11.4.2. Contours for Example 11.4.5

However, the above two equations can not be solved simultaneously for finite value of K_1 and K_2. In fact, if $K_1 \to \infty$ and $K_2 \to 0$ such that $K_1 K_2 = 1$ then the above equations can be solved. Under such conditions, the required torque τ in the system of Fig. 11.3.1 would become infinitely large, hence resulting in an unfeasible design. In order to overcome this undesirable increase in torque, one can redefine the ISE performance index (11.3.9) as

$$I(K_1, K_2, \rho) = \int_0^\infty (e^2(t) + \rho \tau^2(t)) dt \qquad (10)$$

to limit excess variations of torque $\tau(t)$. The parameter ρ is a weighting factor indicating the relative importance between error $e(t)$ and torque $\tau(t)$. Here, two applications of Parseval's Theorem and Table 11.3.1 will be used. The torque $\tau(s)$ can be easily calculated:

$$\tau(s) = K_1 s / (s^2 + K_1 K_2 s + K_1) \qquad (11)$$

In order to apply Parseval's Theorem, one can consider $\sqrt{\rho} \tau(t)$ as a second function $f_2(t)$ of Equation (11.3.1) and proceed to use $a_o = 0, a_1 = \sqrt{\rho} K_1, b_o = K_1, b_1 = K_1 K_2$ and $b_2 = 1$. Thus the second part of the index in (10) would become

$$\int_0^\infty \rho \tau^2(t)\,dt = \rho K_1/2K_2. \tag{12}$$

This value, along with that of Equation (11.3.11) would give the desired expression for $I(K_1, K_2, \rho)$ in (10) i.e.

$$I(K_1, K_2, \rho) = 1/2K_1K_2 + K_2/2 + \rho K_1/2K_2 \tag{13}$$

It is noted that when $\rho = 4$, the index in (13) corresponds to the function $I(x)$ in (7) whose contours are shown in Fig. 1 using program "CONTUR". In fact for any value of ρ the values of K_1 and K_2 corresponding to the minimum value of $I(K_1, K_2, \rho)$ are easily seen to be,

$$K_1 = 1/\sqrt{\rho}, \quad K_2 = \sqrt{2\sqrt{\rho}} \tag{14}$$

with

$$I(K_1, K_2, \rho) = 2\sqrt{\rho} / \sqrt{2\rho \sqrt{\rho}} \tag{15}$$

The following BASIC program, called "FMROS" implements Rosenbrock's [11.4] algorithm. Next, the performance index (13) for $\rho = 4$ will be minimized using program "FMROS". A second example taken from reference [11.11] will follow. The resulting computer outputs are shown then.

7. "FMROS"

```
10    ! PROGRAM NAME : <<FMROS>> PROG
20    PRINT "PROGRAM 'FMROS' FIND THE MINIMUM OF AN UNCONSTRAINED MULTI-"
30    PRINT "VARIABLE FUNCTION VIA THE DIRECT SEARCH METHOD OF ROSENBROCK"
40    OPTION BASE 1
50    DIM Eps(8),Ake(8),V(8,8)
60    COM D(8),Bl(8,8),Blen(8),Aj(8),E(8),Al(8,8),Afk(8)
70    INPUT "Have SUBs  Rosenb and Object BEEN LINKED?(Y/N)",N$
80    IF (N$="Y") OR (N$="y") THEN 130
90    IF (N$="N") OR (N$="n") THEN 110
100   GOTO 70
110   LINK "Rosenb",330,120
120   LINK "Object",1520,130
130   INPUT " No. of Independent Varibles Km",Km
140   REDIM D(Km),Bl(Km,Km),Blen(Km),Aj(Km),E(Km),Al(Km,Km),Afk(Km)
150   INPUT " Maximum No. of Objective Function Evaluations Maxk",Maxk
160   INPUT " Maximum No. of Axes Rotations Mkat",Mkat
170   INPUT " Maximum No. of Successive Failures in Axes Rotations Mcyc",Mcyc
180   INPUT " Nstep:1 use INITIAL step,:2 use previous stage's FINAL step",Nstep
190   PRINT " Km=";Km;"Maxk=";Maxk;"Mkat=";Mkat;"Nstep=";Nstep
200   INPUT "Objective Function Stopping Criterion Epsy",Epsy
210   INPUT "Step Size Scaling Factor for INCREASE Alpha",Alpha
220   INPUT "Step Size Scaling Factor for DECREASE Beta",Beta
230   PRINT "Input Vector of Initial STEP SIZE Eps(*) i.e. a ";Km;"x1 Vector"
240   REDIM Ake(Km),Eps(Km)
250   DISP "Eps";
260   MAT INPUT Eps
270   MAT PRINT Eps;
280   PRINT "Input Vector of Initial VARIABLE  Ake(*), i.e. a ";Km;"x1 Vector"
```

```
290  DISP "Ake";
300  MAT INPUT Ake
310  MAT PRINT Ake;
320  CALL Rosenb(Ake(*),Eps(*),Km,Maxk,Mkat,Mcyc,Alpha,Beta,V(*),Nstep,Epsy)
330  END
```

8. Example

PROGRAM ´FMROS´ FINDS THE MINIMUM OF AN UNCONSTRAINED MULTI-VARIABLE FUNCTION VIA THE DIRECT SEARCH METHOD OF ROSENBROCK
Km= 2 Maxk= 500 Mkat= 50 Nstep= 2
Input Vector of Initial STEP SIZE Eps(*) i.e. a 2 x1 Vector
 .2   .2

Input Vector of Initial VARIABLE Ake(*), i.e. a 2 x1 Vector
 .25   1.25

Stage Number= 1
Value of the Objective function= 2.01208708709
Values of the Independent Variables:
  x(1)= .45
  x(2)= 1.85
Stage Number= 2
Value of the Objective function= 2.00000099145
Values of the Independent Variables:
  x(1)= .499781167854
  x(2)= 1.99810823885
Total Number of Stages= 3
Total Number of Function Evaluations= 32
Value of Objective function = 2.00000099145
  x(1)= .499283357731
  x(2)= 1.99662716109

9. Example

PROGRAM ´FMROS´ FINDS THE MINIMUM OF AN UNCONSTRAINED MULTI-VARIABLE FUNCTION VIA THE DIRECT SEARCH METHOD OF ROSENBROCK
Km= 2 Maxk= 1000 Mkat= 30 Nstep= 2
Input Vector of Initial STEP SIZE Eps(*) i.e. a 2 x1 Vector
 .1   .1

Input Vector of Initial VARIABLE Ake(*), i.e. a 2 x1 Vector
 1   .5

```
Stage Number= 1
Value of the Objective function=-3873.741275
Values of the Independent Variables:
  x( 1 ) = .25
  x( 2 ) = ;.45
Stage Number= 2
Value of the Objective function=-3873.85912884
Values of the Independent Variables:
  x( 1 ) = .224393124268
  x( 2 ) = .45827072653
Stage Number= 3
Value of the Objective function=-3873.91211125
Values of the Independent Variables:
  x( 1 ) = .193990503518
  x( 2 ) = .486150647292
Stage Number= 4
Value of the Objective function=-3873.92071384
Values of the Independent Variables:
  x( 1 ) = .200196781396
  x( 2 ) = ;.483671123203
Stage Number= 5
Value of the Objective function=-3873.92331065
Values of the Independent Variables:
  x( 1 ) = .203973698919
  x( 2 ) = ;.480490203205
Total Number of Stages= 6
Total Number of Function Evaluations= 53
Value of Objective Function =-3873.92345503
  x( 1 ) = .207559052771
  x( 2 ) = .477470618134
```

11.4.2 Variable Metric Method

The variable metric method is the unconstrained functional minimization of Davidon [11.12], Fletcher and Powell [11.13] whose FORTRAN implementation has appeared elsewhere [11.11, 11.14]. The geometric rationale for this method is not clear [11.15], but it has proven to be a very efficient in numerical applications [11.11, 11.12, 11.14, 11.15]. In this section a detail algorithmic representation of the method, followed by some comments, a BASIC program and some examples are given. Suppose that the function to be minimized is represented by $f(x)$. Let its first derivative be the $n \times 1$ vector $g(x)$. In the following algorithm matrices **A**, **B** and **H** are symmetric and it is assumed that for any vector **x** the function value $f(x)$ and its gradient vector $g(x)$ are directly computable.

10. Algorithm

Step 1. Set iteration index $r=0$, start with initial value x^o and compute $f(x^o)$ and $g(x^o)$, let $H^o = I$.

LINEAR CONTROL SYSTEMS

Step 2. Let the direction of search be defined by

$$s^r = -H'g' \tag{16}$$

where s^r represents the direction during the *rth* iteration.

Step 3. Minimize the function $f(x^r + \lambda s^r)$ for the positive scalar λ and let the critical value of λ be β^r. For this purpose any simple method such as Newton's can be utilized.

Step 4. Find the incremental distance d^r and update x^r, i.e.

$$d^r = \beta^r s^r \tag{17a}$$

$$x^{r+1} = x^r + d^r \tag{17b}$$

Step 5. Calculate the next values of function and its gradient $f(x^{r+1})$ and $g(x^{r+1})$ and evaluate the gradient difference

$$y^r = g^{r+1} - g^r \tag{18}$$

Step 6. Calculate the next value of symmetric matrix A^r,

$$A^r \text{'} = \text{'}d^r d^r / d^r \text{'} y^r \tag{19}$$

where the numerator is the "backward" or matrix product of two vectors, the denominator is a scalar product and prime denotes transposition.

Step 7. Calculate the next value of the symmetric matrix B^r,

$$B^r = (H'y')(H'y')' / (y^r{}'H'y^r) \tag{20}$$

where, once again, the numerator represents the matrix product of two vectors, while the denominator is a quadratic form, i.e. a scalar.

Step 8. Update the nel4Evalue of the symmetric matrix H^r,

$$H^{r+1} = H^r + A^r + B^r \tag{21}$$

Step 9. Test whether $r<N$, a limit for the number of iterations. If so, go to *Step 12*.

Step 10. Test if the norms $\|d^r\|$ and $\|s^r\|$ are small, i.e. within a tolerance ϵ. If so, x^r is the minimum point; then go to *Step 12*.

Step 11. Update r = r+1 and go to *Step* 2.

Step 12. STOP

Before a computer code and a few numerical examples are presented, some comments are in order. Fletcher and Power [11.13] have proved that the matrix **H'**, starting as a positive definite matrix will remain so throughout the iteration. The method uses the first derivative of the function **g'**, while the second derivative is estimated by vector **y**^{r} (see Eq. (18)) and incorporated into **H'**, which gradually evolves to become K_{min}^{-1} - the inverse of the matrix inverse of the second partial derivatives, evaluated at the minimum point x_{min}. One can write the Taylor's series expansion of f as [11.15]

$$f(x+v) = f(x) + g'(x)v + \frac{1}{2!}v'K(x)v + \cdots \qquad (22)$$

where **v** is the displacement vector from the arbitrary point of expansion. At the minimum point the gradient vector **g** vanishes. Also it is noted that the gradient at **v** is given by

$$g(v) = g(x) + K(x)v. \qquad (23)$$

If one would want to move to the minimum point, i.e. where **g(v)** is zero from the point **x** where we currently are, then

$$0 = g(x) + K(x)v \qquad (24)$$

or

$$v = -K^{-1}(x)g(x) \qquad (25)$$

Therefore, if **H'** approaches **K**$^{-1}$(**x**), then **s'** is approximately the vector which takes one to the minimum point and hence β^r must be close to unity. The following BASIC program provides a coding for this minimization method.

11. "FMDFP"

```
10    ! PROGRAM NEAME :   <<FMDFP>> PROG
20    PRINT "PROGRAM <<FMDFP>> FINDS THE MINIMUM OF AN UNCONSTRAINED"
30    PRINT "MULTIVARAIBLE  FUNCTION VIA THE VARIABLE METRIC METHOD OF"
40    PRINT " OF DAVIDON-FLECTHER-POWELL"
50    OPTION BASE 1
60    DIM X(9),G(9),H(72)
70    INPUT "Have YOU LINKED SUBs:   <<Fmfp>> &   <<Funct>>",F$
80    IF (F$="Y") OR (F$="y") THEN 130
90    IF (F$="N") OR (F$="n") THEN 110
100   GOTO 70
110   LINK "Fmfp",350,120
120   LINK "Funct",2860,130
130   PRINT "  ***  Davidon-Fletcher-Powell Functional Minimization Algorithm ***"
140   INPUT "Number of unknown variables n",N
150   PRINT "Number of unknown variables n=";N
160   M=N*(N+7)/2
```

```
170  REDIM X(N),G(N),H(M)
180  INPUT "Give Maximum Number of Iterations, Limit",Limit
190  INPUT "Give Estimated Value of the Cost Function : Est",Est
200  INPUT "Give Stopping Parameter : Eps",Eps
210  PRINT "  Limit=";Limit;"Est=";Est;"Eps=";Eps
220  PRINT "Give INITIAL VALUES in the form of a ";N;" x 1 Vector"
230  MAT INPUT X
240  MAT PRINT X;
250  CALL Fmfp(N,M,X(*),F,G(*),Est,Eps,Limit,Ier,H(*),Count)
260  PRINT "         MINIMIZATION Procedure is now COMPLETED"
270  PRINT LIN(1),"Ier=";Ier;" Number of Iterations=";Count
280  PRINT "MINIMUM value of Function f(x1,x2,...,xn)=";F
290  PRINT "FINAL values of x(i),i=1,2,...,n are:"
300  FOR I=1 TO N
310  PRINT "x(";I;")=";X(I)
320  NEXT I
330  PRINT "Gradient of Function g(.):"
340  MAT PRINT G;
350  END
```

12. Example. The function (11.4.7) of Example 11.4.5 was used to test program "FMDFP". The resulting output is shown below:

```
PROGRAM 'FMDFP' FINDS THE MINIMUM OF AN UNCONSTRAINED
MULTIVARAIBLE  FUNCTION VIA THE VARIABLE METRIC METHOD OF
OF DAVIDON-FLECTHER-POWELL
   *** Davidon-Fletcher.-Powell Functional Minimization Algorithm ***
Number of unknown variables n= 2
  Limit= 50 Est= 2.5 Eps= .00001
Give INITIAL VALUES in the form of a  2  x 1 Vector
 1  1


         MINIMIZATION Procedure is now COMPLETED

Ier= 0  Number of Iterations= 3
MINIMUM value of Function f(x1,x2,...,xn)= 1.41421356237
FINAL values of x(i),i=1,2,...,n are:
x( 1 )= 1
x( 2 )= 1.41421359223
Gradient of Function g(.):
 0   .000000021112

PROGRAM <<FMDFP>> FINDS THE MINIMUM OF AN UNCONSTRAINED
MULTIVARAIBLE  FUNCTION VIA THE VARIABLE METRIC METHOD CF
OF DAVIDON-FLECTHER-PCWELL
   *** Davidon-Fletcher.-Powell Functional Minimization Algorithm ***
Number of unknown variables n= 2
  Limit= 150 Est=-2000 Eps= .0001
Give INITIAL VALUES in the form of a  2  x 1 Vector
 1  .5


         MINIMIZATION Procedure is now COMPLETED

Ier= 0  Number of Iterations= 3
MINIMUM value of Function f(x1,x2,...,xn)=-3873.92354779
FINAL values of x(i),i=1,2,...,n are:
x( 1 )= .205658567768
x( 2 )= .479863.3025
```

Gradient of Function g(.):
-.0000000774 -.0000001823

13. Example. This example, taken from reference [11.5], is given below:

$$Maximize\ F = 3803.84 + 138.08x_1 + 232.92x_2 - 123.08x_1^2$$

$$-203.64x_2^2 - 182.25x_1x_2. \quad (26)$$

In order to maximize this function one can minimize - F instead. The resulting computations for this example are shown below:

```
PROGRAM 'FMDFP' FINDS THE MINIMUM OF AN UNCONSTRAINED
MULTIVARAIBLE  FUNCTION VIA THE VARIABLE METRIC METHOD OF
OF DAVIDON-FLECTHER-POWELL
***   Davidon-Fletcher-Powell Functional Minimization Algorithm ***
Number of unknown variables n= 2
  Limit= 50  Est= 3  Eps= .000001
Give INITIAL VALUES in the form of a   2   x 1 Vector
 .25    1.25


           MINIMIZATION Procedure is now COMPLETED

Ier= 0   Number of Iterations= 9
MINIMUM value of Function f(x1,x2,...,xn)= 2
FINAL values of x(i),i=1,2,...,n are:
x( 1 ) = .500000013788
x( 2 ) = 2.00000011622
Gradient of Function g(.):
 .000000005515   .000000058111
```

11.4.3 An Automatic Constrained Method

In this section an "automatic" constrained minimization (maximization) method due to Rosenbrock [11.4, 11.14] is presented. Consider the following minimization problem:

$$minimize:\ I(\mathbf{x}) = f(x_1,...,x_n) \quad (27)$$

$$Subject\ to:\ g(\mathbf{x}) \leqslant \mathbf{x} \leqslant h(\mathbf{x}) \quad (28)$$

where vector $\mathbf{x}' = (x_1 x_2 \cdots x_n)$, $g(\mathbf{x})$ and $h(\mathbf{x})$ are either constant or $\mathbf{x}$ - dependent functions representing the lower and upper constraints on $\mathbf{x}$.

The scheme is a sequential search technique which has proved to be effective for some problems [11.11]. The algorithm proceeds much like the unconstrained direct search method [11.4] described in Section 11.4.1 until either convergence has been achieved or a boundary zone in the vicinity of the constraints is entered. Such zones are defined as follows:

Lower Zone: $g_i(x) \leq x_i \leq (g_i(x) + (h_i(x) - g_i(x)) 10^{-4})$ \hfill (29)

Upper Zone: $h_i(x) \geq x_i \geq (h_i(x) - (h_i(x) - g_i(x)) 10^{-4})$ \hfill (30)

for $i = 1,2,...,n$. In order to initiate this procedure the initial guess must be chosen to be feasible but not within the boundary zones. The search calculations of this scheme are the same as the unconstrained method, except here after each function evaluation, the following steps are carried out:

(i) Let f^o define the current best value of the objective function for a point where the constraints are satisfied, and let f^* denote the current best value of the objective function for a point where the constraints are satisfied and in addition the boundary zone are not violated. At the beginning of the algorithm f^o and f^* are set equal to each other.

(ii) If the current value of the objective function f is worse than f^o or if the constraints are violated, the trial is considered a failure and the unconstrained procedure is continued.

(iii) If the current point is inside a boundary zone, the objective function is modified as follows:

$$f \text{ (new)} = f \text{ (old)} - (f \text{ (old)} - f^*)(3\alpha - 4\alpha^2 + 2\alpha^3) \tag{31}$$

where

$\alpha = $ (distante into boundary zone)/(width of boundary zone)

$= (g_i + (h_i - g_i).10^{-4} - x_i)/(h_i - g_i) 10^{-4} \cdots$ lower zone

$= (x_i - (h_i - (h_i - g_i).10^{-4}))/((h_i - g_i).10^{-4})...$upper zone \hfill (32)

It is noted that at the inner edge of the boundary zone, $\alpha = 0$, i.e. the function is unchanged, $f \text{ (new)} = f \text{ (old)}$. At the constraint, $\alpha = 1$, and thus $f \text{ (new)} = f^*$. Therefore, the best current value of the function in the feasible region would replace the function value and not the value in the boundary zone. For a function which improves as the constraint is approached, the modified function has an optimum in the boundary zone.

(iv) If an improvement in the objective function has been obtained without the boundary zones or constraints, f^* is set equal to f^o and the procedure is continued.

(v) The algorithm (search technique) is terminated when the convergence criterion is satisfied, i.e. when the difference between two consecutive values of the objective function is within a prespecified small value (e.g. 10^{-6}).

Program "HILL" presents a computer implementation of the Rosenbrock's Hill climbing method. The program is followed by a second-order constrained minimization example

14. "HILL"

```
10    ! PROGRAM NAME : <<HILL>> PROG
20    PRINT "PROGRAM 'HILL' FINDS THE MAXIMUM (OR MINIMUM) OF A MULTIVARIABLE"
30    PRINT "NONLINEAR FUNCTION SUBJECT TO NONLINEAR INEQUALITY CONSTRAINTS "
40    PRINT " BY THE HILL ALGORITHM (CONSTRAINED ROSENBROCK'S METHOD)."
50    OPTION BASE 1
60    DIM X(9),E(9),V(9,9),Sa(9),D(9),G(9),H(9),Al(9),Ph(9),A(9,9),B(9,9)
70    DIM Bx(9),Da(1),Vv(9,9),Eint(9),Vm(9),Ev(9)
80    COM Kount
90    INPUT "Problem Controller : M =1 Minimzation & M=-1 Maximization",M
100   INPUT "Number of Variables - P",P
110   INPUT "Number of Variables + Number of Implicit Constraints - L",L
120   INPUT "Maximim Number of Stages to be Calculated - Loopy",Loopy
130   INPUT "Printing Controller i.e Number of Stages Between Outputs- Pr",Pr
140   INPUT "Storage Controller : Nd=1 Storage in Array Da, Nd=0 No Storage",Nd
150   INPUT "Number of Data Points to be Read into Da - Ndata",Ndata
160   INPUT "Step Size Controller Nstep=0 ORIGINAL Step, 1 PREVIOUS  Step",Nstep
170   IF M=1 THEN 200
180   PRINT "    ***      MAXIMIZATION OF A FUNCTION f(x1,x2,...,xn)    ***"
190   GOTO 210
200   PRINT "    ***      MINIMIZATION OF A FUNCTION f(x1,x2,...,xn)    ***"
210   PRINT "Number of Variables - P=";P
220   PRINT "Number of Variables + Number of Implicit Constraints - L=";L
230   PRINT "Maximim Number of Stages to be Calculated - Loopy=";Loopy
240   PRINT "Storage Controller : Nd=1 Storage in Array Da, Nd=0 No Storage =";Nd
250   PRINT "Printing Controller i.e Number of Stages Between Outputs- Pr=";Pr
260   PRINT "Number of Data Points to be Read into Da - Ndata =";Ndata
270   PRINT "Step Size Controller Nstep=0 ORIGINAL Step,1 PREVIOUS  Step =";Nstep
280   PRINT "Give INITIAL GUESS for the Independent Variables :"
290   REDIM X(P),E(P),V(P,P),Sa(P),D(P),G(P),H(P),Al(P),Ph(P),A(P,P),B(P,P)
300   REDIM Bx(P),Da(1),Vv(P,P),Eint(P),Vm(P),Ev(P)
310   FOR Ix=1 TO P
320   DISP "x(";Ix;")";
330   INPUT X(Ix)
340   NEXT Ix
350   MAT PRINT X;
360   PRINT "Give Vector of INITIAL STEP Sizes :"
370   FOR Is=1 TO P
380   DISP "e(";Is;")";
390   INPUT E(Is)
400   NEXT Is
410   MAT PRINT E;
420   PRINT "  ****     ROSENBROCK'S HILLCLIMB ALGORITHM    ****"
430   IF Nd<>1 THEN 470
440   FOR Ka=1 TO Ndata
450   INPUT Da(Ka)
460   NEXT Ka
470   Lap=Pr-1
480   Loop=Isw=Init=Kount=Ierm=Fl=0
490   Dely=1E-10
500   Npar=Ndata
510   N=L
520   FOR K=1 TO L
530   CALL Cxgh(X(*),Da(*),N,Npar,K,Al(*),Xc,Lc,Uc,1)
540   Al(K)=Al(K)*.0001
550   NEXT K
560   MAT V=IDN
570   MAT Eint=E
580   FOR J=1 TO P
590   IF Nstep=0 THEN E(J)=Eint(J)
600   Sa(J)=2.0
610   NEXT J
620   MAT D=ZER
630   Fbest=Fl
640   I=1
650   IF Init=0 THEN 710
660   MAT Ev=E*V
```

```
670  MAT X=X+Ev
680  FOR K=1 TO L
690  H(K)=F0
700  NEXT K
710  CALL Funct(X(*),F1)
720  F1=M*F1
730  IF Isw=0 THEN F0=F1
740  Isw=1
750  IF ABS(Fbest-F1)>Delay THEN 780
760  Term=1
770  GOTO 1650
780  J=1
790  CALL Cxgh(X(*),Da(*),N,Npar,J,Al(*),Xc,Lc,Uc,2.0)
800  CALL Cxgh(X(*),Da(*),N,Npar,J,Al(*),Xc,Lc,Uc,3.0)
810  CALL Cxgh(X(*),Da(*),N,Npar,J,Al(*),Xc,Lc,Uc,4.0)
820  IF (Xc<=Lc) OR (Xc>=Uc) OR (F1<F0) THEN 1560
830  IF (Xc<Lc+Al(J)) OR (Xc>Uc-Al(J)) THEN 860
840  H(J)=F0
850  GOTO 990
860  Bw=Al(J)
870  IF (Xc<=Lc) OR (Uc<=Xc) THEN 920
880  IF (Lc<Xc) AND (Xc<Lc+Bw) THEN 940
890  IF (Uc-Bw<Xc) AND (Xc<Uc) THEN 960
900  Ph(J)=1
910  GOTO 990
920  Ph(J)=0
930  GOTO 980
940  Pw=(Lc+Bw-Xc)/Bw
950  GOTO 970
960  Pw=(Xc-Uc+Bw)/Bw
970  Ph(J)=1-3.0*Pw+4.0*Pw*Pw-2.0+Pw*Pw*Pw
980  F1=H(J)+(F1-H(J))*Ph(J)
990  IF J=L THEN 1020
1000 J=J+1
1010 GOTO 790
1020 Init=1
1030 IF F1<F0 THEN 1560
1040 D(I)=D(I)+E(I)
1050 E(I)=3.0*E(I)
1060 F0=F1
1070 IF Sa(I)>=1.5 THEN Sa(I)=1
1080 FOR Jj=1 TO P
1090 IF Sa(Jj)>=.5 THEN 1620
1100 NEXT Jj
1110 REM    *****        AXES ROTATION       *****
1120 MAT Vv=ZER
1130 FOR R=1 TO P
1140 Kr=R
1150 FOR C=1 TO P
1160 FOR K=Kr TO P
1170 Vv(R,C)=D(K)*V(K,C)+Vv(R,C)
1180 NEXT K
1190 B(R,C)=Vv(R,C)
1200 NEXT C
1210 NEXT R
1220 Bmag=0
1230 FOR C=1 TO P
1240 Bmag=Bmag+B(1,C)*B(1,C)
1250 NEXT C
1260 Bmag=SQR(Bmag)
1270 Bx(1)=Bmag
1280 FOR C=1 TO P
1290 V(1,C)=B(1,C)/Bmag
1300 NEXT C
1310 FOR R=2.0 TO P
1320 Ir=R-1
1330 FOR C=1 TO P
1340 Sumvm=0
```

```
1350 FOR Kk=1 TO Ir
1360 Sumav=0
1370 FOR Kj=1 TO P
1380 Sumav=Sumav+Vv(R,Kj)*V(Kk,Kj)
1390 NEXT Kj
1400 Sumvm=Sumav*V(Kk,C)+Sumvm
1410 NEXT Kk
1420 B(R,C)=Vv(R,C)-Sumvm
1430 NEXT C
1440 FOR R=2.0 TO P
1450 Bbmag=0
1460 FOR K=1 TO P
1470 Bbmag=Bbmag+B(R,K)*B(R,K)
1480 NEXT K
1490 Bbmag=SQR(Bbmag)
1500 FOR C=1 TO P
1510 V(R,C)=B(R,C)/Bbmag
1520 Loop=Loop+1
1530 Lap=Lap+1
1540 IF Lap=Pr THEN 1650
1550 GOTO 580
1560 IF Init=0 THEN 1650
1570 MAT Ev=E*V
1580 MAT X=X-Ev
1590 E(I)=-.5*E(I)
1600 IF Sa(I)<1.5 THEN Sa(I)=0
1610 GOTO 1080
1620 IF I=P THEN 640
1630 I=I+1
1640 GOTO 660
1650 PRINT LIN(1)
1660 PRINT "Stage=";Loop;"Funct.=";F0
1670 PRINT "   Progress=";Bmag;"    Lateral Progress=";Bbmag
1680 PRINT "   Number of Function Evaluations = ";Kount
1690 REM PRINT CURRENT VALUE OF x
1700 PRINT "    Value of Vector x at This Stage"
1710 MAT PRINT X;
1720 Lap=0
1730 IF Init=0 THEN 1760
1740 IF (Term=1) OR (Loop>=Loopy) THEN 1770
1750 GOTO 580
1760 PRINT "   It Appears that the Starting Point has VIOLATED the Constraints"
1770 PRINT "   FINAL Direction Vector Matrix"
1780 MAT PRINT V;
1790 PRINT "       FINAL Step Size  "
1800 MAT PRINT E;
1810 END
1820 SUB Cxgh(X(*),Da(*),N,Npar,K,Al(*),Xc,Lc,Uc,Code)
1830 ON Code GOTO 1840,1910,1930,1950
1840 ON K GOTO 1850,1870
1850 Ch=2.0
1860 GOTO 1880
1870 Ch=2.5
1880 Cg=0
1890 Al(K)=Ch-Cg
1900 SUBEND
1910 Xc=X(K)
1920 SUBEND
1930 Lc=0
1940 SUBEND
1950 ON K GOTO 1960,1980
1960 Ch=2.0
1970 GOTO 1990
1980 Ch=2.5
1990 Uc=Ch
2000 SUBEND
2010 SUB Funct(X(*),F)
2020 COM Kount
```

```
2030 X1=X(1)
2040 X2=X(2)
2050 X12=X1*X1
2060 X22=X2*X2
2070 F=3803.84+138.08*X1+232.92*X2-123.08*X12-203.64*X22-182.25*X1*X2
2080 Kount=Kount+1
2090 SUBEND
```

15. Example. In this example the maximization problem of Example 12 is reconsidered with the additional constraints, i.e.

Maximize

$$f(x) = 3803.84 + 138.08x_1 + 232.92x_2 - 123.08x_1^2 - 203.64x_2^2 - 182.25x_1x_2 \qquad (33)$$

subject to:

$$\begin{bmatrix} 0 \\ 0 \end{bmatrix} \leqslant \begin{bmatrix} x_1 \\ x_2 \end{bmatrix} \leqslant \begin{bmatrix} 2.0 \\ 2.5 \end{bmatrix} \qquad (34)$$

The above problem was implemented on the program "HILL". The function (31) is implemented in subroutine "Funct" and the conditions (29)-(30) and (34) are included in subroutine "Cxgh". The computational results of this example are given below.

```
PROGRAM 'HILL' FINDS THE MAXIMUM (OR MINIMUM) OF A MULTIVARIABLE
NONLINEAR FUNCTION SUBJECT TO NONLINEAR INEQUALITY CONSTRAINTS
BY THE HILL ALGORITHM (CONSTRAINED ROSENBROCK'S METHOD).
***    MINIMIZATION OF A FUNCTION :f(x1,x2,...,xn.)    ***
Number of Variables - P= 2
Number of Variables + Number of Implicit Constraints - L= 2
Maximim Number of Stages to be Calculated - Loopy= 100
Storage Controller : Nd=1 Storage in Array Da, Nd=0 No Storage = 0
Printing Controller i.e Number of Stages Between Outputs- Pr:= 1
Number of Data Points to be Read into Da - Ndata := 0
Step Size Controller Nstep=0 ORIGINAL Step,1 PREVIOUS    Step = 0
Give INITIAL GUESS for the Independent Variables ):
 1   .5

Give Vector of INITIAL STEP Sizes :
 .1   .1

****    ROSENBROCK'S HILLCLIMB ALGORITHM    ****


Stage= 1 Funct.= 3854.189225
   Progress= .20615528128     Lateral Progress= 4.85071250073E-02
   Number of Function Evaluations =  8
   Value of Vector x at This Stage
 .25   .15


Stage= 2 Funct.= 3866.09350825
   Progress= .164034470017    Lateral Progress= :2.12955496158E-02
   Number of Function Evaluations =  14
   Value of Vector x at This Stage
 .31063390626   .248507125007
```

Stage= 3 Funct.= 3873.85619232
  Progress= .15974796907   Lateral Progress= 3.90244585177E-02
  Number of Function Evaluations = 25
  Value of Vector x at This Stage
  .231757858906 .475631122938

Stage= 4 Funct.= 3873.88492022
  Progress= 3.52385378808E-02   Lateral Progress= 4.42905501390E-04
  Number of Function Evaluations = 40
  Value of Vector x at This Stage
  .222211558664 .48134124735

Stage= 5 Funct.= 3873.88586909
  Progress= 2.45315905232E-03   Lateral Progress= 9.70632516140E-07
  Number of Function Evaluations = 69
  Value of Vector x at This Stage
  .222914466009 .480184046915

Stage= 6 Funct.= 3873.89579675
  Progress= 3.57512026684E-02   Lateral Progress= 3.03635865964E-04
  Number of Function Evaluations = 81
  Value of Vector x at This Stage
  .220428122521 .480190449272

Stage= 7 Funct.= 3873.90645826
  Progress= .03573832314   Lateral Progress= 3.97208120508E-04
  Number of Function Evaluations = 93
  Value of Vector x at This Stage
  .217386616422 .479937737118

Stage= 8 Funct.= 3873.91643963
  Progress= 3.57229557198E-02   Lateral Progress= 5.19700824242E-04
  Number of Function Evaluations = 105
  Value of Vector x at This Stage
  .213619082554 .479349712893

Stage= 9 Funct.= 3873.92031277
  Progress= 2.76756144186E-02   Lateral Progress= 1.48836577271E-04
  Number of Function Evaluations = 121
  Value of Vector x at This Stage
  .207963512447 .482538468723

Stage= 10 Funct.= 3873.92114947
   Progress= 3.84500439164E-03   Lateral Progress= 1.54959608137E-05
   Number of Function Evaluations = 142
   Value of Vector x at This Stage
   .210347605339   .479465326555

Stage= 11 Funct.= 3873.92142195
   Progress= 2.13924337530E-03   Lateral Progress= 4.79743240354E-05
   Number of Function Evaluations = 161
   Value of Vector x at This Stage
   .209877998724   .479776192761

Stage= 12 Funct.= 3873.92143848
   Progress= 2.20307922957E-04   Lateral Progress= 2.98329705919E-06
   Number of Function Evaluations = 185
   Value of Vector x at This Stage
   .209476016876   .480268874325

Stage= 13 Funct.= 3873.92337208
   Progress= 4.01417846910E-03   Lateral Progress= 1.87689863812E-04
   Number of Function Evaluations = 206
   Value of Vector x at This Stage
   .207096843933   .479383138178

Stage= 14 Funct.= 3873.92354289
   Progress= 1.64790529952E-03   Lateral Progress= 2.30378850248E-04
   Number of Function Evaluations = 228
   Value of Vector x at This Stage
   .205416721119   .479952984923

Stage= 15 Funct.= 3873.92354664
   Progress= 3.59108225436E-04   Lateral Progress= 9.19392539709E-06
   Number of Function Evaluations = 256
   Value of Vector x at This Stage
   .205578161938   .479953937447

Stage= 16 Funct.= 3873.92354732
   Progress= 3.09853110618E-05   Lateral Progress= 1.85602071987E-05
   Number of Function Evaluations = 290
   Value of Vector x at This Stage
   .20559764763   .479918128366

```
Stage= 17 Funct.= 3873.92354772
  Progress= 2.09069463265E-05      Lateral Progress= 2.41556094425E-05
  Number of Function Evaluations =   318
  Value of Vector x at This Stage
  .205632886903   .479870188909



Stage= 18 Funct.= 3873.92354776
  Progress= 1.75517540394E-05      Lateral Progress= 3.81401136469E-07
  Number of Function Evaluations =   358
  Value of Vector x at This Stage
  .205643518808   .479874680049



Stage= 18 Funct.= 3873.92354776
  Progress= 1.75517540394E-05      Lateral Progress= 3.81401136469E-07
  Number of Function Evaluations =   385
  Value of Vector x at This Stage
  .205631713164   .479862278023



  FINAL Direction Vector Matrix
  .999476344115   3.23579592581E-02

  -.032357959258   1



      FINAL Step Size
  -1.22070312500E-05 .-1.22070312500E-05
```

11.5 DESIGN OF OPTIMAL STATE FEEDBACK

In this section the notion of parameter optimization and functional minimization are used for optimal design of linear systems described by a state model. The use of feedback in system design was already demonstrated previously. In Section 9.9 feedback compensation was used to design a linear SISO system described in frequency domain. The state and output feedbacks were used in Section 10.4 within the context of pole placement. The basic difference in this section is the fact that the feedback gain matrix would minimize (or maximize) an objective function in the design process.

Consider a linear control system described by a state equation,

$$\dot{x} = Ax + Bu \tag{1}$$

LINEAR CONTROL SYSTEMS

$$x(0) = x_o \qquad (2)$$

The feedback control law is assumed to be given by,

$$u = -Kx. \qquad (3)$$

The optimal state feedback design problem is stated as follows: Find a feedback gain matrix K such that (1) - (3) are satisfied while a performance index (or cost function),

$$J = \int_0^T f(x,u,t)\,dt \qquad (4)$$

is minimized. A common special case of this problem is one of regulating (1), i.e. $T \to \infty$ while minimizing a quadratic cost function, i.e. $f(x,u,t) = 1/2(x'\,x + u'\,Ru)$. Using (3), the state equation (1) and cost function (4) can be rewritten as,

$$\dot{x} = Ax - BKx = (A-BK)x \triangleq A_c x \qquad (5)$$

$$J = 1/2 \int_0^\infty (x'x + (-Kx)'R(-Kx))\,dt =$$

$$1/2 \int_0^\infty x'(I+K'RK)x\,dt \triangleq 1/2 \int_0^\infty x'Qx\,dt \qquad (6)$$

In order to find the minimum value of J, one way is to postulate the existence of a positive definite matrix P within the context of the following exact differential, i.e.

$$1/2\,d/dt\,(x'Px) = -x'Qx. \qquad (7)$$

Completing the left-hand side differentiation,

$$\dot{x}'Px + x'P\dot{x} = x'A_c'Px + x'PA_c x = x'(A_c'P + PA_c)x = -x'Qx \qquad (8)$$

which leads to the following Lyapunov equation

$$A_c'(K)P(K) + P(K)A_c(K) + Q(K) = 0 \qquad (9)$$

which can be solved using the subroutine "Lyap" used in program "LYAP" presented in Chapter 7. Using the postulate of Equation (7), one gets

DESIGN OF OPTIMAL STATE FEEDBACK

$$J = 1/2 \int_0^\infty (\frac{-d}{dt}(\mathbf{x}'\mathbf{P}\mathbf{x}) dt = -1/2\mathbf{x}'\mathbf{P}\mathbf{x}|_0^\infty = 1/2(-\mathbf{x}'(\infty)\mathbf{P}\mathbf{x}(\infty) + \mathbf{x}'(0)\mathbf{P}\mathbf{x}(0)) \tag{10}$$

which after assuming that the closed-loop system is stable, i.e. $\mathbf{x}(\infty) = 0$, one can see that $J = 1/2\mathbf{x}_o'\mathbf{P}(\mathbf{K})\mathbf{x}_o$. Thus, the optimal state feedback design can be restated as follows:

$$\underset{\mathbf{K}}{\text{Minimize}} \ J = 1/2\mathbf{x}_o'\mathbf{P}(\mathbf{K})\mathbf{x}_o \tag{11}$$

subject to:

$$\mathbf{A}_c'(\mathbf{K})\mathbf{P}(\mathbf{K}) + \mathbf{P}(\mathbf{K})\mathbf{A}_c(\mathbf{K}) + \mathbf{Q}(\mathbf{K}) = 0 \tag{12}$$

The implementation of the above problem on the computer is rather simple. One can simply use subroutine "Lyap" within subroutine "Funct" of subroutine "Fmfp" of the general Davidon-Fletcher-Powell minimization technique of Section 11.4.2 (see Example 11.4.11). The following two examples illustrate the method.

1. Example. Consider a second order system described by

$$\dot{\mathbf{x}} = \begin{bmatrix} 0 & 1 \\ 0 & 0 \end{bmatrix} \mathbf{x} + \begin{bmatrix} 0 \\ 1 \end{bmatrix} \mathbf{u}, \ \mathbf{x}(0) = \begin{bmatrix} 1 \\ 1 \end{bmatrix} \tag{13}$$

it is desired to find a feedback gain vector $\mathbf{u} = -\mathbf{k}\mathbf{x}$ such that the cost functional

$$J = \int_0^\infty \mathbf{x}'\mathbf{x} \ dt \tag{14}$$

is minimized. Note that the open-loop system is unstable.

The closed-loop matrix $\mathbf{A}_c$ is given by

$$\mathbf{A}_c = \begin{bmatrix} 0 & 1 \\ -k_1 & -k_2 \end{bmatrix}$$

For simplicity, one can let $k_1 = 1$ and try to find minimum of J with respect to k_2. The Lyapunov equation (9) can be solved from

$$\begin{bmatrix} 0 & -1 \\ 1 & -k_2 \end{bmatrix} \begin{bmatrix} p_{11} & p_{12} \\ p_{12} & p_{22} \end{bmatrix} + \begin{bmatrix} p_{11} & p_{12} \\ p_{12} & p_{22} \end{bmatrix} \begin{bmatrix} 0 & 1 \\ -1 & -k_2 \end{bmatrix} + \begin{bmatrix} 1 & 0 \\ 0 & 1 \end{bmatrix} = 0 \tag{15}$$

540   LINEAR CONTROL SYSTEMS

which leads to

$$\mathbf{P}(k_2) = \begin{bmatrix} \dfrac{k_2^2+2}{2k_2} & \dfrac{1}{2} \\ \dfrac{1}{2} & \dfrac{1}{k_2} \end{bmatrix}. \tag{16}$$

Thus the cost functional J is

$$J = \mathbf{x}_o{}'\mathbf{P}\mathbf{x}_o = (k_2^2+2k_2+4)/2k_2 \tag{17}$$

The necessary condition for minimizing (17) is to have

$$dJ/dk_2 = (2k_2^2-8)/4k_2^2 = 0 \tag{18}$$

or $k_2 = 2$, i.e., the optimizing feedback control is given by

$$u^* = -x_1 - 2x_2 \tag{19}$$

and $J = 3$. Note that the system's eigenvalues were changed from $\lambda\{A\} = \{0,0\}$ to $\lambda\{A_c\} = \{-1,-1\}$, i.e. the optimizing control stabilizes the system as well.

2. Example. Consider a second-order optimal state feedback control system problem

$$\dot{\mathbf{x}} = \begin{bmatrix} 0 & 1 \\ 0 & -1 \end{bmatrix}\mathbf{x} + \begin{bmatrix} 0 \\ 1 \end{bmatrix}u, \ \mathbf{x}(0) = \begin{bmatrix} 1 \\ 1 \end{bmatrix} \tag{20}$$

$$J = \int_0^\infty (\mathbf{x}'\mathbf{x} + u^2)dt \tag{21}$$

By assuming a state feedback for the control $u = -k_1x_1 - k_2x_2$, the standard Lyapunov equation (9) will become,

$$\mathbf{A}_c{}'\mathbf{P}+\mathbf{P}\mathbf{A}_c+\mathbf{Q} = 0 \tag{22}$$

where

$$\mathbf{A}_c = \mathbf{A}-\mathbf{B}\mathbf{K} = \begin{bmatrix} 0 & 1 \\ -k_1 & -1-k_2 \end{bmatrix} \tag{23a}$$

DESIGN OF OPTIMAL STATE FEEDBACK

$$Q = I + K'K = \begin{bmatrix} 1+k_1^2 & k_1k_2 \\ k_1k_2 & 1+k_2^2 \end{bmatrix} \quad (23b)$$

This problem was simulated on the computer by utilizing subroutines "Rosenb" of Rosenbrock's functional minimization for

$$J(K) = x_o'Px_o = p_{11}(K) + 2p_{12}(K) + p_{22}(K) \quad (24)$$

along with subroutine "Lyap" of Chapter 7 to find the optimal values of k_1 and k_2. The only important necessary modification lies in the way subroutine "Object" must be written. This is shown below for the benefit of the interested readers.

```
10    SUB Object(Ake(*),Sumn,Km)
20    OPTION BASE 1
21    DIM Ac(2,2),Q(2,2),P(2,2),Xot(1,2),Xo(2,1)
30    K1=Ake(1)
40    K2=Ake(2)
50    Ac(1,1)=0
60    Ac(1,2)=1
70    Ac(2,1)=-K1
80    Ac(2,2)=-(1+K2)
90    Q(1,1)=1+K1*K1
100   Q(1,2)=K1*K2
110   Q(2,1)=K1*K2
120   Q(2,2)=1+K2*K2
130   CALL Lyap(Ac(*),Q(*),Km,100,P(*))
140   REM
150   Sumn=P(1,1)+2*P(1,2)+P(2,2)
160   SUBEND
```

The optimization of this problem was achieved in 3 stages using a total of 26 functions evaluations as shown below.

```
**** SOLUTION OF   OPTIMAL STATE FEEDBACK ****
Minimization of a Multivariable Function via H.ROSENBROCK'S
Method with Solution of an Accompanying Lyapunov Equation.
  Km,Maxk,Mkat,Mcyc,Nstep are:   2   50   30   50   1
Epsy,Alpha,Beta are:   .00001   2   .5
Input Eps(*),Ake(*)  2   2 x1 Vectors
 .1                       .1

0                        0


Lyapunov matrix
  3.28725254482         1.08333330127

  1.08333330127         1.25258861992
```

Stage Number= 1
Value of the Objective function= 5.61666450061
Values of the Independent Variables:
 x(1)= .7
 x(2)= .3

Lyapunov matrix
2.35167889906   1.02150343782

1.02150343782   1.1518987427

Stage Number= 2
Value of the Objective function= 5.54657082085
Values of the Independent Variables:
 x(1)= .820142556187
 x(2)= .36872363625
Lyapunov matrix
2.44334504022   1.00477363675

1.00477363675   1.12264698935

Lyapunov matrix
2.34880812703   1.02112453249

1.02112453249   1.15540471149

Total Number of Stages= 3
Total Number of Function Evaluations= 26
Value of Objective Function = 5.54657082085
 x(1)= .814497333348
 x(2)= .378592606216

For the sake of briefness, only a few intermediate values of the Lyapunov matrix P(k) were printed. The optimal state-feedback control and closed-loop matrix became,

$$u^* = -0.8145x_1 - 0.3786x_2 \tag{25}$$

$$A_c = \begin{bmatrix} 0 & 1 \\ -0.8145 & -1.3786 \end{bmatrix} \tag{26}$$

This indicates that, while minimizing cost function (21), the control (25) stabilizes the system with open- and closed-loop eigenvalues being at {0, -1} and {-0.69 ± j0.582}, respectively.

PROBLEMS

11.1 Consider a SISO system

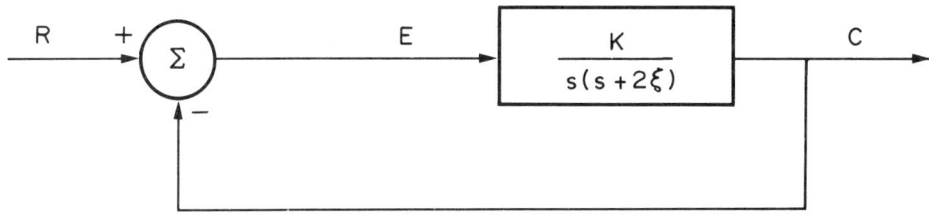

Fig. P11-1

define an ISE and find it as a function of K and ξ.

11.2 If

$$\frac{C(s)}{R(s)} = \frac{G(s)}{1+G(s)} = \frac{(1+\tau_1 s)(1+\tau_2 s)...(1+\tau_m s)}{(1+\tau_a s)(1+\tau_b s)...(1+\tau_n s)}$$

show that

$$\int_0^\infty e(t)dt = (\tau_a+\tau_b+...+\tau_n) - (\tau_1+\tau_2+...+\tau_m)$$

11.3 Compute $\int_0^\infty c^2(t)dt$ with a unit impulse input for

$$\frac{C(s)}{R(s)} = \frac{1}{s^2+4\xi+1}, \quad 0<\xi<1$$

11.4 For the system

$$\ddot{c}+2\xi\dot{c}+c = 0$$

find $\min J = \int_0^\infty (c^2(t)+\dot{c}^2(t))dt$ using a) "FMROS", (b) "FMDFP" and (c) Own calculations

11.5 Evaluate

$$\int_0^\infty |e(t)|dt, \quad \int_0^\infty t|e(t)|dt$$

for system $\dfrac{C(s)}{R(s)} = 1/(s^2+2\xi s+4) \quad (\xi \geq 1).$

11.6 Determine the value of damping factor ξ in

$$\ddot{c}+2\xi\dot{c}+c = 0$$

such that $\int_0^\infty (c^2+5\dot{c}^2)dt$ is minimum.

11.7 Determine a state-space representation of the system of Problem 11.4 and a quadratic cost function for J.

11.8 Use discrete-time version of the Parseval's theorem to find ISE for

$$\mathbf{x}(k+1) = \begin{bmatrix} 0 & 1 \\ -1 & -2\xi \end{bmatrix} \mathbf{x}(k)$$

as a function of ξ.

11.9 Suppose that the parameters K and ξ in Problem 11.1 are constrained by $0<\xi<1$ and $K>0$. Use program "HILL" to find the minimum of ISE.

11.10 For the system

$$\dot{\mathbf{x}} = \begin{bmatrix} 0 & 1 \\ 0 & 0 \end{bmatrix} \mathbf{x} + \begin{bmatrix} 1 \\ 1 \end{bmatrix} u, \quad \mathbf{x}(0) = \begin{bmatrix} 1 \\ 0 \end{bmatrix}$$

let $u = k_1 x_1 + k_2 x_2$ and find the values of k_1 and k_2 such that

$$J = \int_0^\infty (\mathbf{x}'\mathbf{x}+u^2)dt$$

is minimized.

11.11 Find the value of k_1 and k_2 in system

$$\dot{x} = \begin{bmatrix} 1 & 0 \\ -1 & 2 \end{bmatrix} x + \begin{bmatrix} 1 \\ 1 \end{bmatrix} u, \; x(0) = \begin{bmatrix} 1 \\ 1 \end{bmatrix}$$

$u(t) = -k_1 x_1 - k_2 x_2$ such that

$$J = \int_0^\infty (x_1^2 + 2x_2^2 + 4u^2) dt$$

is minimized. Verify the result by using "Lyap", "Object" and "FMROS" programs.

REFERENCES

[11.1] W. R. Perkins and J. B. Cruz, Jr., *Engineering of Dynamic Systems*, John Wiley and Sons, Inc., New York, 1969

[11.2] G. C. Newton, Jr., L. A. Gould and J. Kaiser, *Analytic Design of Linear Feedback Controls*, John Wiley and Sons, Inc., New York, 1957, pp. 372-381

[11.3] E. I. Jury, *Theory and Applications of the z-Transform Methods*, John Wiley and Sons, Inc., New York, 1964, pp. 297-299

[11.4] H. H. Rosenbrock, "An automatic method for finding the greatest or least value of a function," *Comput. J.* Vol 3, (1960) pp 175-184

[11.5] R. Fletcher, "A review of methods for unconstrained optimization," in *Optimization, Symposium of the Institute of Mathematics and its Applications*, Univ. of Keele, England, 1968, London: Academic Press

[11.6] R. Hooke and T. A. Jeeves, "Direct Search" Solution of numerical and statistical Problems," *J. Ass. Comput. Mach.*, Vol 8, p.212, 1961

[11.7] W. H. Swann, "Report on the development of a new direct search method of optimization," *I.C.I. Ltd. Central Instr. Lab. Res. Note 64/3.*, 1964

[11.8] J. A. Nedler and R. Mead, "A Simplex method for function minimization," *Comp. J.*, Vol 7, p. 308, 1965

[11.9] C. S. Smith, "The automatic computation of maximum likelihood estimates," *N.C.B. Sc. Dept. Report S.C. 846/MR/40.*, 1962

[11.10] M.J.D. Powell, "An efficient method of finding the minimum of a function of several variables without calculating derivatives," *Comput.J.*, Vol 7, p.155, 1964

[11.11] J. L. Kuester and J. H. Mize, *Optimization Techniques with FORTRAN*, McGraw-Hill Book Co., New York, 1973

[11.12] W. C. Davidon, "Variable metric method for minimization" AEC Research and Development Report ANL-5990 (Rev.) 1959. See also "Variance algorithm for minimization", *Comp. J.*, Vol. 10, pp. 406-410, 1968

[11.13] R. Fletcher and M. J. D. Powell," A rapidly convergent descent method for minimization, " *Computer J.*, Vol. 6, l963, pp. 1963-1968

[11.14] IBM Corporation, *System/360 Scientific Subroutine Package,* Technical Report no. H20-0205-3, 1966, pp. 221-225

[11.15] F. S. Acton, *Numerical Methods That Work,* Harper & Row Publ., New York, NY. (1970), pp 467-469

APPENDIX

In this appendix the source listing of the general and utility subroutines used throughout the text are listed. It must be noted that several other subroutines which are called only once are not listed here; rather, they are listed along with their respective driver programs. Moreover, the list of the general subroutines are provided in accordance with the section numbers in the text.

LIST OF SUBROUTINES

Routine	Object
Rank	Finds the rank of any $m \times n$ matrix
Norm	Finds Eucleadean norms of a vector or a square matrix
Eigen	Finds eigenvalues and eigenvectors of any real square matrix
Resmat	Finds an analytic expression for the resolvent matrix $(sI-A)^{-1}$
Matpol	Evaluates an mth order polynomial of an $n \times n$ real square matrix
Coeff	Evaluates a coefficient matrix
Polpro	Finds the product of any two polynomials
Poldiv	Finds the dividend of any two polynomials
Expat	Finds $exp(At)$ for any real square matrix
Lyap	Solves the Lyapunov equation through an iterative algorithm (Algorithm 7.9.11)
Cont	Constructs the controllability matrix of a linear MIMO system
Sepcob	Determines the completely controllable or completely observable subsystem of a linear MIMO system
Obs	Constructs the observability matrix of a linear MIMO system
Bdnync	Plots Bode, Nyquist and Nichols diagrams for any linear SISO system

Spoly	Evaluates a polynomial of $s=X+jY$
P_sul	Sets ranges, labels and ticks for a plot
P_sb	A two-level plotter program to set up semi-log plots
P_xslg	Draws semi-log axes for a given plot
P_xlin	Draws linear axes for a given plot
Rosenb	Finds minimum of a function of several variables by Rosenbrock's direct search method
Fmfp	Finds minimum of a function of several variables by Davidon - Fletcher - Powel's variable-metric method
Lambda	Determines distinct eigenvalues and their multiplicities
Vanmat	Constructs a Van Dermonde matrix from distinct eigenvalues
Coemat	Calculates the coefficient matrices for $exp(At)$ in accordance with the Cayely-Hamilton Theorem method
Coeffs	Calculates the coefficients of an nth order polynomial with roots stored in array Lambda (*)
Detinv	Finds the determinant and inverse of a square complex matrix using the Cayley-Hamilton Theorem

APPENDIX

DRIVER PROGRAMS

Name	Object	Section
RANK	Determines rank of any $m \times n$ matrix	2.3
EIGEN	Finds eigenvalues and eigenvectors of any real square matrix	2.6
RESMAT	Finds an analytic expression for the resolvent matrix $(sI-A)^{-1}$	6.5
MATPOL	Evaluates an mth-order polynomial of an $n \times n$ square matrix	2.10
ROOTFD	Finds roots of any polynomial with real and/or complex coefficients	--
STATEQ	Finds state equations in companion form for a $l.t.i.$ continuous- time or discrete-time system from accompanying differential or difference equations	4.4
TRANFN	Finds the transfer matrix of any $l.t.i.$ MIMO system	4.5
TRANSF	Finds the closed-loop transfer function of any SISO linear system	4.5
COMPAN	Transforms a linear SISO system's state equation into a controllable or an observable companion form	4.6
SEPTIM	Performs an iterative separation of the time scales in a singularly- perturbed system	5.2
EXAGR	Finds aggregated models using exact aggregation conditions	5.3
MODAGR	Finds aggregated models using modal (dominant modes) aggregation method	5.3
MOMPAD	Finds aggregated models for SISO systems described in frequency domain using moment matching and/or Pade' approximation	5.3
EXPATA	Finds an analytical expression for the exponential of any real square transition matrix of a $l.t.i.$ continuous-time system) matrix	6.4

APPENDIX

APOWRK	Finds an analytical expression for the transition matrix of a *l.t.i.* discrete-time system, A^k	6.4
COMSCS	Gives a complete solution of a MIMO linear continuous-time system subject to various input signals	6.6
COMSDS	Gives a complete solution of a MIMO linear discrete-time system subject to various input signals	6.6
STABCD	Determines stability of the equilibrium point of any linear continuous-time or discrete-time system	7.3
ROUTH	Stability check of a SISO linear continuous-time system via the Routh-Hurwitz Criterion	7.4
JURY	Stability check of a SISO linear discrete-time system via the Jury-Blanchard-Raible test	7.5
ROOTLC	Draws the root locus of a linear SISO continuous-time or discrete-time system	7.6
BNPLOT	Draws the Bode and Polar (Nyquist) plots of a linear SISO continuous-time or discrete-time system	7.7
LYAP	Solves the matrix Lyapunov equation through an iterative scheme	7.9
CONOBS	Determines whether a linear MIMO continuous-time or discrete-time system is completely controllable or completely observable	8.4
SEPCON	Separates the controllable part of an uncontrollable linear MIMO continuous-time or discrete-time system	8.6
SEPOBS	Separates the observable part of an unobservable linear MIMO continuous-time or discrete-time system	8.6
MACRCL	Plots M and Alpha circles for a linear SISO system superimposed on the polar plot of the open-loop transfer function	9.5
POLPLA	Pole placement in a *l.t.i.* system	10.4

CONTUR	Draws up to 5 closed contours of a function of two variables on the CRT or an X-Y plotter	**11.4**
FMROS	Functional minimization via the unconstrained method of Rosenbrock	**11.4**
FMDFP	Functional minimization via the unconstrained method of Davidon-Fletcher- Powell	**11.4**
HILL	Finds extreme points of a multivariable non-linear inequality constraints by the Hill-climing method	**11.4**

```
10    SUB Rank(A(*),M,N,R)
20    !   "Rank" EVALUATES THE RANK OF AN MxN REAL MATRIX
30    OPTION BASE 1
40    DIM H(N,M)
50    !  Eliminate all the all-zero columns:
60    I=1
70    If=0
80     FOR J=1 TO N
90      IF A(J,I)<.000005 THEN If=If+1
100    NEXT J
110   IF If<N THEN 200
120   M=M-1
130   IF I>=M THEN 230
140    FOR J=I TO M
150     FOR K=1 TO N
160      A(K,J)=A(K,J+1)
170     NEXT K
180    NEXT J
190   GOTO 70
200   I=I+1
210   IF I<=M THEN 70
220   !  Eliminate all the all-zero rows:
230   I=1
240   If=0
250    FOR J=1 TO M
260     IF A(I,J)<.000005 THEN If=If+1
270    NEXT J
280   IF If<M THEN 370
290   N=N-1
300   IF I>=N+1 THEN 400
310    FOR J=I TO N
320     FOR K=1 TO M
330      A(J,K)=A(J+1,K)
340     NEXT K
350    NEXT J
360   GOTO 240
370   I=I+1
380   IF I<=N THEN 240
390   !  Let the smaller dimension become the number of rows:
400   Im=M
410   In=N
420   IF N>=M THEN 490
430   MAT H=TRN(A)
440   REDIM A(N,M)
450   MAT A=F
460   Im=N
470   In=M
480   !  Perform Gaussian elimination:
490    FOR I=1 TO Im
500     IF A(I,I)<>0 THEN Next
```

```
510     IF I=Im THEN End
520       FOR J=I+1 TO Im
530         IF A(J,I)<>0 THEN Found
540       NEXT J
550   GOTO End1
560 Found:   FOR K=1 TO In
570   A=A(J,K)
580   A(J,K)=A(I,K)
590   A(I,K)=A
600     NEXT K
610 Next: FOR L=I TO Im
620   A(I,Im+I-L)=A(I,Im+I-L)/A(I,I)
630     NEXT L
640   IF I=Im THEN End1
650     FOR L=I+1 TO Im
660       FOR J=1 TO In
670         A(L,In+1-J)=A(L,In+1-J)-A(I,In+1-J)*A(L,I)
680       NEXT J
690     NEXT L
700   NEXT I
710 End:   FOR J=I+1 TO In
720   IF A(I,J)<>0 THEN Found1
730   NEXT J
740   GOTO End1
750 Found1:   FOR K=1 TO Im
760   A=A(K,J)
770   A(K,J)=A(K,I)
780   A(K,I)=A
790     NEXT K
800   GOTO Next
810   ! Count the number of nonzero diagonal terms:
820 End1: R=0
830     FOR I=1 TO Im
840       IF A(I,I)<>0 THEN R=R+1
850     NEXT I
860   SUBEND




10    SUB Norm(A(*),N,M,Xnorm)
20    !  "Norm" EVALUATES THE EUCLIDEAN NORM OF A MATRIX OR
      A VECTOR
30    OPTION BASE 1
40    IF (M<>N) AND (M=1) THEN 130
50    Val=0
60    FOR I=1 TO N
70    FOR J=1 TO N
80    Val=Val+A(I,J)^2
90    NEXT J
```

```
100    NEXT I
110    Xnorm=SQR(Val)
120    GOTO 190
130    REDIM A(N,1)
140    Val=0
150    FOR I=1 TO N
160    Val=Val+A(I,1)^2
170    NEXT I
180    Xnorm=SQR(Val)
190    SUBEND




1      SUB Eigen(N,A(*),Evr(*),Evi(*),Vecr(*),Veci(*),Indic(*))
2      Baddta=(N<=0)
3      IF Baddta=0 THEN 7
4      PRINT LIN(2),"ERROR IN SUBPROGRAM Eigen."
5      PRINT "N=";N,LIN(2)
6      PAUSE
7      OPTION BASE 1
8      INTEGER Local(N)
9      DIM Prfact(N),Subdia(N),Work(N)
10     IF N<>1 THEN 17
11     Evr(1)=A(1,1)
12     Evi(1)=0
13     Vecr(1,1)=1
14     Veci(1,1)=0
15     Indic(1)=2
16     GOTO 128
17     CALL Scale(N,A(*),Veci(*),Prfact(*),Enorm)
18     Ex=EXP(-39*LOG(2))
19     CALL Hesqr(N,A(*),Veci(*),Evr(*),Evi(*),Subdia(*),
       Indic(*),Eps,Ex)
20     J=N
21     I=1
22     Local(1)=1
23     IF J=1 THEN 30
24     IF ABS(Subdia(J-1))>Eps THEN 27
25     I=I+1
26     Local(I)=0
27     J=J-1
28     Local(I)=Local(I)+1
29     IF J<>1 THEN 24
30     K=1
31     Kon=0
32     L=Local(1)
33     M=N
34     FOR I=1 TO N
35         Ivec=N-I+1
```

```
36        IF I<=L THEN 40
37        K=K+1
38        M=N-L
39        L=L+Local(K)
40        IF Indic(Ivec)=0 THEN 56
41        IF Evi(Ivec)<>0 THEN 51
42        FOR Kl=1 TO M
43           FOR Ll=Kl TO M
44                A(Kl,Ll)=Veci(Kl,Ll)
45           NEXT Ll
46           IF Kl=1 THEN 48
47           A(Kl,Kl-1)=Subdia(Kl-1)
48        NEXT Kl
49        CALL Realve(N,M,Ivec,A(*),Vecr(*),Evr(*),Evi(*),
          Work(*),Indic(*),Eps,Ex)
50        GOTO 56
51        IF Kon<>0 THEN 55
52        Kon=1
53        CALL Compve(N,M,Ivec,A(*),Vecr(*),Veci(*),Evr(*),
          Evi(*),Indic(*),Subdia( ,Work(*),Eps,Ex)
54        GOTO 56
55        Kon=0
56     NEXT I
57     MAT A=IDN
58     IF N<=2 THEN 73
59     M=N-2
60     FOR K=1 TO M
61        L=K+1
62        FOR J=2 TO N
63           D1=0
64           FOR I=L TO N
65              D2=Veci(I,K)
66              D1=D1+D2*A(J,I)
67           NEXT I
68           FOR I=L TO N
69              A(J,I)=A(J,I)-Veci(I,K)*D1
70           NEXT I
71        NEXT J
72     NEXT K
73     Kon=1
74     FOR I=1 TO N
75        L=0
76        IF Evi(I)=0 THEN 81
77        L=1
78        IF Kon=0 THEN 81
79        Kon=0
80        GOTO 127
81        FOR J=1 TO N
82           D1=D2=0
83           FOR K=1 TO N
```

```
84              D3=A(J,K)
85              D1=D1+D3*Vecr(K,I)
86              IF L=0 THEN 88
87              D2=D2+D3*Vecr(K,I-1)
88           NEXT K
89           Work(J)=D1/Prfact(J)
90           IF L=0 THEN 92
91           Subdia(J)=D2/Prfact(J)
92        NEXT J
93        IF L=1 THEN 105
94        D1=0
95        FOR M=1 TO N
96            D1=D1+Work(M)^2
97        NEXT M
98        D1=SQR(D1)
99        FOR M=1 TO N
100           Veci(M,I)=0
101           Vecr(M,I)=Work(M)/D1
102       NEXT M
103       Evr(I)=Evr(I)*Enorm
104       GOTO 127
105       Kon=1
106       Evr(I)=Evr(I)*Enorm
107       Evr(I-1)=Evr(I)
108       Evi(I)=Evi(I)*Enorm
109       Evi(I-1)=-Evi(I)
110       R=0
111       FOR J=1 TO N
112           R1=Work(J)^2+Subdia(J)^2
113           IF R>=R1 THEN 116
114           R=R1
115           L=J
116       NEXT J
117       D3=Work(L)
118       R1=Subdia(L)
119       FOR J=1 TO N
120           D1=Work(J)
121           D2=Subdia(J)
122           Vecr(J ,I)=(D1*D3+D2*R1)/R
123           Veci(J ,I)=(D2*D3-D1*R1)/R
124           Vecr(J,I-1)=Vecr(J,I)
125           Veci(J,I-1)=-Veci(J,I)
126       NEXT J
127   NEXT I
128   SUBEXIT
129   SUBEND
130   SUB Scale(N,A(*),H(*),Prfact(*),Enorm)
131   OPTION BASE 1
132   INTEGER I,J,Iter,Ncount
133   FOR I=1 TO N
```

```
134         FOR J=1 TO N
135             H(I,J)=A(I,J)
136         NEXT J
137     Prfact(I)=1
138     NEXT I
139     Bound1=.75
140     Bound2=1.33
141     Iter=0
142     Ncount=0
143     FOR I=1 TO N
144         Column=0
145         Row=0
146         FOR J=1 TO N
147             IF I=J THEN 150
148             Column=Column+ABS(A(J,I))
149             Row=Row+ABS(A(I,J))
150         NEXT J
151         IF Column=0 THEN 156
152         IF Row=0 THEN 156
153         Q=Column/Row
154         IF Q<Bound1 THEN 158
155         IF Q>Bound2 THEN 158
156         Ncount=Ncount+1
157         GOTO 165
158         Factor=SQR(Q)
159         FOR J=1 TO N
160             IF I=J THEN 163
161             A(I,J)=A(I,J)*Factor
162             A(J,I)=A(J,I)/Factor
163         NEXT J
164         Prfact(I)=Prfact(I)*Factor
165     NEXT I
166     Iter=Iter+1
167     IF Iter>30 THEN 184
168     IF Ncount<N THEN 142
169     Fnorm=0
170     FOR I=1 TO N
171         FOR J=1 TO N
172             Q=A(I,J)
173             Fnorm=Fnorm+Q*Q
174         NEXT J
175     NEXT I
176     Fnorm=SQR(Fnorm)
177     FOR I=1 TO N
178         FOR J=1 TO N
179             A(I,J)=A(I,J)/Fnorm
180         NEXT J
181     NEXT I
182     Enorm=Fnorm
183     GOTO 191
```

```
184    FOR I=1 TO N
185        Prfact(I)=1
186        FOR J=1 TO N
187            A(I,J)=H(I,J)
188        NEXT J
189    NEXT I
190    Enorm=1
191    SUBEXIT
192    SUBEND
193    SUB Hesqr(N,A(*),H(*),Evr(*),Evi(*),Subdia(*),Indic(*)
       ,Eps,Ex)
194    OPTION BASE 1
195    INTEGER I,J,K,L,M,Maxst,M1,Ns
196    IF N-2<0 THEN 254
197    IF N-2>0 THEN 200
198    Subdia(1)=A(2,1)
199    GOTO 254
200    M=N-2
201    FOR K=1 TO M
202        L=K+1
203        S=0
204        FOR I=L TO N
205            H(I,K)=A(I,K)
206            S=S+ABS(A(I,K))
207        NEXT I
208        IF S<>ABS(A(K+1,K)) THEN 212
209        Subdia(K)=A(K+1,K)
210        H(K+1,K)=0
211        GOTO 249
212        Sr2=0
213        FOR I=L TO N
214            Sr=A(I,K)
215            Sr=Sr/S
216            A(I,K)=Sr
217            Sr2=Sr2+Sr*Sr
218        NEXT I
219        Sr=SQR(Sr2)
220        IF A(L,K)<0 THEN 222
221        Sr=-Sr
222        Sr2=Sr2-Sr*A(L,K)
223        A(L,K)=A(L,K)-Sr
224        H(L,K)=H(L,K)-Sr*S
225        Subdia(K)=Sr*S
226        X=S*SQR(Sr2)
227        FOR I=L TO N
228            H(I,K)=H(I,K)/X
229            Subdia(I)=A(I,K)/Sr2
230        NEXT I
231        FOR J=L TO N
232            Sr=0
```

```
233            FOR I=L TO N
234                Sr=Sr+A(I,K)*A(I,J)
235            NEXT I
236            FOR I=L TO N
237                A(I,J)=A(I,J)-Subdia(I)*Sr
238            NEXT I
239         NEXT J
240         FOR J=1 TO N
241            Sr=0
242            FOR I=L TO N
243                Sr=Sr+A(J,I)*A(I,K)
244            NEXT I
245            FOR I=L TO N
246                A(J,I)=A(J,I)-Subdia(I)*Sr
247            NEXT I
248         NEXT J
249      NEXT K
250      FOR K=1 TO M
251         A(K+1,K)=Subdia(K)
252      NEXT K
253      Subdia(N-1)=A(N,N-1)
254      Eps=0
255      FOR K=1 TO N
256         Indic(K)=0
257         IF K<>N THEN Eps=Eps+Subdia(K)^2
258         FOR I=K TO N
259            H(K,I)=A(K,I)
260            Eps=Eps+A(K,I)^2
261         NEXT I
262      NEXT K
263      Eps=Ex*SQR(Eps)
264      Shift=A(N,N-1)
265      IF N<=2 THEN Shift=0
266      IF A(N,N)<>0 THEN Shift=0
267      IF A(N-1,N)<>0 THEN Shift=0
268      IF A(N-1,N-1)<>0 THEN Shift=0
269      M=N
270      Ns=0
271      Maxst=N*10
272      FOR I=2 TO N
273         FOR K=I TO N
274            IF A(I-1,K)<>0 THEN 283
275         NEXT K
276      NEXT I
277      FOR I=1 TO N
278         Indic(I)=1
279         Evr(I)=A(I,I)
280         Evi(I)=0
281      NEXT I
282      GOTO 386
```

```
283   K=M-1
284   M1=K
285   I=K
286   IF K<0 THEN 386
287   IF K=0 THEN 361
288   IF ABS(A(M,K))<=Eps THEN 361
289   IF M-2=0 THEN 366
290   I=I-1
291   IF ABS(A(K,I))<=Eps THEN 294
292   K=I
293   IF K>1 THEN 290
294   IF K=M1 THEN 366
295   S=A(M,M)+A(M1,M1)+Shift
296   Sr=A(M,M)*A(M1,M1)-A(M,M1)*A(M1,M)+.25*Shift^2
297   A(K+2,K)=0
298   X=A(K,K)*(A(K,K)-S)+A(K,K+1)*A(K+1,K)+Sr
299   Y=A(K+1,K)*(A(K,K)+A(K+1,K+1)-S)
300   R=ABS(X)+ABS(Y)
301   IF R=0 THEN Shift=A(M,M-1)
302   IF R=0 THEN 294
303   Z=A(K+2,K+1)*A(K+1,K)
304   Shift=0
305   Ns=Ns+1
306   FOR I=K TO M1
307      IF I=K THEN 313
308      X=A(I,I-1)
309      Y=A(I+1,I-1)
310      Z=0
311      IF I+2>M THEN 313
312      Z=A(I+2,I-1)
313      Sr2=ABS(X)+ABS(Y)+ABS(Z)
314      IF Sr2=0 THEN 318
315      X=X/Sr2
316      Y=Y/Sr2
317      Z=Z/Sr2
318      S=SQR(X*X+Y*Y+Z*Z)
319      IF X<0 THEN 321
320      S=-S
321      IF I=K THEN 323
322      A(I,I-1)=S*Sr2
323      IF Sr2<>0 THEN 326
324      IF I+3>M THEN 358
325      GOTO 355
326      Sr=1-X/S
327      S=X-S
328      X=Y/S
329      Y=Z/S
330      FOR J=I TO M
331         S=A(I,J)+A(I+1,J)*X
332         IF I+2>M THEN 334
```

```
333         S=S+A(I+2,J)*Y
334         S=S*Sr
335         A(I,J)=A(I,J)-S
336         A(I+1,J)=A(I+1,J)-S*X
337         IF I+2>M THEN 339
338         A(I+2,J)=A(I+2,J)-S*Y
339      NEXT J
340      L=I+2
341      IF I<M1 THEN 343
342      L=M
343      FOR J=K TO L
344         S=A(J,I)+A(J,I+1)*X
345         IF I+2>M THEN 347
346         S=S+A(J,I+2)*Y
347         S=S*Sr
348         A(J,I)=A(J,I)-S
349         A(J,I+1)=A(J,I+1)-S*X
350         IF I+2>M THEN 352
351         A(J,I+2)=A(J,I+2)-S*Y
352      NEXT J
353      IF I+3>M THEN 358
354      S=-A(I+3,I+2)*Y*Sr
355      A(I+3,I)=S
356      A(I+3,I+1)=S*X
357      A(I+3,I+2)=S*Y+A(I+3,I+2)
358   NEXT I
359   IF Ns>Maxst THEN 386
360   GOTO 283
361   Evr(M)=A(M,M)
362   Evi(M)=0
363   Indic(M)=1
364   M=K
365   GOTO 283
366   R=.5*(A(K,K)+A(M,M))
367   S=.5*(A(M,M)-A(K,K))
368   S=S*S+A(K,M)*A(M,K)
369   Indic(K)=1
370   Indic(M)=1
371   IF S<0 THEN 379
372   T=SQR(S)
373   Evr(K)=R-T
374   Evr(M)=R+T
375   Evi(K)=0
376   Evi(M)=0
377   M=M-2
378   GOTO 283
379   T=SQR(-S)
380   Evr(K)=R
381   Evi(K)=T
382   Evr(M)=R
```

```
383  Evi(M)=-T
384  M=M-2
385  GOTO 283
386  SUBEXIT
387  SUBEND
388  SUB Realve(N,M,Ivec,A(*),Vecr(*),Evr(*),Evi(*),Work(*)
     ,Indic(*),Eps,Ex)
389  Baddta=(N<=0) OR (M<=0) OR (Ivec<=0)
390  IF Baddta=0 THEN 394
391  PRINT LIN(2),"ERROR IN SUBPROGRAM Realve."
392  PRINT "N=";N,"M=";M,"Ivec=";Ivec,LIN(2)
393  PAUSE
394  OPTION BASE 1
395  INTEGER Iwork(N)
396  INTEGER I,Iter,J,K,L,Ns
397  Vecr(1,Ivec)=1
398  IF M=1 THEN 493
399  Evalue=Evr(Ivec)
400  IF Ivec=M THEN 409
401  K=Ivec+1
402  R=0
403  FOR I=K TO M
404     IF Evalue<>Evr(I) THEN 407
405     IF Evi(I)<>0 THEN 407
406     R=R+3
407  NEXT I
408  Evalue=Evalue+R*Ex
409  FOR K=1 TO M
410     A(K,K)=A(K,K)-Evalue
411  NEXT K
412  K=M-1
413  FOR I=1 TO K
414     L=I+1
415     Iwork(I)=0
416     IF A(I+1,I)<>0 THEN 420
417     IF A(I,I)<>0 THEN 432
418     A(I,I)=Eps
419     GOTO 432
420     IF ABS(A(I,I))>=ABS(A(I+1,I)) THEN 427
421     Iwork(I)=1
422     FOR J=I TO M
423        R=A(I,J)
424        A(I,J)=A(I+1,J)
425        A(I+1,J)=R
426     NEXT J
427     R=-A(I+1,I)/A(I,I)
428     A(I+1,I)=R
429     FOR J=L TO M
430        A(I+1,J)=A(I+1,J)+R*A(I,J)
431     NEXT J
```

```
432   NEXT I
433   IF A(M,M)<>0 THEN 435
434   A(M,M)=Eps
435   FOR I=1 TO N
436      IF I>M THEN 439
437      Work(I)=1
438      GOTO 440
439      Work(I)=0
440   NEXT I
441   Bound=.01/(Ex*N)
442   Ns=0
443   Iter=1
444   R=0
445   FOR I=1 TO M
446      J=M-I+1
447      S=Work(J)
448      IF J=M THEN 454
449      L=J+1
450      FOR K=L TO M
451         Sr=Work(K)
452         S=S-Sr*A(J,K)
453      NEXT K
454      Work(J)=S/A(J,J)
455      T=ABS(Work(J))
456      IF R>=T THEN 458
457      R=T
458   NEXT I
459   FOR I=1 TO M
460      Work(I)=Work(I)/R
461   NEXT I
462   R1=0
463   FOR I=1 TO M
464      T=0
465      FOR J=I TO M
466         T=T+A(I,J)*Work(J)
467      NEXT J
468      T=ABS(T)
469      IF R1>=T THEN 471
470      R1=T
471   NEXT I
472   IF Iter=1 THEN 474
473   IF Previs<=R1 THEN 493
474   FOR I=1 TO M
475      Vecr(I,Ivec)=Work(I)
476   NEXT I
477   Previs=R1
478   IF Ns=1 THEN 493
479   IF Iter>6 THEN 494
480   Iter=Iter+1
481   IF R<Bound THEN 483
```

```
482   Ns=1
483   K=M-1
484   FOR I=1 TO K
485      R=Work(I+1)
486      IF Iwork(I)=0 THEN 490
487      Work(I+1)=Work(I)+Work(I+1)*A(I+1,I)
488      Work(I)=R
489      GOTO 491
490      Work(I+1)=Work(I)*A(I+1,I)+Work(I+1)
491   NEXT I
492   GOTO 444
493   Indic(Ivec)=2
494   IF M=N THEN 499
495   J=M+1
496   FOR I=J TO N
497      Vecr(I,Ivec)=0
498   NEXT I
499   SUBEXIT
500   SUBEND
501   SUB Compve(N,M,Ivec,A(*),Vecr(*),H(*),Evr(*),Evi(*),
      Indic(*),Subdia(*),Work),Eps,Ex)
502   Baddta=(N<=0) OR (M<=0) OR (Ivec<=0)
503   IF Baddta=0 THEN 507
504   PRINT LIN(2),"ERROR IN SUBPROGRAM Compve."
505   PRINT "N=";N,"M=";M,"Ivec=";Ivec,LIN(2)
506   PAUSE
507   OPTION BASE 1
508   INTEGER Iwork(N)
509   DIM Workl(N),Work2(N)
510   INTEGER I,I1,I2,Iter,J,K,L,Ns
511   Fksi=Evr(Ivec)
512   Eta=Evi(Ivec)
513   IF Ivec=M THEN 524
514   K=Ivec+1
515   R=0
516   FOR I=K TO M
517      IF Fksi<>Evr(I) THEN 520
518      IF ABS(Eta)<>ABS(Evi(I)) THEN 520
519      R=R+3
520   NEXT I
521   R=R*Ex
522   Fksi=Fksi+R
523   Eta=Eta+R
524   R=Fksi*Fksi+Eta*Eta
525   S=2*Fksi
526   L=M-1
527   FOR I=1 TO M
528      FOR J=I TO M
529         D=0
530         A(J,I)=0
```

```
531           FOR K=I TO J
532               D=D+H(I,K)*H(K,J)
533           NEXT K
534           A(I,J)=D-S*H(I,J)
535       NEXT J
536       A(I,I)=A(I,I)+R
537   NEXT I
538   FOR I=1 TO L
539       R=Subdia(I)
540       A(I+1,I)=-S*R
541       I1=I+1
542       FOR J=1 TO I1
543           A(J,I)=A(J,I)+R*H(J,I+1)
544       NEXT J
545       IF I=1 THEN 547
546       A(I+1,I-1)=R*Subdia(I-1)
547       FOR J=I TO M
548           A(I+1,J)=A(I+1,J)+R*H(I,J)
549       NEXT J
550   NEXT I
551   K=M-1
552   FOR I=1 TO K
553       I1=I+1
554       I2=I+2
555       Iwork(I)=0
556       IF I=K THEN 558
557       IF A(I+2,I)<>0 THEN 562
558       IF A(I+1,I)<>0 THEN 562
559       IF A(I,I)<>0 THEN 585
560       A(I,I)=Eps
561       GOTO 585
562       IF I=K THEN 568
563       IF ABS(A(I+1,I))>=ABS(A(I+2,I)) THEN 568
564       IF ABS(A(I,I))>=ABS(A(I+2,I)) THEN 578
565       L=I+2
566       Iwork(I)=2
567       GOTO 571
568       IF ABS(A(I,I))>=ABS(A(I+1,I)) THEN 576
569       L=I+1
570       Iwork(I)=1
571       FOR J=I TO M
572           R=A(I,J)
573           A(I,J)=A(L,J)
574           A(L,J)=R
575       NEXT J
576       IF I<>K THEN 578
577       I2=I1
578       FOR L=I1 TO I2
579           R=-A(L,I)/A(I,I)
580           A(L,I)=R
```

```
581         FOR J=I1 TO M
582             A(L,J)=A(L,J)+R*A(I,J)
583         NEXT J
584       NEXT L
585   NEXT I
586   IF A(M,M)<>0 THEN 588
587   A(M,M)=Eps
588   FOR I=1 TO N
589      IF I>M THEN 593
590      Vecr(I,Ivec)=1
591      Vecr(I,Ivec-1)=1
592      GOTO 595
593      Vecr(I,Ivec)=0
594      Vecr(I,Ivec-1)=0
595   NEXT I
596   Bound=.01/(Ex*N)
597   Ns=0
598   Iter=1
599   FOR I=1 TO M
600      Work(I)=H(I,I)-Fksi
601   NEXT I
602   FOR I=1 TO M
603      D=Work(I)*Vecr(I,Ivec)
604      IF I=1 THEN 606
605      D=D+Subdia(I-1)*Vecr(I-1,Ivec)
606      L=I+1
607      IF L>M THEN 611
608      FOR K=L TO M
609          D=D+H(I,K)*Vecr(K,Ivec)
610      NEXT K
611      Vecr(I,Ivec-1)=D-Eta*Vecr(I,Ivec-1)
612   NEXT I
613   K=M-1
614   FOR I=1 TO K
615      L=I+Iwork(I)
616      R=Vecr(L,Ivec-1)
617      Vecr(L,Ivec-1)=Vecr(I,Ivec-1)
618      Vecr(I,Ivec-1)=R
619      Vecr(I+1,Ivec-1)=Vecr(I+1,Ivec-1)+A(I+1,I)*R
620      IF I=K THEN 622
621      Vecr(I+2,Ivec-1)=Vecr(I+2,Ivec-1)+A(I+2,I)*R
622   NEXT I
623   FOR I=1 TO M
624      J=M-I+1
625      D=Vecr(J,Ivec-1)
626      IF J=M THEN 632
627      L=J+1
628      FOR K=L TO M
629          Dl=A(J,K)
630          D=D-Dl*Vecr(K,Ivec-1)
```

```
631     NEXT K
632     Vecr(J,Ivec-1)=D/A(J,J)
633  NEXT I
634  FOR I=1 TO M
635     D=Work(I)*Vecr(I,Ivec-1)
636     IF I=1 THEN 638
637     D=D+Subdia(I-1)*Vecr(I-1,Ivec-1)
638     L=I+1
639     IF L>M THEN 643
640     FOR K=L TO M
641        D=D+H(I,K)*Vecr(K,Ivec-1)
642     NEXT K
643     Vecr(I,Ivec)=(Vecr(I,Ivec)-D)/Eta
644  NEXT I
645  L=1
646  S=0
647  FOR I=1 TO M
648     R=Vecr(I,Ivec)^2+Vecr(I,Ivec-1)^2
649     IF R<=S THEN 652
650     S=R
651     L=I
652  NEXT I
653  U=Vecr(L,Ivec-1)
654  V=Vecr(L,Ivec)
655  FOR I=1 TO M
656     B=Vecr(I,Ivec)
657     R=Vecr(I,Ivec-1)
658     Vecr(I,Ivec)=(R*U+B*V)/S
659     Vecr(I,Ivec-1)=(B*U-R*V)/S
660  NEXT I
661  B=0
662  FOR I=1 TO M
663     R=Work(I)*Vecr(I,Ivec-1)-Eta*Vecr(I,Ivec)
664     U=Work(I)*Vecr(I,Ivec)+Eta*Vecr(I,Ivec-1)
665     IF I=1 THEN 668
666     R=R+Subdia(I-1)*Vecr(I-1,Ivec-1)
667     U=U+Subdia(I-1)*Vecr(I-1,Ivec)
668     L=I+1
669     IF L>M THEN 674
670     FOR J=L TO M
671        R=R+H(I,J)*Vecr(J,Ivec-1)
672        U=U+H(I,J)*Vecr(J,Ivec)
673     NEXT J
674     U=R*R+U*U
675     IF B>=U THEN 677
676     B=U
677  NEXT I
678  IF Iter=1 THEN 680
679  IF Previs<=B THEN 691
680  FOR I=1 TO N
```

```
681      Work1(I)=Vecr(I,Ivec)
682      Work2(I)=Vecr(I,Ivec-1)
683   NEXT I
684   Previs=B
685   IF Ns=1 THEN 695
686   IF Iter>6 THEN 697
687   Iter=Iter+1
688   IF Bound>SQR(S) THEN 602
689   Ns=1
690   GOTO 602
691   FOR I=1 TO N
692      Vecr(I,Ivec)=Work1(I)
693      Vecr(I,Ivec-1)=Work2(I)
694   NEXT I
695   Indic(Ivec-1)=2
696   Indic(Ivec)=2
697   SUBEND



10    SUB Resmat(N,A(*),P(*),Q(*))
20    !   "Resmat" CALCULATES THE RESOLVENT MATRIX (INVERSE
      OF sI-A)
30    P$="(<1)"
40    IF N<1 THEN 70
50    IF N=INT(N) THEN 110
60    P$="( Not an integer )"
70    PRINT "Error in SUB Resmat: N=";N;P$
80    BEEP
90    PAUSE
100   GOTO 30
110   OPTION BASE 1
120   DIM Idn(N,N),M1(N,N),M2(N,N)
130   MAT Idn=IDN
140   FOR I=1 TO N
150   FOR J=1 TO N
160   Q(N,I,J)=0
170   IF I=J THEN Q(N,I,J)=1
180   NEXT J
190   NEXT I
200   FOR I=N TO 1 STEP -1
210   FOR J=1 TO N
220   FOR K=1 TO N
230   M1(J,K)=Q(I,J,K)
240   NEXT K
250   NEXT J
260   MAT M2=A*M1
270   P(I)=0
280   FOR J=1 TO N
```

```
290   P(I)=P(I)+M2(J,J)
300   NEXT J
310   P(I)=-P(I)/(N-I+1)
320   IF I=1 THEN 410
330   MAT M1=(P(I))*Idn
340   MAT M2=M2+M1
350   FOR J=1 TO N
360   FOR K=1 TO N
370   Q(I-1,J,K)=M2(J,K)
380   NEXT K
390   NEXT J
400   NEXT I
410   SUBEND


10    SUB Matpol(A(*),B(*),N,M,N1,A1(*),R(*))
20    !   "Matpol" CALCULATES ANY POLYNOMIAL OF ANY MATRIX.
30    IF M>=N THEN 60
40    DIM D(N,N),W(N,N)
50    Icode=1
60    DIM C(N1),P(N+1),Q(N,N,N)
70    IF Icode=1 THEN 180
80    CALL Resmat(N,A(*),P(*),Q(*))
90    P(N+1)=1
100   L=(-1)^N
110   MAT P=(L)*P
120    FOR I=1 TO N+1
130     P(I-1)=P(I)
140    NEXT I
150   REDIM P(N)
160   CALL Poldiv(B(*),P(*),M,N,C(*),R(*))
170   GOTO 210
180    FOR I=0 TO M
190     R(I)=B(I)
200    NEXT I
210   MAT D=A
220   MAT A1=ZER
230    FOR K=1 TO N-1
240     L=R(K)
250     MAT W=(L)*A
260     MAT A1=A1+W
270     MAT W=A*D
280     MAT A=W
290    NEXT K
300    FOR I=0 TO N-1
310     A1(I,I)=A1(I,I)+R(0)
320    NEXT I
330   SUBEND
```

APPENDIX

```
10      SUB Coeff(A(*),Aj(*),P,Pj,Nn)
20      FOR I=Nn TO 1 STEP -1
30      R=A(I-1)-P*A(I)+Pj*Aj(I)
40      Aj(I)=Aj(I-1)-P*Aj(I)-Pj*A(I)
50      A(I)=R
60      NEXT I
70      SUBEND
```

```
10      SUB Polpro(A(*),B(*),N,M,R(*))
20      !  "Polpro" CALCULATES THE PRODUCT OF TWO POLYNOMIALS.
30      DIM A1(M+N),B1(M+N)
40      FOR I=0 TO N
50      A1(I)=A(I)
60      NEXT I
70      FOR I=N+1 TO M+N
80      A1(I)=0
90      NEXT I
100     FOR I=0 TO M
110     B1(I)=B(I)
120     NEXT I
130     FOR I=M+1 TO M+N
140     B1(I)=0
150     NEXT I
160     FOR I=0 TO M+N
170     R(I)=0
180     FOR J=0 TO I
190     R(I)=R(I)+A1(J)*B1(I-J)
200     NEXT J
210     NEXT I
220     SUBEND
```

```
10      SUB Poldiv(N(*),D(*),Nn,Nd,C(*),R(*))
20      !  "Poldiv"  DIVIDES TWO POLYNOMIALS
30      Nc=Nn-Nd
40      FOR I=0 TO Nc
50      K=Nc-I
60      C(K)=N(Nn-I)/D(Nd)
70      IF Nc=0 THEN 110
80      FOR J=1 TO Nd
90      N(Nn-I-J)=N(Nn-I-J)-C(K)*D(Nd-J)
100     NEXT J
110     NEXT I
120     IF Nc<>0 THEN 170
130     FOR J=0 TO Nd-1
140     R(J)=N(J)-C(0)*D(J)
```

```
150    NEXT J
160    GOTO 210
170    REDIM R(Nc)
180    FOR I=0 TO Nc-1
190    R(I)=N(I)
200    NEXT I
210    SUBEND




10     SUB Expat(N,A(*),P,Evr(*),Evi(*),M(*),Cr(*),Ci(*))
20     !   "Expat" CALCULATES EXP(A*t) FOR ANY REAL SQUARE
       MATRIX A.
30     P$="(<1)"
40     IF N<1 THEN 70
50     IF N=INT(N) THEN 110
60     P$="( NOT AN INTEGER )"
70     PRINT "ERROR IN SUB Expat: N=";N;P$
80     BEEP
90     PAUSE
100    GOTO 30
110    OPTION BASE 1
120    DIM A1(N,N),Vecr(N,N),Veci(N,N),Indic(N),Vr(N,N),Vi(N,N)
130    MAT A1=A
140    CALL Eigen(N,A1(*),Evr(*),Evi(*),Vecr(*),Veci(*),Indic(*))
150    CALL Lambda(N,Evr(*),Evi(*),P,M(*))
160    CALL Vanmat(N,P,Evr(*),Evi(*),M(*),Vr(*),Vi(*))
170    CALL Detinv(N,Vr(*),Vi(*),Detr,Deti)
180    CALL Coemat(N,A(*),Vr(*),Vi(*),Cr(*),Ci(*))
190    SUBEND




10     SUB Lyap(A(*),C(*),N,Kiter,Ly(*))
20     OPTION BASE 1
30     DIM Asq(N,N),P(N,N),Pt(N,N),Psq(N,N)
40     DIM Ptsq(N,N),Phi1(N,N),Phi2(N,N),Ptemp(N,N),Ptempt(N,N)
50     CALL Norm(A(*),N,N,Norma)
60     Dt=1E-4/(2*Norma)
70     MAT Ly=(Dt)*C
80     MAT P=IDN
90     MAT A=(Dt/2)*A
100    MAT P=P-A
110    MAT A=(2)*A
120    MAT Phi1=P+A
130    MAT Asq=A*A
140    MAT A=(1/12)*Asq
```

```
150    MAT P=P+A
160    MAT Phil=Phil+A
170    MAT A=INV(P)
180    REM Matrix P=πI-h/2.A+h2/12.A^2+-1πI+h/2.A+h2/12.A^2+
190    MAT P=A*Phil
200    REM Evaluate P^2^k & Pt^2^k for k=0,1,2,........
210    MAT Pt=TRN(P)
220    MAT Psq=P*P
230    MAT Ptsq=Pt*Pt
240    MAT Ptemp=IDN
250    MAT Ptempt=IDN
260    FOR K=1 TO Kiter
270    MAT Phil=Ly*Ptemp
280    MAT Phi2=Ptempt*Phil
290    MAT Ly=Phi2+Ly
300    MAT Phil=Psq*Ptemp
310    MAT Psq=Phil
320    MAT Ptemp=Phil
330    MAT Phi2=Ptsq*Ptempt
340    MAT Ptsq=Phi2
350    MAT Ptempt=Phi2
360    NEXT K
370    PRINT "Lyapunov matrix"
380    MAT PRINT Ly;
390    SUBEND




10     SUB Cont(A(*),B(*),N,R,Qc(*))
20     !   "Cont"  CONSTRUCTS THE CONTROLLABILITY MATRIX
30     OPTION BASE 1
40     DIM A1(N,R),A2(N,R)
50     MAT A1=A*B
60     MAT A2=B
70     FOR L=1 TO R
80     FOR J=1 TO N
90     Qc(J,L)=A2(J,L)
100    NEXT J
110    NEXT L
120    FOR I=1 TO N-R
130    N2=0
140    R1=I*R+1
150    R2=R1+R-1
160    FOR L=R1 TO R2
170    N2=N2+1
180    FOR J=1 TO N
190    Qc(J,L)=A1(J,N2)
200    NEXT J
210    NEXT L
```

```
220     MAT A2=A*A1
230     MAT A1=A2
240     NEXT I
250     SUBEND



10      SUB Sepcob(A(*),B(*),N,R,Nc,W2(*),W3(*),W5(*))
20      !   "Sepcob" DETERMINES THE COMPLETELY CONTROLLABLE OR OBSERVABLE
30      !   SUBSYSTEM OF ANY LINEAR MULTIVARIABLE SYSTEM.
40      OPTION BASE 1
50      DIM W(N,N),W1(N,R),W6(N,R)
60      MAT W1=B
70      MAT W2=(0)
80      MAT W3=(0)
90      Rank=0
100     Count=0
110     J=0
120     K=0
130     MAT W=IDN
140     J=J+1
150     MAT W6=W*W1
160     MAT W1=W6
170     Count=Count+R
180     K=K+1
190       FOR I=1 TO N
200        W2(I,J)=W1(I,K)
210       NEXT I
220       MAT W3=W2
230       Nx=N
240       CALL Rank(W3(*),Nx,Nx,Rj)
250       IF Rj<Rank+1 THEN Nextk
260       Rank=Rank+1
270       IF K=R THEN 300
280       J=J+1
290       GOTO 180
300       IF Count=Nc THEN 400
310       MAT W=A
320       K=0
330       GOTO 140
340 Nextk:   IF K=R THEN Nextbl
350       GOTO 180
360 Nextbl: IF Count=Nc THEN 400
370       MAT W=A
380       K=0
390       GOTO 150
400       J=0
410       K=0
420       MAT W5=IDN
```

```
430    K=K+1
435 J=J+1
440    FOR I=1 TO N
450      W2(I,J)=W5(I,K)
460    NEXT I
470    MAT W3=W2
480    Nx=N
490    CALL Rank(W3(*),Nx,Nx,Rj)
500    IF Rj<Rank THEN 430
510    IF J=N THEN Out
530    GOTO 430
540 Out: MAT W5=INV(W2)
550    SUBEND



10     SUB Obs(A(*),C(*),N,M,Qo(*))
20     !    "Obs" CONSTRUCTS THE OBSERVABILITY MATRIX
30     OPTION BASE 1
40     DIM A3(M,N),A4(M,N)
50     MAT A3=C*A
60     MAT A4=C
70     FOR L=1 TO M
80     FOR J=1 TO N
90     Qo(L,J)=A4(L,J)
100    NEXT J
110    NEXT L
120    FOR I=1 TO N-M
130    N2=0
140    M1=I*M+1
150    M2=M1+M-1
160    FOR L=M1 TO M2
170    N2=N2+1
180    FOR J=1 TO N
190    Qo(L,J)=A3(N2,J)
200    NEXT J
210    NEXT L
220    MAT A4=A3*A
230    MAT A3=A4
240    NEXT I
250    SUBEND



10     SUB Bdnync(A(*),N,B(*),M,Type)
20     OPTION BASE 0
30     !
40     COM X0,Y0,X1,Y1,Xmin,Xmax,Ymin,Ymax
50     COM Paxes,P$,Xa,Ya,X$,Y$,Xti,Yti,Xni,Yni,Grid,Print,Pen
       ,Lt,Pt
```

```
60   !
70   ! Plots Bode, Nyquist, or Nichols diagram of G(s)=A(s)
     /B(s)
80   !                                      2            N
90   !   where A(s)=A(0) + A(1)s + A(2)s + ... + A(N)s
100  !   B(s) is similar
110  ! Type = 1-Bode , 2-Nyquist , 3-Nichols
120  !
130  INPUT "Do you want to clear the old graph? (Y/N)",A$
140  IF A$<>"Y" THEN 170
150  PLOTTER IS 13,"GRAPHICS"
160  IF P$="9872A" THEN PLOTTER IS "9872A"
170  PRINT "The frequency F will vary from"
180  PRINT " F(start) to F(end) (Hz.) according to the
     equation:"
190  PRINT "       F(new)=A*F(old)"
200  PRINT
210  DIM X(200),Y(200),Y1(200)
220 I2: INPUT "Enter F(start),F(end),A",Fs,Fe,A
230  Np=INT((LGT(Fe)-LGT(Fs))/LGT(A)+1)      ! number of points
240  REDIM X(Np),Y(Np),Y1(Np)
250  PRINT "There will be ";Np;" points."
260  INPUT "Is this O.K.?(Y/N)",A$
270  IF A$<>"Y" THEN I2
280  !
290  DEG
300  F=Fs
310  I=0
320 Loop: W=F*3.14159/180          ! Convert to radians
330      CALL Spoly(A(*),N,0,W,Ar,Ai,Amag,Aang)
340      CALL Spoly(B(*),M,0,W,Br,Bi,Bmag,Bang)
350      Mag=Amag/Bmag
360      Ang=Aang-Bang         ! value of G(s)
370      Gr=Mag*COS(Ang)
380      Gi=Mag*SIN(Ang)       ! value of G = Gr + j(Gi)
390      ON Type GOTO Bode,Nyq,Nich
400 Bode: X(I)=F
410      Y(I)=20*LGT(Mag)       ! in db
420      Y1(I)=Ang
430      GOTO Cl
440 Nyq: X(I)=Gr
450      Y(I)=Gi
460      GOTO Cl
470 Nich: X(I)=Ang
480      Y(I)=20*LGT(Mag)
490      GOTO Cl
500 Cl: F=A*F             ! next frequency value
510      I=I+1            ! next
520      IF F<=Fe THEN Loop    ! next point
530      PRINT
```

```
540     P$="CRT"
550     INPUT "Do you want the plot on the 9872A plotter?
        (Y/N)",A$
560     IF A$="Y" THEN P$="9872A"
570     ON Type GOTO Bode2,Nyq2,Nich2
580 Bode2: PRINT "BODE plot"
590     CALL P_sul(X(*),Y(*),Np)
600     X$="Frequency,Hz"
610     Y$="Mag,db"
620     ! Magnitude plot
630     Paxes=2              ! semi-log plot
640     Y$="Phase, deg."
650     CALL P_xslg
660     CALL P_sb(X(*),Y(*),Np)
670     ! Phase plot
680     Paxes=2
690     CALL P_sul(X(*),Y1(*),Np)
700     CALL P_xslg
710     CALL P_sb(X(*),Y1(*),Np)
720     GOTO End
730 Nyq2: PRINT "NYQUIST PLOT"
740     CALL P_sul(X(*),Y(*),Np)
750     X$="REAL"
760     Y$="IMAGINARY"
770     Paxes=1
780     CALL P_xlin
790     CALL P_sb(X(*),Y(*),Np)
800     GOTO End
810 Nich2: PRINT "NICHOL'S CHART"
820     CALL P_sul(X(*),Y(*),Np)
830     Paxes=1        ! semilog
840     X$="Phase,deg."
850     Y$="Mag.,db"
860     CALL P_xlin
870     CALL P_sb(X(*),Y(*),Np)
880     GOTO End
890 End: SUBEXIT


2920 ! Spoly - polynomial of s=X+jY
2930 SUB Spoly(P(*),N,Sig,W,Real,Imag,Mag,Phase)
2940 !                              2            N
2950 ! P(s)= p(0) + p(1)s + p(2)s +...+ p(N)s
2960 !       where s=Sig+jW, p(i) are real numbers
2970 ! P(s)=(Real)+j(Imag)=Mag/Phase (Phase in degrees)
2980 DEF FNMod(X,Y)=X-Y*INT(X/Y)
2990 OPTION BASE 0
3000 ! ---------SUM-----------
```

```
3001 DEG
3010 Ms=SQR(Sig^2+W^2)
3020 As=FNAtan(Sig,W)
3030 Real=0
3040 Imag=0
3050 FOR I=0 TO N STEP 1
3060    M=Ms^I*P(I)        ! s^i=Ms^i @ As*i
3070    A=As*I
3080    Real=Real+M*COS(A)
3090    Imag=Imag+M*SIN(A)
3100 NEXT I
3110 ! -----END OF SUM---------
3120 Mag=SQR(Real^2+Imag^2)
3130 Phase=FNAtan(Real,Imag)            ! arctan considering quadrant
3140 SUBEXIT
3150 SUBEND
3160 !
3170 ! Atan - arctan considering quadrant
3180 DEF FNAtan(X,Y)
3190 DEG         ! angle in degrees
3200 IF X<>0 THEN A2
3210    IF Y>0 THEN Ang=90
3220    IF Y<0 THEN Ang=270
3230    IF Y=0 THEN Ang=0
3240    GOTO End
3250 A2:Ang=ATN(Y/X)
3260 IF Y/X>=0 THEN Pos
3270 Neg:IF Y<0 THEN Q4
3280 Q2:Ang=Ang+180           ! quadrant 2
3290    GOTO End
3300 Q4:Ang=Ang+360
3310    GOTO End
3320 Pos:IF X>=0 THEN Q1
3330 Q3:Ang=Ang+180
3340    GOTO End
3350 Q1:Ang=Ang
3360 End:RETURN Ang
3370 FNEND



3390 ! P_sul - plotter setup
3400 !       -   sets ranges, labels, ticks
3410 SUB P_sul(X(*),Y(*),N)
3420 OPTION BASE 0
3430 !
3440 COM X0,Y0,X1,Y1,Xmin,Xmax,Ymin,Ymax
3450 COM Paxes,P$,Xa,Ya,X$,Y$,Xti,Yti,Xni,Yni,Grid,Print,Pen,Lt,Pt
```

```
3460 !
3470 !    LOCATE PLOT ON PAPER
3480 EXIT GRAPHICS
3490 PRINT
3500 PRINT "Please digitize X0,Y0 - then X1,Y1"
3510 WAIT 1000
3520 SETGU
3530 DIGITIZE X0,Y0
3540 MOVE X0,Y0
3550 DRAW X0,Y0        ! Put a point there
3560 WAIT 1000
3570 DIGITIZE X1,Y1
3580 MOVE X1,Y1
3590 DRAW X1,Y1
3600 PENUP
3610 SETUU
3620 !
3630 Maxx=Minx=X(0)
3640 Maxy=Miny=Y(0)
3650 FOR I=1 TO N-1
3660    IF X(I)>Maxx THEN Maxx=X(I)
3670    IF X(I)<Minx THEN Minx=X(I)
3680    IF Y(I)>Maxy THEN Maxy=Y(I)
3690    IF Y(I)<Miny THEN Miny=Y(I)
3700 NEXT I
3710 PRINT
3720 PRINT "Minx,Maxx=";Minx;Maxx
3730 PRINT "Miny,Maxy=";Miny;Maxy
3740 PRINT
3750 !
3760 M1: INPUT "Enter Xmin,Xmax,Ymin,Ymax.",Xmin,Xmax,Ymin
     ,Ymax
3770 PRINT "Xmin=";Xmin;"Xmax=";Xmax;"Ymin=";Ymin;"Ymax="
     ;Ymax
3780    INPUT "Changes ?(Y/N).",A$
3790    IF A$="Y" THEN M1
3800 M2:INPUT "Enter Xti,Xni,Yti,Yni,Xa,Ya.",Xti,Xni,Yti,
     Yni,Xa,Ya ! ti=tick interval
3810 PRINT "Xti=";Xti;"Xni=";Xni;"Yti=";Yti;"Yni=";Yni;"Xa=
     ";Xa;"Ya=";Ya
3820    INPUT "Changes ?(Y/N).",A$
     ! ni=number interval
3830    IF A$="Y" THEN M2
3840 ! DEFAULT VALUES
3850 Paxes=1
3860 Pen=1
3870 Lt=1
3880 Pt=1
3890 Grid=0
3900 Print=1
```

```
3901 INPUT "Would you like a grid?(Y/N)",A$
3902 IF A$="Y" THEN Grid=1
3910 !
3920 SUBEXIT
3930 SUBEND




3960 ! P_sb    - LOW LEVEL PLOTTER PROGRAM
3970 !           will now do semi-log plots
3980 SUB P_sb(X(*),Y(*),N)
3990 !
4000 ! X,Y - INDEPENDENT AND DEPENDENT VARIABLES (sorted in
     ascending X)
4010 ! N - the number of points in X,Y
4020 ! X0,Y0 - lower left corner of plot in GDU's
4030 ! X1,Y1 - upper right corner of plot in GDU's
4040 ! Xmin,Xmax,Ymin,Ymax - desired ranges of plot in UDU's
4050 ! P$ - plotter string
4060 ! Paxes - axes type (0=no axes, 1=lin-lin,2=lin-log)
4070 !
4080 OPTION BASE 0
4090 !
4100 COM X0,Y0,X1,Y1,Xmin,Xmax,Ymin,Ymax
4110 COM Paxes,P$,Xa,Ya,X$,Y$,Xti,Yti,Xni,Yni,Grid,Print,Pen
     ,Lt,Pt
4120 !
4130 ON ERROR GOTO Error1
4140 Plotter$="9872A"
4150 GRAPHICS
4160 SETGU                      ! GDU's
4170 LOCATE X0,X1,Y0,Y1
4180 Xl=Xmin
4190 Xh=Xmax
4200 Yl=Ymin
4210 Yh=Ymax
4220 IF Paxes<>2 THEN P6
4230    Xl=LGT(Xmin)     ! SEMILOG, CHANGE SCALING
4240    Xh=LGT(Xmax)
4250 P6: SCALE Xl,Xh,Yl,Yh       ! UDU's(assumes log values
     not given for semi log)
4260 Plotit: IF Lt>0 THEN C2
4270    IF (Pt<>1) AND (Pt<>2) AND (Pt<>3) THEN 4380
4280 C2: IF P$=Plotter$ THEN PEN Pen
4290 IF Lt<=0 THEN Point    ! NEGATIVE TYPE->POINTS and LINES
4300 Line:LINE TYPE Lt
4310    H=X(0)           ! FIRST POINT
4320    V=Y(0)
4330    IF Paxes<>2 THEN L2
4340    H=LGT(H)
```

```
4350 L2:MOVE H,V                ! LINE PLOT
4360    FOR I=1 TO N-1          ! REST OF POINTS
4370      H=X(I)
4380      V=Y(I)
4390      IF Paxes<>2 THEN L3
4400        H=LGT(H)
4410 L3: DRAW H,V
4420    NEXT I
4430    GOTO End
4440 Point:Ssize=.65            !   POINT PLOT
4450       DEG
4460       LINE TYPE 1
4470       FOR I=0 TO N-1
4480         H=X(I)
4490         V=Y(I)
4500         IF Paxes<>2 THEN P2
4510           H=LGT(H)
4520 P2:     MOVE H,V           ! PUT A POINT
4530         DRAW H,V
4540         MOVE H,V
4550         SETGU
4560         ON Pt GOSUB Circle,Triangl,Box     ! DRAW FIGURE
4570         SETUU
4580       NEXT I
4590       IF Lt=0 THEN End     ! just points
4600       Lt=ABS(Lt)           !  now connect the lines
4610       GOTO Line
4620       SUBEXIT
4630 End:  EXIT GRAPHICS
4640       SUBEND
4650  !
4660 Circle:  Size=Ssize
4670        FOR Arc=0 TO 360 STEP 30
4680          PDIR Arc
4690          RPLOT Size,0         !  RELATIVE PLOT
4700          NEXT Arc
4710          PENUP
4720          RETURN
4730 Triangl:  Size=Ssize
4740        FOR Arc=90 TO 450 STEP 120
4750          PDIR Arc
4760          RPLOT Size,0
4770          NEXT Arc
4780          PENUP
4790          RETURN
4800 Box:  Size=Ssize
4810       FOR Arc=45 TO 405 STEP 90
4820         PDIR Arc
4830         RPLOT Size,0
4840         NEXT Arc
```

```
4850        PENUP
4860        RETURN
4870 ! P_xslg - SEMILOG AXES
4880 SUB-P_xslg
4890 !
4900 ! X0,Y0 - lower left corner of plot in GDU's
4910 ! X1,Y1 - upper right corner of plot in GDU's
4920 ! Xmin,Xmax,Ymin,Ymax - desired ranges of plot in UDU's
4930 ! P$ - name of plotter
4940 ! Paxes - PRINT AXES CODE (0=NO,1=LINEAR,2=SEMILOG)
4950 !       SEMILOG - Xni doesn't matter, Xmin,Xmax=actual
     values
4960 ! Xa,Ya - location of x-axis and yaxis in UDU's
4970 ! X$,Y$ - titles for X,Y axes
4980 ! Yti - y tick interval in UDU's
4990 ! Yni - Y number interval
5000 ! Grid - grid code (0= no grid, 1=grid)
5010 ! Print - print code (0=don't print values , 1=print
     values)
5020 !         prints only in evenly spaced increments
5030 !
5040 OPTION BASE 0
5050 !
5060 COM X0,Y0,X1,Y1,Xmin,Xmax,Ymin,Ymax
5070 COM Paxes,P$,Xa,Ya,X$,Y$,Xti,Yti,Xni,Yni,Grid,Print
     ,Pen,Lt,Pt
5080 !
5090 ON ERROR GOTO Error1
5100 GRAPHICS
5110 SETGU                       ! GDU's
5120 LOCATE X0,X1,Y0,Y1
5130 Xl=LGT(Xmin)
5140 Xh=LGT(Xmax)
5150 Yl=Ymin
5160 Yh=Ymax
5170 P6:Xrg=X1-X0     !  X range in GDU's
5180 Yrg=Y1-Y0        !  Y
5190 Xru=Xh-Xl
5200 Yru=Yh-Yl
5210 SCALE Xl,Xh,Yl,Yh           ! UDU's(assumes log values not
     given for semilog)
5220 Mtcy=Yni/Yti
5230 IF P$<>Plotter$ THEN C4
5240 PEN Pen
5250 C4:LINE TYPE 1              ! AXES FOR SEMILOG
5260  SETUU
5270  Yas=LGT(Ya)                ! Y AXIS SEMILOG CONVERSION
5280  IF Print=1 THEN Printit
5290  SUBEXIT
5300 !
5310 Error1:PRINT ERRM$
```

```
5320 BEEP
5330 PAUSE
5340 GOTO Plotit
5350 !
5360 ! Prints axes and axes labels
5370 !
5380 Printit: CSIZE 2.5
5390          LINE TYPE 1
5400 ! ************AXES NUMBERS*********************
5410    IF P$<>Plotter$ THEN Nopen
5420    PEN Pen
5430 Nopen:    LINE TYPE 1
5440          SETUU                    ! UDU´s
5450 ! ----------------X----------------
5460 Labelx: LORG 6
5470       DEG
5480       LDIR 0
5490 Semilog: FOR D=Xl TO Xh STEP 1
5500  FOR I=1 TO 9      ! DO THESE LINES BETWEEN DECADES
5510   H=LGT(10.0^D*I)      ! Horizontal position
5520   IF Grid<>1 THEN S2
5530   LINE TYPE 3        ! DASHED
5540   IF I=1 THEN LINE TYPE 4 ! DASH-DOT FOR DECADE
5550   MOVE H,Yl         ! GRID LINE
5560   DRAW H,Yh
5570   LINE TYPE 1
5580   GOTO S3
5590 S2:Tl=.5
5600 IF I=1 THEN Tl=1  ! longer tick for decade
5610   MOVE H,Xa+Tl*Yru/Yrg    ! TICK
5620   DRAW H,Xa-Tl*Yru/Yrg
5630 S3:  Yp=1       ! Y POSITION
5640   CSIZE 2.5
5650   V=Xa-Yp*Yru/Yrg
5670   MOVE H,V
5680   X=10.0^H     ! ACTUAL VALUE
5690   IF I=1 THEN LABEL USING "K";X      ! LABEL DECADE ONLY
5700 S4:IF D=Xh THEN Labely          ! STOP AT LAST DECADE
5710  NEXT I
5720 NEXT D
5730 ! ----------------Y----------------
5740 Labely: LORG 8
5750     LDIR 0
5760     FOR Yax=Yl TO Yh STEP Yti         ! SEMILOG TICKS
5770     IF Grid<>1 THEN Ly4
5780     LINE TYPE 3
5790     MOVE Xl,Yax       ! PUT A GRID LINE IF SEMILOG WANTS
5800     DRAW Xh,Yax
5810     LINE TYPE 1
5820     GOTO Ly3
```

```
5830 Ly4: MOVE Yas-1*Xru/Xrg,Yax   ! TICK MARK
5840      DRAW Yas+1*Xru/Xrg,Yax
5850 Ly3: NEXT Yax
5860      MOVE Xl,Xa                ! AXES
5870      DRAW Xh,Xa
5880      MOVE Yas,Yl
5890      DRAW Yas,Yh
5900 Ly2: FOR Yax=Yl TO Yh STEP Yni  ! NUMBERS
5910      MOVE Yas-1.5*Xru/Xrg,Yax   !1.5 GDU's LEFT
5920      IF Yax<>0 THEN Ly1
5930      LABEL USING "K";Yax
5940      GOTO Nexty
5950 Ly1:  IF (ABS(Yax)>=1E5) OR (ABS(Yax)<=1E-5) THEN Toobigy
5960      LABEL USING "K";Yax
5970      GOTO Nexty
5980 Toobigy:   LABEL USING "MZ.DDE";Yax
5990 Nexty: NEXT Yax
6000 ! ------------------------------------------------------
6010 Xtitle:Csize=2
6020      LINE TYPE 1
6030       Tsize=2.5
6040        SETGU
6050        LORG 6
6060        LDIR 0
6070        Xx=X0+Xrg/2
6080        Yx=Y0-5         ! X LABEL POSITION
6090        MOVE Xx,Yx
6100        SETUU
6110        CSIZE Tsize
6120        LABEL USING "K";X$
6130 Ytitle:LORG 4
6140        LDIR 90
6150        SETGU
6160        Xy=X0-9          ! Y LABEL POSITION
6170 ! P_xlin - LOW LEVEL LINEAR   AXES
6180 SUB P_xlin
6190 !
6200 ! X0,Y0 - lower left corner of plot in GDU's
6210 ! Xl,Yl - upper right corner of plot in GDU's
6220 ! Xmin,Xmax,Ymin,Ymax - desired ranges of plot in UDU's
6230 ! P$ - name of plotter
6240 ! Xa,Ya - location of x-axis and yaxis in UDU's
6250 ! X$,Y$ - titles for X,Y axes
6260 ! Xti,Yti - x,y tick interval in UDU's
6270 ! Xni,Yni - X,Y number interval
6280 ! Grid - grid code (0= no grid, 1=grid)
6290 ! Print - print code (0=don't print values , 1=print values)
6300 !           prints only in evenly spaced increments
```

```
6310 !
6320 OPTION BASE 0
6330 !
6340 COM X0,Y0,X1,Y1,Xmin,Xmax,Ymin,Ymax
6350 COM Paxes,P$,Xa,Ya,X$,Y$,Xti,Yti,Xni,Yni,Grid,Print,Pen
     ,Lt,Pt
6360 !
6370 ON ERROR GOTO Error1
6380 Plotter$="9872A"
6390 GRAPHICS
6400 SETGU                           ! GDU's
6410 LOCATE X0,X1,Y0,Y1
6420 P6:Xrg=X1-X0      !  X range in GDU's
6430 Yrg=Y1-Y0         !  Y
6440 Xru=Xmax-Xmin
6450 Yru=Ymax-Ymin
6460 Xl=Xmin
6470 Xh=Xmax
6480 Yl=Ymin
6490 Yh=Ymax
6500 SCALE Xmin,Xmax,Ymin,Ymax  ! UDU's
6510 Mtcx=Xni/Xti
6520 Mtcy=Yni/Yti
6530 AXES Xti,Yti,Ya,Xa,Mtcx,Mtcy
6540 IF Grid=1 THEN GRID Xni,Yni,Ya,Xa,Mtcx,Mtcy
6550 GOTO Printit
6560 !
6570 Error1:PRINT ERRM$
6580 BEEP
6590 PAUSE
6600 !
6610 ! Prints axes and axes labels
6620 !
6630 Printit:  CSIZE 2.5    ! FOR GRAPH PAPER
6640           LINE TYPE 1
6650 ! *************AXES NUMBERS*********************
6660    IF P$<>Plotter$ THEN Nopen
6670    PEN Pen
6680 Nopen:    LINE TYPE 1
6690          SETUU                     ! UDU's
6700 Labelx:LORG 6
6710         DEG
6720         LDIR 0
6730    FOR Xax=Xmin TO Xmax STEP Xni
6740    H=Xax
6750    V=Xa-2*Yru/Yrg              ! 2 GDU's BELOW BOTTOM
6760      MOVE H,V
6770      IF Xax<>0 THEN Lx1
6780      LABEL USING "K";Xax   ! SHORTENED ZERO
6790      GOTO Nextx
```

```
6800 Lx1:IF (ABS(Xax)>=1E5) OR (ABS(Xax)<=1E-5) THEN Toobigx
6810 LABEL USING "K";Xax    ! SHORT VERSION
6820 GOTO Nextx
6830 Toobigx: LABEL USING "MZ.DDE";Xax
6840 Nextx: NEXT Xax
6850 ! ------------------Y------------------
6860 Labely: LORG 8
6870       LDIR 0
6880   FOR Yax=Yl TO Yh STEP Yni  ! NUMBERS
6890     MOVE Ya-1*Xru/Xrg,Yax        !2 GDU's LEFT
6900     IF Yax<>0 THEN Lyl
6910     LABEL USING "k";Yax
6920     GOTO Nexty
6930 Lyl:  IF (ABS(Yax)>=1E5) OR (ABS(Yax)<=1E-5) THEN Toobigy
6940     LABEL USING "K";Yax
6950     GOTO Nexty
6960 Toobigy:  LABEL USING "MZ.DDE";Yax
6970 Nexty: NEXT Yax
6980 IF Print=1 THEN Xtitle
6990 GOTO End
7000 ! -------------------------------------------
7010 Xtitle:Csize=2
7020       LINE TYPE 1
7030       Tsize=2.5
7040        SETGU
7050        LORG 6
7060        LDIR 0
7070        Xx=X0+Xrg/2
7080        Yx=Y0-5           ! X LABEL POSITION
7090        MOVE Xx,Yx
7100        SETUU
7110        CSIZE Tsize
7120        LABEL USING "K";X$
7130 Ytitle:LORG 4
7140        LDIR 90
7150        SETGU
7160        Xy=X0-9            ! Y LABEL POSITION
7170        Yy=Y0+Yrg/2
7180        MOVE Xy,Yy
7190        SETUU
7200        LABEL USING "K";Y$
7210        PENUP
7220 End:  EXIT GRAPHICS
7230        SUBEXIT
7240           SUBEND
```

```
10    REM STORED UNDER NAME: "Rosenb" DATA
20    REM SUB Rosenb Minimization via ROSENBROCK'S Method
30    SUB Rosenb(Ake(*),Eps(*),Km,Maxk,Mkat,Mcyc,Alpha,Beta,
      V(*),Nstep,Epsy)
40    OPTION BASE 1
50    COM D(*),Bl(*),Blen(*),Aj(*),E(*),Al(*),Afk(*)
60    Kat=1
70    MAT V=IDN
80    CALL Object(Ake(*),Sumn,Km)
90    Sumo=Sumn
100   MAT Afk=Ake
110   Kkl=1
120   IF Nstep=1 THEN 140
130   MAT E=Eps
140   FOR I=1 TO Km
150   Fbest=Sumn
160   Aj(I)=2
170   IF Nstep<>1 THEN 190
180   E(I)=Eps(I)
190   D(I)=0
200   NEXT I
210   Iii=0
220   Iii=Iii+1
230   I=1
240   FOR J=1 TO Km
250   Ake(J)=Ake(J)+E(I)*V(I,J)
260   NEXT J
270   CALL Object(Ake(*),Sumn,Km)
280   Kat=Kat+1
290   Sumdif=Fbest-Sumn
300   IF ABS(Sumdif-Epsy)<=1E-4 THEN 1140
310   IF Kat>=Maxk THEN 1140
320   IF Sumn<=Sumo THEN 390
330   FOR J=1 TO Km
340   Ake(J)=Ake(J)-E(I)*V(I,J)
350   NEXT J
360   E(I)=-Beta*E(I)
370   IF Aj(I)<1.5 THEN Aj(I)=0
380   GOTO 450
390   D(I)=D(I)+E(I)
400   E(I)=Alpha*E(I)
410   Sumo=Sumn
420   MAT Afk=Ake
430   IF Aj(I)<=1.5 THEN 450
440   Aj(I)=1
450   FOR J=1 TO Km
460   IF Aj(J)>.5 THEN 490
470   NEXT J
480   GOTO 570
490   IF I=Km THEN 520
```

```
500     I=I+1
510     GOTO 240
520     FOR J=1 TO Km
530     IF Aj(J)<2 THEN 230
540     NEXT J
550     IF Iii<Mcyc THEN 220
560     GOTO 1991
570     MAT Al=ZER
580     REM
590     REM PRINT VALUES OF STAGE, FUNCTION, INDEPENDENT
        VARIABLES
600     REM
610     PRINT "Stage Number=";Kkl
620     PRINT "Value of the Objective function=";Sumo
630     PRINT "Values of the Independent Variables:"
640     FOR Ix=1 TO Km
650     PRINT "   x(";Ix;")=";Ake(Ix)
660     NEXT Ix
670     REM
680     REM ROTATE AXES
690     REM
700     FOR I=1 TO Km
710     Kl=I
720     FOR J=1 TO Km
730     FOR K=Kl TO Km
740     Al(I,J)=D(K)*V(K,J)+Al(I,J)
750     NEXT K
760     Bl(I,J)=Al(I,J)
770     NEXT J
780     NEXT I
790     Blen(1)=0
800     FOR K=1 TO Km
810     Blen(1)=Blen(1)+Bl(1,K)*Bl(1,K)
820     NEXT K
830     Blen(1)=SQR(Blen(1))
840     FOR J=1 TO Km
850     V(1,J)=Bl(1,J)/Blen(1)
860     NEXT J
870     FOR I=2 TO Km
880     Ii=I-1
890     FOR J=1 TO Km
900     Sumavv=0
910     FOR Kk=1 TO Ii
920     Sumav=0
930     FOR K=1 TO Km
940     Sumav=Sumav+Al(I,K)*V(Kk,K)
950     NEXT K
960     Sumavv=Sumav*V(Kk,J)+Sumavv
970     NEXT Kk
980     Bl(I,J)=Al(I,J)-Sumavv
```

```
990   NEXT J
1000  NEXT I
1010  FOR I=2 TO Km
1020  Blen(I)=0
1030  FOR K=1 TO Km
1040  Blen(I)=Blen(I)+Bl(I,K)*Bl(I,K)
1050  NEXT K
1060  Blen(I)=SQR(Blen(I))
1070  FOR J=1 TO Km
1080  V(I,J)=Bl(I,J)/Blen(I)
1090  NEXT J
1100  NEXT I
1110  Kkl=Kkl+1
1120  IF Kkl>=Mkat THEN 1653
1130  GOTO 140
1140  PRINT "Total Number of Stages=";Kkl
1150  PRINT "Total Number of Function Evaluations=";Kat
1160  PRINT "Value of Objective Function =";Sumo
1170  FOR Ix=1 TO Km
1180  PRINT " x(";Ix;")=";Ake(Ix)
1190  NEXT Ix
1200  SUBEND
```

```
10    SUB Object(Ake(*),Sumn,Km)
20    OPTION BASE 1
30    X1=Ake(1)
40    X2=Ake(2)
50    Sumn=3803.84+138.08*X1+232.92*X2-123.08*X1*X1-203.64*X2*X2-182.25*X1*X2
60    Sumn=-Sumn
70    SUBEND
```

```
10    SUB Fmfp(N,M,X(*),F,G(*),Est,Eps,Limit,Ier,H(*),Count)
20    REM COMPUTE FUNCTIONM VALUE AND GRADIENT VECTOR FOR INITIAL ARGUEMENT
30    CALL Funct(N,X(*),F,G(*))
40    REM RESET ITERATION COUNTER AND GENERATE IDENTITY MATRIX
50    REM
60    Ier=0
70    Count=0
80    N2=N+N
90    N3=N2+N
100   N31=N3+1
110   K=N31
```

APPENDIX

```
120     FOR J=1 TO N
130     H(K)=1
140     IF N<=J THEN 240
150     FOR L=1 TO N-J
160     K1=K+L
170     H(K1)=0
180     NEXT L
190     K=K1+1
200     NEXT J
210     REM
220     REM     START ITERATION LOOP
230     REM
240     Count=Count+1
250     REM
260     REM     SAVE FUNCTION VALUE, ARGUMENT VECTOR AND GRADIENT
        VECTOR
270     Oldf=F
280     FOR J=1 TO N
290     K=N+J
300     H(K)=G(J)
310     K=K+N
320     H(K)=X(J)
330     REM
340     REM     DETERMINE DIRECTION VECTOR H
350     REM
360     K=J+N3
370     T=0
380     FOR L=1 TO N
390     T=T-G(L)*H(K)
400     IF L>=J THEN 430
410     K=K+N-L
420     GOTO 440
430     K=K+1
440     NEXT L
450     H(J)=T
460     NEXT J
470     REM
480     REM     CHECK WHETHER FUNCTION WILL DECREASE STEPPING
        ALONG H
490     REM
500     Dy=Hnrm=Gnrm=0
510     REM
520     REM     CALCULATE DIRECTIONAL DERIVATIVES AND TEST VALUES
        FOR DIRECTION
530     REM     VECTOR H AND GRADIENT VECTOR G
540     FOR J=1 TO N
550     Hnrm=Hnrm+ABS(H(J))
560     Gnrm=Gnrm+ABS(G(J))
570     Dy=Dy+H(J)*G(J)
580     NEXT J
```

```
590  REM
600  REM REPEAT SEARCH IN DIRECTION OF STEEPEST DESCENT IF
     DIRECTIONAL DERIVATIV t
610  REM APEARS TO BE POSITIVE OR ZERO
620  IF Dy>=1E-3 THEN 2360
630  REM
640  REM REPEAT SEARCH IN DIRECTION OF STEEPEST DESCENT IF
     DIRECTION VECTOR H IS
650  REM SMALL COMPARED TO GRADIENT VECTOR G
660  IF Hnrm/Gnrm<=Eps THEN 2360
670  REM
680  REM SEARCH MINIMUM ALONG DIRECTION H
690  REM
700  REM SEARCH ALONG H FOR POSITIVE DIRECTIONAL DERIVATIVE
710  Fy=F
720  Alfa=2*(Est-F)/Dy
730  Ambda=1
740  REM
750  REM USE ESTIMATE FOR STEP SIZE ONLY IF IT IS POSITIVE
     AND 1. OTHERWISE
760  REM TAKE 1 AS STEP SIZE.
770  IF Alfa<=1E-3 THEN 800
780  IF Alfa-Ambda>=1E-3 THEN 800
790  Ambda=Alfa
800  Alfa=0
810  REM
820  REM SAVE FUNCTION AND DERIVATIVE VALUES FOR OLD ARGUMENT
830  Fx=Fy
840  Dx=Dy
850  REM
860  REM    STEP ARGUMENT ALONG H
870  FOR I=1 TO N
880  X(I)=X(I)+Ambda*H(I)
890  NEXT I
900  REM
910  REM  COMPUTE FUNCTION VALUE AND GRADIENT FOR NEW
     ARGUMENT
920  CALL Funct(N,X(*),F,G(*))
930  Fy=F
940  REM
950  REM COMPUTE DIRECTIONAL DERIVATIVE Dy FOE NEW ARGUMENT,
     TERMINATE
960  REM SEARCH, IF Dy IS POSITIVE. IF Dy IS ZERO THE MINIMUM
     IS FOUND
970  Dy=0
980  FOR I=1 TO N
990  Dy=Dy+G(I)*H(I)
1000 NEXT I
1010 IF Dy<1E-3 THEN 1060
1020 IF Dy>1E-3 THEN 1230
```

```
1030 GOTO 1670
1040 REM TERMINATE  SEARCH ALSO IF THE FUNCTION VALUE
     INDICATES THAT
1050 REM      A MINIMUM HAS BEEN PASSED
1060 IF Fy>=Fx THEN 1230
1070 REM
1080 REM  REPEAT SEARCH AND DOUBLE STEP SIZE FOR FURTHER
     SEARCHES
1090 Ambda=Ambda+Alfa
1100 Alfa=Ambda
1110 REM END OF SARCH LOOP
1120 REM
1130 REM   TERMINATE IF CHANGE IN ARGUMENT GETS VERY LARGE
1140 IF Hnrm*Ambda<=1E10 THEN 830
1150 REM
1160 REM LINEAR SEARCH TECHNIQUE INDICATES THAT NO MINIMUM
     EXISTS
1170 Ier=2
1180 SUBEND
1190 REM
1200 REM INTERPOLATE CUBICALLY IN THE INTERVAL DEFINED BY
     THE SEARCH
1210 REM ABOVE AND COMPUTE THE ARGUMENT X FOR WHICH THE
     INTERPOLATION
1220 REM POLYNOMIAL IS MINIMIZED
1230 T=0
1240 IF Ambda<>1E-3 THEN 1260
1250 GOTO 1670
1260 Z=3*((Fx-Fy)/Ambda)+Dx+Dy
1270 Alfa=MAX(ABS(Z),ABS(Dx),ABS(Dy))
1280 Dalfa=Z/Alfa
1290 Dalfa=Dalfa*Dalfa-Dx/Alfa*Dy/Alfa
1300 IF Dalfa<1E-3 THEN 2360
1310 W=Alfa*SQR(Dalfa)
1320 Alfa=(Dy+W-Z)*Ambda/(Dy+2*W-Dx)
1330 FOR I=1 TO N
1340 X(I)=X(I)+(T-Alfa)*H(I)
1350 NEXT I
1360 REM
1370 REM TERMINATE IF THE VALUE OF THE ACTUAL FUNCTION AT X
     IS LESS THAN
1380 REM THE FUNCTION VALUES AT THE INTERVAL ENDS. OTHERWISE
     REDUCE THE
1390 REM INTERVAL BY CHOOSING ONE END-POINT EQUAL TO X AND
     REPEAT THE
1400 REM INTERPOLATION. WHICH END-POINT IS CHOSEN DEPENDS ON
     THE VALUE OF
1410 REM THE FUNCTION AND ITS GRADIENT AT X
1420 REM
1430 CALL Funct(N,X(*),F,G(*))
```

```
1440 IF F>Fx THEN 1460
1450 IF F<=Fy THEN 1670
1460 Dalfa=0
1470 FOR I=1 TO N
1480 Dalfa=Dalfa+G(I)*H(I)
1490 IF Dalfa>=1E-3 THEN 1580
1500 IF F>Fx THEN 1580
1510 IF F<Fx THEN 1530
1520 IF Dx=Dalfa THEN 1670
1530 Fx=F
1540 Dx=Dalfa
1550 T=Alfa
1560 Ambda=Alfa
1570 GOTO 1240
1580 IF Fy<>F THEN 1600
1590 IF Dy-Dalfa=1E-3 THEN 1670
1600 Fy=F
1610 Dy=Dalfa
1620 Ambda=Ambda-Alfa
1630 GOTO 1230
1640 REM
1650 REM COMPUTE DIFFERENCE VECTORS OF ARGUMENT AND GRADIENT
     FROM TWO
1660 REM CONSECUTIVE ITERATIONS
1670 FOR J=1 TO N
1680 K=N+J
1690 H(K)=G(J)-H(K)
1700 K=N+K
1710 H(K)=X(J)-H(K)
1720 NEXT J
1730 REM
1740 REM TERMINATE IF FUNCTION HAS NOT DECREASED DURING LAST
     ITERATION
1750 IF F-Oldf>Eps THEN 2360
1760 REM
1770 REM TEST LENGTH OF ARGUEMENT VECTOR AND DIRECTION VECTOR
     IF AT LEAST N
1780 REM ITERATIONS HAVE BEEN EXECUTED. TERMINATE IF BOTH ARE
     LESS THAN Eps
1790 Ier=0
1800 IF Count<N THEN 1930
1810 T=Z=0
1820 FOR J=1 TO N
1830 K=N+J
1840 W=H(K)
1850 K=K+N
1860 T=T+ABS(H(K))
1870 Z=Z+W*H(K)
1880 NEXT J
1890 IF Hnrm>Eps THEN 1930
```

APPENDIX

```
1900 IF T<=Eps THEN 2500
1910 REM
1920 REM TERMINATE IF NUMBER OF ITERATIONS WOULD EXCEED Limit
1930 IF Count>=Limit THEN 2320
1940 REM
1950 REM    PREPARE TO UPDATE MATRIX H
1960 Alfa=0
1970 FOR J=1 TO N
1980 K=J+N3
1990 W=0
2000 FOR L=1 TO N
2010 K1=N+L
2020 W=W+H(K1)*H(K)
2030 IF L>=J THEN 2060
2040 K=K+N-L
2050 GOTO 2070
2060 K=K+1
2070 NEXT L
2080 K=N+J
2090 Alfa=Alfa+W*H(K)
2100 H(J)=W
2110 NEXT J
2120 REM
2130 REM REPEAT SEARCH IN DIRECTION OF STEEPEST DESACENT IF RESULTS ARE
2140 REM     NOT SATISFACTORY
2150 IF Z*Alfa=1E-3 THEN 110
2160 REM
2170 REM   UPDATE MATRIX H
2180 K=N31
2190 FOR L=1 TO N
2200 K1=N2+L
2210 FOR J=L TO N
2220 Nj=N2+J
2230 H(K)=H(K)+H(K1)*H(Nj)/Z-H(L)*H(J)/Alfa
2240 K=K+1
2250 NEXT J
2260 NEXT L
2270 GOTO 240
2280 REM
2290 REM      END OF ITERATION LOOP
2300 REM
2310 REM    NO CONVERGENCE AFTER Limit ITERATIONS
2320 Ier=1
2330 SUBEND
2340 REM
2350 REM   RESTORE OLD VALUES OF FUNCTION AND ARGUMENTS
2360 FOR J=1 TO N
2370 K=N2+J
2380 X(J)=H(K)
```

```
2390 NEXT J
2400 CALL Funct(N,X(*),F,G(*))
2410 REM
2420 REM    REPEAT IN DIRECTIO OF STEEPEST DESCENT IF
     DERIVATIVE
2430 REM      FAILS TO BE SUFFICIENTLY SMALL
2440 IF Gnrm<=Eps THEN 2490
2450 REM    TEST REPEATED FAILURE OF ITERATIONS
2460 IF Ier<0 THEN 2500
2470 Ier=-1
2480 GOTO 110
2490 Ier=0
2500 SUBEND

10    SUB Funct(N,Arg(*),Val,Grad(*))
20    REM
30    REM ARGUMENT LIST
40    REM
50    REM Arg   = VECTOR OF X VALUES
60    REM Val   = OBJECTIVE FUNCTION EQUATION
70    REM Grad  = VECTOR OF OBJECTIVE FUNCTION DERIVATIVES
      (Dim(N))
80    X1=Arg(1)
90    X2=Arg(2)
100   Rho=1
110   Val=Rho*X1/(2*X2)+X2/2+1/(2*X1*X2)
120   Grad(1)=Rho/(2*X2)-1/(2*X1^2*X2)
130   Grad(2)=-Rho*X1/(2*X2^2)-1/(2*X1*X2^2)+.5
140   SUBEND




10    SUB Lambda(N,Evr(*),Evi(*),P,M(*))
20    !   "Lambda" DETERMINES DISTINCT EIGENVALUES AND THEIR
      MULTIPLICITIES
30    P$="(<1)"
40    IF N<1 THEN 70
50    IF N=INT(N) THEN 110
60    P$="( NOT AN INTEGER )"
70    PRINT "ERROR IN SUB Lambda: N=";N;P$
80    BEEP
90    PAUSE
100   GOTO 30
110   P=N
120   MAT M=CON
130   FOR I=1 TO P-1
140   FOR J=1 TO P-I
150   IF (Evr(I+J)<>Evr(I)) OR (Evi(I+J)<>Evi(I)) THEN 240
160   M(I)=M(I)+1
```

```
170    P=P-1
180    IF J>P-I THEN 250
190    FOR K=1 TO P-I-J+1
200    Evr(I+J+K-1)=Evr(I+J+K)
210    Evi(I+J+K-1)=Evi(I+J+K)
220    NEXT K
230    GOTO 150
240    NEXT J
250    NEXT I
260    OPTION BASE 1
270    REDIM Evr(P),Evi(P),M(P)
280    SUBEND



10     SUB Vanmat(N,P,Evr(*),Evi(*),M(*),Vr(*),Vi(*))
20     !   "Vanmat" CONSTUCTS A VANDERMONDE MATRIX FROM DISTINCT EIGENVALUES.
30     P$="(<1)"
40     IF N<1 THEN 70
50     IF N=INT(N) THEN 110
60     P$="( NOT AN INTEGER )"
70     PRINT "ERROR IN SUB Vanmat: N= ";N;P$
80     BEEP
90     PAUSE
100    GOTO 30
110    Nm=0
120    FOR I=1 TO P
130    Nm=Nm+M(I)
140    NEXT I
150    IF N=Nm THEN 180
160    P$="(<>M(1)+...+M(P))"
170    GOTO 70
180    OPTION BASE 1
190    DIM Dr(P,N),Di(P,N)
200    FOR I=1 TO P
210    Dr(I,1)=1
220    Di(I,1)=0
230    FOR J=2 TO N
240    Dr(I,J)=Dr(I,J-1)*Evr(I)-Di(I,J-1)*Evi(I)
250    Di(I,J)=Dr(I,J-1)*Evi(I)+Di(I,J-1)*Evr(I)
260    NEXT J
270    NEXT I
280    R=0
290    FOR I=1 TO P
300    FOR J=1 TO M(I)
310    R=R+1
320    FOR K=1 TO N
330    IF J=1 THEN GOTO 380
340    IF K=N THEN GOTO 410
```

```
350   Vr(R,K)=(N-K)*Vr(R-1,K+1)
360   Vi(R,K)=(N-K)*Vi(R-1,K+1)
370   GOTO 420
380   Vr(R,K)=Dr(I,N-K+1)
390   Vi(R,K)=Di(I,N-K+1)
400   GOTO 420
410   Vr(R,K)=Vi(R,K)=0
420   NEXT K
430   NEXT J
440   NEXT I
450   SUBEND
```

```
10    SUB Coemat(N,A(*),Vir(*),Vii(*),Cr(*),Ci(*))
20    ! "Coemat" CALCULATES THE COEFFICIENT MATRICES FOR
      EXP(A*t)
30    P$="(<1)"
40    IF N<1 THEN 70
50    IF N=INT(N) THEN 110
60    P$="( NOT AN INTEGER )"
70    PRINT "ERROR IN SUB Coemat: N= ";N;P$
30    BEEP
90    PAUSE
100   GOTO 30
110   OPTION BASE 1
120   DIM T1(N,N),T2(N,N),Tr(N,N),Ti(N,N)
130   MAT Cr=ZER
140   MAT Ci=ZER
150   MAT T1=IDN
160   I=1
170   GOTO 210
180   FOR I=2 TO N
190   MAT T2=T1*A
200   MAT T1=T2
210   FOR J=1 TO N
220   MAT Tr=(Vir(N-I+1,J))*T1
230   MAT Ti=(Vii(N-I+1,J))*T1
240   FOR K=1 TO N
250   FOR L=1 TO N
260   Cr(J,K,L)=Cr(J,K,L)+Tr(K,L)
270   Ci(J,K,L)=Ci(J,K,L)+Ti(K,L)
280   NEXT L
290   NEXT K
300   NEXT J
310   IF I=1 THEN 180
320   NEXT I
330   SUBEND
```

```
10      SUB Coeffs(N,W(*),Lambda(*))
20      !    "Coeffs" CALCULATES THE COEFFICIENTS W(i) OF AN
        Nth-ORDER POLYNOMIAL
30      !    WHOSE ROOTS ARE THE REAL NUMBERS Lambda(i),
        i=1,2,...,N
40      OPTION BASE 1
50       FOR In=1 TO N
60        W(1,In)=0
70         FOR I1=1 TO N
80          IF In=1 THEN One
90           FOR I2=I1+1 TO N
100           IF In=2 THEN Two
110            FOR I3=I2+1 TO N
120             IF In=3 THEN Three
130              FOR I4=I3+1 TO N
140               IF In=4 THEN Four
150                FOR I5=I4+1 TO N
160                 GOTO Five
170 One: W(1,In)=W(1,In)+Lambda(I1,1)
180    GOTO 300
190 Two: W(1,In)=W(1,In)+Lambda(I1,1)*Lambda(I2,1)
200    GOTO 290
210 Three: W(1,In)=W(1,In)+Lambda(I1,1)*Lambda(I2,1)
       *Lambda(I3,1)
220    GOTO 280
230 Four:  W(1,In)=W(1,In)+Lambda(I1,1)*Lambda(I2,1)*Lambda
       (I3,1)*Lambda(I4,1)
240    GOTO 270
250 Five:  W(1,In)=W(1,In)+Lambda(I1,1)*Lambda(I2,1)*Lambda
       (I3,1)*Lambda(I4,1)*L bda(I5,1)
260          NEXT I5
270         NEXT I4
280        NEXT I3
290       NEXT I2
300      NEXT I1
310     NEXT In
320    SUBEND




10      SUB Detinv(N,Vr(*),Vi(*),Detr,Deti)
20      ! "Detinv" FIND THE DETERMINANT AND THE INVERSE OF A
        SQUARE COMPLEX MATRIX.
30      P$="(<1)"
40      IF N<1 THEN 70
50      IF N=INT(N) THEN 110
60      P$="( NOT AN INTEGER )"
70      PRINT "ERROR IN SUB Detinv: N=";N;P$
80      BEEP
```

```
90      PAUSE
100     GOTO 30
110     OPTION BASE 1
120     DIM Qr(N,N),Qi(N,N),Q1r(N,N),Q1i(N,N),Q2r(N,N),Q2i(N,N)
        ,Q3r(N,N),Q3i(N,N)
130     Pr=Pi=0
140     FOR I=1 TO N
150     Pr=Pr+Vr(I,I)
160     Pi=Pi+Vi(I,I)
170     NEXT I
180     MAT Qr=IDN
190     MAT Qi=ZER
200     Pr=-Pr
210     Pi=-Pi
220     FOR I=2 TO N
230     MAT Q3r=Qr*Vr
240     MAT Q2r=Qi*Vi
250     MAT Q1r=Q3r-Q2r
260     MAT Q3r=IDN
270     MAT Q2r=(Pr)*Q3r
280     MAT Q3i=Qr*Vi
290     MAT Q2i=Qi*Vr
300     MAT Q1i=Q2i+Q3i
310     MAT Q3i=IDN
320     MAT Q2i=(Pi)*Q3i
330     MAT Qr=Q1r+Q2r
340     MAT Qi=Q1i+Q2i
350     MAT Q3r=Qr*Vr
360     MAT Q2r=Qi*Vi
370     MAT Q3i=Qr*Vi
380     MAT Q2i=Qi*Vr
390     MAT Q1r=Q3r-Q2r
400     MAT Q1i=Q3i+Q2i
410     Pr=Pi=0
420     FOR J=1 TO N
430     Pr=Pr+Q1r(J,J)
440     Pi=Pi+Q1i(J,J)
450     NEXT J
460     Pr=-Pr/I
470     Pi=-Pi/I
480     NEXT I
490     Detr=Pr
500     Deti=Pi
510     IF (Pr<>0) OR (Pi<>0) THEN 530
520     SUBEXIT
530     IF N/2=INT(N/2) THEN 560
540     Detr=-Pr
550     Deti=-Pi
560     Pir=-Pr/(Pr*Pr+Pi*Pi)
570     Pii=Pi/(Pr*Pr+Pi*Pi)
```

```
580    MAT Q3r=(Pir)*Qr
590    MAT Q2r=(Pii)*Qi
600    MAT Q3i=(Pii)*Qr
610    MAT Q2i=(Pir)*Qi
620    MAT Vr=Q3r-Q2r
630    MAT Vi=Q3i+Q2i
640    SUBEND
```

APPENDIX
Utility Routines

Mat	Inputs any nxm matrix or vector in Option Base 1.
Vec	Inputs any vector in Option Base 0.
Prtmat	Prints an nxm matrix
ROOTFD	Driver program to find roots of any polynomial
Rootfd	Finds roots of a polynominal with real or complex coefficients
Inpmat	Inputs any nxm matrix
Prtplt	Subroutine to print state and output vectors of a linear system and driver to plot them
Kutta	Solve an nth order ordinary differential equation via the 4th order Runge-Kutta method
Cheby	Performs a Chebyschev sense curve fitting of a tabulated function
Print	A vector print and plot driver routine
Plot	A CRT/XY plotter plotting routine for any number of functions

APPENDIX

```
10      SUB Mat(Aa(*),Nn,Mm,Dd$)
20      !    "Mat" INPUTS AN NnxMm MATRIX
30      OPTION BASE 1
40      CALL Aread(Aa(*),Nn,Mm,Dd$)
50      SUBEND
60      SUB Aread(A(*),N,M,D$)
70      OPTION BASE 1
80      FIXED 0
90      IF (N=1) AND (M=1) THEN 120
100     IF N=1 THEN Rowread
110     IF M=1 THEN Colread
120     REDIM A(N,M)
130     PRINT LIN(1),"Matrix ";D$;"(";N;"x";M;"):";
140     PRINT LIN(1)
150     FOR I=1 TO N
160     FOR J=1 TO M
170     DISP D$;"(";I;",";J;")";
180     INPUT A(I,J)
190     NEXT J
200     NEXT I
210     IF M>5 THEN 250
220     FLOAT 4
230     MAT PRINT A;
240     GOTO 400
250     L=4
260     FOR J=1 TO M STEP 5
270     IF J+5>M THEN L=M-J
280     FOR K=0 TO L
290     PRINT TAB(13*K+13-INT(LGT(J+K)));J+K;
300     NEXT K
310     FOR I=1 TO N
320     PRINT TAB(0);SPA(INT(LGT(N))-INT(LGT(I)));I;SPA(2);
330     FLOAT 4
340     FOR K=0 TO L
350     PRINT A(I,J+K);
360     NEXT K
370     FIXED 0
380     NEXT I
390     NEXT J
400     INPUT "Any CHANGES (Y/N)?",C$
410     IF (C$="Y") OR (C$="y") THEN 440
420     IF (C$="N") OR (C$="n") THEN 970
430     GOTO 400
440     PRINT "COORDINATES of the element of";D$;" (*),[row,
        column]?"
450     INPUT I,J
460     IF (I<1) OR (I>N) OR (J<1) OR (J>M) OR (I<>INT(I)) OR
        (J<>INT(J)) THEN 440
470     FIXED 0
480     DISP D$;"(";I;",";J;")";
```

```
490  INPUT A(I,J)
500  PRINT LIN(2)
510  FLOAT 4
520  MAT PRINT A;
530  GOTO 400
540 Rowread: Ndim=M
550  PRINT LIN(1),"Row Vector ";D$;"(";Ndim;"):";LIN(1)
560  FOR I=1 TO M
570  FIXED 0
580  DISP D$;"(";I;")";
590  INPUT A(1,I)
600  NEXT I
610  FLOAT 4
620  MAT PRINT A;
630  INPUT "Any CHANGES (Y/N)?",C$
640  IF (C$="Y") OR (C$="y") THEN 670
650  IF (C$="N") OR (C$="n") THEN 970
660  GOTO 630
670  INPUT "Element number to be changed?",In
680  IF (In<1) OR (In>M) THEN 670
690  FIXED 0
700  DISP D$;"(";In;
710  INPUT A(1,In)
720  FLOAT 4
730  MAT PRINT A;
740  GOTO 630
750  PRINT LIN(1)
760 Colread: Ndim=N
770  PRINT LIN(1),"Column Vector ";D$;"(";Ndim;"):";LIN(1)
780  FOR I=1 TO N
790  FIXED 0
800  DISP D$;"(";I;")";
810  INPUT A(I,1)
820  NEXT I
830  FLOAT 4
840  MAT PRINT A;
850  INPUT "Any CHANGES (Y/N)?",C$
860  IF (C$="Y") OR (C$="y") THEN 890
870  IF (C$="N") OR (C$="n") THEN 970
880  GOTO 850
890  INPUT "Element number to be changed?",In
900  IF (In<1) OR (In>N) THEN 890
910  FIXED 0
920  DISP D$;"(";In;
930  INPUT A(In,1)
940  FLOAT 4
950  MAT PRINT A;
960  GOTO 850
970  SUBEND
```

```
10    SUB Vec(A(*),N,A$)
20    !    "Vec"   INPUTS A VECTOR IN OPTION BASE 0
30    FIXED 0
40    FOR I=0 TO N
50    DISP A$;"(";I;")=";
60     INPUT A(I)
70    NEXT I
80    FOR I=0 TO N
90    PRINT A$;"(";I;")=";
100   FLOAT 4
110   PRINT A(I)
120   FIXED 0
130   NEXT I
140   PRINT LIN(1)
150   INPUT "Any Changes(Y/N)",X$
160   IF (X$="Y") OR (X$="y") THEN 190
170   IF (X$="N") OR (X$="n") THEN 270
180   GOTO 150
190   INPUT "Element Number to be Changed ",In
200   FIXED 0
210   DISP A$;"(";In;")=";
220   INPUT A(In)
230   PRINT A$;"(";In;")=";
240   FLOAT 4
250   PRINT A(In)
260   GOTO 150
270   SUBEND
```

```
10    SUB Prtmat(A(*),N,M)
20    !    "Prtmat" PRINTS AN NxM MATRIX
30    Nm$="N"
40    Nm=N
50    FOR I=1 TO 2
60    P$=" (<1) "
70    IF Nm<1 THEN 100
80    P$="( NOT AN INTEGER )"
90    IF Nm=INT(Nm) THEN 140
100   PRINT "ERROR IN SUB Prtmat: ";Nm$;" =";Nm;P$
110   BEEP
120   PAUSE
130   GOTO 30
140   Nm$="M"
150   Nm=M
160   NEXT I
170   IF M>5 THEN 210
180   FLOAT 4
190   MAT PRINT A;
```

```
200   GOTO 360
210   L=4
220   FOR J=1 TO M STEP 5
230   IF J+5>M THEN L=M-J
240   FOR K=0 TO L
250   PRINT TAB(13*K+13-INT(LGT(J+K)));J+K;
260   NEXT K
270   FOR I=1 TO N
280   PRINT TAB(0);SPA(INT(LGT(N))-INT(LGT(I)));I;SPA(2);
290   FLOAT 4
300   FOR K=0 TO L
310   PRINT A(I,J+K);
320   NEXT K
330   FIXED 0
340   NEXT I
350   NEXT J
360   SUBEND



10    !       PROGRAM NAME : "ROOTFD" PROG
20    !       THIS PROGRAM FINDS THE ROOTS OF ANY GENERAL
              POLYNOMIAL
30    !       WITH COMPLEX COEFFICIENTS.
40    INPUT "HAVE YOU LINKED SUB <<Rootfd>> (Y/N)?",C$
50    IF (C$="Y") OR (C$="y") THEN 90
60    IF (C$="N") OR (C$="n") THEN 80
70    GOTO 40
80    LINK "Rootfd",1010,90
90    INPUT "Degree of Polynomial?",N
100   PRINT LIN(1),"Degree of Polynomial=";N
110   IF N<=0 THEN 90
120   CALL Rootfd1(N)
130   END
140   SUB Rootfd1(N)
150   DIM Rroot(1:N),Iroot(1:N),Rcoef(0:N),Icoef(0:N)
160   INPUT "Max # of Iterations?",Itmax
170   PRINT "Max # of Iterations=";Itmax
180   IF Itmax<=0 THEN 160
190   INPUT "Tolerance for Roots?",Tola
200   PRINT "Tolerance for Roots=";Tola
210   IF Tola<=0 THEN 190
220   INPUT "Tolerance for Functional Evaluations?",Tolf
230   PRINT "Tolerance for Functional Evaluations=";Tolf
240   IF Tolf<=0 THEN 220
250   PRINT LIN(1),"Coeffs: [(Rcf(0)+Icf(0)*I)+(Rcf(1)+Icf
      (1)*I)*x^1+.....]"
260   PRINT LIN(1),SPA(8),"REAL","    IMAGINARY",LIN(1)
270   FOR I=0 TO N
```

```
280       DISP "Rcoef(";I;")=";
290       INPUT Rcoef(I)
300       DISP "Icoef(";I;")=";
310       INPUT Icoef(I)
320       PRINT USING 340;Rcoef(I),Icoef(I)
330    NEXT I
340    IMAGE 3X,MZ.6DE,5X,MZ.6DE
350    LINPUT "CHANGES (Y/N)",C$
360    IF (C$="N") OR (C$="n") THEN 480
370    IF (C$="Y") OR (C$="y") THEN 390
380    GOTO 350
390    INPUT "COEFFICIENT NUMBER",I
400    IF (I<0) OR (I>N) THEN 390
410    DISP "Rcoef(";I;")=";
420    INPUT Rcoef(I)
430    DISP "Icoef(";I;")=";
440    INPUT Icoef(I)
450    PRINT USING 460;I,Rcoef(I),I,Icoef(I)
460    IMAGE "Rcoef(",DDD,")=",MZ.6DE,5X,"Icoef(",DDD,")="
       ,MZ.6DE
470    GOTO 350
480    PRINT LIN(2)
490    CALL Rootfd(N,Rcoef(*),Icoef(*),Tola,Tolf,Itmax,Rroot
       (*),Iroot(*))
500    PRINT LIN(2),"ROOTS:",LIN(1),SPA(8),"REAL","
       IMAGINARY",LIN(2)
510    FOR I=1 TO N
520       PRINT USING 540;Rroot(I),Iroot(I)
530    NEXT I
540    IMAGE 3X,MZ.6DE,5X,MZ.6DE
550    PRINT LIN(5)
560    SUBEXIT


1010 SUB Rootfd(N,Rcoef(*),Icoef(*),Tola,Tolf,Itmax,Rroot(*)
     ,Iroot(*))
1020 Baddta=(N<=0) OR (Tola<=0) OR (Tolf<=0) OR (Itmax<=0)
1030 IF Baddta=0 THEN 1100
1040 PRINT LIN(2),"ERROR IN SUBPROGRAM Siljak."
1050 PRINT "N=";N,"Tola=";Tola
1060 PRINT "Tolf=";Tolf,"Itmax=";Itmax,LIN(2)
1070 PAUSE
1080 GOTO 1020
1090 DIM Xsiljak(0:N),Ysiljak(0:N)
1100 MAT Rroot=(9.999999E99)
1110 MAT Iroot=(9.999999E99)
1120 Nn=N
1130 IF N=1 THEN 1720
1140 Y=Ysiljak(1)=Xsiljak(0)=1
```

```
1150 X=Xsiljak(1)=.1
1160 Ysiljak(0)=L=0
1170 GOSUB Siljak
1180 G=F
1190 M=Q=P=0
1200 L=L+1
1210 FOR K=1 TO N
1220     P=P+K*(Rcoef(K)*Xsiljak(K-1)-Icoef(K)*Ysiljak(K-1))
1230     Q=Q+K*(Rcoef(K)*Ysiljak(K-1)+Icoef(K)*Xsiljak(K-1))
1240 NEXT K
1250 Z=P*P+Q*Q
1260 Deltax=-(U*P+V*Q)/Z
1270 Deltay=(U*Q-V*P)/Z
1280 M=M+1
1290 Xsiljak(1)=X+Deltax
1300 Ysiljak(1)=Y+Deltay
1310 GOSUB Siljak
1320 IF F>=G THEN 1380
1330 IF (ABS(Deltax)<Tola) AND (ABS(Deltay)<Tola) THEN 1530
1340 IF L>Itmax THEN 1480
1350 X=Xsiljak(1)
1360 Y=Ysiljak(1)
1370 GOTO 1180
1380 IF M>20 THEN 1420
1390 Deltax=Deltax/4
1400 Deltay=Deltay/4
1410 GOTO 1280
1420 IF (ABS(U)<=Tolf) AND (ABS(V)<=Tolf) THEN 1530
1430 PRINT LIN(2),"ERROR IN SUBPROGRAM Siljak."
1440 PRINT "THE INTERVAL SIZE HAS BEEN QUARTERED 20 TIMES AND "
1450 PRINT "THE TOLERANCE FOR FUNCTIONAL EVALUATIONS IS STILL NOT MET."
1460 PRINT "Tolf=";Tolf,"U=";U,"V=";V,LIN(2)
1470 PAUSE
1480 PRINT LIN(2),"ERROR IN SUBROUTINE Siljak."
1490 PRINT "MAXIMUM # OF ITERATIONS HAS BEEN EXCEEDED."
1500 PRINT "L=";L,"Itmax=";Itmax,LIN(2)
1510 PAUSE
1520 GOTO 1340
1530 Rroot(N)=Xsiljak(1)
1540 Iroot(N)=Ysiljak(1)
1550 A=Rcoef(N)
1560 B=Icoef(N)
1570 Rcoef(N)=Icoef(N)=0
1580 X=Xsiljak(1)
1590 Y=Ysiljak(1)
1600 FOR K=N-1 TO 0 STEP -1
1610     C=Rcoef(K)
1620     D=Icoef(K)
```

```
1630      U=Rcoef(K+1)
1640      V=Icoef(K+1)
1650      Rcoef(K)=A+X*U-Y*V
1660      Icoef(K)=B+X*V+Y*U
1670      A=C
1680      B=D
1690 NEXT K
1700 N=N-1
1710 IF N<>1 THEN 1140
1720 A=Rcoef(0)
1730 U=Rcoef(1)
1740 B=Icoef(0)
1750 V=Icoef(1)
1760 T=U*U+V*V
1770 Rroot(1)=-(A*U+B*V)/T
1780 Iroot(1)=(A*V-U*B)/T
1790 N=Nn
1800 SUBEXIT
1810 Siljak: Z=Xsiljak(1)*Xsiljak(1)+Ysiljak(1)*Ysiljak(1)
1820 T=2*Xsiljak(1)
1830 FOR K=0 TO N-2
1840      Xsiljak(K+2)=T*Xsiljak(K+1)-Z*Xsiljak(K)
1850      Ysiljak(K+2)=T*Ysiljak(K+1)-Z*Ysiljak(K)
1860 NEXT K
1870 U=V=0
1880 FOR K=0 TO N
1890      U=U+Rcoef(K)*Xsiljak(K)-Icoef(K)*Ysiljak(K)
1900      V=V+Rcoef(K)*Ysiljak(K)+Icoef(K)*Xsiljak(K)
1910 NEXT K
1920 F=U*U+V*V
1930 RETURN
1940 SUBEND




10    SUB Inpmat(N,M,Name$,Matrix(*))
20    REM *********************
30    REM * INPUT N BY M MATRIX *
40    REM *********************
50    Nm$="N"
60    Nm=N
70    FOR I=1 TO 2
80    P$="(<1)"
90    IF Nm<1 THEN 120
100   P$="( NOT AN INTEGER )"
110   IF Nm=INT(Nm) THEN 160
120   PRINT "ERROR IN SUB Inpmat: ";Nm$;" =";Nm;P$
130   BEEP
140   PAUSE
```

```
150    GOTO 50
160    Nm$="M"
170    Nm=M
180    NEXT I
190    FOR I=1 TO N
200    FOR J=1 TO M
210    FIXED 0
220    DISP Name$;"(";I;",";J;") =";
230    INPUT Matrix(I,J)
240    NEXT J
250    NEXT I
260    PRINT "DIMENSION OF MATRIX ";Name$;" =";N;"BY";M,LIN(2),
       "MATRIX ";Name$;":" IN(1)
270    IF M>5 THEN 310
280    FLOAT 5
290    MAT PRINT Matrix;
300    GOTO 470
310    L=4
320    FOR J=1 TO M STEP 5
330    IF J+5>M THEN L=M-J
340    FOR K=0 TO L
350    PRINT TAB(13*K+13-INT(LGT(J+K)));J+K;
360    NEXT K
370    FOR I=1 TO N
380    PRINT TAB(0);SPA(INT(LGT(N))-INT(LGT(I)));I;SPA(2);
390    FLOAT 5
400    FOR K=0 TO L
410    PRINT Matrix(I,J+K);
420    NEXT K
430    FIXED 0
440    NEXT I
450    NEXT J
460    A$=" "
470    INPUT "CHANGES (Y/N)?",A$
480    A$=UPC$(A$)
490    IF (A$="Y") OR (A$="YES") THEN 520
500    IF (A$="N") OR (A$="NO") THEN 600
510    GOTO 460
520    DISP "COOREINATES OF ";Name$;" = ( ROW, COLUMN )";
530    INPUT I,J
540    IF (I<1) OR (I>N) OR (J<1) OR (J>M) OR (I<>INT(I)) OR
       (J<>INT(J)) THEN 520
550    FIXED 0
560    DISP Name$;"(";I;",";J;") =";
570    INPUT Matrix(I,J)
580    PRINT LIN(5)
590    GOTO 260
600    SUBEND
```

```
10      SUB Prtplt(N,Nb,To,Dt,Tf,Y(*))
20      OPTION BASE 1
30      Iout=0
40      DIM Er(N),Yplot(Nb+2),X(Nb+2),C(5,N),Yout(5),Outs(5,Nb+2)
50      INPUT "Are you interested in ouputs y(t)=Cx(t)?",C$
60      IF (C$="Y") OR (C$="y") THEN 90
70      IF (C$="N") OR (C$="n") THEN 240
80      GOTO 50
90      INPUT "No. of outputs r(<=5)",Rr
100     REDIM C(Rr,N),Yout(Rr),Outs(Rr,Nb+2)
110     PRINT "Give (rxn) dimensional C matrix row-wise"
120     MAT INPUT C
130     MAT PRINT C
140     Iout=1
150     FOR I=1 TO Nb
160     FOR J=1 TO N
170     Er(J)=Y(J,I)
180     NEXT J
190     MAT Yout=C*Er
200     FOR Ir=1 TO Rr
210     Outs(Ir,I)=Yout(Ir)
220     NEXT Ir
230     NEXT I
240     INPUT "Do you like to PRINT results",C$
250     IF (C$="Y") OR (C$="y") THEN 280
260     IF (C$="N") OR (C$="n") THEN 410
270     GOTO 240
280     FOR I=1 TO Nb
290     FOR J=1 TO N
300     Er(J)=Y(J,I)
310     NEXT J
320     IF Iout=1 THEN 360
330     PRINT "At t=";To+(I-1)*Dt
340     MAT PRINT Er
350     GOTO 400
360     MAT Yout=C*Er
370     PRINT "At t=";To+(I-1)*Dt;"States & Outputs are (in 2 lines)"
380     MAT PRINT Er
390     MAT PRINT Yout
400     NEXT I
410     INPUT "Do you like to PLOT the results",C$
420     IF (C$="Y") OR (C$="y") THEN 450
430     IF (C$="N") OR (C$="n") THEN 740
440     GOTO 410
450     IF Iout=1 THEN 590
460     FOR Ipl=1 TO N
470     FOR I=1 TO Nb
480     Yplot(I)=Y(Ipl,I)
490     X(I)=To+(I-1)*Dt
```

```
500    NEXT I
510    Iplot=0
520    CALL Plot(X(*),Yplot(*),Nb,Iplot,To,Tf)
530    INPUT "Do you like to STOP PLOTTING(Y/N)",C$
540    IF (C$="Y") OR (C$="y") THEN 740
550    IF (C$="N") OR (C$="n") THEN 570
560    GOTO 530
570    NEXT Ipl
580    GOTO 740
590    FOR Ipl=1 TO N+Rr
600    FOR I=1 TO Nb
610    X(I)=To+(I-1)*Dt
620    IF Ipl>N THEN 650
630    Yplot(I)=Y(Ipl,I)
640    GOTO 660
650    Yplot(I)=Outs(Ipl-N,I)
660    NEXT I
670    Iplot=0
680    CALL Plot(X(*),Yplot(*),Nb,Iplot,To,Tf)
690    INPUT "Do you like to STOP PLOTTING any more?",C$
700    IF (C$="Y") OR (C$="y") THEN 740
710    IF (C$="N") OR (C$="n") THEN 730
720    GOTO 690
730    NEXT Ipl
740    SUBEND



10     SUB Kutta(Idm,A,H,B,Maxstp,Ynt(*),Y(*))
20     Baddta=(A>=B) OR (H<=0) OR (Maxstp<=0)
30     IF Baddta=0 THEN 110
40     PRINT LIN(2),"ERROR IN SUBPROGRAM Kutta."
50     PRINT "A=";A;"    B=";B
60     PRINT "H=";H;"    Maxstp=";Maxstp,LIN(2)
70     PAUSE
80     GOTO 20
90     OPTION BASE 1
100    DIM Ysv(Idm),F(Idm),K(4,Idm)
110    Eps=1E-6
120    X=A
130    Hh=H/2
140    N=1
150    FOR I=1 TO Idm
160        Ysv(I)=Ynt(I)
170        Y(I,1)=Ynt(I)
180    NEXT I
190    Xsv=X
200    FOR L=1 TO 4
210        CALL Func(Ysv(*),X,Idm,F(*))
```

```
220      FOR I=1 TO Idm
230          K(L,I)=H*F(I)
240      NEXT I
250      ON L GOTO 260,260,310,350
260      X=Xsv+Hh
270      FOR I=1 TO Idm
280          Ysv(I)=Y(I,N)+K(L,I)*.5
290      NEXT I
300      GOTO 350
310      X=Xsv+H
320      FOR I=1 TO Idm
330          Ysv(I)=Y(I,N)+K(L,I)
340      NEXT I
350   NEXT L
360   Np=N+1
370   FOR I=1 TO Idm
380      Y(I,Np)=Y(I,N)+(K(1,I)+2*(K(2,I)+K(3,I))+K(4,I))/6
390   NEXT I
400   N=Np
410   IF (N>=Maxstp) OR (X>B-Eps) THEN SUBEXIT
420   FOR I=1 TO Idm
430       Ysv(I)=Y(I,Np)
440   NEXT I
450   GOTO 190
460   SUBEND




10    SUB Cheby(M,N,X(*),Y(*),A(*))
100   Baddta=(M<=N+1)
110   IF Baddta=0 THEN 210
140   PRINT LIN(2),"ERROR IN SUBPROGRAM Cheby."
150   PRINT "# OF DATA POINTS MUST BE GREATER THAN DEG. OF
      POLY. +1.",LIN(2)
160   PAUSE
170   GOTO 100
210   DIM T(1:M),Ax(1:N+2),Ay(1:N+2),Ah(1:N+2),By(1:N+2)
      ,Bh(1:N+2)
220   INTEGER In(1:N+2)
230   K=(M-1)/(N+1)
240   FOR I=1 TO N+1
250       In(I)=(I-1)*K+1
260   NEXT I
270   In(N+2)=M
310 Start:  Sign=1
320 FOR I=1 TO N+2
330      Ax(I)=X(In(I))
340      Ay(I)=Y(In(I))
350      Ah(I)=Sign
```

```
360       Sign=-Sign
370    NEXT I
410 Difference:   FOR I=2 TO N+2
420       FOR J=I-1 TO N+2
430          By(J)=Ay(J)
440          Bh(J)=Ah(J)
450       NEXT J
460       FOR J=I TO N+2
470          Diff=Ax(J)-Ax(J-I+1)
480          Ay(J)=(By(J)-By(J-1))/Diff
490          Ah(J)=(Bh(J)-Bh(J-1))/Diff
500       NEXT J
510    NEXT I
520    H=-Ay(N+2)/Ah(N+2)
560 Poly:   FOR I=0 TO N
570       A(I)=Ay(I+1)+Ah(I+1)*H
580       By(I+1)=0
590    NEXT I
600    By(1)=1
610    Tmax=ABS(H)
620    Imax=In(1)
630    FOR I=1 TO N
640       FOR J=0 TO I-1
650          By(I+1-J)=By(I+1-J)-By(I-J)*X(In(I))
660          A(J)=A(J)+A(I)*By(I+1-J)
670       NEXT J
680    NEXT I
720 Error:   FOR I=1 TO M
730       T(I)=A(N)
740       FOR J=0 TO N
750          T(I)=T(I)*X(I)+A(N-J)
760       NEXT J
770       T(I)=T(I)-Y(I)
780       IF ABS(T(I))<=Tmax THEN L1
790       Tmax=ABS(T(I))
800       Imax=I
810 L1:   NEXT I
820    FOR I=1 TO N+2
830       IF Imax<In(I) THEN L2
840       IF Imax=In(I) THEN Fit
850    NEXT I
860 L2:   IF T(Imax)*T(In(I))<0 THEN L3
870    In(I)=Imax
880    GOTO Start
890 L3:   IF In(1)<Imax THEN L4
900    FOR I=1 TO N+1
910       In(N+3-I)=In(N+2-I)
920    NEXT I
930    In(I)=Imax
940    GOTO Start
```

```
950 L4:   IF In(N+2)<=Imax THEN L5
960    In(I-2)=Imax
970    GOTO Start
980 L5:   FOR I=1 TO N+1
990       In(I)=In(I+1)
1000 NEXT I
1010 In(N+2)=Imax
1020 GOTO Start
1030 Fit:SUBEND


10     SUB Print(N,Nb,To,Dt,Tf,Y(*))
20     OPTION BASE 1
30     DIM Er(N),Yplot(Nb+2),X(Nb+2)
40     INPUT "Do you like to PRINT results",C$
50     IF (C$="Y") OR (C$="y") THEN 80
60     IF (C$="N") OR (C$="n") THEN 150
70     GOTO 40
80     FOR I=1 TO Nb
90     FOR J=1 TO N
100    Er(J)=Y(J,I)
110    NEXT J
120    PRINT "At t=";To+(I-1)*Dt;"X(t) Is:"
130    MAT PRINT Er;
140    NEXT I
150    INPUT "Do you like to PLOT X(t)",C$
160    IF (C$="Y") OR (C$="y") THEN 190
170    IF (C$="N") OR (C$="n") THEN 310
180    GOTO 150
190    FOR Ipl=1 TO N
200    FOR I=1 TO Nb
210    Yplot(I)=Y(Ipl,I)
220    X(I)=To+(I-1)*Dt
230    NEXT I
240    Iplot=0
250    CALL Plot(X(*),Yplot(*),Nb,Iplot,To,Tf)
260    INPUT "Do you like to STOP PLOTTING(Y/N)",C$
270    IF (C$="Y") OR (C$="y") THEN 310
280    IF (C$="N") OR (C$="n") THEN 300
290    GOTO 260
300    NEXT Ipl
310    SUBEND


10     SUB Plot(X(*),Y(*),Nmax,Iplot,A,B)
20     PLOTTER IS "GRAPHICS"
30     EXIT GRAPHICS
40     INPUT "Enter the LINE TYPE No (Solid=1,Dashed=3,..,Tic Marks=9)",Line
```

```
50      IF (Line<1) OR (Line>9) THEN 40
60      OUTPUT 705;"VS4;"
70      IF Iaxis=1 THEN 380
80      INPUT "Please enter the title of the plot",Title$
90          DISP "WORKING..."
100         INPUT "Scaling Factor : 1-PROGRAM & 2-USER´S"
            ,Mmcode
110         IF Mmcode=1 THEN 170
120         INPUT "Give MIN. y(t)?",Miny
130         INPUT "Give MAX. y(t)?",Maxy
140         Minx=A
150         Maxx=B
160         GOTO 360
170         Miny=9.99999999999E99
180         Maxy=-9.99999999999E99
190         Minx=A
200         Maxx=B
210         ON ERROR GOTO Bug
220         FOR I=1 TO Nmax
230             IF Y(I)<Miny THEN Miny=Y(I)
240             IF Y(I)>Maxy THEN Maxy=Y(I)
250             GOTO 350
260 Bug:        IF ERRN=31 THEN L31
270             IF ERRN=29 THEN L29
280             BEEP
290             PRINT ERRM$
300             PAUSE
310             GOTO 350
320 L31:    Y(I)=9.99999999999E99
330         GOTO 350
340 L29:    Y(I)=-9.9999999999E99
350         NEXT I
360         INPUT "X-Axis Title?",Xtit$
370         INPUT "Y-Axis Title?",Ytit$
380         Lx=LGT(Maxx-Minx)
390         Ly=LGT(Maxy-Miny)
400         Xfudge=.2*(Maxx-Minx)
410         Yfudge=.2*(Maxy-Miny)
420         Testxtic=FRACT(Lx)+(Lx<0)
430         Testytic=FRACT(Ly)+(Ly<0)
440         Xtic=10^(INT(Lx)-1)*(1+1.5*((Testxtic>.39794) AND
            (Testxtic<.69897))+4 8(Testxtic>=.69897) AND
            (Testxtic<=.87506))+6.5*(Testxtic>.87506))
450         Ytic=10^(INT(Ly)-1)*(1+1.5*((Testytic>.39794) AND
            (Testytic<.69897))+4 q(Testytic>=.69897) AND
            (Testytic<=.87506))+6.5*(Testytic>.87506))
460         LOCATE 0,123,0,95
470         SCALE Minx-Xfudge,Maxx,Miny-Yfudge,Maxy
480         GRAPHICS
490         IF (Iaxis=1) AND (Igo=1) THEN 510
```

```
500        CALL Laxes(Xtic,Ytic,PROUND(Minx,-3),PROUND(Miny,
           -3),1,1,2,Minx-Xfudge _axx,Miny-Yfudge,Maxy,Xtit$
           ,Ytit$)
510        LINE TYPE Line
520        MOVE X(1),Y(1)
530        FOR I=2 TO Nmax
540        DRAW X(I),Y(I)
550        NEXT I
560        IF Iaxis=1 THEN 640
570        LINE TYPE 1
580        SETGU
590        CLIP 0,123,0,100
600        MOVE 61.5,97
610        LORG 5
620        LABEL USING 630;Title$
630        IMAGE #,K
640        BEEP
650        DISP "PROGRAM TERMINATED"
660        IF Iplot=0 THEN 690
670        PENUP
680        GOTO Xnd
690        INPUT "Do you like to plot on the X-Y PLOTTER?
           (Y/N)",A$
700        IF (A$="N") OR (A$="n") THEN Xnd
710        PLOTTER IS "9872A"
720        INPUT "ELIMINATE Axes?(Y/N)",X$
730        IF (X$="Y") OR (X$="y") THEN 760
740        IF (X$="N") OR (X$="n") THEN 770
750        GOTO 720
760        Iaxis=1
770        Iplot=1
780        Igo=1
790        INPUT "Which PEN do you like to use BLACK=1,GREEN
           =2,etc.",Ipen
800        PEN Ipen
810        GOTO 40
820 Xnd:        SUBEND
830   SUB Laxes(Xtic,Ytic,Xorg,Yorg,Xmaj,Ymaj,Minticsize,Xmin,
      Xmax,Ymin,Ymax,Xt$,$)
840   LINE TYPE 1
850   IF (Xmin>=Xmax) OR (Ymin>=Ymax) THEN SUBEXIT
860   Xfudge=.02*(Xmax-Xmin)
870   Yfudge=.02*(Ymax-Ymin)
880   AXES Xtic,Ytic,Xorg,Yorg,Xmaj,Ymaj,Minticsize
890   DEG
900   IF NOT Xtic THEN Labely
910 Labelx: IF SGN(Xtic)=-1 THEN Parx
920   LDIR 90
930   LORG 8
940   GOTO 970
```

```
950 Parx:   LDIR 0
960     LORG 6
970     FOR I=Xorg TO Xmax STEP ABS(Xtic)
980     MOVE I,Yorg-Yfudge
990     LABEL USING 1000;I
1000    IMAGE #,K
1010    NEXT I
1020    LDIR 0
1030    LORG 5
1040    MOVE (Xmax+Xmin)/2,(Yorg-Yfuge)*1.2
1050    LABEL USING 1000;Xt$
1060 Labely: IF NOT Ytic THEN SUBEXIT
1070    IF SGN(Ytic)=-1 THEN Pary
1080    LDIR 0
1090    LORG 8
1100    GOTO 1130
1110 Pary: LDIR -90
1120    LORG 6
1130    FOR I=Yorg TO Ymax STEP ABS(Ytic)
1140    MOVE Xorg-Xfudge,I
1150    LABEL USING 1000;I
1160    NEXT I
1170    LDIR 90
1180    LORG 5
1190    MOVE -(12/100)*(Xmax-Xmin),(Ymax+Ymin)/2
1200    LABEL USING 1000;Yt$
1210    LDIR 0
1220    SUBEND
```

INDEX

α (ω) circles   439
Adjoints   263-266
Aggregation   161, 163, 176, 181, 183
   matrix   176
   methods
      frequency-domain   185
      time-domain   175
Algebraically equivalent   133, 413, 504
Algorithm   171, 517, 525
   solution of the Lyapunov equation   368
Analog-digital converter   251
Angle of departure   314
Angle of entry   314
Asymptotes   312, 485
   asymptotically stable   275, 497
Automatic constrained method   529

Bandwidth   435
Bibo stable   288
Block-diagram representation   145
Block-Jordan form   36
Bode diagrams   338-339
Boundary layer correction   167-168
Break-away points   312
Break frequency   339
Break-in points   312

Canonical decomposition   397
Canonical structure   408
Caunchy's residue theorem   95
Causality   237, 253

Cayley-Hamilton theorem   39, 41, 206-209, 215-217, 233, 398, 399
Characteristic polynomial   23, 225, 414, 485, 488, 489, 493, 494, 499
Circle criterion   361
Closed-loop
   poles   139, 481, 489, 494, 504
   system   487, 493-494
   transfer function   146, 156
Cofactor   151
Companion forms   156, 392
Compensation   419, 426, 494
Complex
   conjugate poles or zero   342
   eigenvalues   35
   functions   382
   $H(2)$ 94
Computer control system   251
Conjugate transpose   19, 265
Constant gain K,   341
Continuous-time   115, 205-206, 267, 397
   Lyapunov function   364
   case   234
   systems   4, 122, 140, 201, 234, 266, 504, 510
Control schemes   161
Control systems   1, 381
Control variables   115
Controllability   381, 384-389, 396-400, 403, 409, 411-414, 483-484, 486-487, 494
   matrix   388-389, 412, 485
Controllable   384
   companion   413

form   123, 125, 130, 134, 157
   state   483
Convolution property   64, 65, 93, 253
Corner frequency   339
Cost function   507
Coupled system   164, 165, 167
Crossover frequency   346

Damped natural frequency   431
Damping ratio   431
Decomposition   133, 399, 404
   canonical   397
   property   236, 250
Definition   276, 364-365
   of "design"   419
   z-transform   90
Delay   92
Design
   of optimal state feedback   537
   via Bode and Nyquist diagrams   460
   via parameter optimization   507
   via root locus   451
Determinant property   205
Deterministic systems   4
Diagonalization   397, 413
   of a square matrix   30
Differential equations   115, 117, 118, 122, 139, 140, 202, 411
Differentiation   153
   Rule   60
Digital simulation   120
Digital-analog converter   251
Dimension   15
Dirac delta function   63
Direct search method   515
Direct sum   15
Direction of search   516
Discrete   371
   impulse response   253
   state equation   252
   systems   251
   time   113, 201, 234, 397
      Lyapunov equation, A'PA-P = -Q   371
      Lyapunov function   365
      multivariable systems   141
      systems   4, 118, 122, 125, 140-141, 203, 205, 205-206, 249, 253, 265, 271, 282, 413, 504, 513
      transition matrices   267
      version   240
Discretization   251
Distributed-parameter systems   3
Disturbance input   482
Domain of convergence   58

Dominant poles   176, 420
   assumption   420
Dot product   16
Dual systems   201, 263
Duality   266, 396
   theorem   396, 413

Ebers-moll equations   153
Eigenvalue   23
   method   206, 208, 215-216
Eigenvector   23, 27, 31
Equilibrium state   274
Error function   146
Estimator   494
Euclidean norm   20
Euler's method   82
Exactness   176
Existence   202
Exponentially stable   278
External stability   273

Fadeeva's method   226
Fast states   168
Feedback
   compensation   472
   control   481
   system   151
   gain   482-483
   matrices   484, 487, 489
   transfer function   146
Field   9
Final value property   64, 93
Forward path   150
Frequency   338, 343
   domain   162
   specifications   435, 440
   responses
      lag compensator   428
      lag-lead compensator   430
      lead compensator   427
   methods   325
Full-dimension observer   495-499
Full rank   13
Functional minimization techniques   515
Functions   276, 363
   generator   121
   square matrix   38
Fundamental matrix   202-205, 264-267, 269, 271, 386, 392, 396

Gain crossover   328, 346
   frequency   328, 346
Gain margin   325, 328, 346-347

Gain matrix   503
Generalized eigenvector   31
Generalized inverse   177
Global stability   275
Gradient vector   153
Gram
  determinant   387
  matrices   382, 411
Grammian   382
Graph determinant   151

Half planes   280
Hermitian   19
Hessian matrix   158
Homogeneous   202, 270
  differential equation   203, 263
  state equations   215, 223, 266

Ideal sampling   89
Identity observer   499
Imaginary-axis crossing   312
Impulse response   139, 271
  matrix   237-238, 253
Infinite-dimensional vector space   15
Initial conditions   141, 254
Initial states   201-203, 240, 250, 265, 390
Initial value property   64, 93
Inner product   16, 263
Input   115-116
  node   150
  space   116
  variables   248
Integral
  absolute value of error (IAE)   509
  square of the error signal (ISE)   508
  squared time multiplied by squared error (ISTSE)   509
  squared time multiplied by absolute value of error (ISTAE)   509
  time multiplied by the absolute value of the error signal (ITAE)   509
  time multiplied by squared error (ITSE)   509
Integration Rule   61
Integrator   121
Internal stability   273
Inverse Laplace transform   59, 65, 234
  partial fraction expansion   65
Inversion of the z-transform   93
Inversion property   205

Jacobian matrix   153
Jordan block   34, 207, 216, 268

Jordan canonical form   31, 156, 206, 217
Jury-Blanchard stability criterion   299

Kalman   396, 408
Kirchhoff's laws   113-114, 120
Kronecker delta   18, 252
Kutta   610

Lag compensation   428
  frequency-response scheme   472
  root locus scheme   470
Lag network   428
Lag-lead
  compensator   429
  network   429
Laplace transform   125, 139, 145, 157, 208, 239
  method   206-207, 209
  numerical inversion   81
  one-sided   57
  pairs   79
  properties   59
  solution of differential equations   88
  translation   63
Large-scale systems   161
  model   162
Lead compensation   426
  frequency-response schemes   471
  root-locus scheme   469
Lead network   426
Lead-lag compensation
  frequency-response schemes   472
  root locus scheme   470
Left eigenvector   23
Leftside boundary layer   167
Leverrier's algorithm   226
Linear
  combination of vectors   11
  discrete-time systems   385
  independence   381, 383, 411
  space   10
  system   4, 381
  time-invariant   115
  time-varying   119
    systems   201, 223
  transformation   501
  vector space   10
Linearity   59, 92
Linearization   152
Lipschitz condition   202
Luenberger   500
Lumped linear circuit   155
Lumped-parameter systems   3
Lyapunov   363
  equation   368
  method   363

M(ω) circles   437
Mason's gain formula   156
Matrix   13, 15, 19, 46
  continued fraction   193
  modal   30, 182
  residues   234
Minimal realizations   409, 414
Mixed methods   191
Modal aggregation   181
Modal matrix   30, 182
Moment matching   185
Moment of inertia   115
Multi-input
  case   488
  system   487
Multiplicity   206, 234
Multivariable   116
  systems   191, 240
    discrete-time   253-254

Node   150
Nonlinear systems   4, 123, 153
Nonlinearities   114
Nonstationary   4
Normal   19
Normed vector space   19
Nov equation   371
Null space   15
Numerical inversion of the Laplace
  transform   81
Nyquist
  diagram   326
  path   330
  stability criterion   328, 331

Obscissa of convergence   58
  403-405, 408, 409, 411, 413, 486, 498, 503
Observability   381, 389-390, 396-397, 412, 414, 484, 487,
  complete   390, 500-501, 504
  matrix   403, 485, 496
Observable   389, 410
  companion form   123, 124-125, 130, 134, 157, 413
Observer   494, 499-501
  poles   502, 504
Open-loop
  state estimator   496-497
  system   494
  transfer function   146, 157
Operating point   154

Optimal control   489
Optimum   507
Orthogonal   18-19
Orthonormal   18-19
Output   115
  equations   117, 154, 201
  error vector   497
  feedback   482, 493
    gain   503
  node   150
  space   116
  variables   249
  vector   116, 201

Pade'
  approximation   187
  modal   191
Parseval's Theorem   510
  discrete-time systems   514
  optimization   510
Partial fraction expansion method   96
Passband   435
Peak
  frequency   436
  magnitude   436
  time   433
Perfect aggregation   183
Performance   431, 507
Perturbation   161, 167
Phase
  crossover frequency   328, 346
    point   328
  margins   325, 328, 346
Polar plot   327
Poles   66, 341
  assignment   487
  placement   486-487, 489
Power series method   94
Pulse response   141
  matrix   252-253

Quadratic form   46
Quasi-steady-state   169

Radius of convergence   90
Raible's table   302
Range space   15
Rational   66
  function   140
Real-axis sub-branches   311
Realization   140
  of minimal-order   409

INDEX

Reduced-dimension 500
   observer 495, 502, 505
Reduced-order observer 499
Reference input 482
Region of convergence 58
Regular perturbation 164
Residue method 94
Residues 66, 94, 234
Resolvent matrix 140, 208, 224-225, 233
Resonant 343
Response 339
Riccati differential 368
Right eigenvector 23, 182
Root locus
   method 310
   plotting 316
   symmetry 312
Routh-Hurwitz stability criterion 290
Runge-Kutta 82
   method 240

Scalar 115
   differential equations 235
   multiplication 10, 121
   product 16
Schur-Cohn
   determinants 300
   criterion 300
Separation
   property 204, 223
   ratio 165
   time-scales 171
Settling time 433
Shifting property 238
Signal flow graphs 150-151, 157
Similarity transformation 30, 397, 488, 502
   matrix 30
Simulation diagrams 120
Singular 301
   peturbation 167
SISO 499
Skew symmetric 19
Slow states 168
Small-signal analysis 153
Solvability condition of Penrose 176
Spectral representation 234
Stability 274, 278, 485
   conditional 335
   criteria 277
   exponential 278
   external 273
   global 275
   "in the large" 275

   internal 273
   margins 335
   uniform 275
Stabilization 419, 486
Standard forms 123
State 116
   diagram 157
   equations 117, 123, 202, 265
   equivalent 408
   error vector 497
   estimation 481, 494-495
   feedback 481, 484-485, 487, 489, 494, 503
      matrix 504
   space 116, 381, 397, 399, 404, 412
      model 119
      representation 117
      transformations 130, 499
         matrix 131-132
   variables 116, 248, 483, 499, 502
   vector 116, 501-503
Static error coefficients 422-424
Steady-state error 421
Step
   inputs 240, 254
   response 240
   size determination 515
Stochastic systems 4
Stretched time-scale 168
Strictly proper 66, 116
   linear system 156
   proper rational function 225
Subspaces 15
Superposition property 236, 270
Systems 1, 4, 115, 278
   matrix 118, 225, 240, 484
   observability 389
   representation 157, 410
   stability 273

Time-domain 162
   performance specifications 434
   specifications 431
Time-function translation 61
Time-invariant 4, 155
   discrete case 215
Time-scale separation 168
Time-varying 4, 118, 155
   system 270
Trajectories 262, 265, 483
Transfer function 138, 141, 145, 151, 240, 410, 503
   matrix 140, 239, 254, 399, 409
Transform method 215, 217

Transformation matrix   400, 403, 414
Transistor circuits   153
Transition
  matrices   202-206, 223, 235, 249, 254, 267, 269-270, 272, 386
  property   205, 235
Transmittance   150-151

Uncontrollable   384, 399, 410, 487, 504
  modes   504
Undamped natural frequency   431
  state equations   202
Uniform asymptotic stability   289
Uniqueness   92, 202
Unit
  delay   121
  impulse   63
Unit step function   59
Unitary   19
Unobservability   389, 404, 504

Vandermonde matrix   28, 207
Variable metric method   525
Vectors   10-11, 16, 116
  addition   10

functions   382, 384
gradient   153
space   10, 15
two x   18
X_1, X_2 18

Weak equivalence   133
Weakly coupled   165

Z-transform   89, 140-141
  method   216
  one-sided   90
  pairs   105
  properties   92
solution of differential equations   106
Zero-input
  response   236, 250, 390
  (z.i.)   237
  stability   273-274
  system behavior   381
Zero-state
  equivalent   398, 400, 403, 405, 409
  response   236, 239, 250, 252-253, 399
  stability   273, 276
Zeros   66, 341